Linear Algebra

Jim Hefferon

Edition: 2014-Dec-25
See: http://joshua.smcvt.edu/linearalgebra

Notation

$\mathbb{R}$, $\mathbb{R}^+$, $\mathbb{R}^n$	real numbers, positive reals, n-tuples of reals
$\mathbb{N}$, $\mathbb{C}$	natural numbers $\{0, 1, 2, \ldots\}$, complex numbers
$(a \mathinner{\ldotp\ldotp} b)$, $[a \mathinner{\ldotp\ldotp} b]$	open interval, closed interval
$\langle \ldots \rangle$	sequence (a list in which order matters)
$h_{i,j}$	row i and column j entry of matrix H
V, W, U	vector spaces
$\vec{v}$, $\vec{0}$, $\vec{0}_V$	vector, zero vector, zero vector of a space V
$\mathcal{P}_n$, $\mathcal{M}_{n\times m}$	space of degree n polynomials, $n\times m$ matrices
$[S]$	span of a set
$\langle B, D\rangle$, $\vec{\beta}, \vec{\delta}$	basis, basis vectors
$\mathcal{E}_n = \langle \vec{e}_1, \ldots, \vec{e}_n \rangle$	standard basis for $\mathbb{R}^n$
$V \cong W$	isomorphic spaces
$M \oplus N$	direct sum of subspaces
h, g	homomorphisms (linear maps)
t, s	transformations (linear maps from a space to itself)
$\text{Rep}_B(\vec{v})$, $\text{Rep}_{B,D}(h)$	representation of a vector, a map
$Z_{n\times m}$ or Z, $I_{n\times n}$ or I	zero matrix, identity matrix
$\lvert T\rvert$	determinant of the matrix
$\mathscr{R}(h)$, $\mathscr{N}(h)$	range space, null space of the map
$\mathscr{R}_\infty(h)$, $\mathscr{N}_\infty(h)$	generalized range space and null space

Greek letters with pronounciation

character	name	character	name
α	alpha *AL-fuh*	ν	nu *NEW*
β	beta *BAY-tuh*	ξ, Ξ	xi *KSIGH*
γ, Γ	gamma *GAM-muh*	o	omicron *OM-uh-CRON*
δ, Δ	delta *DEL-tuh*	π, Π	pi *PIE*
ϵ	epsilon *EP-suh-lon*	ρ	rho *ROW*
ζ	zeta *ZAY-tuh*	σ, Σ	sigma *SIG-muh*
η	eta *AY-tuh*	τ	tau *TOW (as in cow)*
θ, Θ	theta *THAY-tuh*	υ, Υ	upsilon *OOP-suh-LON*
ι	iota *eye-OH-tuh*	ϕ, Φ	phi *FEE, or FI (as in hi)*
κ	kappa *KAP-uh*	χ	chi *KI (as in hi)*
λ, Λ	lambda *LAM-duh*	ψ, Ψ	psi *SIGH, or PSIGH*
μ	mu *MEW*	ω, Ω	omega *oh-MAY-guh*

Capitals shown are the ones that differ from Roman capitals.

Preface

This book helps students to master the material of a standard US undergraduate first course in Linear Algebra.

The material is standard in that the subjects covered are Gaussian reduction, vector spaces, linear maps, determinants, and eigenvalues and eigenvectors. Another standard is book's audience: sophomores or juniors, usually with a background of at least one semester of calculus. The help that it gives to students comes from taking a developmental approach — this book's presentation emphasizes motivation and naturalness, using many examples as well as extensive and careful exercises.

The developmental approach is what most recommends this book so I will elaborate. Courses at the beginning of a mathematics program focus less on theory and more on calculating. Later courses ask for mathematical maturity: the ability to follow different types of arguments, a familiarity with the themes that underlie many mathematical investigations such as elementary set and function facts, and a capacity for some independent reading and thinking. Some programs have a separate course devoted to developing maturity but in any case a Linear Algebra course is an ideal spot to work on this transition to more rigor. It comes early in a program so that progress made here pays off later but also comes late enough so that the students in the class are serious about mathematics. The material is accessible, coherent, and elegant. There are a variety of argument styles, including proofs by contradiction and proofs by induction. And, examples are plentiful.

Helping readers start the transition to being serious students of mathematics requires taking the mathematics seriously so all of the results here are proved. On the other hand, we cannot assume that students have already arrived and so in contrast with more advanced texts this book is filled with examples, often quite detailed.

Some texts that assume a not-yet sophisticated reader begin with extensive computations, including matrix multiplication and determinants. Then, when vector spaces and linear maps finally appear and definitions and proofs start, the abrupt change brings students to an abrupt stop. While this book begins with linear reduction, from the start we do more than compute. The first chapter includes proofs, such as that linear reduction gives a correct and complete solution set. With that as motivation the second chapter begins with real vector spaces. In the schedule below this happens at the start of the third week.

A student progresses most in mathematics while doing exercises. The problem sets start with routine checks and range up to reasonably involved proofs. I have aimed to typically put two dozen in each set, thereby giving a selection. There are even a few that are puzzles taken from various journals, competitions, or problems collections. These are marked with a '?' and as part of the fun I have kept the original wording as much as possible.

That is, as with the rest of the book, the exercises are aimed to both build an ability at, and help students experience the pleasure of, *doing* mathematics. Students should see how the ideas arise and should be able to picture themselves doing the same type of work.

Applications. Applications and computing are interesting and vital aspects of the subject. Consequently, each chapter closes with a selection of topics in those areas. These give a reader a taste of the subject, discuss how Linear Algebra comes in, point to some further reading, and give a few exercises. They are brief enough that an instructor can do one in a day's class or can assign them as projects for individuals or small groups. Whether they figure formally in a course or not, they help readers see for themselves that Linear Algebra is a tool that a professional must have.

Availability. This book is Free. In particular, instructors can run off copies for students and sell them at the bookstore. See this book's web page http://joshua.smcvt.edu/linearalgebra for the license details. That page also has the latest version, exercise answers, beamer slides, lab manual, additional material, and LaTeX source.

Acknowledgments. A lesson of software development is that complex projects need a process for bug fixes. I am grateful for such reports from both instructors and students and I periodically issue revisions. My contact information is on the web page.

I thank Gabriel S Santiago for the cover colors. I am also grateful to Saint Michael's College for supporting this project over many years.

And, I thank my wife Lynne for her unflagging encouragement.

Advice. This book's emphasis on motivation and development, and its availability, make it widely used for self-study. If you are an independent student then good for you, I admire your industry. However, you may find some advice useful.

While an experienced instructor knows what subjects and pace suit their class, this semester's timetable (graciously shared by George Ashline) may help you plan a sensible rate. It presumes Section One.II, the elements of vectors.

week	*Monday*	*Wednesday*	*Friday*
1	One.I.1	One.I.1, 2	One.I.2, 3
2	One.I.3	One.III.1	One.III.2
3	Two.I.1	Two.I.1, 2	Two.I.2
4	Two.II.1	Two.III.1	Two.III.2
5	Two.III.2	Two.III.2, 3	Two.III.3
6	EXAM	Three.I.1	Three.I.1
7	Three.I.2	Three.I.2	Three.II.1
8	Three.II.1	Three.II.2	Three.II.2
9	Three.III.1	Three.III.2	Three.IV.1, 2
10	Three.IV.2, 3	Three.IV.4	Three.V.1
11	Three.V.1	Three.V.2	Four.I.1
12	EXAM	Four.I.2	Four.III.1
13	Five.II.1	–THANKSGIVING BREAK–	
14	Five.II.1, 2	Five.II.2	Five.II.3

(Using this schedule as a target, I find that I have room for a lecture or two on an application or from the lab manual.) Note that in addition to the in-class exams, students in this course do take-home problems that include arguments, for instance showing that a set is a vector space. Computations are important but so are the proofs.

In the table of contents I have marked subsections as optional if some instructors will pass over them in favor of spending more time elsewhere.

As enrichment, you might pick one or two topics that appeal to you from the end of each chapter or from the lab manual. You'll get more from these if you have access to software for calculations. I recommend *Sage*, freely available from `http://sagemath.org`.

My main advice is: do many exercises. I have marked a good sample with ✓'s in the margin. Do not simply read the answers — you must try the problems and possibly struggle with them. For all of the exercises, you must justify your answer either with a computation or with a proof. Be aware that few people can write correct proofs without training; try to find a knowledgeable person to work with you.

Finally, a caution for all students, independent or not: I cannot overemphasize that the statement, "I understand the material but it is only that I have trouble with the problems" shows a misconception. Being able to do things with the ideas is their entire point. The quotes below express this sentiment admirably (I have taken the liberty of formatting them as poetry). They capture the essence of both the beauty and the power of mathematics and science in general, and of Linear Algebra in particular.

I know of no better tactic
than the illustration of exciting principles
by well-chosen particulars.
–Stephen Jay Gould

If you really wish to learn
then you must mount the machine
and become acquainted with its tricks
by actual trial.
–Wilbur Wright

Jim Hefferon
Mathematics, Saint Michael's College
Colchester, Vermont USA 05439
http://joshua.smcvt.edu/linearalgebra
2014-Dec-25

Author's Note. Inventing a good exercise, one that enlightens as well as tests, is a creative act, and hard work. The inventor deserves recognition. But texts have traditionally not given attributions for questions. I have changed that here where I was sure of the source. I would be glad to hear from anyone who can help me to correctly attribute others of the questions.

Contents

*Starred subsections are optional.

Chapter One

Linear Systems

I Solving Linear Systems

Systems of linear equations are common in science and mathematics. These two examples from high school science [Onan] give a sense of how they arise.

The first example is from Statics. Suppose that we have three objects, we know that one has a mass of 2 kg, and we want to find the two unknown masses. Suppose further that experimentation with a meter stick produces these two balances.

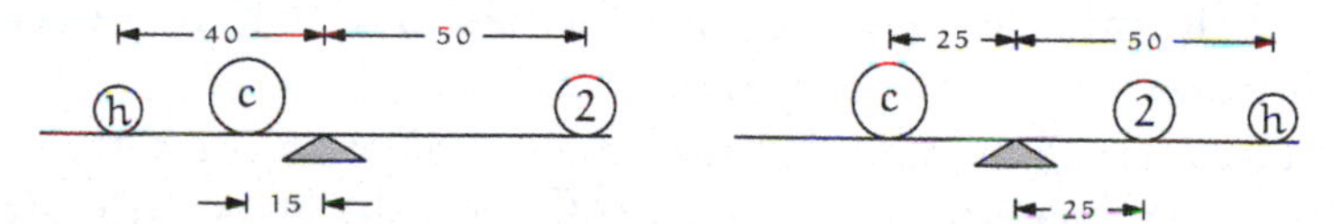

For the masses to balance we must have that the sum of moments on the left equals the sum of moments on the right, where the moment of an object is its mass times its distance from the balance point. That gives a system of two linear equations.

$$\begin{aligned} 40h + 15c &= 100 \\ 25c &= 50 + 50h \end{aligned}$$

The second example is from Chemistry. We can mix, under controlled conditions, toluene C_7H_8 and nitric acid HNO_3 to produce trinitrotoluene $C_7H_5O_6N_3$ along with the byproduct water (conditions have to be very well controlled — trinitrotoluene is better known as TNT). In what proportion should we mix them? The number of atoms of each element present before the reaction

$$x\,C_7H_8 \;+\; y\,HNO_3 \quad\longrightarrow\quad z\,C_7H_5O_6N_3 \;+\; w\,H_2O$$

must equal the number present afterward. Applying that in turn to the elements C, H, N, and O gives this system.

$$\begin{aligned} 7x &= 7z \\ 8x + 1y &= 5z + 2w \\ 1y &= 3z \\ 3y &= 6z + 1w \end{aligned}$$

Both examples come down to solving a system of equations. In each system, the equations involve only the first power of each variable. This chapter shows how to solve any such system.

I.1 Gauss's Method

1.1 Definition A *linear combination* of $x_1, \ldots, x_n$ has the form

$$a_1x_1 + a_2x_2 + a_3x_3 + \cdots + a_nx_n$$

where the numbers $a_1, \ldots, a_n \in \mathbb{R}$ are the combination's *coefficients*. A *linear equation* in the variables $x_1, \ldots, x_n$ has the form $a_1x_1 + a_2x_2 + a_3x_3 + \cdots + a_nx_n = d$ where $d \in \mathbb{R}$ is the *constant*.

An n-tuple $(s_1, s_2, \ldots, s_n) \in \mathbb{R}^n$ is a *solution* of, or *satisfies*, that equation if substituting the numbers $s_1, \ldots, s_n$ for the variables gives a true statement: $a_1s_1 + a_2s_2 + \cdots + a_ns_n = d$. A *system of linear equations*

$$\begin{aligned} a_{1,1}x_1 + a_{1,2}x_2 + \cdots + a_{1,n}x_n &= d_1 \\ a_{2,1}x_1 + a_{2,2}x_2 + \cdots + a_{2,n}x_n &= d_2 \\ &\vdots \\ a_{m,1}x_1 + a_{m,2}x_2 + \cdots + a_{m,n}x_n &= d_m \end{aligned}$$

has the solution $(s_1, s_2, \ldots, s_n)$ if that n-tuple is a solution of all of the equations.

1.2 Example The combination $3x_1 + 2x_2$ of x_1 and x_2 is linear. The combination $3x_1^2 + 2\sin(x_2)$ is not linear, nor is $3x_1^2 + 2x_2$.

1.3 Example The ordered pair $(-1, 5)$ is a solution of this system.

$$\begin{aligned} 3x_1 + 2x_2 &= 7 \\ -x_1 + x_2 &= 6 \end{aligned}$$

In contrast, $(5, -1)$ is not a solution.

Finding the set of all solutions is *solving* the system. We don't need guesswork or good luck, there is an algorithm that always works. This algorithm is *Gauss's Method* (or *Gaussian elimination* or *linear elimination*).

1.4 Example To solve this system

$$\begin{aligned} 3x_3 &= 9 \\ x_1 + 5x_2 - 2x_3 &= 2 \\ \tfrac{1}{3}x_1 + 2x_2 \qquad &= 3 \end{aligned}$$

we transform it, step by step, until it is in a form that we can easily solve.

The first transformation rewrites the system by interchanging the first and third row.

$$\xrightarrow{\text{swap row 1 with row 3}} \begin{aligned} \tfrac{1}{3}x_1 + 2x_2 \qquad &= 3 \\ x_1 + 5x_2 - 2x_3 &= 2 \\ 3x_3 &= 9 \end{aligned}$$

The second transformation rescales the first row by a factor of 3.

$$\xrightarrow{\text{multiply row 1 by 3}} \begin{aligned} x_1 + 6x_2 \qquad &= 9 \\ x_1 + 5x_2 - 2x_3 &= 2 \\ 3x_3 &= 9 \end{aligned}$$

The third transformation is the only nontrivial one in this example. We mentally multiply both sides of the first row by -1, mentally add that to the second row, and write the result in as the new second row.

$$\xrightarrow{\text{add } -1 \text{ times row 1 to row 2}} \begin{aligned} x_1 + 6x_2 \qquad &= 9 \\ -x_2 - 2x_3 &= -7 \\ 3x_3 &= 9 \end{aligned}$$

These steps have brought the system to a form where we can easily find the value of each variable. The bottom equation shows that $x_3 = 3$. Substituting 3 for x_3 in the middle equation shows that $x_2 = 1$. Substituting those two into the top equation gives that $x_1 = 3$. Thus the system has a unique solution; the solution set is $\{(3, 1, 3)\}$.

Most of this subsection and the next one consists of examples of solving linear systems by Gauss's Method, which we will use throughout the book. It is fast and easy. But before we do those examples we will first show that it is also safe: Gauss's Method never loses solutions (any tuple that is a solution to the system before you apply the method is also a solution after), nor does it ever pick up extraneous solutions (any tuple that is not a solution before is also not a solution after).

1.5 Theorem (Gauss's Method) If a linear system is changed to another by one of these operations

(1) an equation is swapped with another

(2) an equation has both sides multiplied by a nonzero constant

(3) an equation is replaced by the sum of itself and a multiple of another

then the two systems have the same set of solutions.

Each of the three Gauss's Method operations has a restriction. Multiplying a row by 0 is not allowed because obviously that can change the solution set. Similarly, adding a multiple of a row to itself is not allowed because adding -1 times the row to itself has the effect of multiplying the row by 0. We disallow swapping a row with itself to make some results in the fourth chapter easier, and also because it's pointless.

PROOF We will cover the equation swap operation here. The other two cases are Exercise 31.

Consider a linear system.

$$\begin{aligned} a_{1,1}x_1 + a_{1,2}x_2 + \cdots + a_{1,n}x_n &= d_1 \\ &\vdots \\ a_{i,1}x_1 + a_{i,2}x_2 + \cdots + a_{i,n}x_n &= d_i \\ &\vdots \\ a_{j,1}x_1 + a_{j,2}x_2 + \cdots + a_{j,n}x_n &= d_j \\ &\vdots \\ a_{m,1}x_1 + a_{m,2}x_2 + \cdots + a_{m,n}x_n &= d_m \end{aligned}$$

The tuple $(s_1, \ldots, s_n)$ satisfies this system if and only if substituting the values for the variables, the s's for the x's, gives a conjunction of true statements: $a_{1,1}s_1 + a_{1,2}s_2 + \cdots + a_{1,n}s_n = d_1$ and ... $a_{i,1}s_1 + a_{i,2}s_2 + \cdots + a_{i,n}s_n = d_i$ and ... $a_{j,1}s_1 + a_{j,2}s_2 + \cdots + a_{j,n}s_n = d_j$ and ... $a_{m,1}s_1 + a_{m,2}s_2 + \cdots + a_{m,n}s_n = d_m$.

In a list of statements joined with 'and' we can rearrange the order of the statements. Thus this requirement is met if and only if $a_{1,1}s_1 + a_{1,2}s_2 + \cdots + a_{1,n}s_n = d_1$ and ... $a_{j,1}s_1 + a_{j,2}s_2 + \cdots + a_{j,n}s_n = d_j$ and ... $a_{i,1}s_1 + a_{i,2}s_2 + \cdots + a_{i,n}s_n = d_i$ and ... $a_{m,1}s_1 + a_{m,2}s_2 + \cdots + a_{m,n}s_n = d_m$. This is exactly the requirement that $(s_1, \ldots, s_n)$ solves the system after the row swap. QED

1.6 Definition The three operations from Theorem 1.5 are the *elementary reduction operations*, or *row operations*, or *Gaussian operations*. They are *swapping*, *multiplying by a scalar* (or *rescaling*), and *row combination*.

When writing out the calculations, we will abbreviate 'row i' by 'ρ_i'. For instance, we will denote a row combination operation by $k\rho_i + \rho_j$, with the row that changes written second. To save writing we will often combine addition steps when they use the same ρ_i as in the next example.

1.7 Example Gauss's Method systematically applies the row operations to solve a system. Here is a typical case.

$$\begin{aligned} x + y \phantom{{}+3z} &= 0 \\ 2x - y + 3z &= 3 \\ x - 2y - z &= 3 \end{aligned}$$

We begin by using the first row to eliminate the $2x$ in the second row and the x in the third. To get rid of the $2x$ we mentally multiply the entire first row by -2, add that to the second row, and write the result in as the new second row. To eliminate the x in the third row we multiply the first row by -1, add that to the third row, and write the result in as the new third row.

$$\xrightarrow[-\rho_1+\rho_3]{-2\rho_1+\rho_2} \begin{aligned} x + y \phantom{{}+3z} &= 0 \\ -3y + 3z &= 3 \\ -3y - z &= 3 \end{aligned}$$

We finish by transforming the second system into a third, where the bottom equation involves only one unknown. We do that by using the second row to eliminate the y term from the third row.

$$\xrightarrow{-\rho_2+\rho_3} \begin{aligned} x + y \phantom{{}+3z} &= 0 \\ -3y + 3z &= 3 \\ -4z &= 0 \end{aligned}$$

Now finding the system's solution is easy. The third row gives $z = 0$. Substitute that back into the second row to get $y = -1$. Then substitute back into the first row to get $x = 1$.

1.8 Example For the Physics problem from the start of this chapter, Gauss's Method gives this.

$$\begin{aligned} 40h + 15c &= 100 \\ -50h + 25c &= 50 \end{aligned} \xrightarrow{5/4\rho_1+\rho_2} \begin{aligned} 40h + 15c &= 100 \\ (175/4)c &= 175 \end{aligned}$$

So $c = 4$, and back-substitution gives that $h = 1$. (We will solve the Chemistry problem later.)

1.9 Example The reduction

$$\begin{array}{rcl} x+y+z=9 \\ 2x+4y-3z=1 \\ 3x+6y-5z=0 \end{array} \xrightarrow[-3\rho_1+\rho_3]{-2\rho_1+\rho_2} \begin{array}{rcl} x+y+z=9 \\ 2y-5z=-17 \\ 3y-8z=-27 \end{array}$$

$$\xrightarrow{-(3/2)\rho_2+\rho_3} \begin{array}{rcl} x+y+z=9 \\ 2y-5z=-17 \\ -(1/2)z=-(3/2) \end{array}$$

shows that $z=3$, $y=-1$, and $x=7$.

As illustrated above, the point of Gauss's Method is to use the elementary reduction operations to set up back-substitution.

1.10 Definition In each row of a system, the first variable with a nonzero coefficient is the row's *leading variable*. A system is in *echelon form* if each leading variable is to the right of the leading variable in the row above it, except for the leading variable in the first row, and any all-zero rows are at the bottom.

1.11 Example The prior three examples only used the operation of row combination. This linear system requires the swap operation to get it into echelon form because after the first combination

$$\begin{array}{rcl} x-y=0 \\ 2x-2y+z+2w=4 \\ y+w=0 \\ 2z+w=5 \end{array} \xrightarrow{-2\rho_1+\rho_2} \begin{array}{rcl} x-y=0 \\ z+2w=4 \\ y+w=0 \\ 2z+w=5 \end{array}$$

the second equation has no leading y. We exchange it for a lower-down row that has a leading y.

$$\xrightarrow{\rho_2\leftrightarrow\rho_3} \begin{array}{rcl} x-y=0 \\ y+w=0 \\ z+2w=4 \\ 2z+w=5 \end{array}$$

(Had there been more than one suitable row below the second then we could have used any one.) With that, Gauss's Method proceeds as before.

$$\xrightarrow{-2\rho_3+\rho_4} \begin{array}{rcl} x-y=0 \\ y+w=0 \\ z+2w=4 \\ -3w=-3 \end{array}$$

Back-substitution gives $w=1$, $z=2$, $y=-1$, and $x=-1$.

Strictly speaking, to solve linear systems we don't need the row rescaling operation. We have introduced it here because it is convenient and because we will use it later in this chapter as part of a variation of Gauss's Method, the Gauss-Jordan Method.

All of the systems so far have the same number of equations as unknowns. All of them have a solution and for all of them there is only one solution. We finish this subsection by seeing other things that can happen.

1.12 Example This system has more equations than variables.

$$\begin{aligned} x + 3y &= 1 \\ 2x + y &= -3 \\ 2x + 2y &= -2 \end{aligned}$$

Gauss's Method helps us understand this system also, since this

$$\xrightarrow[-2\rho_1+\rho_3]{-2\rho_1+\rho_2} \begin{aligned} x + 3y &= 1 \\ -5y &= -5 \\ -4y &= -4 \end{aligned}$$

shows that one of the equations is redundant. Echelon form

$$\xrightarrow{-(4/5)\rho_2+\rho_3} \begin{aligned} x + 3y &= 1 \\ -5y &= -5 \\ 0 &= 0 \end{aligned}$$

gives that $y = 1$ and $x = -2$. The '$0 = 0$' reflects the redundancy.

Gauss's Method is also useful on systems with more variables than equations. The next subsection has many examples.

Another way that linear systems can differ from the examples shown above is that some linear systems do not have a unique solution. This can happen in two ways. The first is that a system can fail to have any solution at all.

1.13 Example Contrast the system in the last example with this one.

$$\begin{aligned} x + 3y &= 1 \\ 2x + y &= -3 \\ 2x + 2y &= 0 \end{aligned} \quad \xrightarrow[-2\rho_1+\rho_3]{-2\rho_1+\rho_2} \quad \begin{aligned} x + 3y &= 1 \\ -5y &= -5 \\ -4y &= -2 \end{aligned}$$

Here the system is inconsistent: no pair of numbers (s_1, s_2) satisfies all three equations simultaneously. Echelon form makes the inconsistency obvious.

$$\xrightarrow{-(4/5)\rho_2+\rho_3} \begin{aligned} x + 3y &= 1 \\ -5y &= -5 \\ 0 &= 2 \end{aligned}$$

The solution set is empty.

1.14 Example The prior system has more equations than unknowns but that is not what causes the inconsistency — Example 1.12 has more equations than unknowns and yet is consistent. Nor is having more equations than unknowns necessary for inconsistency, as we see with this inconsistent system that has the same number of equations as unknowns.

$$\begin{array}{rl} x+2y=8 \\ 2x+4y=8 \end{array} \xrightarrow{-2\rho_1+\rho_2} \begin{array}{rl} x+2y= & 8 \\ 0= & -8 \end{array}$$

Instead, inconsistency has to do with the interaction of the left and right sides; in the first system above the left side's second equation is twice the first but the right side's second constant is not twice the first. Later we will have more to say about dependencies between a system's parts.

The other way that a linear system can fail to have a unique solution, besides having no solutions, is to have many solutions.

1.15 Example In this system

$$\begin{array}{rl} x+y=4 \\ 2x+2y=8 \end{array}$$

any pair of numbers satisfying the first equation also satisfies the second. The solution set $\{(x,y) \mid x+y=4\}$ is infinite; some example member pairs are $(0,4)$, $(-1,5)$, and $(2.5,1.5)$.

The result of applying Gauss's Method here contrasts with the prior example because we do not get a contradictory equation.

$$\xrightarrow{-2\rho_1+\rho_2} \begin{array}{rl} x+y=4 \\ 0=0 \end{array}$$

Don't be fooled by that example: a $0=0$ equation is not the signal that a system has many solutions.

1.16 Example The absence of a $0=0$ equation does not keep a system from having many different solutions. This system is in echelon form, has no $0=0$, but has infinitely many solutions, including $(0,1,-1)$, $(0,1/2,-1/2)$, $(0,0,0)$, and $(0,-\pi,\pi)$ (any triple whose first component is 0 and whose second component is the negative of the third is a solution).

$$\begin{array}{rl} x+y+z=0 \\ y+z=0 \end{array}$$

Nor does the presence of $0=0$ mean that the system must have many solutions. Example 1.12 shows that. So does this system, which does not have any solutions at all despite that in echelon form it has a $0=0$ row.

$$\begin{array}{rcl} \begin{array}{rcrcr} 2x & & -2z & = & 6 \\ & y+ & z & = & 1 \\ 2x+ & y- & z & = & 7 \\ & 3y+ & 3z & = & 0 \end{array} & \xrightarrow{-\rho_1+\rho_3} & \begin{array}{rcrcr} 2x & & -2z & = & 6 \\ & y+ & z & = & 1 \\ & y+ & z & = & 1 \\ & 3y+ & 3z & = & 0 \end{array} \\ & \xrightarrow[-3\rho_2+\rho_4]{-\rho_2+\rho_3} & \begin{array}{rcrcr} 2x & & -2z & = & 6 \\ & y+ & z & = & 1 \\ & & 0 & = & 0 \\ & & 0 & = & -3 \end{array} \end{array}$$

In summary, Gauss's Method uses the row operations to set a system up for back substitution. If any step shows a contradictory equation then we can stop with the conclusion that the system has no solutions. If we reach echelon form without a contradictory equation, and each variable is a leading variable in its row, then the system has a unique solution and we find it by back substitution. Finally, if we reach echelon form without a contradictory equation, and there is not a unique solution — that is, at least one variable is not a leading variable — then the system has many solutions.

The next subsection explores the third case. We will see that such a system must have infinitely many solutions and we will describe the solution set.

Note. *Here, and in the rest of the book, you must justify all of your exercise answers. For instance, if a question asks whether a system has a solution then you must justify a yes response by producing the solution and must justify a no response by showing that no solution exists.*

Exercises

✓ **1.17** Use Gauss's Method to find the unique solution for each system.

(a) $\begin{array}{rcr} 2x+3y & = & 13 \\ x-\ y & = & -1 \end{array}$ (b) $\begin{array}{rcrcrcr} x & & & - & z & = & 0 \\ 3x & + & y & & & = & 1 \\ -x & + & y & + & z & = & 4 \end{array}$

✓ **1.18** Use Gauss's Method to solve each system or conclude 'many solutions' or 'no solutions'.

(a) $\begin{array}{rcr} 2x+2y & = & 5 \\ x-4y & = & 0 \end{array}$ (b) $\begin{array}{rcr} -x+y & = & 1 \\ x+y & = & 2 \end{array}$ (c) $\begin{array}{rcr} x-3y+\ z & = & 1 \\ x+\ y+2z & = & 14 \end{array}$ (d) $\begin{array}{rcr} -x-\ y & = & 1 \\ -3x-3y & = & 2 \end{array}$

(e) $\begin{array}{rcrcrcr} & & 4y & + & z & = & 20 \\ 2x & - & 2y & + & z & = & 0 \\ x & & & + & z & = & 5 \\ x & + & y & - & z & = & 10 \end{array}$ (f) $\begin{array}{rcrcrcrcr} 2x & & & + & z & + & w & = & 5 \\ & & y & & & - & w & = & -1 \\ 3x & & & - & z & - & w & = & 0 \\ 4x & + & y & + & 2z & + & w & = & 9 \end{array}$

✓ **1.19** We can solve linear systems by methods other than Gauss's. One often taught in high school is to solve one of the equations for a variable, then substitute the resulting expression into other equations. Then we repeat that step until there is an equation with only one variable. From that we get the first number in the solution and then we get the rest with back-substitution. This method takes longer than Gauss's Method, since it involves more arithmetic operations, and is also more likely to lead to errors. To illustrate how it can lead to wrong conclusions, we will use the system

$$\begin{aligned} x + 3y &= 1 \\ 2x + y &= -3 \\ 2x + 2y &= 0 \end{aligned}$$

from Example 1.13.

(a) Solve the first equation for x and substitute that expression into the second equation. Find the resulting y.

(b) Again solve the first equation for x, but this time substitute that expression into the third equation. Find this y.

What extra step must a user of this method take to avoid erroneously concluding a system has a solution?

✓ **1.20** For which values of k are there no solutions, many solutions, or a unique solution to this system?

$$\begin{aligned} x - y &= 1 \\ 3x - 3y &= k \end{aligned}$$

✓ **1.21** This system is not linear in that it says $\sin\alpha$ instead of α

$$\begin{aligned} 2\sin\alpha - \cos\beta + 3\tan\gamma &= 3 \\ 4\sin\alpha + 2\cos\beta - 2\tan\gamma &= 10 \\ 6\sin\alpha - 3\cos\beta + \tan\gamma &= 9 \end{aligned}$$

and yet we can apply Gauss's Method. Do so. Does the system have a solution?

✓ **1.22** What conditions must the constants, the b's, satisfy so that each of these systems has a solution? *Hint.* Apply Gauss's Method and see what happens to the right side.

(a) $$\begin{aligned} x - 3y &= b_1 \\ 3x + y &= b_2 \\ x + 7y &= b_3 \\ 2x + 4y &= b_4 \end{aligned}$$

(b) $$\begin{aligned} x_1 + 2x_2 + 3x_3 &= b_1 \\ 2x_1 + 5x_2 + 3x_3 &= b_2 \\ x_1 \qquad + 8x_3 &= b_3 \end{aligned}$$

1.23 True or false: a system with more unknowns than equations has at least one solution. (As always, to say 'true' you must prove it, while to say 'false' you must produce a counterexample.)

1.24 Must any Chemistry problem like the one that starts this subsection — a balance the reaction problem — have infinitely many solutions?

✓ **1.25** Find the coefficients a, b, and c so that the graph of $f(x) = ax^2 + bx + c$ passes through the points $(1, 2)$, $(-1, 6)$, and $(2, 3)$.

1.26 After Theorem 1.5 we note that multiplying a row by 0 is not allowed because that could change a solution set. Give an example of a system with solution set S_0 where after multiplying a row by 0 the new system has a solution set S_1 and S_0 is a proper subset of S_1. Give an example where $S_0 = S_1$.

1.27 Gauss's Method works by combining the equations in a system to make new equations.

(a) Can we derive the equation $3x - 2y = 5$ by a sequence of Gaussian reduction steps from the equations in this system?

$$\begin{aligned} x + y &= 1 \\ 4x - y &= 6 \end{aligned}$$

(b) Can we derive the equation $5x - 3y = 2$ with a sequence of Gaussian reduction steps from the equations in this system?

$$\begin{aligned} 2x + 2y &= 5 \\ 3x + y &= 4 \end{aligned}$$

(c) Can we derive $6x - 9y + 5z = -2$ by a sequence of Gaussian reduction steps from the equations in the system?

$$\begin{aligned} 2x + y - z &= 4 \\ 6x - 3y + z &= 5 \end{aligned}$$

1.28 Prove that, where $a, b, \ldots, e$ are real numbers and $a \neq 0$, if

$$ax + by = c$$

has the same solution set as

$$ax + dy = e$$

then they are the same equation. What if $a = 0$?

✓ 1.29 Show that if $ad - bc \neq 0$ then

$$\begin{aligned} ax + by &= j \\ cx + dy &= k \end{aligned}$$

has a unique solution.

✓ 1.30 In the system

$$\begin{aligned} ax + by &= c \\ dx + ey &= f \end{aligned}$$

each of the equations describes a line in the xy-plane. By geometrical reasoning, show that there are three possibilities: there is a unique solution, there is no solution, and there are infinitely many solutions.

1.31 Finish the proof of Theorem 1.5.

1.32 Is there a two-unknowns linear system whose solution set is all of $\mathbb{R}^2$?

✓ 1.33 Are any of the operations used in Gauss's Method redundant? That is, can we make any of the operations from a combination of the others?

1.34 Prove that each operation of Gauss's Method is reversible. That is, show that if two systems are related by a row operation $S_1 \to S_2$ then there is a row operation to go back $S_2 \to S_1$.

? 1.35 [Anton] A box holding pennies, nickels and dimes contains thirteen coins with a total value of 83 cents. How many coins of each type are in the box? (These are US coins; a penny is 1 cent, a nickel is 5 cents, and a dime is 10 cents.)

? 1.36 [Con. Prob. 1955] Four positive integers are given. Select any three of the integers, find their arithmetic average, and add this result to the fourth integer. Thus the numbers 29, 23, 21, and 17 are obtained. One of the original integers is:

(a) 19 (b) 21 (c) 23 (d) 29 (e) 17

? ✓ **1.37** [Am. Math. Mon., Jan. 1935] Laugh at this: AHAHA + TEHE = TEHAW. It resulted from substituting a code letter for each digit of a simple example in addition, and it is required to identify the letters and prove the solution unique.

? **1.38** [Wohascum no. 2] The Wohascum County Board of Commissioners, which has 20 members, recently had to elect a President. There were three candidates (A, B, and C); on each ballot the three candidates were to be listed in order of preference, with no abstentions. It was found that 11 members, a majority, preferred A over B (thus the other 9 preferred B over A). Similarly, it was found that 12 members preferred C over A. Given these results, it was suggested that B should withdraw, to enable a runoff election between A and C. However, B protested, and it was then found that 14 members preferred B over C! The Board has not yet recovered from the resulting confusion. Given that every possible order of A, B, C appeared on at least one ballot, how many members voted for B as their first choice?

? **1.39** [Am. Math. Mon., Jan. 1963] "This system of n linear equations with n unknowns," said the Great Mathematician, "has a curious property."

"Good heavens!" said the Poor Nut, "What is it?"

"Note," said the Great Mathematician, "that the constants are in arithmetic progression."

"It's all so clear when you explain it!" said the Poor Nut. "Do you mean like $6x + 9y = 12$ and $15x + 18y = 21$?"

"Quite so," said the Great Mathematician, pulling out his bassoon. "Indeed, the system has a unique solution. Can you find it?"

"Good heavens!" cried the Poor Nut, "I am baffled."

Are you?

I.2 Describing the Solution Set

A linear system with a unique solution has a solution set with one element. A linear system with no solution has a solution set that is empty. In these cases the solution set is easy to describe. Solution sets are a challenge to describe only when they contain many elements.

2.1 Example This system has many solutions because in echelon form

$$\begin{array}{rcl} 2x \quad\;\; + z = 3 \\ x - y - z = 1 \\ 3x - y \quad\;\; = 4 \end{array} \xrightarrow[-(3/2)\rho_1+\rho_3]{-(1/2)\rho_1+\rho_2} \begin{array}{rcl} 2x \quad + \qquad z = 3 \\ -y - (3/2)z = -1/2 \\ -y - (3/2)z = -1/2 \end{array}$$

$$\xrightarrow{-\rho_2+\rho_3} \begin{array}{rcl} 2x \quad + \qquad z = 3 \\ -y - (3/2)z = -1/2 \\ 0 = 0 \end{array}$$

not all of the variables are leading variables. Theorem 1.5 shows that an (x, y, z) satisfies the first system if and only if it satisfies the third. So we can describe the solution set $\{(x, y, z) \mid 2x + z = 3 \text{ and } x - y - z = 1 \text{ and } 3x - y = 4\}$ in this way.

$$\{(x, y, z) \mid 2x + z = 3 \text{ and } -y - 3z/2 = -1/2\} \qquad (*)$$

This description is better because it has two equations instead of three but it is not optimal because it still has some hard to understand interactions among the variables.

To improve it, use the variable that does not lead any equation, z, to describe the variables that do lead, x and y. The second equation gives $y = (1/2) - (3/2)z$ and the first equation gives $x = (3/2) - (1/2)z$. Thus we can describe the solution set in this way.

$$\{(x, y, z) = ((3/2) - (1/2)z, (1/2) - (3/2)z, z) \mid z \in \mathbb{R}\} \qquad (**)$$

Compared with $(*)$, the advantage of $(**)$ is that z can be any real number. This makes the job of deciding which tuples are in the solution set much easier. For instance, taking $z = 2$ shows that $(1/2, -5/2, 2)$ is a solution.

2.2 Definition In an echelon form linear system the variables that are not leading are *free.*

2.3 Example Reduction of a linear system can end with more than one variable free. Gauss's Method on this system

$$\begin{array}{rrrrr} x+ & y+ & z- & w= & 1 \\ & y- & z+ & w= & -1 \\ 3x & & +6z- & 6w= & 6 \\ & -y+ & z- & w= & 1 \end{array} \xrightarrow{-3\rho_1+\rho_3} \begin{array}{rrrrr} x+ & y+ & z- & w= & 1 \\ & y- & z+ & w= & -1 \\ & -3y+ & 3z- & 3w= & 3 \\ & -y+ & z- & w= & 1 \end{array}$$

$$\xrightarrow[\rho_2+\rho_4]{3\rho_2+\rho_3} \begin{array}{rr} x+y+z-w= & 1 \\ y-z+w= & -1 \\ 0= & 0 \\ 0= & 0 \end{array}$$

leaves x and y leading and both z and w free. To get the description that we prefer, we work from the bottom. We first express the leading variable y in terms of z and w, as $y = -1 + z - w$. Moving up to the top equation, substituting for y gives $x + (-1 + z - w) + z - w = 1$ and solving for x leaves $x = 2 - 2z + 2w$. The solution set

$$\{(2 - 2z + 2w, -1 + z - w, z, w) \mid z, w \in \mathbb{R}\} \qquad (**)$$

has the leading variables in terms of the variables that are free.

2.4 Example The list of leading variables may skip over some columns. After this reduction

$$\begin{array}{rcl} 2x-2y &=0 \\ z+3w &=2 \\ 3x-3y &=0 \\ x-y+2z+6w &=4 \end{array} \xrightarrow[-(1/2)\rho_1+\rho_4]{-(3/2)\rho_1+\rho_3} \begin{array}{rcl} 2x-2y &=0 \\ z+3w &=2 \\ 0 &=0 \\ 2z+6w &=4 \end{array}$$

$$\xrightarrow{-2\rho_2+\rho_4} \begin{array}{rcl} 2x-2y &=0 \\ z+3w &=2 \\ 0 &=0 \\ 0 &=0 \end{array}$$

x and z are the leading variables, not x and y. The free variables are y and w and so we can describe the solution set as $\{(y, y, 2-3w, w) \mid y, w \in \mathbb{R}\}$. For instance, $(1, 1, 2, 0)$ satisfies the system — take $y = 1$ and $w = 0$. The four-tuple $(1, 0, 5, 4)$ is not a solution since its first coordinate does not equal its second.

A variable that we use to describe a family of solutions is a *parameter*. We say that the solution set in the prior example is *parametrized* with y and w.

(The terms 'parameter' and 'free variable' do not mean the same thing. In the prior example y and w are free because in the echelon form system they do not lead while they are parameters because of how we used them to describe the set of solutions. Had we instead rewritten the second equation as $w = 2/3 - (1/3)z$ then the free variables would still be y and w but the parameters would be y and z.)

In the rest of this book we will solve linear systems by bringing them to echelon form and then parametrizing with the free variables.

2.5 Example This is another system with infinitely many solutions.

$$\begin{array}{rcl} x+2y &=1 \\ 2x+z &=2 \\ 3x+2y+z-w &=4 \end{array} \xrightarrow[-3\rho_1+\rho_3]{-2\rho_1+\rho_2} \begin{array}{rcl} x+2y &=1 \\ -4y+z &=0 \\ -4y+z-w &=1 \end{array}$$

$$\xrightarrow{-\rho_2+\rho_3} \begin{array}{rcl} x+2y &=1 \\ -4y+z &=0 \\ -w &=1 \end{array}$$

The leading variables are x, y, and w. The variable z is free. Notice that, although there are infinitely many solutions, the value of w doesn't vary but is constant at -1. To parametrize, write w in terms of z with $w = -1 + 0z$. Then $y = (1/4)z$. Substitute for y in the first equation to get $x = 1 - (1/2)z$. The solution set is $\{(1-(1/2)z, (1/4)z, z, -1) \mid z \in \mathbb{R}\}$.

Parametrizing solution sets shows that systems with free variables have infinitely many solutions. For instance, above z takes on all of infinitely many real number values, each associated with a different solution.

We finish this subsection by developing a streamlined notation for linear systems and their solution sets.

2.6 Definition An $m \times n$ *matrix* is a rectangular array of numbers with m *rows* and n *columns*. Each number in the matrix is an *entry*.

We usually denote a matrix with an upper case roman letters. For instance,

$$A = \begin{pmatrix} 1 & 2.2 & 5 \\ 3 & 4 & -7 \end{pmatrix}$$

has 2 rows and 3 columns and so is a 2×3 matrix. Read that aloud as "two-by-three"; the number of rows is always given first. (The matrix has parentheses on either side so that when two matrices are adjacent we can tell where one ends and the other begins.) We name matrix entries with the corresponding lower-case letter so that $a_{2,1} = 3$ is the entry in the second row and first column of the above array. Note that the order of the subscripts matters: $a_{1,2} \neq a_{2,1}$ since $a_{1,2} = 2.2$. We denote the set of all $n \times m$ matrices by $\mathcal{M}_{n \times m}$.

We use matrices to do Gauss's Method in essentially the same way that we did it for systems of equations: where a row's *leading entry*. is its first nonzero entry (if it has one), we perform row operations to arrive at *matrix echelon form*,where the leading entry in lower rows are to the right of those in the rows above. We switch to this notation because it lightens the clerical load of Gauss's Method—the copying of variables and the writing of $+$'s and $=$'s.

2.7 Example We can abbreviate this linear system

$$\begin{aligned} x + 2y \quad\quad &= 4 \\ y - \ z &= 0 \\ x \quad\quad + 2z &= 4 \end{aligned}$$

with this matrix.

$$\left(\begin{array}{ccc|c} 1 & 2 & 0 & 4 \\ 0 & 1 & -1 & 0 \\ 1 & 0 & 2 & 4 \end{array}\right)$$

The vertical bar reminds a reader of the difference between the coefficients on the system's left hand side and the constants on the right. With a bar, this is an *augmented* matrix.

$$\left(\begin{array}{ccc|c} 1 & 2 & 0 & 4 \\ 0 & 1 & -1 & 0 \\ 1 & 0 & 2 & 4 \end{array}\right) \xrightarrow{-\rho_1+\rho_3} \left(\begin{array}{ccc|c} 1 & 2 & 0 & 4 \\ 0 & 1 & -1 & 0 \\ 0 & -2 & 2 & 0 \end{array}\right) \xrightarrow{2\rho_2+\rho_3} \left(\begin{array}{ccc|c} 1 & 2 & 0 & 4 \\ 0 & 1 & -1 & 0 \\ 0 & 0 & 0 & 0 \end{array}\right)$$

The second row stands for $y - z = 0$ and the first row stands for $x + 2y = 4$ so the solution set is $\{(4 - 2z, z, z) \mid z \in \mathbb{R}\}$.

Matrix notation also clarifies the descriptions of solution sets. Example 2.3's $\{(2 - 2z + 2w, -1 + z - w, z, w) \mid z, w \in \mathbb{R}\}$ is hard to read. We will rewrite it to group all of the constants together, all of the coefficients of z together, and all of the coefficients of w together. We write them vertically, in one-column matrices.

$$\{\begin{pmatrix}2\\-1\\0\\0\end{pmatrix} + \begin{pmatrix}-2\\1\\1\\0\end{pmatrix} \cdot z + \begin{pmatrix}2\\-1\\0\\1\end{pmatrix} \cdot w \mid z, w \in \mathbb{R}\}$$

For instance, the top line says that $x = 2 - 2z + 2w$ and the second line says that $y = -1 + z - w$. (Our next section gives a geometric interpretation that will help picture the solution sets.)

2.8 Definition A *vector* (or *column vector*) is a matrix with a single column. A matrix with a single row is a *row vector*. The entries of a vector are its *components*. A column or row vector whose components are all zeros is a *zero vector*.

Vectors are an exception to the convention of representing matrices with capital roman letters. We use lower-case roman or greek letters overlined with an arrow: $\vec{a}$, $\vec{b}$, ... or $\vec{\alpha}$, $\vec{\beta}$, ... (boldface is also common: $\mathbf{a}$ or $\boldsymbol{\alpha}$). For instance, this is a column vector with a third component of 7.

$$\vec{v} = \begin{pmatrix}1\\3\\7\end{pmatrix}$$

A zero vector is denoted $\vec{0}$. There are many different zero vectors—the one-tall zero vector, the two-tall zero vector, etc.—but nonetheless we will often say "the" zero vector, expecting that the size will be clear from the context.

2.9 Definition The linear equation $a_1x_1 + a_2x_2 + \cdots + a_nx_n = d$ with unknowns $x_1, \ldots, x_n$ is *satisfied* by

$$\vec{s} = \begin{pmatrix}s_1\\\vdots\\s_n\end{pmatrix}$$

if $a_1s_1 + a_2s_2 + \cdots + a_ns_n = d$. A vector satisfies a linear system if it satisfies each equation in the system.

The style of description of solution sets that we use involves adding the vectors, and also multiplying them by real numbers. Before we give the examples showing the style we first need to define these operations.

2.10 Definition The *vector sum* of $\vec{u}$ and $\vec{v}$ is the vector of the sums.

$$\vec{u}+\vec{v}=\begin{pmatrix}u_1\\ \vdots\\ u_n\end{pmatrix}+\begin{pmatrix}v_1\\ \vdots\\ v_n\end{pmatrix}=\begin{pmatrix}u_1+v_1\\ \vdots\\ u_n+v_n\end{pmatrix}$$

Note that for the addition to be defined the vectors must have the same number of entries. This entry-by-entry addition works for any pair of matrices, not just vectors, provided that they have the same number of rows and columns.

2.11 Definition The *scalar multiplication* of the real number r and the vector $\vec{v}$ is the vector of the multiples.

$$r\cdot\vec{v}=r\cdot\begin{pmatrix}v_1\\ \vdots\\ v_n\end{pmatrix}=\begin{pmatrix}rv_1\\ \vdots\\ rv_n\end{pmatrix}$$

As with the addition operation, the entry-by-entry scalar multiplication operation extends beyond vectors to apply to any matrix.

We write scalar multiplication either as $r\cdot\vec{v}$ or $\vec{v}\cdot r$, or even without the '$\cdot$' symbol: $r\vec{v}$. (Do not refer to scalar multiplication as 'scalar product' because we will use that name for a different operation.)

2.12 Example

$$\begin{pmatrix}2\\3\\1\end{pmatrix}+\begin{pmatrix}3\\-1\\4\end{pmatrix}=\begin{pmatrix}2+3\\3-1\\1+4\end{pmatrix}=\begin{pmatrix}5\\2\\5\end{pmatrix}\qquad 7\cdot\begin{pmatrix}1\\4\\-1\\-3\end{pmatrix}=\begin{pmatrix}7\\28\\-7\\-21\end{pmatrix}$$

Observe that the definitions of addition and scalar multiplication agree where they overlap; for instance, $\vec{v}+\vec{v}=2\vec{v}$.

With these definitions, we are set to use matrix and vector notation to both solve systems and express the solution.

2.13 Example This system

$$\begin{array}{rcrcrcrcl}2x&+&y&&&-&w&&=4\\&&y&&&+&w&+u&=4\\x&&&-&z&+&2w&&=0\end{array}$$

reduces in this way.

$$\left(\begin{array}{ccccc|c} 2 & 1 & 0 & -1 & 0 & 4 \\ 0 & 1 & 0 & 1 & 1 & 4 \\ 1 & 0 & -1 & 2 & 0 & 0 \end{array}\right) \xrightarrow{-(1/2)\rho_1+\rho_3} \left(\begin{array}{ccccc|c} 2 & 1 & 0 & -1 & 0 & 4 \\ 0 & 1 & 0 & 1 & 1 & 4 \\ 0 & -1/2 & -1 & 5/2 & 0 & -2 \end{array}\right)$$

$$\xrightarrow{(1/2)\rho_2+\rho_3} \left(\begin{array}{ccccc|c} 2 & 1 & 0 & -1 & 0 & 4 \\ 0 & 1 & 0 & 1 & 1 & 4 \\ 0 & 0 & -1 & 3 & 1/2 & 0 \end{array}\right)$$

The solution set is $\{(w+(1/2)u, 4-w-u, 3w+(1/2)u, w, u) \mid w, u \in \mathbb{R}\}$. We write that in vector form.

$$\{\begin{pmatrix} x \\ y \\ z \\ w \\ u \end{pmatrix} = \begin{pmatrix} 0 \\ 4 \\ 0 \\ 0 \\ 0 \end{pmatrix} + \begin{pmatrix} 1 \\ -1 \\ 3 \\ 1 \\ 0 \end{pmatrix} w + \begin{pmatrix} 1/2 \\ -1 \\ 1/2 \\ 0 \\ 1 \end{pmatrix} u \mid w, u \in \mathbb{R}\}$$

Note how well vector notation sets off the coefficients of each parameter. For instance, the third row of the vector form shows plainly that if u is fixed then z increases three times as fast as w. Another thing shown plainly is that setting both w and u to zero gives that

$$\begin{pmatrix} x \\ y \\ z \\ w \\ u \end{pmatrix} = \begin{pmatrix} 0 \\ 4 \\ 0 \\ 0 \\ 0 \end{pmatrix}$$

is a particular solution of the linear system.

2.14 Example In the same way, the system

$$\begin{aligned} x - y + z &= 1 \\ 3x + z &= 3 \\ 5x - 2y + 3z &= 5 \end{aligned}$$

reduces

$$\left(\begin{array}{ccc|c} 1 & -1 & 1 & 1 \\ 3 & 0 & 1 & 3 \\ 5 & -2 & 3 & 5 \end{array}\right) \xrightarrow[-5\rho_1+\rho_3]{-3\rho_1+\rho_2} \left(\begin{array}{ccc|c} 1 & -1 & 1 & 1 \\ 0 & 3 & -2 & 0 \\ 0 & 3 & -2 & 0 \end{array}\right)$$

$$\xrightarrow{-\rho_2+\rho_3} \left(\begin{array}{ccc|c} 1 & -1 & 1 & 1 \\ 0 & 3 & -2 & 0 \\ 0 & 0 & 0 & 0 \end{array}\right)$$

to give a one-parameter solution set.

$$\{\begin{pmatrix}1\\0\\0\end{pmatrix}+\begin{pmatrix}-1/3\\2/3\\1\end{pmatrix}z \mid z\in\mathbb{R}\}$$

As in the prior example, the vector not associated with the parameter

$$\begin{pmatrix}1\\0\\0\end{pmatrix}$$

is a particular solution of the system.

Before the exercises, we will consider what we have accomplished and what we have yet to do.

So far we have done the mechanics of Gauss's Method. We have not stopped to consider any of the interesting questions that arise, except for proving Theorem 1.5 — which justifies the method by showing that it gives the right answers.

For example, can we always describe solution sets as above, with a particular solution vector added to an unrestricted linear combination of some other vectors? We've noted that the solution sets we described in this way have infinitely many solutions so an answer to this question would tell us about the size of solution sets.

Many questions arise from our observation that we can do Gauss's Method in more than one way (for instance, when swapping rows we may have a choice of more than one row). Theorem 1.5 says that we must get the same solution set no matter how we proceed but if we do Gauss's Method in two ways must we get the same number of free variables in each echelon form system? Must those be the same variables, that is, is it impossible to solve a problem one way to get y and w free and solve it another way to get y and z free?

In the rest of this chapter we will answer these questions. The answer to each is 'yes'. In the next subsection we do the first one: we will prove that we can always describe solution sets in that way. Then, in this chapter's second section, we will use that understanding to describe the geometry of solution sets. In this chapter's final section, we will settle the questions about the parameters.

When we are done, we will not only have a solid grounding in the practice of Gauss's Method but we will also have a solid grounding in the theory. We will know exactly what can and cannot happen in a reduction.

Exercises

✓ **2.15** Find the indicated entry of the matrix, if it is defined.

$$A=\begin{pmatrix}1&3&1\\2&-1&4\end{pmatrix}$$

(a) $a_{2,1}$ (b) $a_{1,2}$ (c) $a_{2,2}$ (d) $a_{3,1}$

✓ **2.16** Give the size of each matrix.

(a) $\begin{pmatrix}1&0&4\\2&1&5\end{pmatrix}$ (b) $\begin{pmatrix}1&1\\-1&1\\3&-1\end{pmatrix}$ (c) $\begin{pmatrix}5&10\\10&5\end{pmatrix}$

✓ **2.17** Do the indicated vector operation, if it is defined.

(a) $\begin{pmatrix}2\\1\\1\end{pmatrix}+\begin{pmatrix}3\\0\\4\end{pmatrix}$ (b) $5\begin{pmatrix}4\\-1\end{pmatrix}$ (c) $\begin{pmatrix}1\\5\\1\end{pmatrix}-\begin{pmatrix}3\\1\\1\end{pmatrix}$ (d) $7\begin{pmatrix}2\\1\end{pmatrix}+9\begin{pmatrix}3\\5\end{pmatrix}$

(e) $\begin{pmatrix}1\\2\end{pmatrix}+\begin{pmatrix}1\\2\\3\end{pmatrix}$ (f) $6\begin{pmatrix}3\\1\\1\end{pmatrix}-4\begin{pmatrix}2\\0\\3\end{pmatrix}+2\begin{pmatrix}1\\1\\5\end{pmatrix}$

✓ **2.18** Solve each system using matrix notation. Express the solution using vectors.

(a) $\begin{aligned}3x+6y&=18\\x+2y&=6\end{aligned}$ (b) $\begin{aligned}x+y&=1\\x-y&=-1\end{aligned}$ (c) $\begin{aligned}x_1\qquad\quad+x_3&=4\\x_1-x_2+2x_3&=5\\4x_1-x_2+5x_3&=17\end{aligned}$

(d) $\begin{aligned}2a+b-c&=2\\2a\quad\;+c&=3\\a-b\qquad&=0\end{aligned}$ (e) $\begin{aligned}x+2y-z\qquad&=3\\2x+y\qquad+w&=4\\x-y+z+w&=1\end{aligned}$ (f) $\begin{aligned}x\qquad+z+w&=4\\2x+y\qquad-w&=2\\3x+y+z\qquad&=7\end{aligned}$

✓ **2.19** Solve each system using matrix notation. Give each solution set in vector notation.

(a) $\begin{aligned}2x+y-z&=1\\4x-y\qquad&=3\end{aligned}$ (b) $\begin{aligned}x\qquad-z\qquad&=1\\y+2z-w&=3\\x+2y+3z-w&=7\end{aligned}$ (c) $\begin{aligned}x-y+z\qquad&=0\\y\qquad+w&=0\\3x-2y+3z+w&=0\\-y\qquad-w&=0\end{aligned}$

(d) $\begin{aligned}a+2b+3c+d-e&=1\\3a-b+c+d+e&=3\end{aligned}$

2.20 Solve each system using matrix notation. Express the solution set using vectors.

(a) $\begin{aligned}3x+2y+z&=1\\x-y+z&=2\\5x+5y+z&=0\end{aligned}$ (b) $\begin{aligned}x+y-2z&=0\\x-y\qquad&=-3\\3x-y-2z&=-6\\2y-2z&=3\end{aligned}$ (c) $\begin{aligned}2x-y-z+w&=4\\x+y+z\qquad&=-1\end{aligned}$

(d) $\begin{aligned}x+y-2z&=0\\x-y\qquad&=-3\\3x-y-2z&=0\end{aligned}$

✓ **2.21** The vector is in the set. What value of the parameters produces that vector?

(a) $\begin{pmatrix}5\\-5\end{pmatrix}$, $\{\begin{pmatrix}1\\-1\end{pmatrix}k \mid k\in\mathbb{R}\}$

(b) $\begin{pmatrix}-1\\2\\1\end{pmatrix}$, $\{\begin{pmatrix}-2\\1\\0\end{pmatrix}i+\begin{pmatrix}3\\0\\1\end{pmatrix}j \mid i,j\in\mathbb{R}\}$

(c) $\begin{pmatrix}0\\-4\\2\end{pmatrix}$, $\{\begin{pmatrix}1\\1\\0\end{pmatrix}m+\begin{pmatrix}2\\0\\1\end{pmatrix}n \mid m,n\in\mathbb{R}\}$

2.22 Decide if the vector is in the set.

(a) $\begin{pmatrix}3\\-1\end{pmatrix}$, $\{\begin{pmatrix}-6\\2\end{pmatrix}k \mid k\in\mathbb{R}\}$

(b) $\begin{pmatrix}5\\4\end{pmatrix}$, $\{\begin{pmatrix}5\\-4\end{pmatrix}j \mid j\in\mathbb{R}\}$

(c) $\begin{pmatrix}2\\1\\-1\end{pmatrix}$, $\{\begin{pmatrix}0\\3\\-7\end{pmatrix}+\begin{pmatrix}1\\-1\\3\end{pmatrix}r \mid r\in\mathbb{R}\}$

(d) $\begin{pmatrix}1\\0\\1\end{pmatrix}$, $\{\begin{pmatrix}2\\0\\1\end{pmatrix}j+\begin{pmatrix}-3\\-1\\1\end{pmatrix}k \mid j,k\in\mathbb{R}\}$

2.23 [Cleary] A farmer with 1200 acres is considering planting three different crops, corn, soybeans, and oats. The farmer wants to use all 1200 acres. Seed corn costs \$20 per acre, while soybean and oat seed cost \$50 and \$12 per acre respectively. The farmer has \$40 000 available to buy seed and intends to spend it all.

(a) Use the information above to formulate two linear equations with three unknowns and solve it.

(b) Solutions to the system are choices that the farmer can make. Write down two reasonable solutions.

(c) Suppose that in the fall when the crops mature, the farmer can bring in revenue of \$100 per acre for corn, \$300 per acre for soybeans and \$80 per acre for oats. Which of your two solutions in the prior part would have resulted in a larger revenue?

2.24 Parametrize the solution set of this one-equation system.

$$x_1+x_2+\cdots+x_n=0$$

✓ **2.25** (a) Apply Gauss's Method to the left-hand side to solve

$$\begin{aligned}x+2y-w&=a\\2x+z&=b\\x+y+2w&=c\end{aligned}$$

for x, y, z, and w, in terms of the constants a, b, and c.

(b) Use your answer from the prior part to solve this.

$$\begin{aligned}x+2y-w&=3\\2x+z&=1\\x+y+2w&=-2\end{aligned}$$

✓ **2.26** Why is the comma needed in the notation '$a_{i,j}$' for matrix entries?

✓ **2.27** Give the 4×4 matrix whose i,j-th entry is

(a) $i+j$; (b) -1 to the $i+j$ power.

2.28 For any matrix A, the *transpose* of A, written A^{T}, is the matrix whose columns are the rows of A. Find the transpose of each of these.

(a) $\begin{pmatrix} 1 & 2 & 3 \\ 4 & 5 & 6 \end{pmatrix}$ (b) $\begin{pmatrix} 2 & -3 \\ 1 & 1 \end{pmatrix}$ (c) $\begin{pmatrix} 5 & 10 \\ 10 & 5 \end{pmatrix}$ (d) $\begin{pmatrix} 1 \\ 1 \\ 0 \end{pmatrix}$

✓ **2.29** (a) Describe all functions $f(x) = ax^2 + bx + c$ such that $f(1) = 2$ and $f(-1) = 6$.
(b) Describe all functions $f(x) = ax^2 + bx + c$ such that $f(1) = 2$.

2.30 Show that any set of five points from the plane $\mathbb{R}^2$ lie on a common conic section, that is, they all satisfy some equation of the form $ax^2 + by^2 + cxy + dx + ey + f = 0$ where some of $a, \ldots, f$ are nonzero.

2.31 Make up a four equations/four unknowns system having
(a) a one-parameter solution set;
(b) a two-parameter solution set;
(c) a three-parameter solution set.

? **2.32** [Shepelev] This puzzle is from a Russian web-site `http://www.arbuz.uz/` and there are many solutions to it, but mine uses linear algebra and is very naive. There's a planet inhabited by arbuzoids (watermeloners, to translate from Russian). Those creatures are found in three colors: red, green and blue. There are 13 red arbuzoids, 15 blue ones, and 17 green. When two differently colored arbuzoids meet, they both change to the third color.

The question is, can it ever happen that all of them assume the same color?

? **2.33** [USSR Olympiad no. 174]
(a) Solve the system of equations.

$$\begin{aligned} ax + y &= a^2 \\ x + ay &= 1 \end{aligned}$$

For what values of a does the system fail to have solutions, and for what values of a are there infinitely many solutions?
(b) Answer the above question for the system.

$$\begin{aligned} ax + y &= a^3 \\ x + ay &= 1 \end{aligned}$$

? **2.34** [Math. Mag., Sept. 1952] In air a gold-surfaced sphere weighs 7588 grams. It is known that it may contain one or more of the metals aluminum, copper, silver, or lead. When weighed successively under standard conditions in water, benzene, alcohol, and glycerin its respective weights are 6588, 6688, 6778, and 6328 grams. How much, if any, of the forenamed metals does it contain if the specific gravities of the designated substances are taken to be as follows?

Aluminum	2.7	Alcohol	0.81
Copper	8.9	Benzene	0.90
Gold	19.3	Glycerin	1.26
Lead	11.3	Water	1.00
Silver	10.8		

I.3 General = Particular + Homogeneous

In the prior subsection the descriptions of solution sets all fit a pattern. They have a vector that is a particular solution of the system added to an unrestricted combination of some other vectors. The solution set from Example 2.13 illustrates.

$$\{\underbrace{\begin{pmatrix}0\\4\\0\\0\\0\end{pmatrix}}_{\text{particular solution}}+\underbrace{w\begin{pmatrix}1\\-1\\3\\1\\0\end{pmatrix}+u\begin{pmatrix}1/2\\-1\\1/2\\0\\1\end{pmatrix}}_{\text{unrestricted combination}}\mid w,u\in\mathbb{R}\}$$

The combination is unrestricted in that w and u can be any real numbers—there is no condition like "such that $2w-u=0$" to restrict which pairs w, u we can use.

That example shows an infinite solution set fitting the pattern. The other two kinds of solution sets also fit. A one-element solution set fits because it has a particular solution, and the unrestricted combination part is trivial. (That is, instead of being a combination of two vectors or of one vector, it is a combination of no vectors. By convention the sum of an empty set of vectors is the zero vector.) An empty solution set fits the pattern because there is no particular solution and thus there are no sums of that form at all.

3.1 Theorem Any linear system's solution set has the form

$$\{\vec{p}+c_1\vec{\beta}_1+\cdots+c_k\vec{\beta}_k \mid c_1,\ldots,c_k\in\mathbb{R}\}$$

where $\vec{p}$ is any particular solution and where the number of vectors $\vec{\beta}_1, \ldots, \vec{\beta}_k$ equals the number of free variables that the system has after a Gaussian reduction.

The solution description has two parts, the particular solution $\vec{p}$ and the unrestricted linear combination of the $\vec{\beta}$'s. We shall prove the theorem with two corresponding lemmas.

We will focus first on the unrestricted combination. For that we consider systems that have the vector of zeroes as a particular solution so that we can shorten $\vec{p}+c_1\vec{\beta}_1+\cdots+c_k\vec{\beta}_k$ to $c_1\vec{\beta}_1+\cdots+c_k\vec{\beta}_k$.

3.2 Definition A linear equation is *homogeneous* if it has a constant of zero, so that it can be written as $a_1x_1+a_2x_2+\cdots+a_nx_n=0$.

3.3 Example With any linear system like

$$\begin{aligned} 3x + 4y &= 3 \\ 2x - \ y &= 1 \end{aligned}$$

we associate a system of homogeneous equations by setting the right side to zeros.

$$\begin{aligned} 3x + 4y &= 0 \\ 2x - \ y &= 0 \end{aligned}$$

Compare the reduction of the original system

$$\begin{aligned} 3x + 4y &= 3 \\ 2x - \ y &= 1 \end{aligned} \quad \xrightarrow{-(2/3)\rho_1+\rho_2} \quad \begin{aligned} 3x + \qquad 4y &= 3 \\ -(11/3)y &= -1 \end{aligned}$$

with the reduction of the associated homogeneous system.

$$\begin{aligned} 3x + 4y &= 0 \\ 2x - \ y &= 0 \end{aligned} \quad \xrightarrow{-(2/3)\rho_1+\rho_2} \quad \begin{aligned} 3x + \qquad 4y &= 0 \\ -(11/3)y &= 0 \end{aligned}$$

Obviously the two reductions go in the same way. We can study how to reduce a linear systems by instead studying how to reduce the associated homogeneous system.

Studying the associated homogeneous system has a great advantage over studying the original system. Nonhomogeneous systems can be inconsistent. But a homogeneous system must be consistent since there is always at least one solution, the zero vector.

3.4 Example Some homogeneous systems have the zero vector as their only solution.

$$\begin{aligned} 3x + 2y + z &= 0 \\ 6x + 4y \qquad &= 0 \\ y + z &= 0 \end{aligned} \quad \xrightarrow{-2\rho_1+\rho_2} \quad \begin{aligned} 3x + 2y + \quad z &= 0 \\ -2z &= 0 \\ y + \quad z &= 0 \end{aligned} \quad \xrightarrow{\rho_2\leftrightarrow\rho_3} \quad \begin{aligned} 3x + 2y + \quad z &= 0 \\ y + \quad z &= 0 \\ -2z &= 0 \end{aligned}$$

3.5 Example Some homogeneous systems have many solutions. One is the

Chemistry problem from the first page of the first subsection.

$$\begin{array}{rcl} 7x \qquad -7z \qquad\quad =0 \\ 8x+\ y-5z-2w=0 \\ y-3z \qquad\quad =0 \\ 3y-6z-\ w=0 \end{array} \xrightarrow{-(8/7)\rho_1+\rho_2} \begin{array}{r} 7x \qquad -7z \qquad\quad =0 \\ y+3z-2w=0 \\ y-3z \qquad\quad =0 \\ 3y-6z-\ w=0 \end{array}$$

$$\xrightarrow[-3\rho_2+\rho_4]{-\rho_2+\rho_3} \begin{array}{r} 7x \quad - \quad 7z \qquad\quad =0 \\ y+\quad 3z-2w=0 \\ -6z+2w=0 \\ -15z+5w=0 \end{array}$$

$$\xrightarrow{-(5/2)\rho_3+\rho_4} \begin{array}{r} 7x \quad - \quad 7z \qquad\quad =0 \\ y+\quad 3z-2w=0 \\ -6z+2w=0 \\ 0=0 \end{array}$$

The solution set

$$\{\begin{pmatrix} 1/3 \\ 1 \\ 1/3 \\ 1 \end{pmatrix} w \mid w \in \mathbb{R}\}$$

has many vectors besides the zero vector (if we interpret w as a number of molecules then solutions make sense only when w is a nonnegative multiple of 3).

3.6 Lemma For any homogeneous linear system there exist vectors $\vec{\beta}_1, \ldots, \vec{\beta}_k$ such that the solution set of the system is

$$\{c_1\vec{\beta}_1 + \cdots + c_k\vec{\beta}_k \mid c_1, \ldots, c_k \in \mathbb{R}\}$$

where k is the number of free variables in an echelon form version of the system.

We will make two points before the proof. The first is that the basic idea of the proof is straightforward. Consider this system of homogeneous equations in echelon form.

$$\begin{array}{r} x+y+2z+u+v=0 \\ y+\ \ z+u-v=0 \\ u+v=0 \end{array}$$

Start with the bottom equation. Express its leading variable in terms of the free variables with $u=-v$. For the next row up, substitute for the leading variable u of the row below $y+z+(-v)-v=0$ and solve for this row's leading variable $y=-z+2v$. Iterate: on the next row up, substitute expressions found in lower rows $x+(-z+2v)+2z+(-v)+v=0$ and solve for the leading variable $x=-z-2v$. To finish, write the solution in vector notation

$$\begin{pmatrix} x \\ y \\ z \\ u \\ v \end{pmatrix} = \begin{pmatrix} -1 \\ -1 \\ 1 \\ 0 \\ 0 \end{pmatrix} z + \begin{pmatrix} -2 \\ 2 \\ 0 \\ -1 \\ 1 \end{pmatrix} v \qquad \text{for } z, v \in \mathbb{R}$$

and recognize that the $\vec{\beta}_1$ and $\vec{\beta}_2$ of the lemma are the vectors associated with the free variables z and v.

The prior paragraph is an example, not a proof. But it does suggest the second point about the proof, its approach. The example moves row-by-row up the system, using the equations from lower rows to do the next row. This points to doing the proof by mathematical induction.*

Induction is an important and non-obvious proof technique that we shall use a number of times in this book. We will do proofs by induction in two steps, a base step and an inductive step. In the base step we verify that the statement is true for some first instance, here that for the bottom equation we can write the leading variable in terms of free variables. In the inductive step we must establish an implication, that if the statement is true for all prior cases then it follows for the present case also. Here we will establish that if for the bottom-most t rows we can express the leading variables in terms of the free variables, then for the $t+1$-th row from the bottom we can also express the leading variable in terms of those that are free.

Those two steps together prove the statement for all the rows because by the base step it is true for the bottom equation, and by the inductive step the fact that it is true for the bottom equation shows that it is true for the next one up. Then another application of the inductive step implies that it is true for the third equation up, etc.

Proof Apply Gauss's Method to get to echelon form. There may be some $0=0$ equations; we ignore these (if the system consists only of $0=0$ equations then the lemma is trivially true because there are no leading variables). But because the system is homogeneous there are no contradictory equations.

We will use induction to verify that each leading variable can be expressed in terms of free variables. That will finish the proof because we can use the free variables as parameters and the $\vec{\beta}$'s are the vectors of coefficients of those free variables.

For the base step consider the bottom-most equation

$$a_{m,\ell_m}x_{\ell_m} + a_{m,\ell_m+1}x_{\ell_m+1} + \cdots + a_{m,n}x_n = 0 \qquad (*)$$

* More information on mathematical induction is in the appendix.

where $a_{m,\ell_m} \neq 0$. (The 'ℓ' means "leading" so that x_{ℓ_m} is the leading variable in row m.) This is the bottom row so any variables after the leading one must be free. Move these to the right hand side and divide by a_{m,ℓ_m}

$$x_{\ell_m} = (-a_{m,\ell_m+1}/a_{m,\ell_m})x_{\ell_m+1} + \cdots + (-a_{m,n}/a_{m,\ell_m})x_n$$

to express the leading variable in terms of free variables. (There is a tricky technical point here: if in the bottom equation $(*)$ there are no variables to the right of x_{l_m} then $x_{\ell_m} = 0$. This satisfies the statement we are verifying because, as alluded to at the start of this subsection, it has x_{ℓ_m} written as a sum of a number of the free variables, namely as the sum of zero many, under the convention that a trivial sum totals to 0.)

For the inductive step assume that the statement holds for the bottom-most t rows, with $0 \leqslant t < m-1$. That is, assume that for the m-th equation, and the $(m-1)$-th equation, etc., up to and including the $(m-t)$-th equation, we can express the leading variable in terms of free ones. We must verify that this then also holds for the next equation up, the $(m-(t+1))$-th equation. For that, take each variable that leads in a lower equation $x_{\ell_m}, \ldots, x_{\ell_{m-t}}$ and substitute its expression in terms of free variables. We only need expressions for leading variables from lower equations because the system is in echelon form, so the leading variables in equations above this one do not appear in this equation. The result has a leading term of $a_{m-(t+1),\ell_{m-(t+1)}}x_{\ell_{m-(t+1)}}$ with $a_{m-(t+1),\ell_{m-(t+1)}} \neq 0$, and the rest of the left hand side is a linear combination of free variables. Move the free variables to the right side and divide by $a_{m-(t+1),\ell_{m-(t+1)}}$ to end with this equation's leading variable $x_{\ell_{m-(t+1)}}$ in terms of free variables.

We have done both the base step and the inductive step so by the principle of mathematical induction the proposition is true. QED

This shows, as discussed between the lemma and its proof, that we can parametrize solution sets using the free variables. We say that the set of vectors $\{c_1\vec{\beta}_1 + \cdots + c_k\vec{\beta}_k \mid c_1, \ldots, c_k \in \mathbb{R}\}$ is *generated by* or *spanned by* the set $\{\vec{\beta}_1, \ldots, \vec{\beta}_k\}$.

To finish the proof of Theorem 3.1 the next lemma considers the particular solution part of the solution set's description.

3.7 Lemma For a linear system and for any particular solution $\vec{p}$, the solution set equals $\{\vec{p} + \vec{h} \mid \vec{h} \text{ satisfies the associated homogeneous system}\}$.

PROOF We will show mutual set inclusion, that any solution to the system is in the above set and that anything in the set is a solution of the system.*

* More information on set equality is in the appendix.

For set inclusion the first way, that if a vector solves the system then it is in the set described above, assume that $\vec{s}$ solves the system. Then $\vec{s}-\vec{p}$ solves the associated homogeneous system since for each equation index i,

$$\begin{aligned}
&a_{i,1}(s_1-p_1)+\cdots+a_{i,n}(s_n-p_n) \\
&\qquad=(a_{i,1}s_1+\cdots+a_{i,n}s_n)-(a_{i,1}p_1+\cdots+a_{i,n}p_n)=d_i-d_i=0
\end{aligned}$$

where p_j and s_j are the j-th components of $\vec{p}$ and $\vec{s}$. Express $\vec{s}$ in the required $\vec{p}+\vec{h}$ form by writing $\vec{s}-\vec{p}$ as $\vec{h}$.

For set inclusion the other way, take a vector of the form $\vec{p}+\vec{h}$, where $\vec{p}$ solves the system and $\vec{h}$ solves the associated homogeneous system and note that $\vec{p}+\vec{h}$ solves the given system since for any equation index i,

$$\begin{aligned}
&a_{i,1}(p_1+h_1)+\cdots+a_{i,n}(p_n+h_n) \\
&\qquad=(a_{i,1}p_1+\cdots+a_{i,n}p_n)+(a_{i,1}h_1+\cdots+a_{i,n}h_n)=d_i+0=d_i
\end{aligned}$$

where as earlier p_j and h_j are the j-th components of $\vec{p}$ and $\vec{h}$. QED

The two lemmas together establish Theorem 3.1. Remember that theorem with the slogan, "General = Particular + Homogeneous".

3.8 Example This system illustrates Theorem 3.1.

$$\begin{aligned}
x+2y-z&=1 \\
2x+4y&=2 \\
y-3z&=0
\end{aligned}$$

Gauss's Method

$$\xrightarrow{-2\rho_1+\rho_2}
\begin{aligned}
x+2y-z&=1 \\
2z&=0 \\
y-3z&=0
\end{aligned}
\xrightarrow{\rho_2\leftrightarrow\rho_3}
\begin{aligned}
x+2y-z&=1 \\
y-3z&=0 \\
2z&=0
\end{aligned}$$

shows that the general solution is a singleton set.

$$\{\begin{pmatrix}1\\0\\0\end{pmatrix}\}$$

That single vector is obviously a particular solution. The associated homogeneous system reduces via the same row operations

$$\begin{aligned}
x+2y-z&=0 \\
2x+4y&=0 \\
y-3z&=0
\end{aligned}
\xrightarrow{-2\rho_1+\rho_2}\xrightarrow{\rho_2\leftrightarrow\rho_3}
\begin{aligned}
x+2y-z&=0 \\
y-3z&=0 \\
2z&=0
\end{aligned}$$

to also give a singleton set.

$$\{\begin{pmatrix}0\\0\\0\end{pmatrix}\}$$

So, as discussed at the start of this subsection, in this single-solution case the general solution results from taking the particular solution and adding to it the unique solution of the associated homogeneous system.

3.9 Example The start of this subsection also discusses that the case where the general solution set is empty fits the General = Particular + Homogeneous pattern too. This system illustrates.

$$\begin{array}{rcl} x \quad + z + w = -1 \\ 2x - y \quad + w = 3 \\ x + y + 3z + 2w = 1 \end{array} \xrightarrow[-\rho_1+\rho_3]{-2\rho_1+\rho_2} \begin{array}{rcl} x \quad + z + w = -1 \\ -y - 2z - w = 5 \\ y + 2z + w = 2 \end{array}$$

It has no solutions because the final two equations conflict. But the associated homogeneous system does have a solution, as do all homogeneous systems.

$$\begin{array}{rcl} x \quad + z + w = 0 \\ 2x - y \quad + w = 0 \\ x + y + 3z + 2w = 0 \end{array} \xrightarrow[-\rho_1+\rho_3]{-2\rho_1+\rho_2} \xrightarrow{\rho_2+\rho_3} \begin{array}{rcl} x \quad + z + w = 0 \\ -y - 2z - w = 0 \\ 0 = 0 \end{array}$$

In fact, the solution set is infinite.

$$\{\begin{pmatrix}-1\\-2\\1\\0\end{pmatrix}z + \begin{pmatrix}-1\\-1\\0\\1\end{pmatrix}w \mid z, w \in \mathbb{R}\}$$

Nonetheless, because the original system has no particular solution, its general solution set is empty—there are no vectors of the form $\vec{p} + \vec{h}$ because there are no $\vec{p}$'s.

3.10 Corollary Solution sets of linear systems are either empty, have one element, or have infinitely many elements.

PROOF We've seen examples of all three happening so we need only prove that there are no other possibilities.

First observe a homogeneous system with at least one non-$\vec{0}$ solution $\vec{v}$ has infinitely many solutions. This is because any scalar multiple of $\vec{v}$ also solves the homogeneous system and there are infinitely many vectors in the set of scalar multiples of $\vec{v}$: if $s, t \in \mathbb{R}$ are unequal then $s\vec{v} \neq t\vec{v}$, since $s\vec{v} - t\vec{v} = (s - t)\vec{v}$ is non-$\vec{0}$ as any non-0 component of $\vec{v}$, when rescaled by the non-0 factor $s-t$, will give a non-0 value.

Now apply Lemma 3.7 to conclude that a solution set

$$\{\vec{p}+\vec{h} \mid \vec{h} \text{ solves the associated homogeneous system}\}$$

is either empty (if there is no particular solution $\vec{p}$), or has one element (if there is a $\vec{p}$ and the homogeneous system has the unique solution $\vec{0}$), or is infinite (if there is a $\vec{p}$ and the homogeneous system has a non-$\vec{0}$ solution, and thus by the prior paragraph has infinitely many solutions). QED

This table summarizes the factors affecting the size of a general solution.

		number of solutions of the homogeneous system	
		one	*infinitely many*
particular solution exists?	*yes*	unique solution	infinitely many solutions
	no	no solutions	no solutions

The dimension on the top of the table is the simpler one. When we perform Gauss's Method on a linear system, ignoring the constants on the right side and so paying attention only to the coefficients on the left-hand side, we either end with every variable leading some row or else we find some variable that does not lead a row, that is, we find some variable that is free. (We formalize "ignoring the constants on the right" by considering the associated homogeneous system.)

A notable special case is systems having the same number of equations as unknowns. Such a system will have a solution, and that solution will be unique, if and only if it reduces to an echelon form system where every variable leads its row (since there are the same number of variables as rows), which will happen if and only if the associated homogeneous system has a unique solution.

3.11 Definition A square matrix is *nonsingular* if it is the matrix of coefficients of a homogeneous system with a unique solution. It is *singular* otherwise, that is, if it is the matrix of coefficients of a homogeneous system with infinitely many solutions.

3.12 Example The first of these matrices is nonsingular while the second is singular

$$\begin{pmatrix} 1 & 2 \\ 3 & 4 \end{pmatrix} \qquad \begin{pmatrix} 1 & 2 \\ 3 & 6 \end{pmatrix}$$

because the first of these homogeneous systems has a unique solution while the second has infinitely many solutions.

$$\begin{aligned} x + 2y &= 0 \\ 3x + 4y &= 0 \end{aligned} \qquad \begin{aligned} x + 2y &= 0 \\ 3x + 6y &= 0 \end{aligned}$$

We have made the distinction in the definition because a system with the same number of equations as variables behaves in one of two ways, depending on whether its matrix of coefficients is nonsingular or singular. Where the matrix of coefficients is nonsingular the system has a unique solution for any constants on the right side: for instance, Gauss's Method shows that this system

$$\begin{aligned} x + 2y &= a \\ 3x + 4y &= b \end{aligned}$$

has the unique solution $x = b - 2a$ and $y = (3a - b)/2$. On the other hand, where the matrix of coefficients is singular the system never has a unique solution — it has either no solutions or else has infinitely many, as with these.

$$\begin{aligned} x + 2y &= 1 \\ 3x + 6y &= 2 \end{aligned} \qquad \begin{aligned} x + 2y &= 1 \\ 3x + 6y &= 3 \end{aligned}$$

The definition uses the word 'singular' because it means "departing from general expectation." People often, naively, expect that systems with the same number of variables as equations will have a unique solution. Thus, we can think of the word as connoting "troublesome," or at least "not ideal." (That 'singular' applies to those systems that never have exactly one solution is ironic, but it is the standard term.)

3.13 Example The systems from Example 3.3, Example 3.4, and Example 3.8 each have an associated homogeneous system with a unique solution. Thus these matrices are nonsingular.

$$\begin{pmatrix} 3 & 4 \\ 2 & -1 \end{pmatrix} \qquad \begin{pmatrix} 3 & 2 & 1 \\ 6 & -4 & 0 \\ 0 & 1 & 1 \end{pmatrix} \qquad \begin{pmatrix} 1 & 2 & -1 \\ 2 & 4 & 0 \\ 0 & 1 & -3 \end{pmatrix}$$

The Chemistry problem from Example 3.5 is a homogeneous system with more than one solution so its matrix is singular.

$$\begin{pmatrix} 7 & 0 & -7 & 0 \\ 8 & 1 & -5 & -2 \\ 0 & 1 & -3 & 0 \\ 0 & 3 & -6 & -1 \end{pmatrix}$$

The table above has two dimensions. We have considered the one on top: we can tell into which column a given linear system goes solely by considering the system's left-hand side; the constants on the right-hand side play no role in this.

The table's other dimension, determining whether a particular solution exists, is tougher. Consider these two systems with the same left side but different right sides.

$$\begin{aligned} 3x + 2y &= 5 \\ 3x + 2y &= 5 \end{aligned} \qquad \begin{aligned} 3x + 2y &= 5 \\ 3x + 2y &= 4 \end{aligned}$$

The first has a solution while the second does not, so here the constants on the right side decide if the system has a solution. We could conjecture that the left side of a linear system determines the number of solutions while the right side determines if solutions exist but that guess is not correct. Compare these two, with the same right sides but different left sides.

$$\begin{aligned} 3x + 2y &= 5 \\ 4x + 2y &= 4 \end{aligned} \qquad \begin{aligned} 3x + 2y &= 5 \\ 3x + 2y &= 4 \end{aligned}$$

The first has a solution but the second does not. Thus the constants on the right side of a system don't alone determine whether a solution exists. Rather, that depends on some interaction between the left and right.

For some intuition about that interaction, consider this system with one of the coefficients left unspecified, as the variable c.

$$\begin{aligned} x + 2y + 3z &= 1 \\ x + y + z &= 1 \\ cx + 3y + 4z &= 0 \end{aligned}$$

If $c = 2$ then this system has no solution because the left-hand side has the third row as the sum of the first two, while the right-hand does not. If $c \neq 2$ then this system has a unique solution (try it with $c = 1$). For a system to have a solution, if one row of the matrix of coefficients on the left is a linear combination of other rows then on the right the constant from that row must be the same combination of constants from the same rows.

More intuition about the interaction comes from studying linear combinations. That will be our focus in the second chapter, after we finish the study of Gauss's Method itself in the rest of this chapter.

Exercises

3.14 Solve this system. Then solve the associated homogeneous system.

$$\begin{aligned} x + y - 2z &= 0 \\ x - y &= -3 \\ 3x - y - 2z &= -6 \\ 2y - 2z &= 3 \end{aligned}$$

✓ **3.15** Solve each system. Express the solution set using vectors. Identify the particular solution and the solution set of the homogeneous system. (These systems also appear in Exercise 18.)

(a) $\begin{array}{rl} 3x+6y &= 18 \\ x+2y &= 6 \end{array}$
(b) $\begin{array}{rl} x+y &= 1 \\ x-y &= -1 \end{array}$
(c) $\begin{array}{rl} x_1 \phantom{{}-x_2} + x_3 &= 4 \\ x_1-x_2+2x_3 &= 5 \\ 4x_1-x_2+5x_3 &= 17 \end{array}$

(d) $\begin{array}{rl} 2a+b-c &= 2 \\ 2a \phantom{{}+b} +c &= 3 \\ a-b \phantom{{}+c} &= 0 \end{array}$
(e) $\begin{array}{rl} x+2y-z \phantom{{}+w} &= 3 \\ 2x+y \phantom{{}+z} +w &= 4 \\ x-y+z+w &= 1 \end{array}$
(f) $\begin{array}{rl} x \phantom{{}+y} +z+w &= 4 \\ 2x+y \phantom{{}+z} -w &= 2 \\ 3x+y+z \phantom{{}+w} &= 7 \end{array}$

3.16 Solve each system, giving the solution set in vector notation. Identify the particular solution and the solution of the homogeneous system.

(a) $\begin{array}{rl} 2x+y-z &= 1 \\ 4x-y \phantom{{}-z} &= 3 \end{array}$
(b) $\begin{array}{rl} x \phantom{{}+2y} -z \phantom{{}-w} &= 1 \\ y+2z-w &= 3 \\ x+2y+3z-w &= 7 \end{array}$
(c) $\begin{array}{rl} x-y+z \phantom{{}+w} &= 0 \\ y \phantom{{}+z} +w &= 0 \\ 3x-2y+3z+w &= 0 \\ -y \phantom{{}+z} -w &= 0 \end{array}$

(d) $\begin{array}{rl} a+2b+3c+d-e &= 1 \\ 3a-b+c+d+e &= 3 \end{array}$

✓ **3.17** For the system

$$\begin{array}{rl} 2x-y \phantom{{}+z} -w &= 3 \\ y+z+2w &= 2 \\ x-2y-z \phantom{{}+2w} &= -1 \end{array}$$

which of these can be used as the particular solution part of some general solution?

(a) $\begin{pmatrix} 0 \\ -3 \\ 5 \\ 0 \end{pmatrix}$ (b) $\begin{pmatrix} 2 \\ 1 \\ 1 \\ 0 \end{pmatrix}$ (c) $\begin{pmatrix} -1 \\ -4 \\ 8 \\ -1 \end{pmatrix}$

✓ **3.18** Lemma 3.7 says that we can use any particular solution for $\vec{p}$. Find, if possible, a general solution to this system

$$\begin{array}{rl} x-y \phantom{{}-z} +w &= 4 \\ 2x+3y-z \phantom{{}+w} &= 0 \\ y+z+w &= 4 \end{array}$$

that uses the given vector as its particular solution.

(a) $\begin{pmatrix} 0 \\ 0 \\ 0 \\ 4 \end{pmatrix}$ (b) $\begin{pmatrix} -5 \\ 1 \\ -7 \\ 10 \end{pmatrix}$ (c) $\begin{pmatrix} 2 \\ -1 \\ 1 \\ 1 \end{pmatrix}$

3.19 One is nonsingular while the other is singular. Which is which?

(a) $\begin{pmatrix} 1 & 3 \\ 4 & -12 \end{pmatrix}$ (b) $\begin{pmatrix} 1 & 3 \\ 4 & 12 \end{pmatrix}$

✓ **3.20** Singular or nonsingular?

(a) $\begin{pmatrix} 1 & 2 \\ 1 & 3 \end{pmatrix}$ (b) $\begin{pmatrix} 1 & 2 \\ -3 & -6 \end{pmatrix}$ (c) $\begin{pmatrix} 1 & 2 & 1 \\ 1 & 3 & 1 \end{pmatrix}$ (Careful!)

(d) $\begin{pmatrix} 1 & 2 & 1 \\ 1 & 1 & 3 \\ 3 & 4 & 7 \end{pmatrix}$ (e) $\begin{pmatrix} 2 & 2 & 1 \\ 1 & 0 & 5 \\ -1 & 1 & 4 \end{pmatrix}$

✓ **3.21** Is the given vector in the set generated by the given set?

(a) $\begin{pmatrix}2\\3\end{pmatrix}, \{\begin{pmatrix}1\\4\end{pmatrix}, \begin{pmatrix}1\\5\end{pmatrix}\}$

(b) $\begin{pmatrix}-1\\0\\1\end{pmatrix}, \{\begin{pmatrix}2\\1\\0\end{pmatrix}, \begin{pmatrix}1\\0\\1\end{pmatrix}\}$

(c) $\begin{pmatrix}1\\3\\0\end{pmatrix}, \{\begin{pmatrix}1\\0\\4\end{pmatrix}, \begin{pmatrix}2\\1\\5\end{pmatrix}, \begin{pmatrix}3\\3\\0\end{pmatrix}, \begin{pmatrix}4\\2\\1\end{pmatrix}\}$

(d) $\begin{pmatrix}1\\0\\1\\1\end{pmatrix}, \{\begin{pmatrix}2\\1\\0\\1\end{pmatrix}, \begin{pmatrix}3\\0\\0\\2\end{pmatrix}\}$

3.22 Prove that any linear system with a nonsingular matrix of coefficients has a solution, and that the solution is unique.

3.23 In the proof of Lemma 3.6, what happens if there are no non-$0 = 0$ equations?

✓ **3.24** Prove that if $\vec{s}$ and $\vec{t}$ satisfy a homogeneous system then so do these vectors.

(a) $\vec{s}+\vec{t}$ (b) $3\vec{s}$ (c) $k\vec{s}+m\vec{t}$ for $k, m \in \mathbb{R}$

What's wrong with this argument: "These three show that if a homogeneous system has one solution then it has many solutions — any multiple of a solution is another solution, and any sum of solutions is a solution also — so there are no homogeneous systems with exactly one solution."?

3.25 Prove that if a system with only rational coefficients and constants has a solution then it has at least one all-rational solution. Must it have infinitely many?

II Linear Geometry

If you have seen the elements of vectors then this section is an optional review. However, later work will refer to this material so if this is not a review then it is not optional.

In the first section we had to do a bit of work to show that there are only three types of solution sets — singleton, empty, and infinite. But this is easy to see geometrically in the case of systems with two equations and two unknowns. Draw each two-unknowns equation as a line in the plane, and then the two lines could have a unique intersection, be parallel, or be the same line.

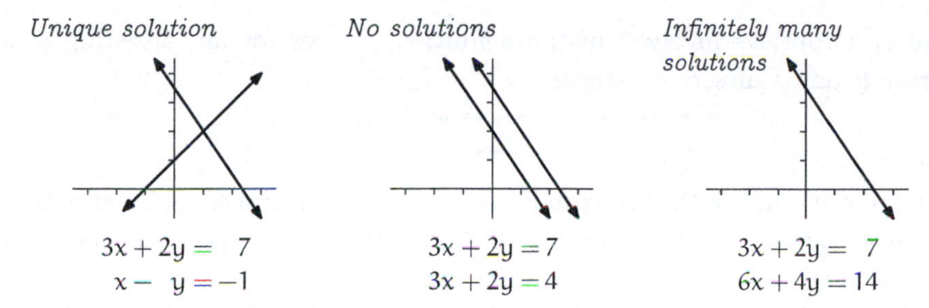

These pictures aren't a short way to prove the results from the prior section, because those results apply to linear systems of any size. But they do broaden our understanding of those results.

This section develops what we need to express our results geometrically. In particular, while the two-dimensional case is familiar, to extend to systems with more than two unknowns we shall need some higher-dimensional geometry.

II.1 Vectors in Space

"Higher-dimensional geometry" sounds exotic. It is exotic — interesting and eye-opening. But it isn't distant or unreachable.

We begin by defining one-dimensional space to be $\mathbb{R}$. To see that the definition is reasonable, picture a one-dimensional space

and pick a point to label 0 and another to label 1.

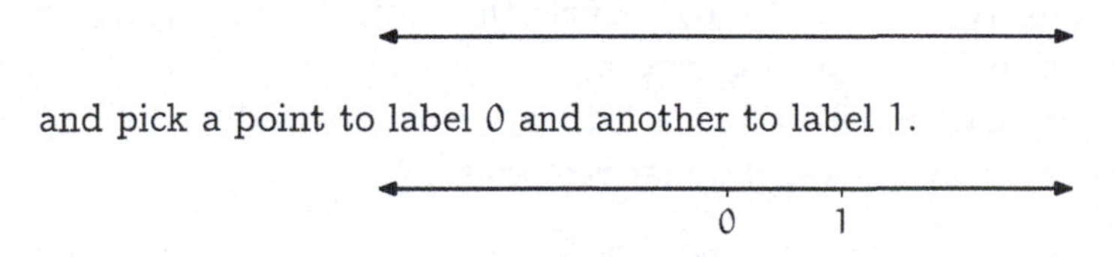

Now, with a scale and a direction, we have a correspondence with $\mathbb{R}$. For instance, to find the point matching $+2.17$, start at 0 and head in the direction of 1, and go 2.17 times as far.

The basic idea here, combining magnitude with direction, is the key to extending to higher dimensions.

An object comprised of a magnitude and a direction is a *vector* (we use the same word as in the prior section because we shall show below how to describe such an object with a column vector). We can draw a vector as having some length and pointing in some direction.

There is a subtlety involved in the definition of a vector as consisting of a magnitude and a direction—these

are equal, even though they start in different places They are equal because they have equal lengths and equal directions. Again: those vectors are not just alike, they are equal.

How can things that are in different places be equal? Think of a vector as representing a displacement (the word 'vector' is Latin for "carrier" or "traveler"). These two squares undergo displacements that are equal despite that they start in different places.

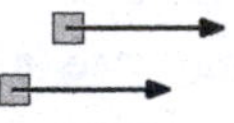

When we want to emphasize this property vectors have of not being anchored we refer to them as *free* vectors. Thus, these free vectors are equal, as each is a displacement of one over and two up.

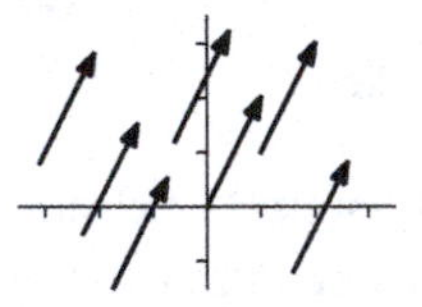

More generally, vectors in the plane are the same if and only if they have the same change in first components and the same change in second components: the vector extending from (a_1, a_2) to (b_1, b_2) equals the vector from (c_1, c_2) to (d_1, d_2) if and only if $b_1 - a_1 = d_1 - c_1$ and $b_2 - a_2 = d_2 - c_2$.

Saying 'the vector that, were it to start at (a_1, a_2), would extend to (b_1, b_2)' would be unwieldy. We instead describe that vector as

$$\begin{pmatrix} b_1 - a_1 \\ b_2 - a_2 \end{pmatrix}$$

so that we represent the 'one over and two up' arrows shown above in this way.

$$\begin{pmatrix}1\\2\end{pmatrix}$$

We often draw the arrow as starting at the origin, and we then say it is in the *canonical position* (or *natural position* or *standard position*). When

$$\vec{v} = \begin{pmatrix}v_1\\v_2\end{pmatrix}$$

is in canonical position then it extends from the origin to the endpoint (v_1, v_2).

We will typically say "the point

$$\begin{pmatrix}1\\2\end{pmatrix}\text{"}$$

rather than "the endpoint of the canonical position of" that vector. Thus, we will call each of these $\mathbb{R}^2$.

$$\{(x_1, x_2) \mid x_1, x_2 \in \mathbb{R}\} \qquad \{\begin{pmatrix}x_1\\x_2\end{pmatrix} \mid x_1, x_2 \in \mathbb{R}\}$$

In the prior section we defined vectors and vector operations with an algebraic motivation;

$$r \cdot \begin{pmatrix}v_1\\v_2\end{pmatrix} = \begin{pmatrix}rv_1\\rv_2\end{pmatrix} \qquad \begin{pmatrix}v_1\\v_2\end{pmatrix} + \begin{pmatrix}w_1\\w_2\end{pmatrix} = \begin{pmatrix}v_1 + w_1\\v_2 + w_2\end{pmatrix}$$

we can now understand those operations geometrically. For instance, if $\vec{v}$ represents a displacement then $3\vec{v}$ represents a displacement in the same direction but three times as far and $-1\vec{v}$ represents a displacement of the same distance as $\vec{v}$ but in the opposite direction.

$\vec{v}$

$3\vec{v}$

$-\vec{v}$

And, where $\vec{v}$ and $\vec{w}$ represent displacements, $\vec{v}+\vec{w}$ represents those displacements combined.

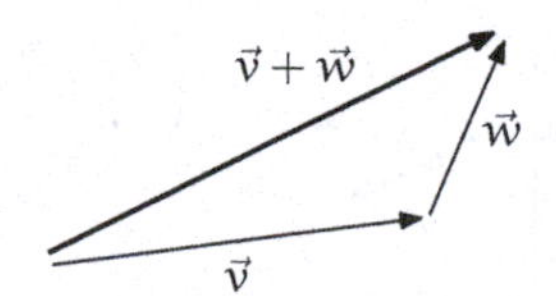

The long arrow is the combined displacement in this sense: imagine that you are walking on a ship's deck. Suppose that in one minute the ship's motion gives it a displacement relative to the sea of $\vec{v}$, and in the same minute your walking gives you a displacement relative to the ship's deck of $\vec{w}$. Then $\vec{v}+\vec{w}$ is your displacement relative to the sea.

Another way to understand the vector sum is with the *parallelogram rule.* Draw the parallelogram formed by the vectors $\vec{v}$ and $\vec{w}$. Then the sum $\vec{v}+\vec{w}$ extends along the diagonal to the far corner.

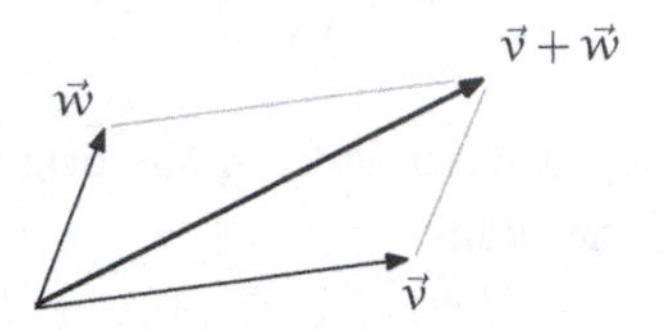

The above drawings show how vectors and vector operations behave in $\mathbb{R}^2$. We can extend to $\mathbb{R}^3$, or to even higher-dimensional spaces where we have no pictures, with the obvious generalization: the free vector that, if it starts at $(a_1,\ldots,a_n)$, ends at $(b_1,\ldots,b_n)$, is represented by this column.

$$\begin{pmatrix} b_1-a_1 \\ \vdots \\ b_n-a_n \end{pmatrix}$$

Vectors are equal if they have the same representation. We aren't too careful about distinguishing between a point and the vector whose canonical representation ends at that point.

$$\mathbb{R}^n=\{\begin{pmatrix} v_1 \\ \vdots \\ v_n \end{pmatrix} \mid v_1,\ldots,v_n\in\mathbb{R}\}$$

And, we do addition and scalar multiplication component-wise.

Having considered points, we next turn to lines. In $\mathbb{R}^2$, the line through $(1,2)$ and $(3,1)$ is comprised of (the endpoints of) the vectors in this set.

$$\{\begin{pmatrix} 1 \\ 2 \end{pmatrix}+t\begin{pmatrix} 2 \\ -1 \end{pmatrix} \mid t\in\mathbb{R}\}$$

That description expresses this picture.

$$\begin{pmatrix} 2 \\ -1 \end{pmatrix}=\begin{pmatrix} 3 \\ 1 \end{pmatrix}-\begin{pmatrix} 1 \\ 2 \end{pmatrix}$$

The vector that in the description is associated with the parameter t

$$\begin{pmatrix}2\\-1\end{pmatrix}=\begin{pmatrix}3\\1\end{pmatrix}-\begin{pmatrix}1\\2\end{pmatrix}$$

is the one shown in the picture as having its whole body in the line — it is a *direction vector* for the line. Note that points on the line to the left of $x = 1$ are described using negative values of t.

In $\mathbb{R}^3$, the line through $(1, 2, 1)$ and $(2, 3, 2)$ is the set of (endpoints of) vectors of this form

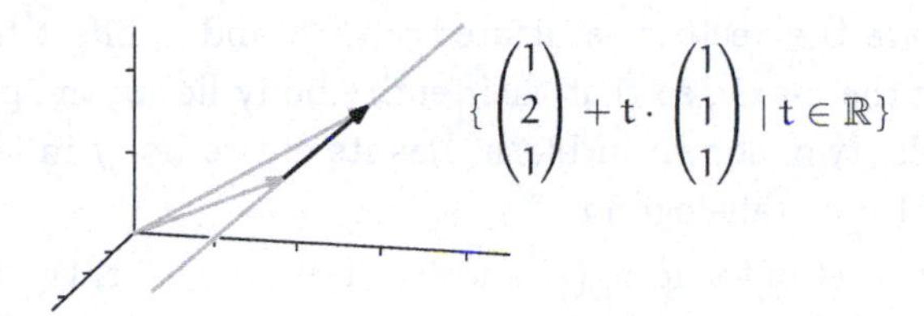

$$\{\begin{pmatrix}1\\2\\1\end{pmatrix}+t\cdot\begin{pmatrix}1\\1\\1\end{pmatrix}\mid t\in\mathbb{R}\}$$

and lines in even higher-dimensional spaces work in the same way.

In $\mathbb{R}^3$, a line uses one parameter so that a particle on that line would be free to move back and forth in one dimension. A plane involves two parameters. For example, the plane through the points $(1, 0, 5)$, $(2, 1, -3)$, and $(-2, 4, 0.5)$ consists of (endpoints of) the vectors in this set.

$$\{\begin{pmatrix}1\\0\\5\end{pmatrix}+t\begin{pmatrix}1\\1\\-8\end{pmatrix}+s\begin{pmatrix}-3\\4\\-4.5\end{pmatrix}\mid t,s\in\mathbb{R}\}$$

The column vectors associated with the parameters come from these calculations.

$$\begin{pmatrix}1\\1\\-8\end{pmatrix}=\begin{pmatrix}2\\1\\-3\end{pmatrix}-\begin{pmatrix}1\\0\\5\end{pmatrix}\qquad\begin{pmatrix}-3\\4\\-4.5\end{pmatrix}=\begin{pmatrix}-2\\4\\0.5\end{pmatrix}-\begin{pmatrix}1\\0\\5\end{pmatrix}$$

As with the line, note that we describe some points in this plane with negative t's or negative s's or both.

Calculus books often describe a plane by using a single linear equation.

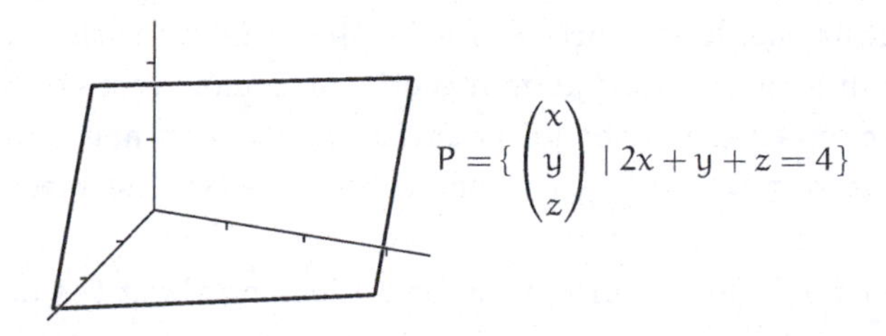

$$P=\{\begin{pmatrix}x\\y\\z\end{pmatrix}\mid 2x+y+z=4\}$$

To translate from this to the vector description, think of this as a one-equation linear system and parametrize: $x = 2 - y/2 - z/2$.

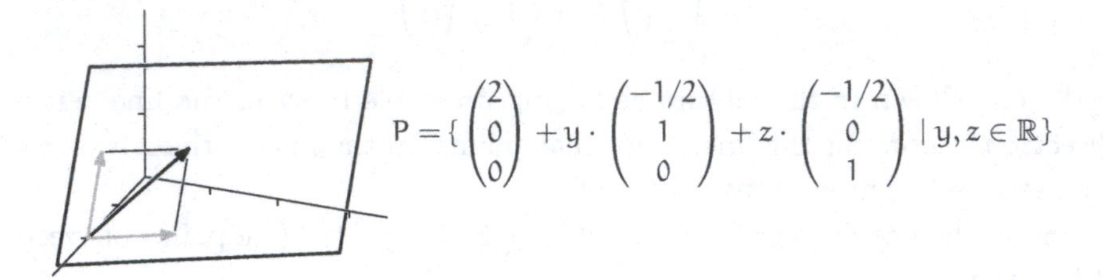

$$P = \{ \begin{pmatrix} 2 \\ 0 \\ 0 \end{pmatrix} + y \cdot \begin{pmatrix} -1/2 \\ 1 \\ 0 \end{pmatrix} + z \cdot \begin{pmatrix} -1/2 \\ 0 \\ 1 \end{pmatrix} \mid y, z \in \mathbb{R} \}$$

Shown in grey are the vectors associated with y and z, offset from the origin by 2 units along the x-axis, so that their entire body lies in the plane. Thus the vector sum of the two, shown in black, has its entire body in the plane along with the rest of the parallelogram.

Generalizing, a set of the form $\{\vec{p} + t_1\vec{v}_1 + t_2\vec{v}_2 + \cdots + t_k\vec{v}_k \mid t_1, \ldots, t_k \in \mathbb{R}\}$ where $\vec{v}_1, \ldots, \vec{v}_k \in \mathbb{R}^n$ and $k \leqslant n$ is a k-*dimensional linear surface* (or k-*flat*). For example, in $\mathbb{R}^4$

$$\{ \begin{pmatrix} 2 \\ \pi \\ 3 \\ -0.5 \end{pmatrix} + t \begin{pmatrix} 1 \\ 0 \\ 0 \\ 0 \end{pmatrix} \mid t \in \mathbb{R} \}$$

is a line,

$$\{ \begin{pmatrix} 0 \\ 0 \\ 0 \\ 0 \end{pmatrix} + t \begin{pmatrix} 1 \\ 1 \\ 0 \\ -1 \end{pmatrix} + s \begin{pmatrix} 2 \\ 0 \\ 1 \\ 0 \end{pmatrix} \mid t, s \in \mathbb{R} \}$$

is a plane, and

$$\{ \begin{pmatrix} 3 \\ 1 \\ -2 \\ 0.5 \end{pmatrix} + r \begin{pmatrix} 0 \\ 0 \\ 0 \\ -1 \end{pmatrix} + s \begin{pmatrix} 1 \\ 0 \\ 1 \\ 0 \end{pmatrix} + t \begin{pmatrix} 2 \\ 0 \\ 1 \\ 0 \end{pmatrix} \mid r, s, t \in \mathbb{R} \}$$

is a three-dimensional linear surface. Again, the intuition is that a line permits motion in one direction, a plane permits motion in combinations of two directions, etc. When the dimension of the linear surface is one less than the dimension of the space, that is, when in $\mathbb{R}^n$ we have an $(n-1)$-flat, the surface is called a *hyperplane*.

A description of a linear surface can be misleading about the dimension. For example, this

$$L = \{ \begin{pmatrix} 1 \\ 0 \\ -1 \\ -2 \end{pmatrix} + t \begin{pmatrix} 1 \\ 1 \\ 0 \\ -1 \end{pmatrix} + s \begin{pmatrix} 2 \\ 2 \\ 0 \\ -2 \end{pmatrix} \mid t, s \in \mathbb{R} \}$$

is a *degenerate* plane because it is actually a line, since the vectors are multiples of each other and we can omit one.

$$L = \{ \begin{pmatrix} 1 \\ 0 \\ -1 \\ -2 \end{pmatrix} + r \begin{pmatrix} 1 \\ 1 \\ 0 \\ -1 \end{pmatrix} \mid r \in \mathbb{R} \}$$

We shall see in the Linear Independence section of Chapter Two what relationships among vectors causes the linear surface they generate to be degenerate.

We now can restate in geometric terms our conclusions from earlier. First, the solution set of a linear system with n unknowns is a linear surface in $\mathbb{R}^n$. Specifically, it is a k-dimensional linear surface, where k is the number of free variables in an echelon form version of the system. For instance, in the single equation case the solution set is an $n-1$-dimensional hyperplane in $\mathbb{R}^n$ (where $n > 0$). Second, the solution set of a homogeneous linear system is a linear surface passing through the origin. Finally, we can view the general solution set of any linear system as being the solution set of its associated homogeneous system offset from the origin by a vector, namely by any particular solution.

Exercises

✓ **1.1** Find the canonical name for each vector.

(a) the vector from $(2, 1)$ to $(4, 2)$ in $\mathbb{R}^2$
(b) the vector from $(3, 3)$ to $(2, 5)$ in $\mathbb{R}^2$
(c) the vector from $(1, 0, 6)$ to $(5, 0, 3)$ in $\mathbb{R}^3$
(d) the vector from $(6, 8, 8)$ to $(6, 8, 8)$ in $\mathbb{R}^3$

✓ **1.2** Decide if the two vectors are equal.

(a) the vector from $(5, 3)$ to $(6, 2)$ and the vector from $(1, -2)$ to $(1, 1)$
(b) the vector from $(2, 1, 1)$ to $(3, 0, 4)$ and the vector from $(5, 1, 4)$ to $(6, 0, 7)$

✓ **1.3** Does $(1, 0, 2, 1)$ lie on the line through $(-2, 1, 1, 0)$ and $(5, 10, -1, 4)$?

✓ **1.4** (a) Describe the plane through $(1, 1, 5, -1)$, $(2, 2, 2, 0)$, and $(3, 1, 0, 4)$.
(b) Is the origin in that plane?

1.5 Describe the plane that contains this point and line.

$$\begin{pmatrix} 2 \\ 0 \\ 3 \end{pmatrix} \qquad \{ \begin{pmatrix} -1 \\ 0 \\ -4 \end{pmatrix} + \begin{pmatrix} 1 \\ 1 \\ 2 \end{pmatrix} t \mid t \in \mathbb{R} \}$$

✓ **1.6** Intersect these planes.

$$\{\begin{pmatrix}1\\1\\1\end{pmatrix}t+\begin{pmatrix}0\\1\\3\end{pmatrix}s \mid t,s\in\mathbb{R}\}\qquad \{\begin{pmatrix}1\\1\\0\end{pmatrix}+\begin{pmatrix}0\\3\\0\end{pmatrix}k+\begin{pmatrix}2\\0\\4\end{pmatrix}m \mid k,m\in\mathbb{R}\}$$

✓ **1.7** Intersect each pair, if possible.

(a) $\{\begin{pmatrix}1\\1\\2\end{pmatrix}+t\begin{pmatrix}0\\1\\1\end{pmatrix} \mid t\in\mathbb{R}\}$, $\{\begin{pmatrix}1\\3\\-2\end{pmatrix}+s\begin{pmatrix}0\\1\\2\end{pmatrix} \mid s\in\mathbb{R}\}$

(b) $\{\begin{pmatrix}2\\0\\1\end{pmatrix}+t\begin{pmatrix}1\\1\\-1\end{pmatrix} \mid t\in\mathbb{R}\}$, $\{s\begin{pmatrix}0\\1\\2\end{pmatrix}+w\begin{pmatrix}0\\4\\1\end{pmatrix} \mid s,w\in\mathbb{R}\}$

1.8 When a plane does not pass through the origin, performing operations on vectors whose bodies lie in it is more complicated than when the plane passes through the origin. Consider the picture in this subsection of the plane

$$\{\begin{pmatrix}2\\0\\0\end{pmatrix}+\begin{pmatrix}-0.5\\1\\0\end{pmatrix}y+\begin{pmatrix}-0.5\\0\\1\end{pmatrix}z \mid y,z\in\mathbb{R}\}$$

and the three vectors with endpoints $(2,0,0)$, $(1.5,1,0)$, and $(1.5,0,1)$.

(a) Redraw the picture, including the vector in the plane that is twice as long as the one with endpoint $(1.5,1,0)$. The endpoint of your vector is not $(3,2,0)$; what is it?

(b) Redraw the picture, including the parallelogram in the plane that shows the sum of the vectors ending at $(1.5,0,1)$ and $(1.5,1,0)$. The endpoint of the sum, on the diagonal, is not $(3,1,1)$; what is it?

1.9 Show that the line segments $\overline{(a_1,a_2)(b_1,b_2)}$ and $\overline{(c_1,c_2)(d_1,d_2)}$ have the same lengths and slopes if $b_1-a_1=d_1-c_1$ and $b_2-a_2=d_2-c_2$. Is that only if?

1.10 How should we define $\mathbb{R}^0$?

? ✓ **1.11** [Math. Mag., Jan. 1957] A person traveling eastward at a rate of 3 miles per hour finds that the wind appears to blow directly from the north. On doubling his speed it appears to come from the north east. What was the wind's velocity?

1.12 Euclid describes a plane as "a surface which lies evenly with the straight lines on itself". Commentators such as Heron have interpreted this to mean, "(A plane surface is) such that, if a straight line pass through two points on it, the line coincides wholly with it at every spot, all ways". (Translations from [Heath], pp. 171-172.) Do planes, as described in this section, have that property? Does this description adequately define planes?

II.2 Length and Angle Measures

We've translated the first section's results about solution sets into geometric terms, to better understand those sets. But we must be careful not to be misled by our own terms — labeling subsets of $\mathbb{R}^k$ of the forms $\{\vec{p}+t\vec{v} \mid t\in\mathbb{R}\}$ and $\{\vec{p}+t\vec{v}+s\vec{w} \mid t,s\in\mathbb{R}\}$ as 'lines' and 'planes' doesn't make them act like the lines and planes of our past experience. Rather, we must ensure that the names suit the sets. While we can't prove that the sets satisfy our intuition — we can't prove anything about intuition — in this subsection we'll observe that a result familiar from $\mathbb{R}^2$ and $\mathbb{R}^3$, when generalized to arbitrary $\mathbb{R}^n$, supports the idea that a line is straight and a plane is flat. Specifically, we'll see how to do Euclidean geometry in a 'plane' by giving a definition of the angle between two $\mathbb{R}^n$ vectors, in the plane that they generate.

2.1 Definition The *length* of a vector $\vec{v}\in\mathbb{R}^n$ is the square root of the sum of the squares of its components.

$$|\vec{v}\,| = \sqrt{v_1^2+\cdots+v_n^2}$$

2.2 Remark This is a natural generalization of the Pythagorean Theorem. A classic motivating discussion is in [Polya].

For any nonzero $\vec{v}$, the vector $\vec{v}/|\vec{v}|$ has length one. We say that the second *normalizes* $\vec{v}$ to length one.

We can use that to get a formula for the angle between two vectors. Consider two vectors in $\mathbb{R}^3$ where neither is a multiple of the other

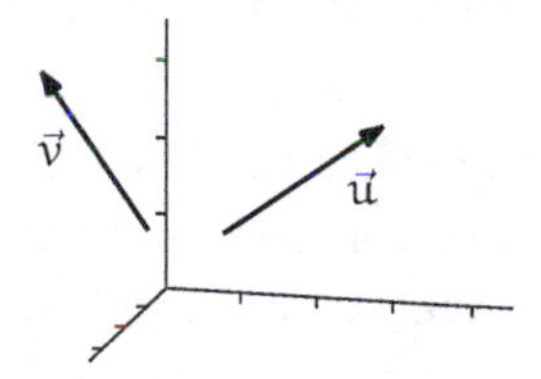

(the special case of multiples will turn out below not to be an exception). They determine a two-dimensional plane — for instance, put them in canonical position and take the plane formed by the origin and the endpoints. In that plane consider the triangle with sides $\vec{u}$, $\vec{v}$, and $\vec{u}-\vec{v}$.

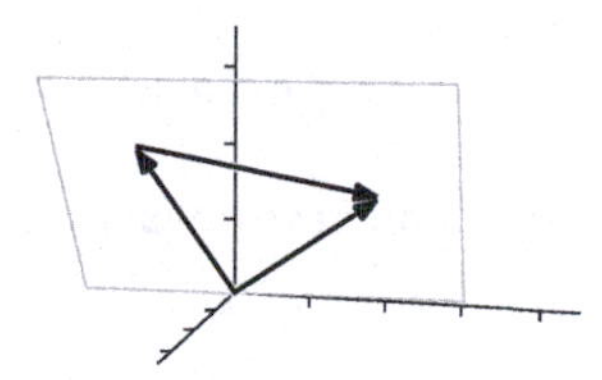

Apply the Law of Cosines: $|\vec{u}-\vec{v}\,|^2 = |\vec{u}\,|^2+|\vec{v}\,|^2-2\,|\vec{u}\,|\,|\vec{v}\,|\cos\theta$ where θ is the angle between the vectors. The left side gives

$$\begin{aligned}(u_1-v_1)^2+(u_2-v_2)^2+(u_3-v_3)^2\\ =(u_1^2-2u_1v_1+v_1^2)+(u_2^2-2u_2v_2+v_2^2)+(u_3^2-2u_3v_3+v_3^2)\end{aligned}$$

while the right side gives this.

$$(u_1^2+u_2^2+u_3^2)+(v_1^2+v_2^2+v_3^2)-2\,|\vec{u}\,|\,|\vec{v}\,|\cos\theta$$

Canceling squares u_1^2, ..., v_3^2 and dividing by 2 gives a formula for the angle.

$$\theta=\arccos(\,\frac{u_1v_1+u_2v_2+u_3v_3}{|\vec{u}\,|\,|\vec{v}\,|}\,)$$

In higher dimensions we cannot draw pictures as above but we can instead make the argument analytically. First, the form of the numerator is clear; it comes from the middle terms of $(u_i-v_i)^2$.

2.3 Definition The *dot product* (or *inner product* or *scalar product*) of two n-component real vectors is the linear combination of their components.

$$\vec{u}\cdot\vec{v}=u_1v_1+u_2v_2+\cdots+u_nv_n$$

Note that the dot product of two vectors is a real number, not a vector, and that the dot product is only defined if the two vectors have the same number of components. Note also that dot product is related to length: $\vec{u}\cdot\vec{u}=u_1u_1+\cdots+u_nu_n=|\vec{u}\,|^2$.

2.4 Remark Some authors require that the first vector be a row vector and that the second vector be a column vector. We shall not be that strict and will allow the dot product operation between two column vectors.

Still reasoning analytically but guided by the pictures, we use the next theorem to argue that the triangle formed by the line segments making the bodies of $\vec{u}$, $\vec{v}$, and $\vec{u}+\vec{v}$ in $\mathbb{R}^n$ lies in the planar subset of $\mathbb{R}^n$ generated by $\vec{u}$ and $\vec{v}$ (see the figure below).

2.5 Theorem (Triangle Inequality) For any $\vec{u},\vec{v}\in\mathbb{R}^n$,

$$|\vec{u}+\vec{v}\,|\leqslant|\vec{u}\,|+|\vec{v}\,|$$

with equality if and only if one of the vectors is a nonnegative scalar multiple of the other one.

This is the source of the familiar saying, "The shortest distance between two points is in a straight line."

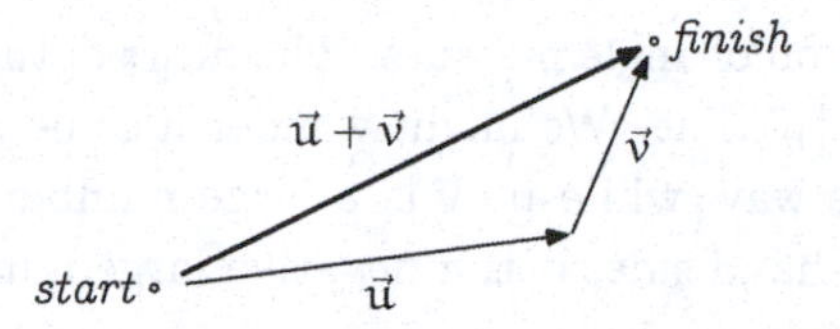

Proof (We'll use some algebraic properties of dot product that we have not yet checked, for instance that $\vec{u}\cdot(\vec{a}+\vec{b})=\vec{u}\cdot\vec{a}+\vec{u}\cdot\vec{b}$ and that $\vec{u}\cdot\vec{v}=\vec{v}\cdot\vec{u}$. See Exercise 18.) Since all the numbers are positive, the inequality holds if and only if its square holds.

$$\begin{aligned}
|\vec{u}+\vec{v}|^2 &\leqslant (\,|\vec{u}\,|+|\vec{v}\,|\,)^2 \\
(\,\vec{u}+\vec{v}\,)\cdot(\,\vec{u}+\vec{v}\,) &\leqslant |\vec{u}|^2+2|\vec{u}\,|\,|\vec{v}|+|\vec{v}|^2 \\
\vec{u}\cdot\vec{u}+\vec{u}\cdot\vec{v}+\vec{v}\cdot\vec{u}+\vec{v}\cdot\vec{v} &\leqslant \vec{u}\cdot\vec{u}+2|\vec{u}\,|\,|\vec{v}|+\vec{v}\cdot\vec{v} \\
2\,\vec{u}\cdot\vec{v} &\leqslant 2\,|\vec{u}\,|\,|\vec{v}\,|
\end{aligned}$$

That, in turn, holds if and only if the relationship obtained by multiplying both sides by the nonnegative numbers $|\vec{u}\,|$ and $|\vec{v}\,|$

$$2\,(\,|\vec{v}\,|\,\vec{u}\,)\cdot(\,|\vec{u}\,|\,\vec{v}\,)\leqslant 2\,|\vec{u}\,|^2\,|\vec{v}\,|^2$$

and rewriting

$$0\leqslant|\vec{u}\,|^2\,|\vec{v}\,|^2-2\,(\,|\vec{v}\,|\,\vec{u}\,)\cdot(\,|\vec{u}\,|\,\vec{v}\,)+|\vec{u}\,|^2\,|\vec{v}\,|^2$$

is true. But factoring shows that it is true

$$0\leqslant(\,|\vec{u}\,|\,\vec{v}-|\vec{v}\,|\,\vec{u}\,)\cdot(\,|\vec{u}\,|\,\vec{v}-|\vec{v}\,|\,\vec{u}\,)$$

since it only says that the square of the length of the vector $|\vec{u}\,|\,\vec{v}-|\vec{v}\,|\,\vec{u}$ is not negative. As for equality, it holds when, and only when, $|\vec{u}\,|\,\vec{v}-|\vec{v}\,|\,\vec{u}$ is $\vec{0}$. The check that $|\vec{u}\,|\,\vec{v}=|\vec{v}\,|\,\vec{u}$ if and only if one vector is a nonnegative real scalar multiple of the other is easy. QED

This result supports the intuition that even in higher-dimensional spaces, lines are straight and planes are flat. We can easily check from the definition that linear surfaces have the property that for any two points in that surface, the line segment between them is contained in that surface. But if the linear surface were not flat then that would allow for a shortcut.

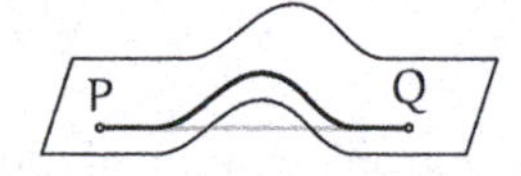

Because the Triangle Inequality says that in any $\mathbb{R}^n$ the shortest cut between two endpoints is simply the line segment connecting them, linear surfaces have no bends.

Back to the definition of angle measure. The heart of the Triangle Inequality's proof is the $\vec{u}\cdot\vec{v} \leqslant |\vec{u}\,||\vec{v}\,|$ line. We might wonder if some pairs of vectors satisfy the inequality in this way: while $\vec{u}\cdot\vec{v}$ is a large number, with absolute value bigger than the right-hand side, it is a negative large number. The next result says that does not happen.

2.6 Corollary (Cauchy-Schwarz Inequality) For any $\vec{u},\vec{v}\in\mathbb{R}^n$,

$$|\,\vec{u}\cdot\vec{v}\,| \leqslant |\,\vec{u}\,|\,|\vec{v}\,|$$

with equality if and only if one vector is a scalar multiple of the other.

Proof The Triangle Inequality's proof shows that $\vec{u}\cdot\vec{v} \leqslant |\vec{u}\,||\vec{v}\,|$ so if $\vec{u}\cdot\vec{v}$ is positive or zero then we are done. If $\vec{u}\cdot\vec{v}$ is negative then this holds.

$$|\,\vec{u}\cdot\vec{v}\,| = -(\,\vec{u}\cdot\vec{v}\,) = (-\vec{u}\,)\cdot\vec{v} \leqslant |-\vec{u}\,||\vec{v}\,| = |\vec{u}\,||\vec{v}\,|$$

The equality condition is Exercise 19. QED

The Cauchy-Schwarz inequality assures us that the next definition makes sense because the fraction has absolute value less than or equal to one.

2.7 Definition The *angle* between two nonzero vectors $\vec{u},\vec{v}\in\mathbb{R}^n$ is

$$\theta = \arccos(\,\frac{\vec{u}\cdot\vec{v}}{|\vec{u}\,||\vec{v}\,|}\,)$$

(if either is the zero vector, we take the angle to be right).

2.8 Corollary Vectors from $\mathbb{R}^n$ are orthogonal, that is, perpendicular, if and only if their dot product is zero. They are parallel if and only if their dot product equals the product of their lengths.

2.9 Example These vectors are orthogonal.

$$\begin{pmatrix}1\\-1\end{pmatrix}\cdot\begin{pmatrix}1\\1\end{pmatrix} = 0$$

We've drawn the arrows away from canonical position but nevertheless the vectors are orthogonal.

2.10 Example The $\mathbb{R}^3$ angle formula given at the start of this subsection is a special case of the definition. Between these two

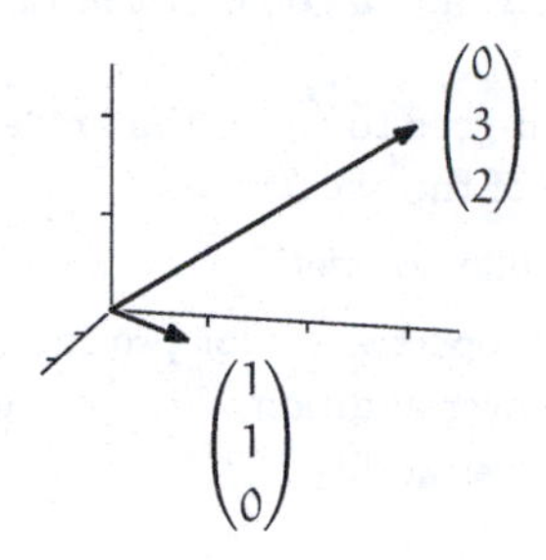

the angle is

$$\arccos\left(\frac{(1)(0)+(1)(3)+(0)(2)}{\sqrt{1^2+1^2+0^2}\sqrt{0^2+3^2+2^2}}\right) = \arccos\left(\frac{3}{\sqrt{2}\sqrt{13}}\right)$$

approximately 0.94 radians. Notice that these vectors are not orthogonal. Although the yz-plane may appear to be perpendicular to the xy-plane, in fact the two planes are that way only in the weak sense that there are vectors in each orthogonal to all vectors in the other. Not every vector in each is orthogonal to all vectors in the other.

Exercises

✓ **2.11** Find the length of each vector.

(a) $\begin{pmatrix}3\\1\end{pmatrix}$ (b) $\begin{pmatrix}-1\\2\end{pmatrix}$ (c) $\begin{pmatrix}4\\1\\1\end{pmatrix}$ (d) $\begin{pmatrix}0\\0\\0\end{pmatrix}$ (e) $\begin{pmatrix}1\\-1\\1\\0\end{pmatrix}$

✓ **2.12** Find the angle between each two, if it is defined.

(a) $\begin{pmatrix}1\\2\end{pmatrix}, \begin{pmatrix}1\\4\end{pmatrix}$ (b) $\begin{pmatrix}1\\2\\0\end{pmatrix}, \begin{pmatrix}0\\4\\1\end{pmatrix}$ (c) $\begin{pmatrix}1\\2\end{pmatrix}, \begin{pmatrix}1\\4\\-1\end{pmatrix}$

✓ **2.13** [Ohanian] During maneuvers preceding the Battle of Jutland, the British battle cruiser *Lion* moved as follows (in nautical miles): 1.2 miles north, 6.1 miles 38 degrees east of south, 4.0 miles at 89 degrees east of north, and 6.5 miles at 31 degrees east of north. Find the distance between starting and ending positions. (Ignore the earth's curvature.)

2.14 Find k so that these two vectors are perpendicular.

$$\begin{pmatrix}k\\1\end{pmatrix} \quad \begin{pmatrix}4\\3\end{pmatrix}$$

2.15 Describe the set of vectors in $\mathbb{R}^3$ orthogonal to the one with entires 1, 3, and -1.

✓ **2.16** (a) Find the angle between the diagonal of the unit square in $\mathbb{R}^2$ and one of the axes.

(b) Find the angle between the diagonal of the unit cube in $\mathbb{R}^3$ and one of the axes.

(c) Find the angle between the diagonal of the unit cube in $\mathbb{R}^n$ and one of the axes.

(d) What is the limit, as n goes to ∞, of the angle between the diagonal of the unit cube in $\mathbb{R}^n$ and one of the axes?

2.17 Is any vector perpendicular to itself?

✓ **2.18** Describe the algebraic properties of dot product.

(a) Is it right-distributive over addition: $(\vec{u}+\vec{v})\cdot\vec{w}=\vec{u}\cdot\vec{w}+\vec{v}\cdot\vec{w}$?

(b) Is it left-distributive (over addition)?

(c) Does it commute?

(d) Associate?

(e) How does it interact with scalar multiplication?

As always, you must back any assertion with either a proof or an example.

2.19 Verify the equality condition in Corollary 2.6, the Cauchy-Schwarz Inequality.

(a) Show that if $\vec{u}$ is a negative scalar multiple of $\vec{v}$ then $\vec{u}\cdot\vec{v}$ and $\vec{v}\cdot\vec{u}$ are less than or equal to zero.

(b) Show that $|\vec{u}\cdot\vec{v}|=|\vec{u}\,|\,|\vec{v}\,|$ if and only if one vector is a scalar multiple of the other.

2.20 Suppose that $\vec{u}\cdot\vec{v}=\vec{u}\cdot\vec{w}$ and $\vec{u}\neq\vec{0}$. Must $\vec{v}=\vec{w}$?

✓ **2.21** Does any vector have length zero except a zero vector? (If "yes", produce an example. If "no", prove it.)

✓ **2.22** Find the midpoint of the line segment connecting (x_1,y_1) with (x_2,y_2) in $\mathbb{R}^2$. Generalize to $\mathbb{R}^n$.

2.23 Show that if $\vec{v}\neq\vec{0}$ then $\vec{v}/|\vec{v}\,|$ has length one. What if $\vec{v}=\vec{0}$?

2.24 Show that if $r\geqslant 0$ then $r\vec{v}$ is r times as long as $\vec{v}$. What if $r<0$?

✓ **2.25** A vector $\vec{v}\in\mathbb{R}^n$ of length one is a *unit* vector. Show that the dot product of two unit vectors has absolute value less than or equal to one. Can 'less than' happen? Can 'equal to'?

2.26 Prove that $|\vec{u}+\vec{v}\,|^2+|\vec{u}-\vec{v}\,|^2=2|\vec{u}\,|^2+2|\vec{v}\,|^2$.

2.27 Show that if $\vec{x}\cdot\vec{y}=0$ for every $\vec{y}$ then $\vec{x}=\vec{0}$.

2.28 Is $|\vec{u}_1+\cdots+\vec{u}_n|\leqslant|\vec{u}_1|+\cdots+|\vec{u}_n|$? If it is true then it would generalize the Triangle Inequality.

2.29 What is the ratio between the sides in the Cauchy-Schwarz inequality?

2.30 Why is the zero vector defined to be perpendicular to every vector?

2.31 Describe the angle between two vectors in $\mathbb{R}^1$.

2.32 Give a simple necessary and sufficient condition to determine whether the angle between two vectors is acute, right, or obtuse.

✓ **2.33** Generalize to $\mathbb{R}^n$ the converse of the Pythagorean Theorem, that if $\vec{u}$ and $\vec{v}$ are perpendicular then $|\vec{u}+\vec{v}\,|^2=|\vec{u}\,|^2+|\vec{v}\,|^2$.

2.34 Show that $|\vec{u}\,|=|\vec{v}\,|$ if and only if $\vec{u}+\vec{v}$ and $\vec{u}-\vec{v}$ are perpendicular. Give an example in $\mathbb{R}^2$.

2.35 Show that if a vector is perpendicular to each of two others then it is perpendicular to each vector in the plane they generate. (*Remark.* They could generate a degenerate plane — a line or a point — but the statement remains true.)

2.36 Prove that, where $\vec{u}, \vec{v} \in \mathbb{R}^n$ are nonzero vectors, the vector

$$\frac{\vec{u}}{|\vec{u}|} + \frac{\vec{v}}{|\vec{v}|}$$

bisects the angle between them. Illustrate in $\mathbb{R}^2$.

2.37 Verify that the definition of angle is dimensionally correct: (1) if $k > 0$ then the cosine of the angle between $k\vec{u}$ and $\vec{v}$ equals the cosine of the angle between $\vec{u}$ and $\vec{v}$, and (2) if $k < 0$ then the cosine of the angle between $k\vec{u}$ and $\vec{v}$ is the negative of the cosine of the angle between $\vec{u}$ and $\vec{v}$.

✓ **2.38** Show that the inner product operation is *linear*: for $\vec{u}, \vec{v}, \vec{w} \in \mathbb{R}^n$ and $k, m \in \mathbb{R}$, $\vec{u} \cdot (k\vec{v} + m\vec{w}) = k(\vec{u} \cdot \vec{v}) + m(\vec{u} \cdot \vec{w})$.

✓ **2.39** The *geometric mean* of two positive reals x, y is $\sqrt{xy}$. It is analogous to the *arithmetic mean* $(x + y)/2$. Use the Cauchy-Schwarz inequality to show that the geometric mean of any $x, y \in \mathbb{R}$ is less than or equal to the arithmetic mean.

? **2.40** [Cleary] Astrologers claim to be able to recognize trends in personality and fortune that depend on an individual's birthday by somehow incorporating where the stars were 2000 years ago. Suppose that instead of star-gazers coming up with stuff, math teachers who like linear algebra (we'll call them vectologers) had come up with a similar system as follows: Consider your birthday as a row vector $(\text{month} \;\; \text{day})$. For instance, I was born on July 12 so my vector would be $(7 \;\; 12)$. Vectologers have made the rule that how well individuals get along with each other depends on the angle between vectors. The smaller the angle, the more harmonious the relationship.

(a) Find the angle between your vector and mine, in radians.

(b) Would you get along better with me, or with a professor born on September 19?

(c) For maximum harmony in a relationship, when should the other person be born?

(d) Is there a person with whom you have a "worst case" relationship, i.e., your vector and theirs are orthogonal? If so, what are the birthdate(s) for such people? If not, explain why not.

? **2.41** [Am. Math. Mon., Feb. 1933] A ship is sailing with speed and direction $\vec{v}_1$; the wind blows apparently (judging by the vane on the mast) in the direction of a vector $\vec{a}$; on changing the direction and speed of the ship from $\vec{v}_1$ to $\vec{v}_2$ the apparent wind is in the direction of a vector $\vec{b}$.

Find the vector velocity of the wind.

2.42 Verify the Cauchy-Schwarz inequality by first proving Lagrange's identity:

$$\left(\sum_{1\leqslant j\leqslant n} a_j b_j\right)^2 = \left(\sum_{1\leqslant j\leqslant n} a_j^2\right)\left(\sum_{1\leqslant j\leqslant n} b_j^2\right) - \sum_{1\leqslant k<j\leqslant n} (a_k b_j - a_j b_k)^2$$

and then noting that the final term is positive. This result is an improvement over Cauchy-Schwarz because it gives a formula for the difference between the two sides. Interpret that difference in $\mathbb{R}^2$.

III Reduced Echelon Form

After developing the mechanics of Gauss's Method, we observed that it can be done in more than one way. For example, from this matrix

$$\begin{pmatrix} 2 & 2 \\ 4 & 3 \end{pmatrix}$$

we could derive any of these three echelon form matrices.

$$\begin{pmatrix} 2 & 2 \\ 0 & -1 \end{pmatrix} \qquad \begin{pmatrix} 1 & 1 \\ 0 & -1 \end{pmatrix} \qquad \begin{pmatrix} 2 & 0 \\ 0 & -1 \end{pmatrix}$$

The first results from $-2\rho_1 + \rho_2$. The second comes from doing $(1/2)\rho_1$ and then $-4\rho_1 + \rho_2$. The third comes from $-2\rho_1 + \rho_2$ followed by $2\rho_2 + \rho_1$ (after the first row combination the matrix is already in echelon form so the second one is extra work but it is nonetheless a legal row operation).

The fact that echelon form is not unique raises questions. Will any two echelon form versions of a linear system have the same number of free variables? If yes, will the two have exactly the same free variables? In this section we will give a way to decide if one linear system can be derived from another by row operations. The answers to both questions, both 'yes', will follow from this.

III.1 Gauss-Jordan Reduction

Here is an extension of Gauss's Method that has some advantages.

1.1 Example To solve

$$\begin{aligned} x + y - 2z &= -2 \\ y + 3z &= 7 \\ x \qquad - z &= -1 \end{aligned}$$

we can start as usual by reducing it to echelon form.

$$\xrightarrow{-\rho_1+\rho_3} \left(\begin{array}{ccc|c} 1 & 1 & -2 & -2 \\ 0 & 1 & 3 & 7 \\ 0 & -1 & 1 & 1 \end{array}\right) \xrightarrow{\rho_2+\rho_3} \left(\begin{array}{ccc|c} 1 & 1 & -2 & -2 \\ 0 & 1 & 3 & 7 \\ 0 & 0 & 4 & 8 \end{array}\right)$$

We can keep going to a second stage by making the leading entries into 1's

$$\xrightarrow{(1/4)\rho_3} \left(\begin{array}{ccc|c} 1 & 1 & -2 & -2 \\ 0 & 1 & 3 & 7 \\ 0 & 0 & 1 & 2 \end{array}\right)$$

and then to a third stage that uses the leading entries to eliminate all of the other entries in each column by combining upwards.

$$\xrightarrow[2\rho_3+\rho_1]{-3\rho_3+\rho_2}\left(\begin{array}{ccc|c}1&1&0&2\\0&1&0&1\\0&0&1&2\end{array}\right)\xrightarrow{-\rho_2+\rho_1}\left(\begin{array}{ccc|c}1&0&0&1\\0&1&0&1\\0&0&1&2\end{array}\right)$$

The answer is $x=1$, $y=1$, and $z=2$.

Using one entry to clear out the rest of a column is *pivoting* on that entry.

Note that the row combination operations in the first stage move left to right, from column one to column three, while the combination operations in the third stage move right to left.

1.2 Example The middle stage operations that turn the leading entries into 1's don't interact so we can combine multiple ones into a single step.

$$\left(\begin{array}{cc|c}2&1&7\\4&-2&6\end{array}\right)\xrightarrow{-2\rho_1+\rho_2}\left(\begin{array}{cc|c}2&1&7\\0&-4&-8\end{array}\right)$$
$$\xrightarrow[(-1/4)\rho_2]{(1/2)\rho_1}\left(\begin{array}{cc|c}1&1/2&7/2\\0&1&2\end{array}\right)$$
$$\xrightarrow{-(1/2)\rho_2+\rho_1}\left(\begin{array}{cc|c}1&0&5/2\\0&1&2\end{array}\right)$$

The answer is $x=5/2$ and $y=2$.

This extension of Gauss's Method is the *Gauss-Jordan Method* or *Gauss-Jordan reduction*.

1.3 Definition A matrix or linear system is in *reduced echelon form* if, in addition to being in echelon form, each leading entry is a 1 and is the only nonzero entry in its column.

The cost of using Gauss-Jordan reduction to solve a system is the additional arithmetic. The benefit is that we can just read off the solution set description.

In any echelon form system, reduced or not, we can read off when the system has an empty solution set because there is a contradictory equation. We can read off when the system has a one-element solution set because there is no contradiction and every variable is the leading variable in some row. And, we can read off when the system has an infinite solution set because there is no contradiction and at least one variable is free.

However, in reduced echelon form we can read off not just the size of the solution set but also its description. We have no trouble describing the solution set when it is empty, of course. Example 1.1 and 1.2 show how in a single element solution set case the single element is in the column of constants. The next example shows how to read the parametrization of an infinite solution set.

1.4 Example

$$\left(\begin{array}{cccc|c} 2 & 6 & 1 & 2 & 5 \\ 0 & 3 & 1 & 4 & 1 \\ 0 & 3 & 1 & 2 & 5 \end{array}\right) \xrightarrow{-\rho_2+\rho_3} \left(\begin{array}{cccc|c} 2 & 6 & 1 & 2 & 5 \\ 0 & 3 & 1 & 4 & 1 \\ 0 & 0 & 0 & -2 & 4 \end{array}\right)$$

$$\xrightarrow[\substack{(1/3)\rho_2 \\ -(1/2)\rho_3}]{(1/2)\rho_1} \xrightarrow[-\rho_3+\rho_1]{-(4/3)\rho_3+\rho_2} \xrightarrow{-3\rho_2+\rho_1} \left(\begin{array}{cccc|c} 1 & 0 & -1/2 & 0 & -9/2 \\ 0 & 1 & 1/3 & 0 & 3 \\ 0 & 0 & 0 & 1 & -2 \end{array}\right)$$

As a linear system this is

$$\begin{array}{rcl} x_1 \quad - 1/2x_3 & = & -9/2 \\ x_2 + 1/3x_3 & = & 3 \\ x_4 & = & -2 \end{array}$$

so a solution set description is this.

$$S = \{ \begin{pmatrix} x_1 \\ x_2 \\ x_3 \\ x_4 \end{pmatrix} = \begin{pmatrix} -9/2 \\ 3 \\ 0 \\ -2 \end{pmatrix} + \begin{pmatrix} 1/2 \\ -1/3 \\ 1 \\ 0 \end{pmatrix} x_3 \mid x_3 \in \mathbb{R} \}$$

Thus echelon form isn't some kind of one best form for systems. Other forms, such as reduced echelon form, have advantages and disadvantages. Instead of picturing linear systems (and the associated matrices) as things we operate on, always directed toward the goal of echelon form, we can think of them as interrelated when we can get from one to another by row operations. The rest of this subsection develops this relationship.

1.5 Lemma Elementary row operations are reversible.

PROOF For any matrix A, the effect of swapping rows is reversed by swapping them back, multiplying a row by a nonzero k is undone by multiplying by $1/k$, and adding a multiple of row i to row j (with $i \neq j$) is undone by subtracting the same multiple of row i from row j.

$$A \xrightarrow{\rho_i \leftrightarrow \rho_j} \xrightarrow{\rho_j \leftrightarrow \rho_i} A \qquad A \xrightarrow{k\rho_i} \xrightarrow{(1/k)\rho_i} A \qquad A \xrightarrow{k\rho_i+\rho_j} \xrightarrow{-k\rho_i+\rho_j} A$$

(We need the $i \neq j$ condition; see Exercise 17.) QED

Again, the point of view that we are developing, supported now by the lemma, is that the term 'reduces to' is misleading: where $A \longrightarrow B$, we shouldn't think of B as after A or simpler than A. Instead we should think of the two matrices as interrelated. Below is a picture. It shows the matrices from the start of this section and their reduced echelon form version in a cluster, as inter-reducible.

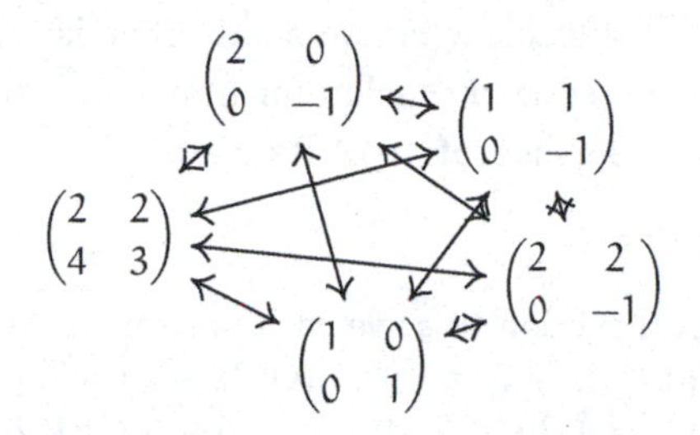

We say that matrices that reduce to each other are equivalent with respect to the relationship of row reducibility. The next result justifies this, using the definition of an equivalence.*

1.6 Lemma Between matrices, 'reduces to' is an equivalence relation.

Proof We must check the conditions (i) reflexivity, that any matrix reduces to itself, (ii) symmetry, that if A reduces to B then B reduces to A, and (iii) transitivity, that if A reduces to B and B reduces to C then A reduces to C.

Reflexivity is easy; any matrix reduces to itself in zero-many operations.

The relationship is symmetric by the prior lemma—if A reduces to B by some row operations then also B reduces to A by reversing those operations.

For transitivity, suppose that A reduces to B and that B reduces to C. Following the reduction steps from $A \to \cdots \to B$ with those from $B \to \cdots \to C$ gives a reduction from A to C. QED

1.7 Definition Two matrices that are interreducible by elementary row operations are *row equivalent*.

The diagram below shows the collection of all matrices as a box. Inside that box each matrix lies in a class. Matrices are in the same class if and only if they are interreducible. The classes are disjoint—no matrix is in two distinct classes. We have partitioned the collection of matrices into *row equivalence classes*.†

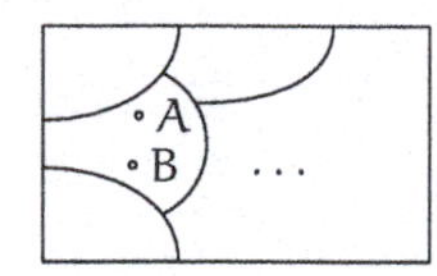

* More information on equivalence relations is in the appendix.

† More information on partitions and class representatives is in the appendix.

One of the classes is the cluster of interrelated matrices from the start of this section pictured earlier, expanded to include all of the nonsingular 2×2 matrices.

The next subsection proves that the reduced echelon form of a matrix is unique. Rephrased in terms of the row-equivalence relationship, we shall prove that every matrix is row equivalent to one and only one reduced echelon form matrix. In terms of the partition what we shall prove is: every equivalence class contains one and only one reduced echelon form matrix. So each reduced echelon form matrix serves as a representative of its class.

Exercises

✓ **1.8** Use Gauss-Jordan reduction to solve each system.

(a) $\begin{aligned} x+y&=2 \\ x-y&=0 \end{aligned}$ (b) $\begin{aligned} x \quad\quad -z&=4 \\ 2x+2y \quad\quad &=1 \end{aligned}$ (c) $\begin{aligned} 3x-2y&=1 \\ 6x+y&=1/2 \end{aligned}$

(d) $\begin{aligned} 2x-y \quad\quad &=-1 \\ x+3y-z&=5 \\ y+2z&=5 \end{aligned}$

1.9 Do Gauss-Jordan reduction.

(a) $\begin{aligned} x+y-z&=3 \\ 2x-y-z&=1 \\ 3x+y+2z&=0 \end{aligned}$ (b) $\begin{aligned} x+y+2z&=0 \\ 2x-y+z&=1 \\ 4x+y+5z&=1 \end{aligned}$

✓ **1.10** Find the reduced echelon form of each matrix.

(a) $\begin{pmatrix} 2 & 1 \\ 1 & 3 \end{pmatrix}$ (b) $\begin{pmatrix} 1 & 3 & 1 \\ 2 & 0 & 4 \\ -1 & -3 & -3 \end{pmatrix}$ (c) $\begin{pmatrix} 1 & 0 & 3 & 1 & 2 \\ 1 & 4 & 2 & 1 & 5 \\ 3 & 4 & 8 & 1 & 2 \end{pmatrix}$

(d) $\begin{pmatrix} 0 & 1 & 3 & 2 \\ 0 & 0 & 5 & 6 \\ 1 & 5 & 1 & 5 \end{pmatrix}$

1.11 Get the reduced echelon form of each.

(a) $\begin{pmatrix} 0 & 2 & 1 \\ 2 & -1 & 1 \\ -2 & -1 & 0 \end{pmatrix}$ (b) $\begin{pmatrix} 1 & 3 & 1 \\ 2 & 6 & 2 \\ -1 & 0 & 0 \end{pmatrix}$

✓ **1.12** Find each solution set by using Gauss-Jordan reduction and then reading off the parametrization.

(a) $\begin{aligned} 2x+y-z&=1 \\ 4x-y \quad\quad &=3 \end{aligned}$

(b) $\begin{aligned} x \quad\quad -z \quad\quad &=1 \\ y+2z-w&=3 \\ x+2y+3z-w&=7 \end{aligned}$

(c) $\begin{aligned} x-y+z \quad\quad &=0 \\ y \quad\quad +w&=0 \\ 3x-2y+3z+w&=0 \\ -y \quad\quad -w&=0 \end{aligned}$

(d) $\begin{aligned} a+2b+3c+d-e&=1 \\ 3a-b+c+d+e&=3 \end{aligned}$

1.13 Give two distinct echelon form versions of this matrix.

$$\begin{pmatrix} 2 & 1 & 1 & 3 \\ 6 & 4 & 1 & 2 \\ 1 & 5 & 1 & 5 \end{pmatrix}$$

✓ **1.14** List the reduced echelon forms possible for each size.
(a) $2{\times}2$ (b) $2{\times}3$ (c) $3{\times}2$ (d) $3{\times}3$

✓ **1.15** What results from applying Gauss-Jordan reduction to a nonsingular matrix?

1.16 [Cleary] Consider the following relationship on the set of $2{\times}2$ matrices: we say that A is *sum-what like* B if the sum of all of the entries in A is the same as the sum of all the entries in B. For instance, the zero matrix would be sum-what like the matrix whose first row had two sevens, and whose second row had two negative sevens. Prove or disprove that this is an equivalence relation on the set of $2{\times}2$ matrices.

1.17 The proof of Lemma 1.5 contains a reference to the $i \neq j$ condition on the row combination operation.
(a) Write down a $2{\times}2$ matrix with nonzero entries, and show that the $-1\cdot\rho_1+\rho_1$ operation is not reversed by $1\cdot\rho_1+\rho_1$.
(b) Expand the proof of that lemma to make explicit exactly where it uses the $i \neq j$ condition on combining.

1.18 [Cleary] Consider the set of students in a class. Which of the following relationships are equivalence relations? Explain each answer in at least a sentence.
(a) Two students x, y are related if x has taken at least as many math classes as y.
(b) Students x, y are related if they have names that start with the same letter.

1.19 Show that each of these is an equivalence on the set of $2{\times}2$ matrices. Describe the equivalence classes.
(a) Two matrices are related if they have the same product down the diagonal, that is, if the product of the entries in the upper left and lower right are equal.
(b) Two matrices are related if they both have at least one entry that is a 1, or if neither does.

1.20 Show that each is not an equivalence on the set of $2{\times}2$ matrices.
(a) Two matrices A, B are related if $a_{1,1} = -b_{1,1}$.
(b) Two matrices are related if the sum of their entries are within 5, that is, A is related to B if $|(a_{1,1}+\cdots+a_{2,2})-(b_{1,1}+\cdots+b_{2,2})| < 5$.

III.2 The Linear Combination Lemma

We will close this chapter by proving that every matrix is row equivalent to one and only one reduced echelon form matrix. The ideas here will reappear, and be further developed, in the next chapter.

The crucial observation concerns how row operations act to transform one matrix into another: the new rows are linear combinations of the old.

2.1 Example Consider this Gauss-Jordan reduction.

$$\left(\begin{array}{cc|c} 2 & 1 & 0 \\ 1 & 3 & 5 \end{array}\right) \xrightarrow{-(1/2)\rho_1+\rho_2} \left(\begin{array}{cc|c} 2 & 1 & 0 \\ 0 & 5/2 & 5 \end{array}\right)$$
$$\xrightarrow[(2/5)\rho_2]{(1/2)\rho_1} \left(\begin{array}{cc|c} 1 & 1/2 & 0 \\ 0 & 1 & 2 \end{array}\right)$$
$$\xrightarrow{-(1/2)\rho_2+\rho_1} \left(\begin{array}{cc|c} 1 & 0 & -1 \\ 0 & 1 & 2 \end{array}\right)$$

Denoting those matrices $A \to D \to G \to B$ and writing the rows of A as α_1 and α_2, etc., we have this.

$$\begin{pmatrix} \alpha_1 \\ \alpha_2 \end{pmatrix} \xrightarrow{-(1/2)\rho_1+\rho_2} \begin{pmatrix} \delta_1 = \alpha_1 \\ \delta_2 = -(1/2)\alpha_1 + \alpha_2 \end{pmatrix}$$
$$\xrightarrow[(2/5)\rho_2]{(1/2)\rho_1} \begin{pmatrix} \gamma_1 = (1/2)\alpha_1 \\ \gamma_2 = -(1/5)\alpha_1 + (2/5)\alpha_2 \end{pmatrix}$$
$$\xrightarrow{-(1/2)\rho_2+\rho_1} \begin{pmatrix} \beta_1 = (3/5)\alpha_1 - (1/5)\alpha_2 \\ \beta_2 = -(1/5)\alpha_1 + (2/5)\alpha_2 \end{pmatrix}$$

2.2 Example The fact that Gaussian operations combine rows linearly also holds if there is a row swap. With this A, D, G, and B

$$\begin{pmatrix} 0 & 2 \\ 1 & 1 \end{pmatrix} \xrightarrow{\rho_1 \leftrightarrow \rho_2} \begin{pmatrix} 1 & 1 \\ 0 & 2 \end{pmatrix} \xrightarrow{(1/2)\rho_2} \begin{pmatrix} 1 & 1 \\ 0 & 1 \end{pmatrix} \xrightarrow{-\rho_2+\rho_1} \begin{pmatrix} 1 & 0 \\ 0 & 1 \end{pmatrix}$$

we get these linear relationships.

$$\begin{pmatrix} \vec{\alpha}_1 \\ \vec{\alpha}_2 \end{pmatrix} \xrightarrow{\rho_1 \leftrightarrow \rho_2} \begin{pmatrix} \vec{\delta}_1 = \vec{\alpha}_2 \\ \vec{\delta}_2 = \vec{\alpha}_1 \end{pmatrix} \xrightarrow{(1/2)\rho_2} \begin{pmatrix} \vec{\gamma}_1 = \vec{\alpha}_2 \\ \vec{\gamma}_2 = (1/2)\vec{\alpha}_1 \end{pmatrix}$$
$$\xrightarrow{-\rho_2+\rho_1} \begin{pmatrix} \vec{\beta}_1 = (-1/2)\vec{\alpha}_1 + 1 \cdot \vec{\alpha}_2 \\ \vec{\beta}_2 = (1/2)\vec{\alpha}_1 \end{pmatrix}$$

In summary, Gauss's Method systematically finds a suitable sequence of linear combinations of the rows.

2.3 Lemma (Linear Combination Lemma) A linear combination of linear combinations is a linear combination.

PROOF Given the set $c_{1,1}x_1 + \cdots + c_{1,n}x_n$ through $c_{m,1}x_1 + \cdots + c_{m,n}x_n$ of linear combinations of the x's, consider a combination of those

$$d_1(c_{1,1}x_1 + \cdots + c_{1,n}x_n) + \cdots + d_m(c_{m,1}x_1 + \cdots + c_{m,n}x_n)$$

where the d's are scalars along with the c's. Distributing those d's and regrouping gives

$$= (d_1c_{1,1} + \cdots + d_mc_{m,1})x_1 + \cdots + (d_1c_{1,n} + \cdots + d_mc_{m,n})x_n$$

which is also a linear combination of the x's. QED

2.4 Corollary Where one matrix reduces to another, each row of the second is a linear combination of the rows of the first.

PROOF For any two interreducible matrices A and B there is some minimum number of row operations that will take one to the other. We proceed by induction on that number.

In the base step, that we can go from the first to the second using zero reduction operations, the two matrices are equal. Then each row of B is trivially a combination of A's rows $\vec{\beta}_i = 0 \cdot \vec{\alpha}_1 + \cdots + 1 \cdot \vec{\alpha}_i + \cdots + 0 \cdot \vec{\alpha}_m$.

For the inductive step assume the inductive hypothesis: with $k \geqslant 0$, any matrix that can be derived from A in k or fewer operations has rows that are linear combinations of A's rows. Consider a matrix B such that reducing A to B requires $k+1$ operations. In that reduction there is a next-to-last matrix G, so that $A \longrightarrow \cdots \longrightarrow G \longrightarrow B$. The inductive hypothesis applies to this G because it is only k steps away from A. That is, each row of G is a linear combination of the rows of A.

We will verify that the rows of B are linear combinations of the rows of G. Then the Linear Combination Lemma, Lemma 2.3, applies to show that the rows of B are linear combinations of the rows of A.

If the row operation taking G to B is a swap then the rows of B are just the rows of G reordered and each row of B is a linear combination of the rows of G. If the operation taking G to B is multiplication of a row by a scalar $c\rho_i$ then $\vec{\beta}_i = c\vec{\gamma}_i$ and the other rows are unchanged. Finally, if the row operation is adding a multiple of one row to another $r\rho_i + \rho_j$ then only row j of B differs from the matching row of G, and $\vec{\beta}_j = r\gamma_i + \gamma_j$, which is indeed a linear combinations of the rows of G.

Because we have proved both a base step and an inductive step, the proposition follows by the principle of mathematical induction. QED

We now have the insight that Gauss's Method builds linear combinations of the rows. But of course its goal is to end in echelon form, since that is a particularly basic version of a linear system, as it has isolated the variables. For instance, in this matrix

$$R = \begin{pmatrix} 2 & 3 & 7 & 8 & 0 & 0 \\ 0 & 0 & 1 & 5 & 1 & 1 \\ 0 & 0 & 0 & 3 & 3 & 0 \\ 0 & 0 & 0 & 0 & 2 & 1 \end{pmatrix}$$

x_1 has been removed from x_5's equation. That is, Gauss's Method has made x_5's row in some way independent of x_1's row.

The following result makes this intuition precise. What Gauss's Method eliminates is linear relationships among the rows.

2.5 Lemma In an echelon form matrix, no nonzero row is a linear combination of the other nonzero rows.

Proof Let R be an echelon form matrix and consider its non-$\vec{0}$ rows. First observe that if we have a row written as a combination of the others $\vec{\rho}_i = c_1\vec{\rho}_1 + \cdots + c_{i-1}\vec{\rho}_{i-1} + c_{i+1}\vec{\rho}_{i+1} + \cdots + c_m\vec{\rho}_m$ then we can rewrite that equation as

$$\vec{0} = c_1\vec{\rho}_1 + \cdots + c_{i-1}\vec{\rho}_{i-1} + c_i\vec{\rho}_i + c_{i+1}\vec{\rho}_{i+1} + \cdots + c_m\vec{\rho}_m \qquad (*)$$

where not all the coefficients are zero; specifically, $c_i = -1$. The converse holds also: given equation $(*)$ where some $c_i \neq 0$ we could express $\vec{\rho}_i$ as a combination of the other rows by moving $c_i\vec{\rho}_i$ to the left and dividing by $-c_i$. Therefore we will have proved the theorem if we show that in $(*)$ all of the coefficients are 0. For that we use induction on the row number i.

The base case is the first row $i = 1$ (if there is no such nonzero row, so that R is the zero matrix, then the lemma holds vacuously). Let ℓ_i be the column number of the leading entry in row i. Consider the entry of each row that is in column ℓ_1. Equation $(*)$ gives this.

$$0 = c_1 r_{1,\ell_1} + c_2 r_{2,\ell_1} + \cdots + c_m r_{m,\ell_1} \qquad (**)$$

The matrix is in echelon form so every row after the first has a zero entry in that column $r_{2,\ell_1} = \cdots = r_{m,\ell_1} = 0$. Thus equation $(**)$ shows that $c_1 = 0$, because $r_{1,\ell_1} \neq 0$ as it leads the row.

The inductive step is much the same as the base step. Again consider equation $(*)$. We will prove that if the coefficient c_i is 0 for each row index $i \in \{1, \ldots, k\}$ then c_{k+1} is also 0. We focus on the entries from column ℓ_{k+1}.

$$0 = c_1 r_{1,\ell_{k+1}} + \cdots + c_{k+1} r_{k+1,\ell_{k+1}} + \cdots + c_m r_{m,\ell_{k+1}}$$

By the inductive hypothesis $c_1, \ldots c_k$ are all 0 so this reduces to the equation $0 = c_{k+1}r_{k+1,\ell_{k+1}} + \cdots + c_m r_{m,\ell_{k+1}}$. The matrix is in echelon form so the entries $r_{k+2,\ell_{k+1}}, \ldots, r_{m,\ell_{k+1}}$ are all 0. Thus $c_{k+1} = 0$, because $r_{k+1,\ell_{k+1}} \neq 0$ as it is the leading entry. QED

2.6 Theorem Each matrix is row equivalent to a unique reduced echelon form matrix.

PROOF [Yuster] Fix a number of rows m. We will proceed by induction on the number of columns n.

The base case is that the matrix has $n = 1$ column. If this is the zero matrix then its echelon form is the zero matrix. If instead it has any nonzero entries then when the matrix is brought to reduced echelon form it must have at least one nonzero entry, which must be a 1 in the first row. Either way, its reduced echelon form is unique.

For the inductive step we assume that $n > 1$ and that all m row matrices having fewer than n columns have a unique reduced echelon form. Consider an $m \times n$ matrix A and suppose that B and C are two reduced echelon form matrices derived from A. We will show that these two must be equal.

Let $\hat{A}$ be the matrix consisting of the first $n - 1$ columns of A. Observe that any sequence of row operations that bring A to reduced echelon form will also bring $\hat{A}$ to reduced echelon form. By the inductive hypothesis this reduced echelon form of $\hat{A}$ is unique, so if B and C differ then the difference must occur in column n.

We finish the inductive step, and the argument, by showing that the two cannot differ only in that column. Consider a homogeneous system of equations for which A is the matrix of coefficients.

$$\begin{aligned} a_{1,1}x_1 + a_{1,2}x_2 + \cdots + a_{1,n}x_n &= 0 \\ a_{2,1}x_1 + a_{2,2}x_2 + \cdots + a_{2,n}x_n &= 0 \\ &\vdots \\ a_{m,1}x_1 + a_{m,2}x_2 + \cdots + a_{m,n}x_n &= 0 \end{aligned} \qquad (*)$$

By Theorem One.I.1.5 the set of solutions to that system is the same as the set of solutions to B's system

$$\begin{aligned} b_{1,1}x_1 + b_{1,2}x_2 + \cdots + b_{1,n}x_n &= 0 \\ b_{2,1}x_1 + b_{2,2}x_2 + \cdots + b_{2,n}x_n &= 0 \\ &\vdots \\ b_{m,1}x_1 + b_{m,2}x_2 + \cdots + b_{m,n}x_n &= 0 \end{aligned} \qquad (**)$$

and to C's.

$$\begin{aligned} c_{1,1}x_1 + \ c_{1,2}x_2 + \cdots + \ c_{1,n}x_n &= 0 \\ c_{2,1}x_1 + \ c_{2,2}x_2 + \cdots + \ c_{2,n}x_n &= 0 \\ &\vdots \\ c_{m,1}x_1 + c_{m,2}x_2 + \cdots + c_{m,n}x_n &= 0 \end{aligned} \qquad (***)$$

With B and C different only in column n, suppose that they differ in row i. Subtract row i of (***) from row i of (**) to get the equation $(b_{i,n}-c_{i,n})\cdot x_n = 0$. We've assumed that $b_{i,n} \neq c_{i,n}$ so $x_n = 0$. Thus in (**) and (***) the n-th column contains a leading entry, or else the variable x_n would be free. That's a contradiction because with B and C equal on the first $n-1$ columns, the leading entries in the n-th column would have to be in the same row, and with both matrices in reduced echelon form, both leading entries would have to be 1, and would have to be the only nonzero entries in that column. So $B = C$. QED

That result answers the two questions from this section's introduction: do any two echelon form versions of a linear system have the same number of free variables, and if so are they exactly the same variables? We get from any echelon form version to the reduced echelon form by eliminating up, so any echelon form version of a system has the same free variables as the reduced echelon form, and therefore uniqueness of reduced echelon form gives that the same variables are free in all echelon form version of a system. Thus both questions are answered "yes." There is no linear system and no combination of row operations such that, say, we could solve the system one way and get y and z free but solve it another way and get y and w free.

We close with a recap. In Gauss's Method we start with a matrix and then derive a sequence of other matrices. We defined two matrices to be related if we can derive one from the other. That relation is an equivalence relation, called row equivalence, and so partitions the set of all matrices into row equivalence classes.

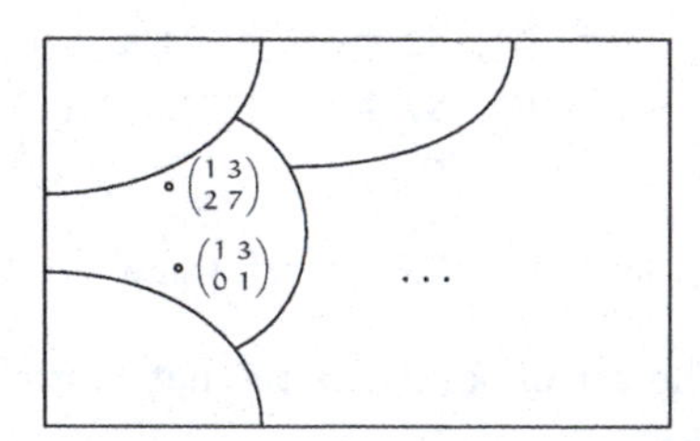

(There are infinitely many matrices in the pictured class, but we've only got room to show two.) We have proved there is one and only one reduced echelon form matrix in each row equivalence class. So the reduced echelon form is a canonical form* for row equivalence: the reduced echelon form matrices are

* More information on canonical representatives is in the appendix.

representatives of the classes.

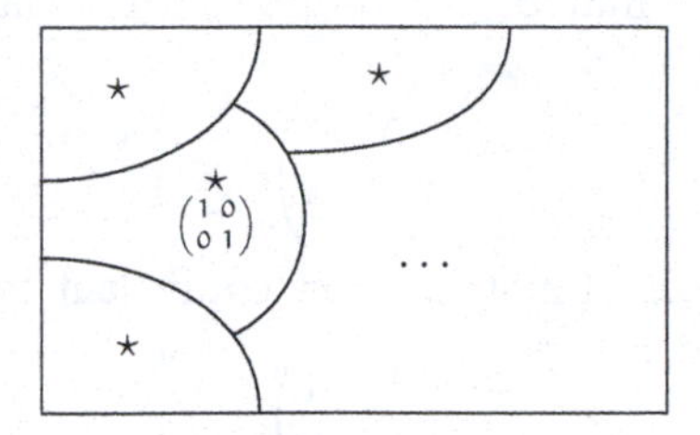

The idea here is that one way to understand a mathematical situation is by being able to classify the cases that can happen. This is a theme in this book and we have seen this several times already. We classified solution sets of linear systems into the no-elements, one-element, and infinitely-many elements cases. We also classified linear systems with the same number of equations as unknowns into the nonsingular and singular cases.

Here, where we are investigating row equivalence, we know that the set of all matrices breaks into the row equivalence classes and we now have a way to put our finger on each of those classes — we can think of the matrices in a class as derived by row operations from the unique reduced echelon form matrix in that class.

Put in more operational terms, uniqueness of reduced echelon form lets us answer questions about the classes by translating them into questions about the representatives. For instance, as promised in this section's opening, we now can decide whether one matrix can be derived from another by row reduction. We apply the Gauss-Jordan procedure to both and see if they yield the same reduced echelon form.

2.7 Example These matrices are not row equivalent

$$\begin{pmatrix} 1 & -3 \\ -2 & 6 \end{pmatrix} \qquad \begin{pmatrix} 1 & -3 \\ -2 & 5 \end{pmatrix}$$

because their reduced echelon forms are not equal.

$$\begin{pmatrix} 1 & -3 \\ 0 & 0 \end{pmatrix} \qquad \begin{pmatrix} 1 & 0 \\ 0 & 1 \end{pmatrix}$$

2.8 Example Any nonsingular 3×3 matrix Gauss-Jordan reduces to this.

$$\begin{pmatrix} 1 & 0 & 0 \\ 0 & 1 & 0 \\ 0 & 0 & 1 \end{pmatrix}$$

2.9 Example We can describe all the classes by listing all possible reduced echelon form matrices. Any 2×2 matrix lies in one of these: the class of matrices row equivalent to this,

$$\begin{pmatrix} 0 & 0 \\ 0 & 0 \end{pmatrix}$$

the infinitely many classes of matrices row equivalent to one of this type

$$\begin{pmatrix} 1 & a \\ 0 & 0 \end{pmatrix}$$

where $a \in \mathbb{R}$ (including $a = 0$), the class of matrices row equivalent to this,

$$\begin{pmatrix} 0 & 1 \\ 0 & 0 \end{pmatrix}$$

and the class of matrices row equivalent to this

$$\begin{pmatrix} 1 & 0 \\ 0 & 1 \end{pmatrix}$$

(this is the class of nonsingular 2×2 matrices).

Exercises

✓ **2.10** Decide if the matrices are row equivalent.

(a) $\begin{pmatrix} 1 & 2 \\ 4 & 8 \end{pmatrix}, \begin{pmatrix} 0 & 1 \\ 1 & 2 \end{pmatrix}$ (b) $\begin{pmatrix} 1 & 0 & 2 \\ 3 & -1 & 1 \\ 5 & -1 & 5 \end{pmatrix}, \begin{pmatrix} 1 & 0 & 2 \\ 0 & 2 & 10 \\ 2 & 0 & 4 \end{pmatrix}$

(c) $\begin{pmatrix} 2 & 1 & -1 \\ 1 & 1 & 0 \\ 4 & 3 & -1 \end{pmatrix}, \begin{pmatrix} 1 & 0 & 2 \\ 0 & 2 & 10 \end{pmatrix}$ (d) $\begin{pmatrix} 1 & 1 & 1 \\ -1 & 2 & 2 \end{pmatrix}, \begin{pmatrix} 0 & 3 & -1 \\ 2 & 2 & 5 \end{pmatrix}$

(e) $\begin{pmatrix} 1 & 1 & 1 \\ 0 & 0 & 3 \end{pmatrix}, \begin{pmatrix} 0 & 1 & 2 \\ 1 & -1 & 1 \end{pmatrix}$

2.11 Describe the matrices in each of the classes represented in Example 2.9.

2.12 Describe all matrices in the row equivalence class of these.

(a) $\begin{pmatrix} 1 & 0 \\ 0 & 0 \end{pmatrix}$ (b) $\begin{pmatrix} 1 & 2 \\ 2 & 4 \end{pmatrix}$ (c) $\begin{pmatrix} 1 & 1 \\ 1 & 3 \end{pmatrix}$

2.13 How many row equivalence classes are there?

2.14 Can row equivalence classes contain different-sized matrices?

2.15 How big are the row equivalence classes?

(a) Show that for any matrix of all zeros, the class is finite.

(b) Do any other classes contain only finitely many members?

✓ **2.16** Give two reduced echelon form matrices that have their leading entries in the same columns, but that are not row equivalent.

✓ **2.17** Show that any two $n\times n$ nonsingular matrices are row equivalent. Are any two singular matrices row equivalent?

✓ **2.18** Describe all of the row equivalence classes containing these.

(a) 2×2 matrices (b) 2×3 matrices (c) 3×2 matrices
(d) 3×3 matrices

2.19 (a) Show that a vector $\vec{\beta}_0$ is a linear combination of members of the set $\{\vec{\beta}_1, \ldots, \vec{\beta}_n\}$ if and only if there is a linear relationship $\vec{0} = c_0\vec{\beta}_0 + \cdots + c_n\vec{\beta}_n$ where c_0 is not zero. (*Hint.* Watch out for the $\vec{\beta}_0 = \vec{0}$ case.)
(b) Use that to simplify the proof of Lemma 2.5.

✓ **2.20** [Trono] Three truck drivers went into a roadside cafe. One truck driver purchased four sandwiches, a cup of coffee, and ten doughnuts for \$8.45. Another driver purchased three sandwiches, a cup of coffee, and seven doughnuts for \$6.30. What did the third truck driver pay for a sandwich, a cup of coffee, and a doughnut?

✓ **2.21** The Linear Combination Lemma says which equations can be gotten from Gaussian reduction of a given linear system.
(1) Produce an equation not implied by this system.

$$\begin{aligned} 3x + 4y &= 8 \\ 2x + y &= 3 \end{aligned}$$

(2) Can any equation be derived from an inconsistent system?

2.22 [Hoffman & Kunze] Extend the definition of row equivalence to linear systems. Under your definition, do equivalent systems have the same solution set?

✓ **2.23** In this matrix

$$\begin{pmatrix} 1 & 2 & 3 \\ 3 & 0 & 3 \\ 1 & 4 & 5 \end{pmatrix}$$

the first and second columns add to the third.
(a) Show that remains true under any row operation.
(b) Make a conjecture.
(c) Prove that it holds.

Topic

Computer Algebra Systems

The linear systems in this chapter are small enough that their solution by hand is easy. For large systems, including those involving thousands of equations, we need a computer. There are special purpose programs such as LINPACK for this. Also popular are general purpose computer algebra systems including *Maple*, *Mathematica*, or *MATLAB*, and *Sage*.

For example, in the Topic on Networks, we need to solve this.

$$\begin{array}{rcl} i_0 - i_1 - i_2 & = & 0 \\ i_1 - i_3 - i_5 & = & 0 \\ i_2 - i_4 + i_5 & = & 0 \\ i_3 + i_4 - i_6 & = & 0 \\ 5i_1 + 10i_3 & = & 10 \\ 2i_2 + 4i_4 & = & 10 \\ 5i_1 - 2i_2 + 50i_5 & = & 0 \end{array}$$

Doing this by hand would take time and be error-prone. A computer is better.

Here is that system solved with *Sage*. (There are many ways to do this; the one here has the advantage of simplicity.)

```
sage: var('i0,i1,i2,i3,i4,i5,i6')
(i0, i1, i2, i3, i4, i5, i6)
sage: network_system=[i0-i1-i2==0, i1-i3-i5==0,
....:          i2-i4+i5==0, i3+i4-i6==0, 5*i1+10*i3==10,
....:          2*i2+4*i4==10, 5*i1-2*i2+50*i5==0]
sage: solve(network_system, i0,i1,i2,i3,i4,i5,i6)
[[i0 == (7/3), i1 == (2/3), i2 == (5/3), i3 == (2/3),
    i4 == (5/3), i5 == 0, i6 == (7/3)]]
```

Magic.

Here is the same system solved under Maple. We enter the array of coefficients and the vector of constants, and then we get the solution.

```
> A:=array( [[1,-1,-1,0,0,0,0],
             [0,1,0,-1,0,-1,0],
             [0,0,1,0,-1,1,0],
             [0,0,0,1,1,0,-1],
             [0,5,0,10,0,0,0],
```

```
           [0,0,2,0,4,0,0],
           [0,5,-2,0,0,50,0]] );
> u:=array( [0,0,0,0,10,10,0] );
> linsolve(A,u);
     7  2  5  2  5     7
  [ -, -, -, -, -, 0, - ]
     3  3  3  3  3     3
```

If a system has infinitely many solutions then the program will return a parametrization.

Exercises

1 Use the computer to solve the two problems that opened this chapter.

(a) This is the Statics problem.

$$\begin{aligned} 40h + 15c &= 100 \\ 25c &= 50 + 50h \end{aligned}$$

(b) This is the Chemistry problem.

$$\begin{aligned} 7h &= 7j \\ 8h + 1i &= 5j + 2k \\ 1i &= 3j \\ 3i &= 6j + 1k \end{aligned}$$

2 Use the computer to solve these systems from the first subsection, or conclude 'many solutions' or 'no solutions'.

(a) $\begin{aligned} 2x + 2y &= 5 \\ x - 4y &= 0 \end{aligned}$ (b) $\begin{aligned} -x + y &= 1 \\ x + y &= 2 \end{aligned}$ (c) $\begin{aligned} x - 3y + z &= 1 \\ x + y + 2z &= 14 \end{aligned}$ (d) $\begin{aligned} -x - y &= 1 \\ -3x - 3y &= 2 \end{aligned}$

(e) $\begin{aligned} 4y + z &= 20 \\ 2x - 2y + z &= 0 \\ x + z &= 5 \\ x + y - z &= 10 \end{aligned}$ (f) $\begin{aligned} 2x + z + w &= 5 \\ y - w &= -1 \\ 3x - z - w &= 0 \\ 4x + y + 2z + w &= 9 \end{aligned}$

3 Use the computer to solve these systems from the second subsection.

(a) $\begin{aligned} 3x + 6y &= 18 \\ x + 2y &= 6 \end{aligned}$ (b) $\begin{aligned} x + y &= 1 \\ x - y &= -1 \end{aligned}$ (c) $\begin{aligned} x_1 + x_3 &= 4 \\ x_1 - x_2 + 2x_3 &= 5 \\ 4x_1 - x_2 + 5x_3 &= 17 \end{aligned}$

(d) $\begin{aligned} 2a + b - c &= 2 \\ 2a + c &= 3 \\ a - b &= 0 \end{aligned}$ (e) $\begin{aligned} x + 2y - z &= 3 \\ 2x + y + w &= 4 \\ x - y + z + w &= 1 \end{aligned}$ (f) $\begin{aligned} x + z + w &= 4 \\ 2x + y - w &= 2 \\ 3x + y + z &= 7 \end{aligned}$

4 What does the computer give for the solution of the general 2×2 system?

$$\begin{aligned} ax + cy &= p \\ bx + dy &= q \end{aligned}$$

Topic

Accuracy of Computations

Gauss's Method lends itself to computerization. The code below illustrates. It operates on an $n \times n$ matrix named a, doing row combinations using the first row, then the second row, etc.

```
for(row=1; row<=n-1; row++){
  for(row_below=row+1; row_below<=n; row_below++){
    multiplier=a[row_below,row]/a[row,row];
    for(col=row; col<=n; col++){
      a[row_below,col]-=multiplier*a[row,col];
    }
  }
}
```

This is in the C language. The `for(row=1; row<=n-1; row++){ .. }` loop initializes `row` at 1 and then iterates while `row` is less than or equal to $n-1$, each time through incrementing `row` by one with the `++` operation. The other non-obvious language construct is that the `-=` in the innermost loop has the effect of `a[row_below,col]=-1*multiplier*a[row,col]+a[row_below,col]`.

While that code is a first take on mechanizing Gauss's Method, it is naive. For one thing, it assumes that the entry in the `row,row` position is nonzero. So one way that it needs to be extended is to cover the case where finding a zero in that location leads to a row swap or to the conclusion that the matrix is singular.

We could add some `if` statements to cover those cases but we will instead consider another way in which this code is naive. It is prone to pitfalls arising from the computer's reliance on floating point arithmetic.

For example, above we have seen that we must handle a singular system as a separate case. But systems that are nearly singular also require care. Consider this one (the extra digits are in the ninth significant place).

$$\begin{aligned} x + 2y &= 3 \\ 1.000\,000\,01x + 2y &= 3.000\,000\,01 \end{aligned} \qquad (*)$$

By eye we easily spot the solution $x = 1$, $y = 1$. A computer has more trouble. If it represents real numbers to eight significant places, called *single precision*, then it will represent the second equation internally as $1.0000000x + 2y = 3.0000000$, losing the digits in the ninth place. Instead of reporting the correct solution, this computer will think that the two equations are equal and it will report that the system is singular.

For some intuition about how the computer could come up with something that far off, consider this graph of the system.

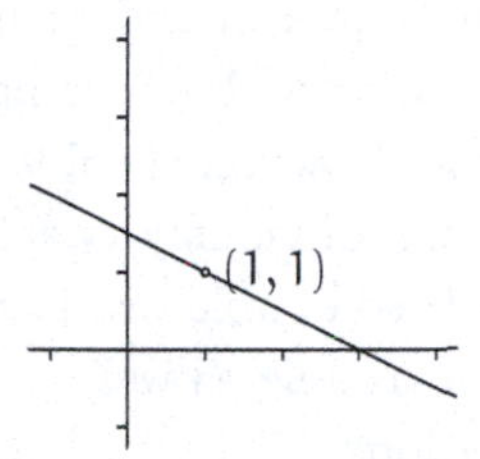

We cannot tell the two lines apart; this system is nearly singular in the sense that the two lines are nearly the same line. This gives the system $(*)$ the property that a small change in an equation can cause a large change in the solution. For instance, changing the 3.00000001 to 3.00000003 changes the intersection point from $(1,1)$ to $(3,0)$. The solution changes radically depending on the ninth digit, which explains why an eight-place computer has trouble. A problem that is very sensitive to inaccuracy or uncertainties in the input values is *ill-conditioned.*

The above example gives one way in which a system can be difficult to solve on a computer. It has the advantage that the picture of nearly-equal lines gives a memorable insight into one way for numerical difficulties to happen. Unfortunately this insight isn't useful when we wish to solve some large system. We typically will not understand the geometry of an arbitrary large system.

There are other ways that a computer's results may be unreliable, besides that the angle between some of the linear surfaces is small. For example, consider this system (from [Hamming]).

$$\begin{aligned} 0.001x + y &= 1 \\ x - y &= 0 \end{aligned} \qquad (**)$$

The second equation gives $x = y$, so $x = y = 1/1.001$ and thus both variables have values that are just less than 1. A computer using two digits represents the system internally in this way (we will do this example in two-digit floating point arithmetic for clarity but inventing a similar one with eight or more digits is easy).

$$\begin{aligned} (1.0 \times 10^{-3}) \cdot x + (1.0 \times 10^{0}) \cdot y &= 1.0 \times 10^{0} \\ (1.0 \times 10^{0}) \cdot x - (1.0 \times 10^{0}) \cdot y &= 0.0 \times 10^{0} \end{aligned}$$

The row reduction step $-1000\rho_1 + \rho_2$ produces a second equation $-1001y = -1000$, which this computer rounds to two places as $(-1.0 \times 10^{3})y = -1.0 \times 10^{3}$.

The computer decides from the second equation that $y = 1$ and with that it concludes from the first equation that $x = 0$. The y value is close but the x is bad — the ratio of the actual answer to the computer's answer is infinite. In short, another cause of unreliable output is the computer's reliance on floating point arithmetic when the system-solving code leads to using leading entries that are small.

An experienced programmer may respond by using *double precision*, which retains sixteen significant digits, or perhaps using some even larger size. This will indeed solve many problems. However, double precision has greater memory requirements and besides we can obviously tweak the above to give the same trouble in the seventeenth digit, so double precision isn't a panacea. We need a strategy to minimize numerical trouble as well as some guidance about how far we can trust the reported solutions.

A basic improvement on the naive code above is to not determine the factor to use for row combinations by simply taking the entry in the `row,row` position, but rather to look at all of the entries in the `row` column below the `row,row` entry and take one that is likely to give reliable results because it is not too small. This is *partial pivoting*.

For example, to solve the troublesome system $(**)$ above we start by looking at both equations for a best entry to use, and take the 1 in the second equation as more likely to give good results. The combination step of $-.001\rho_2 + \rho_1$ gives a first equation of $1.001y = 1$, which the computer will represent as $(1.0 \times 10^0)y = 1.0 \times 10^0$, leading to the conclusion that $y = 1$ and, after back-substitution, that $x = 1$, both of which are close to right. We can adapt the code from above to do this.

```
for(row=1; row<=n-1; row++){
/* find the largest entry in this column (in row max) */
  max=row;
  for(row_below=row+1; row_below<=n; row_below++){
    if (abs(a[row_below,row]) > abs(a[max,row]));
      max = row_below;
  }
/* swap rows to move that best entry up */
  for(col=row; col<=n; col++){
    temp=a[row,col];
    a[row,col]=a[max,col];
    a[max,col]=temp;
  }
/* proceed as before */
  for(row_below=row+1; row_below<=n; row_below++){
    multiplier=a[row_below,row]/a[row,row];
      for(col=row; col<=n; col++){
        a[row_below,col]-=multiplier*a[row,col];
    }
  }
}
```

A full analysis of the best way to implement Gauss's Method is beyond the scope of this book (see [Wilkinson 1965]), but the method recommended by most experts first finds the best entry among the candidates and then scales it to a number that is less likely to give trouble. This is *scaled partial pivoting.*

In addition to returning a result that is likely to be reliable, most well-done code will return a *conditioning number* that describes the factor by which uncertainties in the input numbers could be magnified to become inaccuracies in the results returned (see [Rice]).

The lesson is that just because Gauss's Method always works in theory, and just because computer code correctly implements that method, doesn't mean that the answer is reliable. In practice, always use a package where experts have worked hard to counter what can go wrong.

Exercises

1 Using two decimal places, add 253 and 2/3.

2 This intersect-the-lines problem contrasts with the example discussed above.

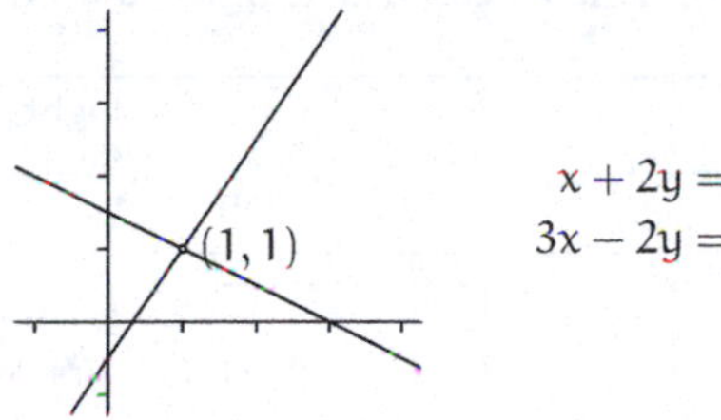

$$\begin{aligned} x + 2y &= 3 \\ 3x - 2y &= 1 \end{aligned}$$

Illustrate that in this system some small change in the numbers will produce only a small change in the solution by changing the constant in the bottom equation to 1.008 and solving. Compare it to the solution of the unchanged system.

3 Consider this system ([Rice]).

$$\begin{aligned} 0.000\,3x + 1.556y &= 1.569 \\ 0.345\,4x - 2.346y &= 1.018 \end{aligned}$$

(a) Solve it. (b) Solve it by rounding at each step to four digits.

4 Rounding inside the computer often has an effect on the result. Assume that your machine has eight significant digits.

(a) Show that the machine will compute $(2/3) + ((2/3) - (1/3))$ as unequal to $((2/3) + (2/3)) - (1/3)$. Thus, computer arithmetic is not associative.

(b) Compare the computer's version of $(1/3)x + y = 0$ and $(2/3)x + 2y = 0$. Is twice the first equation the same as the second?

5 Ill-conditioning is not only dependent on the matrix of coefficients. This example [Hamming] shows that it can arise from an interaction between the left and right sides of the system. Let ε be a small real.

$$\begin{aligned} 3x + 2y + z &= 6 \\ 2x + 2\varepsilon y + 2\varepsilon z &= 2 + 4\varepsilon \\ x + 2\varepsilon y - \varepsilon z &= 1 + \varepsilon \end{aligned}$$

(a) Solve the system by hand. Notice that the ε's divide out only because there is an exact cancellation of the integer parts on the right side as well as on the left.

(b) Solve the system by hand, rounding to two decimal places, and with $\varepsilon = 0.001$.

Topic

Analyzing Networks

The diagram below shows some of a car's electrical network. The battery is on the left, drawn as stacked line segments. The wires are lines, shown straight and with sharp right angles for neatness. Each light is a circle enclosing a loop.

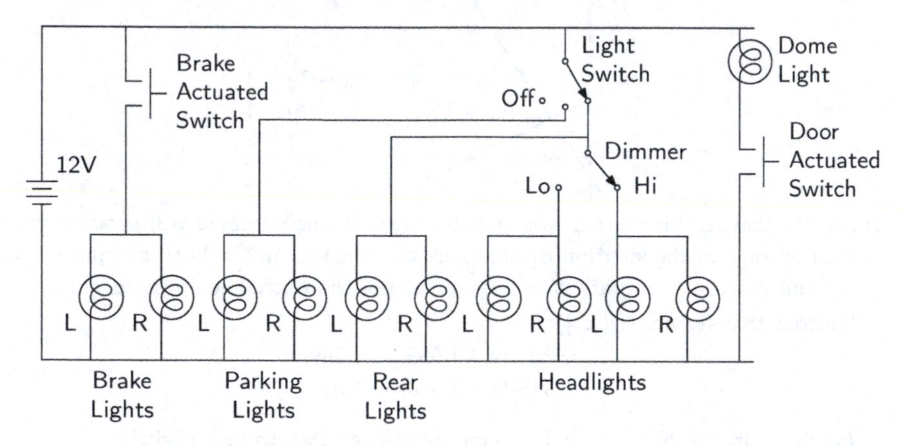

The designer of such a network needs to answer questions such as: how much electricity flows when both the hi-beam headlights and the brake lights are on? We will use linear systems to analyze simple electrical networks.

For the analysis we need two facts about electricity and two facts about electrical networks.

The first fact is that a battery is like a pump, providing a force impelling the electricity to flow, if there is a path. We say that the battery provides a *potential*. For instance, when the driver steps on the brake then the switch makes contact and so makes a circuit on the left side of the diagram, which includes the brake lights. Once the circuit exists, the battery's force creates a current flowing through that circuit, lighting the lights.

The second electrical fact is that in some kinds of network components the amount of flow is proportional to the force provided by the battery. That is, for each such component there is a number, it's *resistance*, such that the potential is equal to the flow times the resistance. Potential is measured in *volts*, the rate of flow is in *amperes*, and resistance to the flow is in *ohms*; these units are defined so that volts = amperes · ohms.

Components with this property, that the voltage-amperage response curve is a line through the origin, are *resistors*. For example, if a resistor measures 2 ohms then wiring it to a 12 volt battery results in a flow of 6 amperes. Conversely, if electrical current of 2 amperes flows through that resistor then there must be a 4 volt potential difference between it's ends. This is the *voltage drop* across the resistor. One way to think of the electrical circuits that we consider here is that the battery provides a voltage rise while the other components are voltage drops.

The facts that we need about networks are *Kirchoff's Current Law*, that for any point in a network the flow in equals the flow out and *Kirchoff's Voltage Law*, that around any circuit the total drop equals the total rise.

We start with the network below. It has a battery that provides the potential to flow and three resistors, shown as zig-zags. When components are wired one after another, as here, they are in *series*.

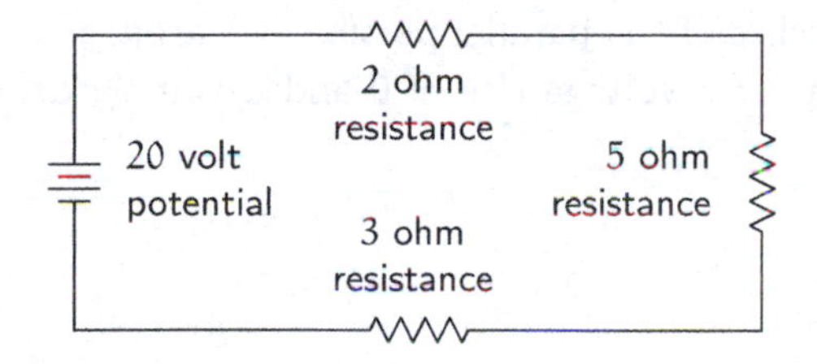

By Kirchoff's Voltage Law, because the voltage rise is 20 volts, the total voltage drop must also be 20 volts. Since the resistance from start to finish is 10 ohms (the resistance of the wire connecting the components is negligible), the current is $(20/10) = 2$ amperes. Now, by Kirchhoff's Current Law, there are 2 amperes through each resistor. Therefore the voltage drops are: 4 volts across the 2 ohm resistor, 10 volts across the 5 ohm resistor, and 6 volts across the 3 ohm resistor.

The prior network is simple enough that we didn't use a linear system but the next one is more complicated. Here the resistors are in *parallel*.

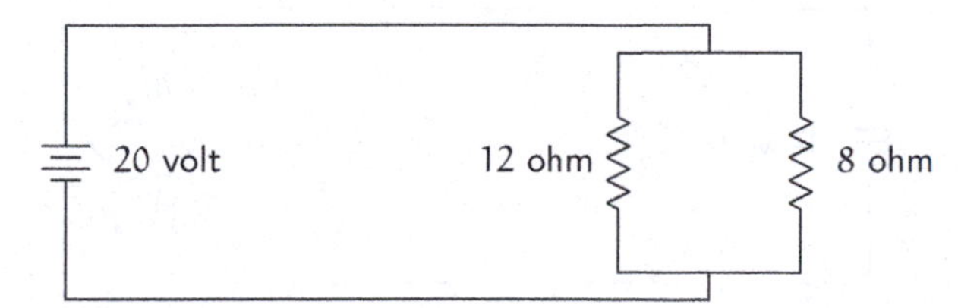

We begin by labeling the branches as below. Let the current through the left branch of the parallel portion be i_1 and that through the right branch be i_2, and also let the current through the battery be i_0. Note that we don't need to know the actual direction of flow — if current flows in the direction opposite to our arrow then we will get a negative number in the solution.

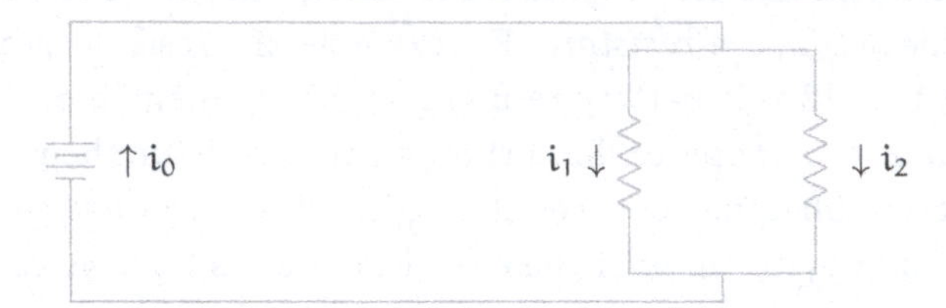

The Current Law, applied to the split point in the upper right, gives that $i_0 = i_1 + i_2$. Applied to the split point lower right it gives $i_1 + i_2 = i_0$. In the circuit that loops out of the top of the battery, down the left branch of the parallel portion, and back into the bottom of the battery, the voltage rise is 20 while the voltage drop is $i_1 \cdot 12$, so the Voltage Law gives that $12i_1 = 20$. Similarly, the circuit from the battery to the right branch and back to the battery gives that $8i_2 = 20$. And, in the circuit that simply loops around in the left and right branches of the parallel portion (we arbitrarily take the direction of clockwise), there is a voltage rise of 0 and a voltage drop of $8i_2 - 12i_1$ so $8i_2 - 12i_1 = 0$.

$$\begin{array}{rcrcrcl} i_0 & - & i_1 & - & i_2 & = & 0 \\ -i_0 & + & i_1 & + & i_2 & = & 0 \\ & & 12i_1 & & & = & 20 \\ & & & & 8i_2 & = & 20 \\ & & -12i_1 & + & 8i_2 & = & 0 \end{array}$$

The solution is $i_0 = 25/6$, $i_1 = 5/3$, and $i_2 = 5/2$, all in amperes. (Incidentally, this illustrates that redundant equations can arise in practice.)

Kirchhoff's laws can establish the electrical properties of very complex networks. The next diagram shows five resistors, whose values are in ohms, wired in *series-parallel*.

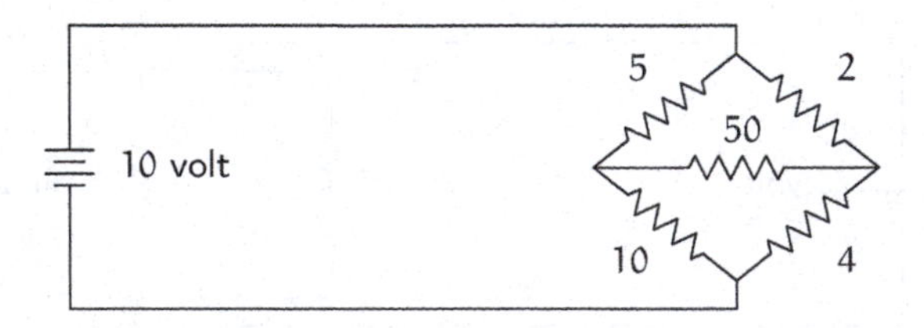

This is a *Wheatstone bridge* (see Exercise 3). To analyze it, we can place the arrows in this way.

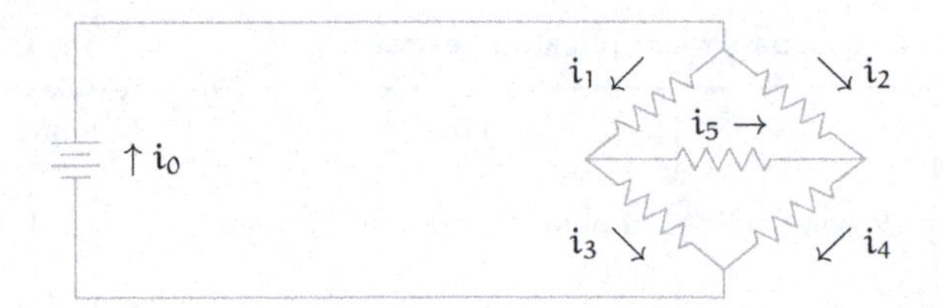

Kirchhoff's Current Law, applied to the top node, the left node, the right node, and the bottom node gives these.

$$\begin{aligned} i_0 &= i_1 + i_2 \\ i_1 &= i_3 + i_5 \\ i_2 + i_5 &= i_4 \\ i_3 + i_4 &= i_0 \end{aligned}$$

Kirchhoff's Voltage Law, applied to the inside loop (the i_0 to i_1 to i_3 to i_0 loop), the outside loop, and the upper loop not involving the battery, gives these.

$$\begin{aligned} 5i_1 + 10i_3 &= 10 \\ 2i_2 + 4i_4 &= 10 \\ 5i_1 + 50i_5 - 2i_2 &= 0 \end{aligned}$$

Those suffice to determine the solution $i_0 = 7/3$, $i_1 = 2/3$, $i_2 = 5/3$, $i_3 = 2/3$, $i_4 = 5/3$, and $i_5 = 0$.

We can understand many kinds of networks in this way. For instance, the exercises analyze some networks of streets.

Exercises

1 Calculate the amperages in each part of each network.

(a) This is a simple network.

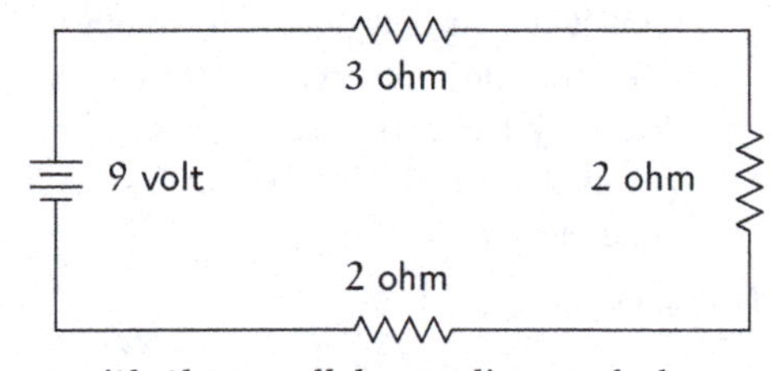

(b) Compare this one with the parallel case discussed above.

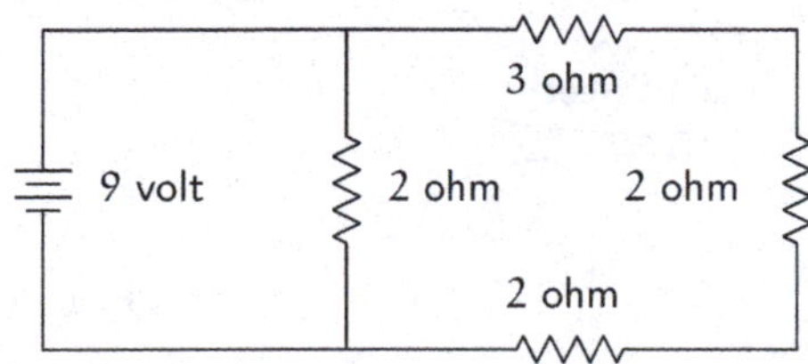

(c) This is a reasonably complicated network.

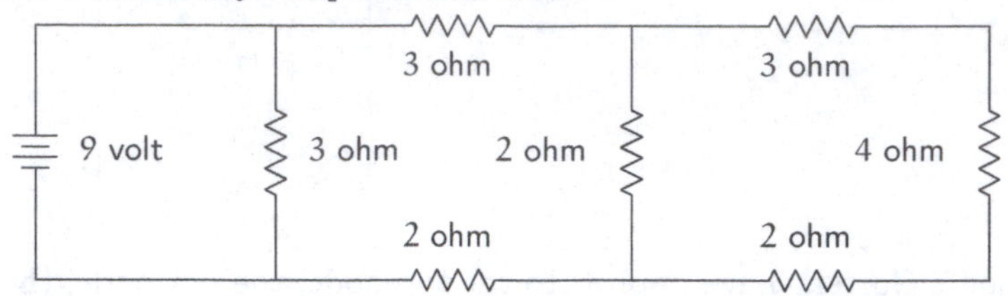

2 In the first network that we analyzed, with the three resistors in series, we just added to get that they acted together like a single resistor of 10 ohms. We can do a similar thing for parallel circuits. In the second circuit analyzed,

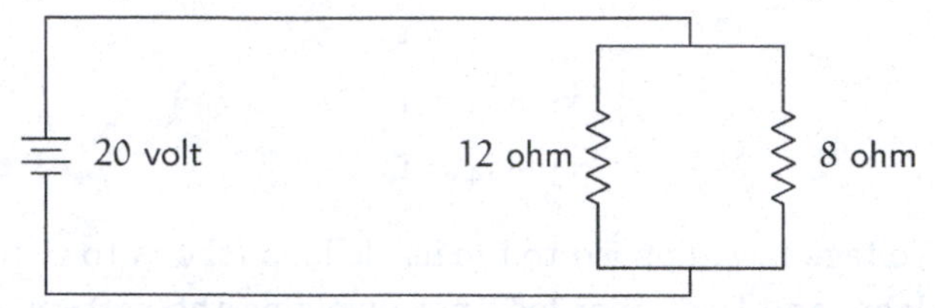

the electric current through the battery is 25/6 amperes. Thus, the parallel portion is *equivalent* to a single resistor of $20/(25/6) = 4.8$ ohms.

(a) What is the equivalent resistance if we change the 12 ohm resistor to 5 ohms?

(b) What is the equivalent resistance if the two are each 8 ohms?

(c) Find the formula for the equivalent resistance if the two resistors in parallel are r_1 ohms and r_2 ohms.

3 A *Wheatstone bridge* is used to measure resistance.

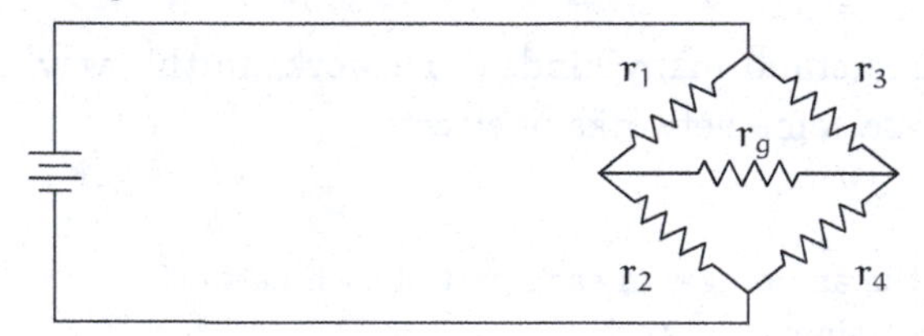

Show that in this circuit if the current flowing through r_g is zero then $r_4 = r_2 r_3 / r_1$. (To operate the device, put the unknown resistance at r_4. At r_g is a meter that shows the current. We vary the three resistances r_1, r_2, and r_3—typically they each have a calibrated knob—until the current in the middle reads 0. Then the equation gives the value of r_4.)

4 Consider this traffic circle.

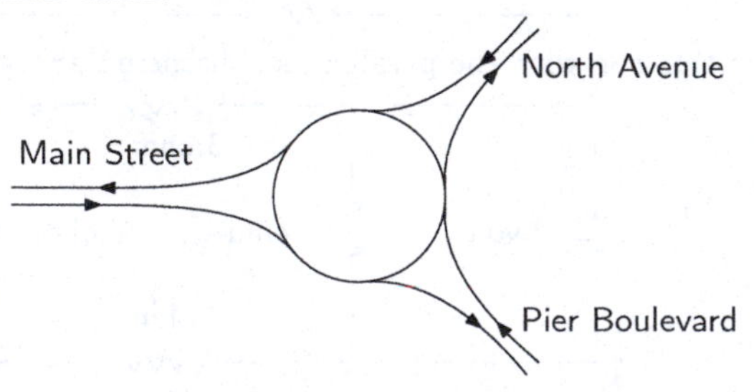

This is the traffic volume, in units of cars per five minutes.

	North	*Pier*	*Main*
into	100	150	25
out of	75	150	50

We can set up equations to model how the traffic flows.

(a) Adapt Kirchhoff's Current Law to this circumstance. Is it a reasonable modeling assumption?

(b) Label the three between-road arcs in the circle with a variable. Using the (adapted) Current Law, for each of the three in-out intersections state an equation describing the traffic flow at that node.

(c) Solve that system.

(d) Interpret your solution.

(e) Restate the Voltage Law for this circumstance. How reasonable is it?

5 This is a network of streets.

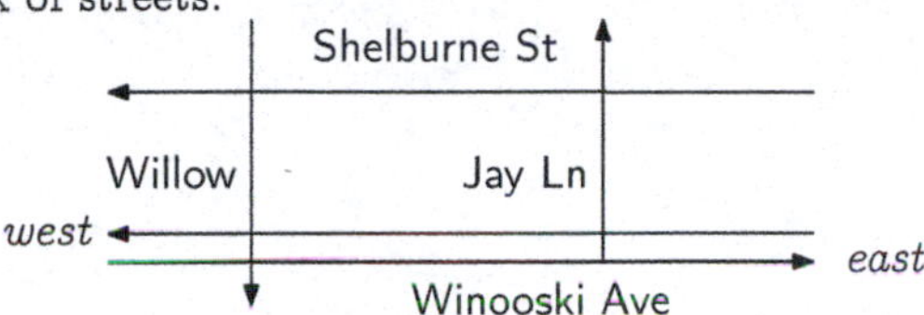

We can observe the hourly flow of cars into this network's entrances, and out of its exits.

	east Winooski	*west Winooski*	*Willow*	*Jay*	*Shelburne*
into	80	50	65	–	40
out of	30	5	70	55	75

(Note that to reach Jay a car must enter the network via some other road first, which is why there is no 'into Jay' entry in the table. Note also that over a long period of time, the total in must approximately equal the total out, which is why both rows add to 235 cars.) Once inside the network, the traffic may flow in different ways, perhaps filling Willow and leaving Jay mostly empty, or perhaps flowing in some other way. Kirchhoff's Laws give the limits on that freedom.

(a) Determine the restrictions on the flow inside this network of streets by setting up a variable for each block, establishing the equations, and solving them. Notice that some streets are one-way only. (*Hint:* this will not yield a unique solution, since traffic can flow through this network in various ways; you should get at least one free variable.)

(b) Suppose that someone proposes construction for Winooski Avenue East between Willow and Jay, and traffic on that block will be reduced. What is the least amount of traffic flow that can we can allow on that block without disrupting the hourly flow into and out of the network?

Chapter Two

Vector Spaces

The first chapter finished with a fair understanding of how Gauss's Method solves a linear system. It systematically takes linear combinations of the rows. Here we move to a general study of linear combinations.

We need a setting. At times in the first chapter we've combined vectors from $\mathbb{R}^2$, at other times vectors from $\mathbb{R}^3$, and at other times vectors from higher-dimensional spaces. So our first impulse might be to work in $\mathbb{R}^n$, leaving n unspecified. This would have the advantage that any of the results would hold for $\mathbb{R}^2$ and for $\mathbb{R}^3$ and for many other spaces, simultaneously.

But if having the results apply to many spaces at once is advantageous then sticking only to $\mathbb{R}^n$'s is overly restrictive. We'd like our results to apply to combinations of row vectors, as in the final section of the first chapter. We've even seen some spaces that are not simply a collection of all of the same-sized column vectors or row vectors. For instance, we've seen a homogeneous system's solution set that is a plane inside of $\mathbb{R}^3$. This set is a closed system in that a linear combination of these solutions is also a solution. But it does not contain all of the three-tall column vectors, only some of them.

We want the results about linear combinations to apply anywhere that linear combinations make sense. We shall call any such set a *vector space*. Our results, instead of being phrased as "Whenever we have a collection in which we can sensibly take linear combinations ...", will be stated "In any vector space ..."

Such a statement describes at once what happens in many spaces. To understand the advantages of moving from studying a single space to studying a class of spaces, consider this analogy. Imagine that the government made laws one person at a time: "Leslie Jones can't jay walk." That would be bad; statements have the virtue of economy when they apply to many cases at once. Or suppose that they said, "Kim Ke must stop when passing an accident." Contrast that with, "Any doctor must stop when passing an accident." More general statements, in some ways, are clearer.

I Definition of Vector Space

We shall study structures with two operations, an addition and a scalar multiplication, that are subject to some simple conditions. We will reflect more on the conditions later but on first reading notice how reasonable they are. For instance, surely any operation that can be called an addition (e.g., column vector addition, row vector addition, or real number addition) will satisfy conditions (1) through (5) below.

I.1 Definition and Examples

1.1 Definition A *vector space* (over $\mathbb{R}$) consists of a set V along with two operations '+' and '·' subject to the conditions that for all vectors $\vec{v}, \vec{w}, \vec{u} \in V$ and all *scalars* $r, s \in \mathbb{R}$:

(1) the set V is closed under vector addition, that is, $\vec{v} + \vec{w} \in V$
(2) vector addition is commutative, $\vec{v} + \vec{w} = \vec{w} + \vec{v}$
(3) vector addition is associative, $(\vec{v} + \vec{w}) + \vec{u} = \vec{v} + (\vec{w} + \vec{u})$
(4) there is a *zero vector* $\vec{0} \in V$ such that $\vec{v} + \vec{0} = \vec{v}$ for all $\vec{v} \in V$
(5) each $\vec{v} \in V$ has an *additive inverse* $\vec{w} \in V$ such that $\vec{w} + \vec{v} = \vec{0}$
(6) the set V is closed under scalar multiplication, that is, $r \cdot \vec{v} \in V$
(7) addition of scalars distributes over scalar multiplication, $(r+s) \cdot \vec{v} = r \cdot \vec{v} + s \cdot \vec{v}$
(8) scalar multiplication distributes over vector addition, $r \cdot (\vec{v} + \vec{w}) = r \cdot \vec{v} + r \cdot \vec{w}$
(9) ordinary multipication of scalars associates with scalar multiplication, $(rs) \cdot \vec{v} = r \cdot (s \cdot \vec{v})$
(10) multiplication by the scalar 1 is the identity operation, $1 \cdot \vec{v} = \vec{v}$.

1.2 Remark The definition involves two kinds of addition and two kinds of multiplication, and so may at first seem confused. For instance, in condition (7) the '+' on the left is addition of two real numbers while the '+' on the right is addition of two vectors in V. These expressions aren't ambiguous because of context; for example, r and s are real numbers so '$r + s$' can only mean real number addition. In the same way, item (9)'s left side 'rs' is ordinary real number multiplication, while its right side '$s \cdot \vec{v}$' is the scalar multipliction defined for this vector space.

The best way to understand the definition is to go through the examples below and for each, check all ten conditions. The first example includes that check, written out at length. Use it as a model for the others. Especially important are the *closure* conditions, (1) and (6). They specify that the addition and scalar multiplication operations are always sensible — they are defined for every pair of vectors and every scalar and vector, and the result of the operation is a member of the set (see Example 1.4).

1.3 Example The set $\mathbb{R}^2$ is a vector space if the operations '+' and '·' have their usual meaning.

$$\begin{pmatrix}x_1\\x_2\end{pmatrix}+\begin{pmatrix}y_1\\y_2\end{pmatrix}=\begin{pmatrix}x_1+y_1\\x_2+y_2\end{pmatrix}\qquad r\cdot\begin{pmatrix}x_1\\x_2\end{pmatrix}=\begin{pmatrix}rx_1\\rx_2\end{pmatrix}$$

We shall check all of the conditions.

There are five conditions in the paragraph having to do with addition. For (1), closure of addition, observe that for any $v_1,v_2,w_1,w_2\in\mathbb{R}$ the result of the vector sum

$$\begin{pmatrix}v_1\\v_2\end{pmatrix}+\begin{pmatrix}w_1\\w_2\end{pmatrix}=\begin{pmatrix}v_1+w_1\\v_2+w_2\end{pmatrix}$$

is a column array with two real entries, and so is in $\mathbb{R}^2$. For (2), that addition of vectors commutes, take all entries to be real numbers and compute

$$\begin{pmatrix}v_1\\v_2\end{pmatrix}+\begin{pmatrix}w_1\\w_2\end{pmatrix}=\begin{pmatrix}v_1+w_1\\v_2+w_2\end{pmatrix}=\begin{pmatrix}w_1+v_1\\w_2+v_2\end{pmatrix}=\begin{pmatrix}w_1\\w_2\end{pmatrix}+\begin{pmatrix}v_1\\v_2\end{pmatrix}$$

(the second equality follows from the fact that the components of the vectors are real numbers, and the addition of real numbers is commutative). Condition (3), associativity of vector addition, is similar.

$$\begin{aligned}(\begin{pmatrix}v_1\\v_2\end{pmatrix}+\begin{pmatrix}w_1\\w_2\end{pmatrix})+\begin{pmatrix}u_1\\u_2\end{pmatrix}&=\begin{pmatrix}(v_1+w_1)+u_1\\(v_2+w_2)+u_2\end{pmatrix}\\&=\begin{pmatrix}v_1+(w_1+u_1)\\v_2+(w_2+u_2)\end{pmatrix}\\&=\begin{pmatrix}v_1\\v_2\end{pmatrix}+(\begin{pmatrix}w_1\\w_2\end{pmatrix}+\begin{pmatrix}u_1\\u_2\end{pmatrix})\end{aligned}$$

For the fourth condition we must produce a zero element — the vector of zeroes is it.

$$\begin{pmatrix}v_1\\v_2\end{pmatrix}+\begin{pmatrix}0\\0\end{pmatrix}=\begin{pmatrix}v_1\\v_2\end{pmatrix}$$

For (5), to produce an additive inverse, note that for any $v_1,v_2\in\mathbb{R}$ we have

$$\begin{pmatrix}-v_1\\-v_2\end{pmatrix}+\begin{pmatrix}v_1\\v_2\end{pmatrix}=\begin{pmatrix}0\\0\end{pmatrix}$$

so the first vector is the desired additive inverse of the second.

The checks for the five conditions having to do with scalar multiplication are similar. For (6), closure under scalar multiplication, where $r, v_1, v_2 \in \mathbb{R}$,

$$r \cdot \begin{pmatrix} v_1 \\ v_2 \end{pmatrix} = \begin{pmatrix} rv_1 \\ rv_2 \end{pmatrix}$$

is a column array with two real entries, and so is in $\mathbb{R}^2$. Next, this checks (7).

$$(r+s) \cdot \begin{pmatrix} v_1 \\ v_2 \end{pmatrix} = \begin{pmatrix} (r+s)v_1 \\ (r+s)v_2 \end{pmatrix} = \begin{pmatrix} rv_1 + sv_1 \\ rv_2 + sv_2 \end{pmatrix} = r \cdot \begin{pmatrix} v_1 \\ v_2 \end{pmatrix} + s \cdot \begin{pmatrix} v_1 \\ v_2 \end{pmatrix}$$

For (8), that scalar multiplication distributes from the left over vector addition, we have this.

$$r \cdot (\begin{pmatrix} v_1 \\ v_2 \end{pmatrix} + \begin{pmatrix} w_1 \\ w_2 \end{pmatrix}) = \begin{pmatrix} r(v_1 + w_1) \\ r(v_2 + w_2) \end{pmatrix} = \begin{pmatrix} rv_1 + rw_1 \\ rv_2 + rw_2 \end{pmatrix} = r \cdot \begin{pmatrix} v_1 \\ v_2 \end{pmatrix} + r \cdot \begin{pmatrix} w_1 \\ w_2 \end{pmatrix}$$

The ninth

$$(rs) \cdot \begin{pmatrix} v_1 \\ v_2 \end{pmatrix} = \begin{pmatrix} (rs)v_1 \\ (rs)v_2 \end{pmatrix} = \begin{pmatrix} r(sv_1) \\ r(sv_2) \end{pmatrix} = r \cdot (s \cdot \begin{pmatrix} v_1 \\ v_2 \end{pmatrix})$$

and tenth conditions are also straightforward.

$$1 \cdot \begin{pmatrix} v_1 \\ v_2 \end{pmatrix} = \begin{pmatrix} 1v_1 \\ 1v_2 \end{pmatrix} = \begin{pmatrix} v_1 \\ v_2 \end{pmatrix}$$

In a similar way, each $\mathbb{R}^n$ is a vector space with the usual operations of vector addition and scalar multiplication. (In $\mathbb{R}^1$, we usually do not write the members as column vectors, i.e., we usually do not write '(π)'. Instead we just write 'π'.)

1.4 Example This subset of $\mathbb{R}^3$ that is a plane through the origin

$$P = \{ \begin{pmatrix} x \\ y \\ z \end{pmatrix} \mid x + y + z = 0 \}$$

is a vector space if '+' and '$\cdot$' are interpreted in this way.

$$\begin{pmatrix} x_1 \\ y_1 \\ z_1 \end{pmatrix} + \begin{pmatrix} x_2 \\ y_2 \\ z_2 \end{pmatrix} = \begin{pmatrix} x_1 + x_2 \\ y_1 + y_2 \\ z_1 + z_2 \end{pmatrix} \qquad r \cdot \begin{pmatrix} x \\ y \\ z \end{pmatrix} = \begin{pmatrix} rx \\ ry \\ rz \end{pmatrix}$$

The addition and scalar multiplication operations here are just the ones of $\mathbb{R}^3$, reused on its subset P. We say that P *inherits* these operations from $\mathbb{R}^3$. This example of an addition in P

$$\begin{pmatrix}1\\1\\-2\end{pmatrix}+\begin{pmatrix}-1\\0\\1\end{pmatrix}=\begin{pmatrix}0\\1\\-1\end{pmatrix}$$

illustrates that P is closed under addition. We've added two vectors from P — that is, with the property that the sum of their three entries is zero — and the result is a vector also in P. Of course, this example is not a proof. For the proof that P is closed under addition, take two elements of P.

$$\begin{pmatrix}x_1\\y_1\\z_1\end{pmatrix}\quad\begin{pmatrix}x_2\\y_2\\z_2\end{pmatrix}$$

Membership in P means that $x_1+y_1+z_1=0$ and $x_2+y_2+z_2=0$. Observe that their sum

$$\begin{pmatrix}x_1+x_2\\y_1+y_2\\z_1+z_2\end{pmatrix}$$

is also in P since its entries add $(x_1+x_2)+(y_1+y_2)+(z_1+z_2)=(x_1+y_1+z_1)+(x_2+y_2+z_2)$ to 0. To show that P is closed under scalar multiplication, start with a vector from P

$$\begin{pmatrix}x\\y\\z\end{pmatrix}$$

where $x+y+z=0$, and then for $r\in\mathbb{R}$ observe that the scalar multiple

$$r\cdot\begin{pmatrix}x\\y\\z\end{pmatrix}=\begin{pmatrix}rx\\ry\\rz\end{pmatrix}$$

gives $rx+ry+rz=r(x+y+z)=0$. Thus the two closure conditions are satisfied. Verification of the other conditions in the definition of a vector space are just as straightforward.

1.5 Example Example 1.3 shows that the set of all two-tall vectors with real entries is a vector space. Example 1.4 gives a subset of an $\mathbb{R}^n$ that is also a vector space. In contrast with those two, consider the set of two-tall columns with entries that are integers (under the usual operations of component-wise addition and scalar multiplication). This is a subset of a vector space but it is not itself a vector space. The reason is that this set is not closed under scalar multiplication, that is, it does not satisfy condition (6). Here is a column with integer entries and a scalar such that the outcome of the operation

$$0.5 \cdot \begin{pmatrix} 4 \\ 3 \end{pmatrix} = \begin{pmatrix} 2 \\ 1.5 \end{pmatrix}$$

is not a member of the set, since its entries are not all integers.

1.6 Example The singleton set

$$\{\begin{pmatrix} 0 \\ 0 \\ 0 \\ 0 \end{pmatrix}\}$$

is a vector space under the operations

$$\begin{pmatrix} 0 \\ 0 \\ 0 \\ 0 \end{pmatrix} + \begin{pmatrix} 0 \\ 0 \\ 0 \\ 0 \end{pmatrix} = \begin{pmatrix} 0 \\ 0 \\ 0 \\ 0 \end{pmatrix} \qquad r \cdot \begin{pmatrix} 0 \\ 0 \\ 0 \\ 0 \end{pmatrix} = \begin{pmatrix} 0 \\ 0 \\ 0 \\ 0 \end{pmatrix}$$

that it inherits from $\mathbb{R}^4$.

A vector space must have at least one element, its zero vector. Thus a one-element vector space is the smallest possible.

1.7 Definition A one-element vector space is a *trivial* space.

The examples so far involve sets of column vectors with the usual operations. But vector spaces need not be collections of column vectors, or even of row vectors. Below are some other types of vector spaces. The term 'vector space' does not mean 'collection of columns of reals'. It means something more like 'collection in which any linear combination is sensible'.

1.8 Example Consider $\mathcal{P}_3 = \{a_0 + a_1x + a_2x^2 + a_3x^3 \mid a_0, \ldots, a_3 \in \mathbb{R}\}$, the set of polynomials of degree three or less (in this book, we'll take constant polynomials, including the zero polynomial, to be of degree zero). It is a vector space under the operations

$$\begin{aligned}(a_0 + a_1x + a_2x^2 + a_3x^3) + (b_0 + b_1x + b_2x^2 + b_3x^3) \\ = (a_0 + b_0) + (a_1 + b_1)x + (a_2 + b_2)x^2 + (a_3 + b_3)x^3\end{aligned}$$

and

$$r \cdot (a_0 + a_1x + a_2x^2 + a_3x^3) = (ra_0) + (ra_1)x + (ra_2)x^2 + (ra_3)x^3$$

(the verification is easy). This vector space is worthy of attention because these are the polynomial operations familiar from high school algebra. For instance, $3 \cdot (1 - 2x + 3x^2 - 4x^3) - 2 \cdot (2 - 3x + x^2 - (1/2)x^3) = -1 + 7x^2 - 11x^3$.

Although this space is not a subset of any $\mathbb{R}^n$, there is a sense in which we can think of $\mathcal{P}_3$ as "the same" as $\mathbb{R}^4$. If we identify these two space's elements in this way

$$a_0 + a_1x + a_2x^2 + a_3x^3 \quad \text{corresponds to} \quad \begin{pmatrix} a_0 \\ a_1 \\ a_2 \\ a_3 \end{pmatrix}$$

then the operations also correspond. Here is an example of corresponding additions.

$$\begin{array}{rr} & 1 - 2x + 0x^2 + 1x^3 \\ + & 2 + 3x + 7x^2 - 4x^3 \\ \hline & 3 + 1x + 7x^2 - 3x^3 \end{array} \quad \text{corresponds to} \quad \begin{pmatrix} 1 \\ -2 \\ 0 \\ 1 \end{pmatrix} + \begin{pmatrix} 2 \\ 3 \\ 7 \\ -4 \end{pmatrix} = \begin{pmatrix} 3 \\ 1 \\ 7 \\ -3 \end{pmatrix}$$

Things we are thinking of as "the same" add to "the same" sum. Chapter Three makes precise this idea of vector space correspondence. For now we shall just leave it as an intuition.

1.9 Example The set $\mathcal{M}_{2\times2}$ of 2×2 matrices with real number entries is a vector space under the natural entry-by-entry operations.

$$\begin{pmatrix} a & b \\ c & d \end{pmatrix} + \begin{pmatrix} w & x \\ y & z \end{pmatrix} = \begin{pmatrix} a+w & b+x \\ c+y & d+z \end{pmatrix} \qquad r \cdot \begin{pmatrix} a & b \\ c & d \end{pmatrix} = \begin{pmatrix} ra & rb \\ rc & rd \end{pmatrix}$$

As in the prior example, we can think of this space as "the same" as $\mathbb{R}^4$.

1.10 Example The set $\{f \mid f\colon \mathbb{N} \to \mathbb{R}\}$ of all real-valued functions of one natural number variable is a vector space under the operations

$$(f_1 + f_2)\,(n) = f_1(n) + f_2(n) \qquad (r \cdot f)\,(n) = r\,f(n)$$

so that if, for example, $f_1(n) = n^2 + 2\sin(n)$ and $f_2(n) = -\sin(n) + 0.5$ then $(f_1 + 2f_2)\,(n) = n^2 + 1$.

We can view this space as a generalization of Example 1.3—instead of 2-tall vectors, these functions are like infinitely-tall vectors.

$$\begin{array}{c|c} n & f(n) = n^2 + 1 \\ \hline 0 & 1 \\ 1 & 2 \\ 2 & 5 \\ 3 & 10 \\ \vdots & \vdots \end{array} \quad \text{corresponds to} \quad \begin{pmatrix} 1 \\ 2 \\ 5 \\ 10 \\ \vdots \end{pmatrix}$$

Addition and scalar multiplication are component-wise, as in Example 1.3. (We can formalize "infinitely-tall" by saying that it means an infinite sequence, or that it means a function from $\mathbb{N}$ to $\mathbb{R}$.)

1.11 Example The set of polynomials with real coefficients

$$\{a_0 + a_1x + \cdots + a_nx^n \mid n \in \mathbb{N} \text{ and } a_0, \ldots, a_n \in \mathbb{R}\}$$

makes a vector space when given the natural '+'

$$\begin{aligned}(a_0 + a_1x + \cdots + a_nx^n) + (b_0 + b_1x + \cdots + b_nx^n)&\\ = (a_0 + b_0) + (a_1 + b_1)x + \cdots + (a_n + b_n)x^n&\end{aligned}$$

and '$\cdot$'.

$$r \cdot (a_0 + a_1x + \ldots a_nx^n) = (ra_0) + (ra_1)x + \ldots (ra_n)x^n$$

This space differs from the space $\mathcal{P}_3$ of Example 1.8. This space contains not just degree three polynomials, but degree thirty polynomials and degree three hundred polynomials, too. Each individual polynomial of course is of a finite degree, but the set has no single bound on the degree of all of its members.

We can think of this example, like the prior one, in terms of infinite-tuples. For instance, we can think of $1 + 3x + 5x^2$ as corresponding to $(1, 3, 5, 0, 0, \ldots)$. However, this space differs from the one in Example 1.10. Here, each member of the set has a finite degree, that is, under the correspondence there is no element from this space matching $(1, 2, 5, 10, \ldots)$. Vectors in this space correspond to infinite-tuples that end in zeroes.

1.12 Example The set $\{f \mid f\colon \mathbb{R} \to \mathbb{R}\}$ of all real-valued functions of one real variable is a vector space under these.

$$(f_1 + f_2)\,(x) = f_1(x) + f_2(x) \qquad (r \cdot f)\,(x) = r\,f(x)$$

The difference between this and Example 1.10 is the domain of the functions.

1.13 Example The set $F = \{a\cos\theta + b\sin\theta \mid a, b \in \mathbb{R}\}$ of real-valued functions of the real variable θ is a vector space under the operations

$$(a_1\cos\theta + b_1\sin\theta) + (a_2\cos\theta + b_2\sin\theta) = (a_1 + a_2)\cos\theta + (b_1 + b_2)\sin\theta$$

and

$$r \cdot (a\cos\theta + b\sin\theta) = (ra)\cos\theta + (rb)\sin\theta$$

inherited from the space in the prior example. (We can think of F as "the same" as $\mathbb{R}^2$ in that $a\cos\theta + b\sin\theta$ corresponds to the vector with components a and b.)

1.14 Example The set

$$\{f\colon \mathbb{R}\to\mathbb{R} \mid \frac{d^2f}{dx^2}+f=0\}$$

is a vector space under the, by now natural, interpretation.

$$(f+g)\,(x)=f(x)+g(x) \qquad (r\cdot f)\,(x)=r\,f(x)$$

In particular, notice that closure is a consequence

$$\frac{d^2(f+g)}{dx^2}+(f+g)=(\frac{d^2f}{dx^2}+f)+(\frac{d^2g}{dx^2}+g)$$

and

$$\frac{d^2(rf)}{dx^2}+(rf)=r(\frac{d^2f}{dx^2}+f)$$

of basic Calculus. This turns out to equal the space from the prior example—functions satisfying this differential equation have the form $a\cos\theta+b\sin\theta$—but this description suggests an extension to solutions sets of other differential equations.

1.15 Example The set of solutions of a homogeneous linear system in n variables is a vector space under the operations inherited from $\mathbb{R}^n$. For example, for closure under addition consider a typical equation in that system $c_1x_1+\cdots+c_nx_n=0$ and suppose that both these vectors

$$\vec{v}=\begin{pmatrix}v_1\\ \vdots\\ v_n\end{pmatrix} \qquad \vec{w}=\begin{pmatrix}w_1\\ \vdots\\ w_n\end{pmatrix}$$

satisfy the equation. Then their sum $\vec{v}+\vec{w}$ also satisfies that equation: $c_1(v_1+w_1)+\cdots+c_n(v_n+w_n)=(c_1v_1+\cdots+c_nv_n)+(c_1w_1+\cdots+c_nw_n)=0$. The checks of the other vector space conditions are just as routine.

We often omit the multiplication symbol '$\cdot$' between the scalar and the vector. We distinguish the multiplication in c_1v_1 from that in $r\vec{v}$ by context, since if both multiplicands are real numbers then it must be real-real multiplication while if one is a vector then it must be scalar-vector multiplication.

Example 1.15 has brought us full circle since it is one of our motivating examples. Now, with some feel for the kinds of structures that satisfy the definition of a vector space, we can reflect on that definition. For example, why specify in the definition the condition that $1\cdot\vec{v}=\vec{v}$ but not a condition that $0\cdot\vec{v}=\vec{0}$?

One answer is that this is just a definition—it gives the rules and you need to follow those rules to continue.

Another answer is perhaps more satisfying. People in this area have worked to develop the right balance of power and generality. This definition is shaped so that it contains the conditions needed to prove all of the interesting and important properties of spaces of linear combinations. As we proceed, we shall derive all of the properties natural to collections of linear combinations from the conditions given in the definition.

The next result is an example. We do not need to include these properties in the definition of vector space because they follow from the properties already listed there.

1.16 Lemma In any vector space V, for any $\vec{v} \in V$ and $r \in \mathbb{R}$, we have (1) $0 \cdot \vec{v} = \vec{0}$, (2) $(-1 \cdot \vec{v}) + \vec{v} = \vec{0}$, and (3) $r \cdot \vec{0} = \vec{0}$.

Proof For (1) note that $\vec{v} = (1+0) \cdot \vec{v} = \vec{v} + (0 \cdot \vec{v})$. Add to both sides the additive inverse of $\vec{v}$, the vector $\vec{w}$ such that $\vec{w} + \vec{v} = \vec{0}$.

$$\begin{aligned} \vec{w} + \vec{v} &= \vec{w} + \vec{v} + 0 \cdot \vec{v} \\ \vec{0} &= \vec{0} + 0 \cdot \vec{v} \\ \vec{0} &= 0 \cdot \vec{v} \end{aligned}$$

Item (2) is easy: $(-1 \cdot \vec{v}) + \vec{v} = (-1+1) \cdot \vec{v} = 0 \cdot \vec{v} = \vec{0}$. For (3), $r \cdot \vec{0} = r \cdot (0 \cdot \vec{0}) = (r \cdot 0) \cdot \vec{0} = \vec{0}$ will do. QED

The second item shows that we can write the additive inverse of $\vec{v}$ as '$-\vec{v}$' without worrying about any confusion with $(-1) \cdot \vec{v}$.

A recap: our study in Chapter One of Gaussian reduction led us to consider collections of linear combinations. So in this chapter we have defined a vector space to be a structure in which we can form such combinations, subject to simple conditions on the addition and scalar multiplication operations. In a phrase: vector spaces are the right context in which to study linearity.

From the fact that it forms a whole chapter, and especially because that chapter is the first one, a reader could suppose that our purpose in this book is the study of linear systems. The truth is that we will not so much use vector spaces in the study of linear systems as we instead have linear systems start us on the study of vector spaces. The wide variety of examples from this subsection shows that the study of vector spaces is interesting and important in its own right. Linear systems won't go away. But from now on our primary objects of study will be vector spaces.

Exercises

1.17 Name the zero vector for each of these vector spaces.

(a) The space of degree three polynomials under the natural operations.

(b) The space of 2×4 matrices.
(c) The space $\{f\colon [0..1]\to\mathbb{R} \mid f \text{ is continuous}\}$.
(d) The space of real-valued functions of one natural number variable.

✓ **1.18** Find the additive inverse, in the vector space, of the vector.
(a) In $\mathcal{P}_3$, the vector $-3-2x+x^2$.
(b) In the space 2×2,

$$\begin{pmatrix}1&-1\\0&3\end{pmatrix}.$$

(c) In $\{ae^x+be^{-x} \mid a,b\in\mathbb{R}\}$, the space of functions of the real variable x under the natural operations, the vector $3e^x-2e^{-x}$.

✓ **1.19** For each, list three elements and then show it is a vector space.
(a) The set of linear polynomials $\mathcal{P}_1=\{a_0+a_1x \mid a_0,a_1\in\mathbb{R}\}$ under the usual polynomial addition and scalar multiplication operations.
(b) The set of linear polynomials $\{a_0+a_1x \mid a_0-2a_1=0\}$, under the usual polynomial addition and scalar multiplication operations.

Hint. Use Example 1.3 as a guide. Most of the ten conditions are just verifications.

1.20 For each, list three elements and then show it is a vector space.
(a) The set of 2×2 matrices with real entries under the usual matrix operations.
(b) The set of 2×2 matrices with real entries where the $2,1$ entry is zero, under the usual matrix operations.

✓ **1.21** For each, list three elements and then show it is a vector space.
(a) The set of three-component row vectors with their usual operations.
(b) The set

$$\{\begin{pmatrix}x\\y\\z\\w\end{pmatrix}\in\mathbb{R}^4 \mid x+y-z+w=0\}$$

under the operations inherited from $\mathbb{R}^4$.

✓ **1.22** Show that each of these is not a vector space. (*Hint.* Check closure by listing two members of each set and trying some operations on them.)
(a) Under the operations inherited from $\mathbb{R}^3$, this set

$$\{\begin{pmatrix}x\\y\\z\end{pmatrix}\in\mathbb{R}^3 \mid x+y+z=1\}$$

(b) Under the operations inherited from $\mathbb{R}^3$, this set

$$\{\begin{pmatrix}x\\y\\z\end{pmatrix}\in\mathbb{R}^3 \mid x^2+y^2+z^2=1\}$$

(c) Under the usual matrix operations,

$$\{\begin{pmatrix}a&1\\b&c\end{pmatrix} \mid a,b,c\in\mathbb{R}\}$$

(d) Under the usual polynomial operations,

$$\{a_0+a_1x+a_2x^2 \mid a_0,a_1,a_2\in\mathbb{R}^+\}$$

where $\mathbb{R}^+$ is the set of reals greater than zero

(e) Under the inherited operations,

$$\{\begin{pmatrix}x\\y\end{pmatrix}\in\mathbb{R}^2 \mid x+3y=4 \text{ and } 2x-y=3 \text{ and } 6x+4y=10\}$$

1.23 Define addition and scalar multiplication operations to make the complex numbers a vector space over $\mathbb{R}$.

✓ **1.24** Is the set of rational numbers a vector space over $\mathbb{R}$ under the usual addition and scalar multiplication operations?

1.25 Show that the set of linear combinations of the variables x, y, z is a vector space under the natural addition and scalar multiplication operations.

1.26 Prove that this is not a vector space: the set of two-tall column vectors with real entries subject to these operations.

$$\begin{pmatrix}x_1\\y_1\end{pmatrix}+\begin{pmatrix}x_2\\y_2\end{pmatrix}=\begin{pmatrix}x_1-x_2\\y_1-y_2\end{pmatrix}\qquad r\cdot\begin{pmatrix}x\\y\end{pmatrix}=\begin{pmatrix}rx\\ry\end{pmatrix}$$

1.27 Prove or disprove that $\mathbb{R}^3$ is a vector space under these operations.

(a) $\begin{pmatrix}x_1\\y_1\\z_1\end{pmatrix}+\begin{pmatrix}x_2\\y_2\\z_2\end{pmatrix}=\begin{pmatrix}0\\0\\0\end{pmatrix}$ and $r\begin{pmatrix}x\\y\\z\end{pmatrix}=\begin{pmatrix}rx\\ry\\rz\end{pmatrix}$

(b) $\begin{pmatrix}x_1\\y_1\\z_1\end{pmatrix}+\begin{pmatrix}x_2\\y_2\\z_2\end{pmatrix}=\begin{pmatrix}0\\0\\0\end{pmatrix}$ and $r\begin{pmatrix}x\\y\\z\end{pmatrix}=\begin{pmatrix}0\\0\\0\end{pmatrix}$

✓ **1.28** For each, decide if it is a vector space; the intended operations are the natural ones.

(a) The *diagonal* 2×2 matrices

$$\{\begin{pmatrix}a&0\\0&b\end{pmatrix}\mid a,b\in\mathbb{R}\}$$

(b) This set of 2×2 matrices

$$\{\begin{pmatrix}x&x+y\\x+y&y\end{pmatrix}\mid x,y\in\mathbb{R}\}$$

(c) This set

$$\{\begin{pmatrix}x\\y\\z\\w\end{pmatrix}\in\mathbb{R}^4\mid x+y+w=1\}$$

(d) The set of functions $\{f\colon\mathbb{R}\to\mathbb{R}\mid df/dx+2f=0\}$

(e) The set of functions $\{f\colon\mathbb{R}\to\mathbb{R}\mid df/dx+2f=1\}$

✓ **1.29** Prove or disprove that this is a vector space: the real-valued functions f of one real variable such that $f(7)=0$.

✓ **1.30** Show that the set $\mathbb{R}^+$ of positive reals is a vector space when we interpret '$x+y$' to mean the product of x and y (so that $2+3$ is 6), and we interpret '$r\cdot x$' as the r-th power of x.

1.31 Is $\{(x,y)\mid x,y\in\mathbb{R}\}$ a vector space under these operations?

(a) $(x_1,y_1)+(x_2,y_2)=(x_1+x_2,y_1+y_2)$ and $r\cdot(x,y)=(rx,y)$

(b) $(x_1,y_1)+(x_2,y_2)=(x_1+x_2,y_1+y_2)$ and $r\cdot(x,y)=(rx,0)$

1.32 Prove or disprove that this is a vector space: the set of polynomials of degree greater than or equal to two, along with the zero polynomial.

1.33 At this point "the same" is only an intuition, but nonetheless for each vector space identify the k for which the space is "the same" as $\mathbb{R}^k$.
(a) The 2×3 matrices under the usual operations
(b) The $n\times m$ matrices (under their usual operations)
(c) This set of 2×2 matrices

$$\{\begin{pmatrix} a & 0 \\ b & c \end{pmatrix} \mid a, b, c \in \mathbb{R}\}$$

(d) This set of 2×2 matrices

$$\{\begin{pmatrix} a & 0 \\ b & c \end{pmatrix} \mid a+b+c=0\}$$

✓ 1.34 Using $\vec{+}$ to represent vector addition and $\vec{\cdot}$ for scalar multiplication, restate the definition of vector space.

✓ 1.35 Prove these.
(a) For any $\vec{v} \in V$, if $\vec{w} \in V$ is an additive inverse of $\vec{v}$, then $\vec{v}$ is an additive inverse of $\vec{w}$. So a vector is an additive inverse of any additive inverse of itself.
(b) Vector addition left-cancels: if $\vec{v}, \vec{s}, \vec{t} \in V$ then $\vec{v}+\vec{s}=\vec{v}+\vec{t}$ implies that $\vec{s}=\vec{t}$.

1.36 The definition of vector spaces does not explicitly say that $\vec{0}+\vec{v}=\vec{v}$ (it instead says that $\vec{v}+\vec{0}=\vec{v}$). Show that it must nonetheless hold in any vector space.

✓ 1.37 Prove or disprove that this is a vector space: the set of all matrices, under the usual operations.

1.38 In a vector space every element has an additive inverse. Can some elements have two or more?

1.39 (a) Prove that every point, line, or plane thru the origin in $\mathbb{R}^3$ is a vector space under the inherited operations.
(b) What if it doesn't contain the origin?

✓ 1.40 Using the idea of a vector space we can easily reprove that the solution set of a homogeneous linear system has either one element or infinitely many elements. Assume that $\vec{v} \in V$ is not $\vec{0}$.
(a) Prove that $r\cdot\vec{v}=\vec{0}$ if and only if $r=0$.
(b) Prove that $r_1\cdot\vec{v}=r_2\cdot\vec{v}$ if and only if $r_1=r_2$.
(c) Prove that any nontrivial vector space is infinite.
(d) Use the fact that a nonempty solution set of a homogeneous linear system is a vector space to draw the conclusion.

1.41 Is this a vector space under the natural operations: the real-valued functions of one real variable that are differentiable?

1.42 A *vector space over the complex numbers* $\mathbb{C}$ has the same definition as a vector space over the reals except that scalars are drawn from $\mathbb{C}$ instead of from $\mathbb{R}$. Show that each of these is a vector space over the complex numbers. (Recall how complex numbers add and multiply: $(a_0+a_1i)+(b_0+b_1i)=(a_0+b_0)+(a_1+b_1)i$ and $(a_0+a_1i)(b_0+b_1i)=(a_0b_0-a_1b_1)+(a_0b_1+a_1b_0)i$.)
(a) The set of degree two polynomials with complex coefficients

(b) This set

$$\{\begin{pmatrix}0 & a\\ b & 0\end{pmatrix} \mid a, b \in \mathbb{C} \text{ and } a+b=0+0i\}$$

1.43 Name a property shared by all of the $\mathbb{R}^n$'s but not listed as a requirement for a vector space.

✓ 1.44 (a) Prove that for any four vectors $\vec{v}_1, \ldots, \vec{v}_4 \in V$ we can associate their sum in any way without changing the result.

$$\begin{aligned}((\vec{v}_1+\vec{v}_2)+\vec{v}_3)+\vec{v}_4 &= (\vec{v}_1+(\vec{v}_2+\vec{v}_3))+\vec{v}_4 = (\vec{v}_1+\vec{v}_2)+(\vec{v}_3+\vec{v}_4)\\ &= \vec{v}_1+((\vec{v}_2+\vec{v}_3)+\vec{v}_4) = \vec{v}_1+(\vec{v}_2+(\vec{v}_3+\vec{v}_4))\end{aligned}$$

This allows us to write '$\vec{v}_1+\vec{v}_2+\vec{v}_3+\vec{v}_4$' without ambiguity.

(b) Prove that any two ways of associating a sum of any number of vectors give the same sum. (*Hint.* Use induction on the number of vectors.)

1.45 Example 1.5 gives a subset of $\mathbb{R}^2$ that is not a vector space, under the obvious operations, because while it is closed under addition, it is not closed under scalar multiplication. Consider the set of vectors in the plane whose components have the same sign or are 0. Show that this set is closed under scalar multiplication but not addition.

I.2 Subspaces and Spanning Sets

One of the examples that led us to define vector spaces was the solution set of a homogeneous system. For instance, we saw in Example 1.4 such a space that is a planar subset of $\mathbb{R}^3$. There, the vector space $\mathbb{R}^3$ contains inside it another vector space, the plane.

2.1 Definition For any vector space, a *subspace* is a subset that is itself a vector space, under the inherited operations.

2.2 Example Example 1.4's plane

$$P = \{\begin{pmatrix}x\\ y\\ z\end{pmatrix} \mid x+y+z=0\}$$

is a subspace of $\mathbb{R}^3$. As required by the definition the plane's operations are inherited from the larger space, that is, vectors add in P as they add in $\mathbb{R}^3$

$$\begin{pmatrix}x_1\\ y_1\\ z_1\end{pmatrix} + \begin{pmatrix}x_2\\ y_2\\ z_2\end{pmatrix} = \begin{pmatrix}x_1+x_2\\ y_1+y_2\\ z_1+z_2\end{pmatrix}$$

and scalar multiplication is also the same as in $\mathbb{R}^3$. To show that P is a subspace we need only note that it is a subset and then verify that it is a space. We have already checked in Example 1.4 that P satisfies the conditions in the definition of a vector space. For instance, for closure under addition we noted that if the summands satisfy that $x_1+y_1+z_1=0$ and $x_2+y_2+z_2=0$ then the sum satisfies that $(x_1+x_2)+(y_1+y_2)+(z_1+z_2)=(x_1+y_1+z_1)+(x_2+y_2+z_2)=0$.

2.3 Example The x-axis in $\mathbb{R}^2$ is a subspace, where the addition and scalar multiplication operations are the inherited ones.

$$\begin{pmatrix}x_1\\0\end{pmatrix}+\begin{pmatrix}x_2\\0\end{pmatrix}=\begin{pmatrix}x_1+x_2\\0\end{pmatrix}\qquad r\cdot\begin{pmatrix}x\\0\end{pmatrix}=\begin{pmatrix}rx\\0\end{pmatrix}$$

As in the prior example, to verify directly from the definition that this is a subspace we simply that it is a subset and then check that it satisfies the conditions in definition of a vector space. For instance the two closure conditions are satisfied: adding two vectors with a second component of zero results in a vector with a second component of zero and multiplying a scalar times a vector with a second component of zero results in a vector with a second component of zero.

2.4 Example Another subspace of $\mathbb{R}^2$ is its trivial subspace.

$$\{\begin{pmatrix}0\\0\end{pmatrix}\}$$

Any vector space has a trivial subspace $\{\vec{0}\,\}$. At the opposite extreme, any vector space has itself for a subspace. These two are the *improper* subspaces. Other subspaces are *proper*.

2.5 Example Vector spaces that are not $\mathbb{R}^n$'s also have subspaces. The space of cubic polynomials $\{a+bx+cx^2+dx^3 \mid a,b,c,d\in\mathbb{R}\}$ has a subspace comprised of all linear polynomials $\{m+nx \mid m,n\in\mathbb{R}\}$.

2.6 Example Another example of a subspace that is not a subset of an $\mathbb{R}^n$ followed the definition of a vector space. The space in Example 1.12 of all real-valued functions of one real variable $\{f \mid f\colon\mathbb{R}\to\mathbb{R}\}$ has the subspace in Example 1.14 of functions satisfying the restriction $(d^2 f/dx^2)+f=0$.

2.7 Example The definition requires that the addition and scalar multiplication operations must be the ones inherited from the larger space. The set $S=\{1\}$ is a subset of $\mathbb{R}^1$. And, under the operations $1+1=1$ and $r\cdot 1=1$ the set S is a vector space, specifically, a trivial space. However, S is not a subspace of $\mathbb{R}^1$ because those aren't the inherited operations, since of course $\mathbb{R}^1$ has $1+1=2$.

2.8 Example Being vector spaces themselves, subspaces must satisfy the closure conditions. The set $\mathbb{R}^+$ is not a subspace of the vector space $\mathbb{R}^1$ because with the inherited operations it is not closed under scalar multiplication: if $\vec{v} = 1$ then $-1 \cdot \vec{v} \notin \mathbb{R}^+$.

The next result says that Example 2.8 is prototypical. The only way that a subset can fail to be a subspace, if it is nonempty and uses the inherited operations, is if it isn't closed.

2.9 Lemma For a nonempty subset S of a vector space, under the inherited operations the following are equivalent statements.*

(1) S is a subspace of that vector space

(2) S is closed under linear combinations of pairs of vectors: for any vectors $\vec{s}_1, \vec{s}_2 \in S$ and scalars r_1, r_2 the vector $r_1\vec{s}_1 + r_2\vec{s}_2$ is in S

(3) S is closed under linear combinations of any number of vectors: for any vectors $\vec{s}_1, \ldots, \vec{s}_n \in S$ and scalars $r_1, \ldots, r_n$ the vector $r_1\vec{s}_1 + \cdots + r_n\vec{s}_n$ is an element of S.

Briefly, a subset is a subspace if and only if it is closed under linear combinations.

Proof 'The following are equivalent' means that each pair of statements are equivalent.

$$(1) \iff (2) \qquad (2) \iff (3) \qquad (3) \iff (1)$$

We will prove the equivalence by establishing that $(1) \implies (3) \implies (2) \implies (1)$. This strategy is suggested by the observation that the implications $(1) \implies (3)$ and $(3) \implies (2)$ are easy and so we need only argue that $(2) \implies (1)$.

Assume that S is a nonempty subset of a vector space V that is S closed under combinations of pairs of vectors. We will show that S is a vector space by checking the conditions.

The vector space definition has five conditions on addition. First, for closure under addition, if $\vec{s}_1, \vec{s}_2 \in S$ then $\vec{s}_1 + \vec{s}_2 \in S$, as it is a combination of a pair of vectors and we are assuming that S is closed under those. Second, for any $\vec{s}_1, \vec{s}_2 \in S$, because addition is inherited from V, the sum $\vec{s}_1 + \vec{s}_2$ in S equals the sum $\vec{s}_1 + \vec{s}_2$ in V, and that equals the sum $\vec{s}_2 + \vec{s}_1$ in V (because V is a vector space, its addition is commutative), and that in turn equals the sum $\vec{s}_2 + \vec{s}_1$ in S. The argument for the third condition is similar to that for the second. For the fourth, consider the zero vector of V and note that closure of S under linear combinations of pairs of vectors gives that $0 \cdot \vec{s} + 0 \cdot \vec{s} = \vec{0}$ is an element of S (where $\vec{s}$ is any member of the nonempty set S); checking that $\vec{0}$ acts under the inherited operations as the additive identity of S is easy. The fifth condition is satisfied because for any $\vec{s} \in S$, closure under linear combinations of pairs of vectors shows that $0 \cdot \vec{0} + (-1) \cdot \vec{s}$ is an element of S, and it is obviously the additive inverse of $\vec{s}$ under the inherited operations.

*More information on equivalence of statements is in the appendix.

The verifications for the scalar multiplication conditions are similar; see Exercise 34. QED

We will usually verify that a subset is a subspace by checking that it satisfies statement (2).

2.10 Remark At the start of this chapter we introduced vector spaces as collections in which linear combinations "make sense." Theorem 2.9's statements (1)-(3) say that we can always make sense of an expression like $r_1\vec{s}_1 + r_2\vec{s}_2$ in that the vector described is in the set S.

As a contrast, consider the set T of two-tall vectors whose entries add to a number greater than or equal to zero. Here we cannot just write any linear combination such as $2\vec{t}_1 - 3\vec{t}_2$ and be confident the result is an element of T.

Lemma 2.9 suggests that a good way to think of a vector space is as a collection of unrestricted linear combinations. The next two examples take some spaces and recasts their descriptions to be in that form.

2.11 Example We can show that this plane through the origin subset of $\mathbb{R}^3$

$$S = \{\begin{pmatrix}x\\y\\z\end{pmatrix} \mid x - 2y + z = 0\}$$

is a subspace under the usual addition and scalar multiplication operations of column vectors by checking that it is nonempty and closed under linear combinations of two vectors. But there is another way. Think of $x - 2y + z = 0$ as a one-equation linear system and parametrize it by expressing the leading variable in terms of the free variables $x = 2y - z$.

$$S = \{\begin{pmatrix}2y-z\\y\\z\end{pmatrix} \mid y, z \in \mathbb{R}\} = \{y\begin{pmatrix}2\\1\\0\end{pmatrix} + z\begin{pmatrix}-1\\0\\1\end{pmatrix} \mid y, z \in \mathbb{R}\} \qquad (*)$$

Now, to show that this is a subspace consider $r_1\vec{s}_1 + r_2\vec{s}_2$. Each $\vec{s}_i$ is a linear combination of the two vectors in $(*)$ so this is a linear combination of linear combinations.

$$r_1 \cdot (y_1\begin{pmatrix}2\\1\\0\end{pmatrix} + z_1\begin{pmatrix}-1\\0\\1\end{pmatrix}) + r_2 \cdot (y_2\begin{pmatrix}2\\1\\0\end{pmatrix} + z_2\begin{pmatrix}-1\\0\\1\end{pmatrix})$$

The Linear Combination Lemma, Lemma One.III.2.3, shows that the total is a linear combination of the two vectors and so Theorem 2.9's statement (2) is satisfied.

2.12 Example This is a subspace of the 2×2 matrices $\mathcal{M}_{2\times2}$.

$$L=\{\begin{pmatrix}a&0\\b&c\end{pmatrix}\mid a+b+c=0\}$$

To parametrize, express the condition as $a=-b-c$.

$$L=\{\begin{pmatrix}-b-c&0\\b&c\end{pmatrix}\mid b,c\in\mathbb{R}\}=\{b\begin{pmatrix}-1&0\\1&0\end{pmatrix}+c\begin{pmatrix}-1&0\\0&1\end{pmatrix}\mid b,c\in\mathbb{R}\}$$

As above, we've described the subspace as a collection of unrestricted linear combinations. To show it is a subspace, note that a linear combination of vectors from L is a linear combination of linear combinations and so statement (2) is true.

2.13 Definition The *span* (or *linear closure*) of a nonempty subset S of a vector space is the set of all linear combinations of vectors from S.

$$[S]=\{c_1\vec{s}_1+\cdots+c_n\vec{s}_n\mid c_1,\ldots,c_n\in\mathbb{R}\text{ and }\vec{s}_1,\ldots,\vec{s}_n\in S\}$$

The span of the empty subset of a vector space is its trivial subspace.

No notation for the span is completely standard. The square brackets used here are common but so are 'span(S)' and 'sp(S)'.

2.14 Remark In Chapter One, after we showed that we can write the solution set of a homogeneous linear system as $\{c_1\vec{\beta}_1+\cdots+c_k\vec{\beta}_k\mid c_1,\ldots,c_k\in\mathbb{R}\}$, we described that as the set 'generated' by the $\vec{\beta}$'s. We now call that the span of $\{\vec{\beta}_1,\ldots,\vec{\beta}_k\}$.

Recall also from that proof that the span of the empty set is defined to be the set $\{\vec{0}\}$ because of the convention that a trivial linear combination, a combination of zero-many vectors, adds to $\vec{0}$. Besides, defining the empty set's span to be the trivial subspace is convenient because it keeps results like the next one from needing exceptions for the empty set.

2.15 Lemma In a vector space, the span of any subset is a subspace.

Proof If the subset S is empty then by definition its span is the trivial subspace. If S is not empty then by Lemma 2.9 we need only check that the span $[S]$ is closed under linear combinations of pairs of elements. For a pair of vectors from that span, $\vec{v}=c_1\vec{s}_1+\cdots+c_n\vec{s}_n$ and $\vec{w}=c_{n+1}\vec{s}_{n+1}+\cdots+c_m\vec{s}_m$, a linear combination

$$\begin{aligned}p\cdot(c_1\vec{s}_1+\cdots+c_n\vec{s}_n)+r\cdot(c_{n+1}\vec{s}_{n+1}+\cdots+c_m\vec{s}_m)\\=pc_1\vec{s}_1+\cdots+pc_n\vec{s}_n+rc_{n+1}\vec{s}_{n+1}+\cdots+rc_m\vec{s}_m\end{aligned}$$

is a linear combination of elements of S and so is an element of [S] (possibly some of the $\vec{s}_i$'s from $\vec{v}$ equal some of the $\vec{s}_j$'s from $\vec{w}$ but that does not matter). QED

The converse of the lemma holds: any subspace is the span of some set, because a subspace is obviously the span of itself, the set of all of its members. Thus a subset of a vector space is a subspace if and only if it is a span. This fits the intuition that a good way to think of a vector space is as a collection in which linear combinations are sensible.

Taken together, Lemma 2.9 and Lemma 2.15 show that the span of a subset S of a vector space is the smallest subspace containing all of the members of S.

2.16 Example In any vector space V, for any vector $\vec{v} \in V$, the set $\{r \cdot \vec{v} \mid r \in \mathbb{R}\}$ is a subspace of V. For instance, for any vector $\vec{v} \in \mathbb{R}^3$ the line through the origin containing that vector $\{k\vec{v} \mid k \in \mathbb{R}\}$ is a subspace of $\mathbb{R}^3$. This is true even if $\vec{v}$ is the zero vector, in which case it is the degenerate line, the trivial subspace.

2.17 Example The span of this set is all of $\mathbb{R}^2$.

$$\{\begin{pmatrix}1\\1\end{pmatrix}, \begin{pmatrix}1\\-1\end{pmatrix}\}$$

We know that the span is some subspace of $\mathbb{R}^2$. To check that it is all of $\mathbb{R}^2$ we must show that any member of $\mathbb{R}^2$ is a linear combination of these two vectors. So we ask: for which vectors with real components x and y are there scalars c_1 and c_2 such that this holds?

$$c_1\begin{pmatrix}1\\1\end{pmatrix} + c_2\begin{pmatrix}1\\-1\end{pmatrix} = \begin{pmatrix}x\\y\end{pmatrix} \qquad (*)$$

Gauss's Method

$$\begin{array}{rl} c_1 + c_2 = x \\ c_1 - c_2 = y \end{array} \xrightarrow{-\rho_1+\rho_2} \begin{array}{rl} c_1 + c_2 = & x \\ -2c_2 = & -x + y \end{array}$$

with back substitution gives $c_2 = (x - y)/2$ and $c_1 = (x + y)/2$. This shows that for any x, y there are appropriate coefficients c_1, c_2 making (*) true — we can write any element of $\mathbb{R}^2$ as a linear combination of the two given ones. For instance, for $x = 1$ and $y = 2$ the coefficients $c_2 = -1/2$ and $c_1 = 3/2$ will do.

Since spans are subspaces, and we know that a good way to understand a subspace is to parametrize its description, we can try to understand a set's span in that way.

2.18 Example Consider, in the vector space of quadratic polynomials $\mathcal{P}_2$, the span of the set $S = \{3x - x^2, 2x\}$. By the definition of span, it is the set of unrestricted linear combinations of the two $\{c_1(3x-x^2)+c_2(2x) \mid c_1, c_2 \in \mathbb{R}\}$. Clearly polynomials in this span must have a constant term of zero. Is that necessary condition also sufficient?

We are asking: for which members $a_2x^2+a_1x+a_0$ of $\mathcal{P}_2$ are there c_1 and c_2 such that $a_2x^2+a_1x+a_0 = c_1(3x-x^2)+c_2(2x)$? Polynomials are equal when their coefficients are equal so we want conditions on a_2, a_1, and a_0 making that triple a solution of this system.

$$\begin{aligned} -c_1 \qquad &= a_2 \\ 3c_1 + 2c_2 &= a_1 \\ 0 &= a_0 \end{aligned}$$

Gauss's Method and back-substitution gives $c_1 = -a_2$, and $c_2 = (3/2)a_2 + (1/2)a_1$, and $0 = a_0$. Thus as long as there is no constant term $a_0 = 0$ we can give coefficients c_1 and c_2 to describe that polynomial as an element of the span. For instance, for the polynomial $0-4x+3x^2$, the coefficients $c_1 = -3$ and $c_2 = 5/2$ will do. So the span of the given set is $[S] = \{a_1x + a_2x^2 \mid a_1, a_2 \in \mathbb{R}\}$.

Incidentally, this shows that the set $\{x, x^2\}$ spans the same subspace. A space can have more than one spanning set. Two other sets spanning this subspace are $\{x, x^2, -x+2x^2\}$ and $\{x, x+x^2, x+2x^2, \ldots\}$.

2.19 Example The picture below shows the subspaces of $\mathbb{R}^3$ that we now know of: the trivial subspace, lines through the origin, planes through the origin, and the whole space. (Of course, the picture shows only a few of the infinitely many cases. Line segments connect subsets with their supersets.) In the next section we will prove that $\mathbb{R}^3$ has no other kind of subspace, so in fact this lists them all.

This describes each subspace as the span of a set with a minimal number of members. With this, the subspaces fall naturally into levels—planes on one level, lines on another, etc.

$$\{x\begin{pmatrix}1\\0\\0\end{pmatrix}+y\begin{pmatrix}0\\1\\0\end{pmatrix}+z\begin{pmatrix}0\\0\\1\end{pmatrix}\}$$

$$\{x\begin{pmatrix}1\\0\\0\end{pmatrix}+y\begin{pmatrix}0\\1\\0\end{pmatrix}\} \quad \{x\begin{pmatrix}1\\0\\0\end{pmatrix}+z\begin{pmatrix}0\\0\\1\end{pmatrix}\} \quad \{x\begin{pmatrix}1\\1\\0\end{pmatrix}+z\begin{pmatrix}0\\0\\1\end{pmatrix}\} \quad \ldots$$

$$\{x\begin{pmatrix}1\\0\\0\end{pmatrix}\} \quad \{y\begin{pmatrix}0\\1\\0\end{pmatrix}\} \quad \{y\begin{pmatrix}2\\1\\0\end{pmatrix}\} \quad \{y\begin{pmatrix}1\\1\\1\end{pmatrix}\} \quad \ldots$$

$$\{\begin{pmatrix}0\\0\\0\end{pmatrix}\}$$

So far in this chapter we have seen that to study the properties of linear combinations, the right setting is a collection that is closed under these combinations. In the first subsection we introduced such collections, vector spaces, and we saw a great variety of examples. In this subsection we saw still more spaces, ones that are subspaces of others. In all of the variety there is a commonality. Example 2.19 above brings it out: vector spaces and subspaces are best understood as a span, and especially as a span of a small number of vectors. The next section studies spanning sets that are minimal.

Exercises

✓ **2.20** Which of these subsets of the vector space of 2×2 matrices are subspaces under the inherited operations? For each one that is a subspace, parametrize its description. For each that is not, give a condition that fails.

(a) $\{\begin{pmatrix}a&0\\0&b\end{pmatrix} \mid a,b\in\mathbb{R}\}$

(b) $\{\begin{pmatrix}a&0\\0&b\end{pmatrix} \mid a+b=0\}$

(c) $\{\begin{pmatrix}a&0\\0&b\end{pmatrix} \mid a+b=5\}$

(d) $\{\begin{pmatrix}a&c\\0&b\end{pmatrix} \mid a+b=0, c\in\mathbb{R}\}$

✓ **2.21** Is this a subspace of $\mathcal{P}_2$: $\{a_0+a_1x+a_2x^2 \mid a_0+2a_1+a_2=4\}$? If it is then parametrize its description.

2.22 Is the vector in the span of the set?

$$\begin{pmatrix}1\\0\\3\end{pmatrix} \qquad \{\begin{pmatrix}2\\1\\-1\end{pmatrix}, \begin{pmatrix}1\\-1\\1\end{pmatrix}\}$$

✓ **2.23** Decide if the vector lies in the span of the set, inside of the space.

(a) $\begin{pmatrix}2\\0\\1\end{pmatrix}$, $\{\begin{pmatrix}1\\0\\0\end{pmatrix}, \begin{pmatrix}0\\0\\1\end{pmatrix}\}$, in $\mathbb{R}^3$

(b) $x-x^3$, $\{x^2, 2x+x^2, x+x^3\}$, in $\mathcal{P}_3$

(c) $\begin{pmatrix}0&1\\4&2\end{pmatrix}$, $\{\begin{pmatrix}1&0\\1&1\end{pmatrix}, \begin{pmatrix}2&0\\2&3\end{pmatrix}\}$, in $\mathcal{M}_{2\times 2}$

2.24 Which of these are members of the span $[\{\cos^2 x, \sin^2 x\}]$ in the vector space of real-valued functions of one real variable?

(a) $f(x)=1$ (b) $f(x)=3+x^2$ (c) $f(x)=\sin x$ (d) $f(x)=\cos(2x)$

✓ **2.25** Which of these sets spans $\mathbb{R}^3$? That is, which of these sets has the property that any three-tall vector can be expressed as a suitable linear combination of the set's elements?

(a) $\{\begin{pmatrix}1\\0\\0\end{pmatrix},\begin{pmatrix}0\\2\\0\end{pmatrix},\begin{pmatrix}0\\0\\3\end{pmatrix}\}$ (b) $\{\begin{pmatrix}2\\0\\1\end{pmatrix},\begin{pmatrix}1\\1\\0\end{pmatrix},\begin{pmatrix}0\\0\\1\end{pmatrix}\}$ (c) $\{\begin{pmatrix}1\\1\\0\end{pmatrix},\begin{pmatrix}3\\0\\0\end{pmatrix}\}$

(d) $\{\begin{pmatrix}1\\0\\1\end{pmatrix},\begin{pmatrix}3\\1\\0\end{pmatrix},\begin{pmatrix}-1\\0\\0\end{pmatrix},\begin{pmatrix}2\\1\\5\end{pmatrix}\}$ (e) $\{\begin{pmatrix}2\\1\\1\end{pmatrix},\begin{pmatrix}3\\0\\1\end{pmatrix},\begin{pmatrix}5\\1\\2\end{pmatrix},\begin{pmatrix}6\\0\\2\end{pmatrix}\}$

✓ **2.26** Parametrize each subspace's description. Then express each subspace as a span.

(a) The subset $\{(a\ \ b\ \ c) \mid a-c=0\}$ of the three-wide row vectors

(b) This subset of $\mathcal{M}_{2\times 2}$

$$\{\begin{pmatrix}a&b\\c&d\end{pmatrix} \mid a+d=0\}$$

(c) This subset of $\mathcal{M}_{2\times 2}$

$$\{\begin{pmatrix}a&b\\c&d\end{pmatrix} \mid 2a-c-d=0 \text{ and } a+3b=0\}$$

(d) The subset $\{a+bx+cx^3 \mid a-2b+c=0\}$ of $\mathcal{P}_3$

(e) The subset of $\mathcal{P}_2$ of quadratic polynomials p such that $p(7)=0$

✓ **2.27** Find a set to span the given subspace of the given space. (*Hint.* Parametrize each.)

(a) the xz-plane in $\mathbb{R}^3$

(b) $\{\begin{pmatrix}x\\y\\z\end{pmatrix} \mid 3x+2y+z=0\}$ in $\mathbb{R}^3$

(c) $\{\begin{pmatrix}x\\y\\z\\w\end{pmatrix} \mid 2x+y+w=0 \text{ and } y+2z=0\}$ in $\mathbb{R}^4$

(d) $\{a_0+a_1x+a_2x^2+a_3x^3 \mid a_0+a_1=0 \text{ and } a_2-a_3=0\}$ in $\mathcal{P}_3$

(e) The set $\mathcal{P}_4$ in the space $\mathcal{P}_4$

(f) $\mathcal{M}_{2\times 2}$ in $\mathcal{M}_{2\times 2}$

2.28 Is $\mathbb{R}^2$ a subspace of $\mathbb{R}^3$?

✓ **2.29** Decide if each is a subspace of the vector space of real-valued functions of one real variable.

(a) The *even* functions $\{f\colon \mathbb{R}\to\mathbb{R} \mid f(-x)=f(x) \text{ for all } x\}$. For example, two members of this set are $f_1(x)=x^2$ and $f_2(x)=\cos(x)$.

(b) The *odd* functions $\{f\colon \mathbb{R}\to\mathbb{R} \mid f(-x)=-f(x) \text{ for all } x\}$. Two members are $f_3(x)=x^3$ and $f_4(x)=\sin(x)$.

2.30 Example 2.16 says that for any vector $\vec{v}$ that is an element of a vector space V, the set $\{r\cdot\vec{v} \mid r\in\mathbb{R}\}$ is a subspace of V. (This is of course, simply the span of the singleton set $\{\vec{v}\}$.) Must any such subspace be a proper subspace, or can it be improper?

2.31 An example following the definition of a vector space shows that the solution set of a homogeneous linear system is a vector space. In the terminology of this subsection, it is a subspace of $\mathbb{R}^n$ where the system has n variables. What about a non-homogeneous linear system; do its solutions form a subspace (under the inherited operations)?

2.32 [Cleary] Give an example of each or explain why it would be impossible to do so.

(a) A nonempty subset of $\mathcal{M}_{2\times 2}$ that is not a subspace.

(b) A set of two vectors in $\mathbb{R}^2$ that does not span the space.

2.33 Example 2.19 shows that $\mathbb{R}^3$ has infinitely many subspaces. Does every nontrivial space have infinitely many subspaces?

2.34 Finish the proof of Lemma 2.9.

2.35 Show that each vector space has only one trivial subspace.

✓ **2.36** Show that for any subset S of a vector space, the span of the span equals the span $[[S]] = [S]$. (*Hint.* Members of $[S]$ are linear combinations of members of S. Members of $[[S]]$ are linear combinations of linear combinations of members of S.)

2.37 All of the subspaces that we've seen in some way use zero in their description. For example, the subspace in Example 2.3 consists of all the vectors from $\mathbb{R}^2$ with a second component of zero. In contrast, the collection of vectors from $\mathbb{R}^2$ with a second component of one does not form a subspace (it is not closed under scalar multiplication). Another example is Example 2.2, where the condition on the vectors is that the three components add to zero. If the condition there were that the three components add to one then it would not be a subspace (again, it would fail to be closed). However, a reliance on zero is not strictly necessary. Consider the set

$$\{\begin{pmatrix} x \\ y \\ z \end{pmatrix} \mid x+y+z=1\}$$

under these operations.

$$\begin{pmatrix} x_1 \\ y_1 \\ z_1 \end{pmatrix} + \begin{pmatrix} x_2 \\ y_2 \\ z_2 \end{pmatrix} = \begin{pmatrix} x_1+x_2-1 \\ y_1+y_2 \\ z_1+z_2 \end{pmatrix} \qquad r\begin{pmatrix} x \\ y \\ z \end{pmatrix} = \begin{pmatrix} rx-r+1 \\ ry \\ rz \end{pmatrix}$$

(a) Show that it is not a subspace of $\mathbb{R}^3$. (*Hint.* See Example 2.7).

(b) Show that it is a vector space. Note that by the prior item, Lemma 2.9 can not apply.

(c) Show that any subspace of $\mathbb{R}^3$ must pass through the origin, and so any subspace of $\mathbb{R}^3$ must involve zero in its description. Does the converse hold? Does any subset of $\mathbb{R}^3$ that contains the origin become a subspace when given the inherited operations?

2.38 We can give a justification for the convention that the sum of zero-many vectors equals the zero vector. Consider this sum of three vectors $\vec{v}_1 + \vec{v}_2 + \vec{v}_3$.

(a) What is the difference between this sum of three vectors and the sum of the first two of these three?

(b) What is the difference between the prior sum and the sum of just the first one vector?

(c) What should be the difference between the prior sum of one vector and the sum of no vectors?

(d) So what should be the definition of the sum of no vectors?

2.39 Is a space determined by its subspaces? That is, if two vector spaces have the same subspaces, must the two be equal?

2.40 (a) Give a set that is closed under scalar multiplication but not addition.
(b) Give a set closed under addition but not scalar multiplication.
(c) Give a set closed under neither.

2.41 Show that the span of a set of vectors does not depend on the order in which the vectors are listed in that set.

2.42 Which trivial subspace is the span of the empty set? Is it

$$\{\begin{pmatrix}0\\0\\0\end{pmatrix}\}\subseteq\mathbb{R}^3,\quad\text{or}\quad\{0+0x\}\subseteq\mathcal{P}_1,$$

or some other subspace?

2.43 Show that if a vector is in the span of a set then adding that vector to the set won't make the span any bigger. Is that also 'only if'?

✓ **2.44** Subspaces are subsets and so we naturally consider how 'is a subspace of' interacts with the usual set operations.
(a) If A, B are subspaces of a vector space, must their intersection $A\cap B$ be a subspace? Always? Sometimes? Never?
(b) Must the union $A\cup B$ be a subspace?
(c) If A is a subspace, must its complement be a subspace?
(*Hint.* Try some test subspaces from Example 2.19.)

✓ **2.45** Does the span of a set depend on the enclosing space? That is, if W is a subspace of V and S is a subset of W (and so also a subset of V), might the span of S in W differ from the span of S in V?

2.46 Is the relation 'is a subspace of' transitive? That is, if V is a subspace of W and W is a subspace of X, must V be a subspace of X?

✓ **2.47** Because 'span of' is an operation on sets we naturally consider how it interacts with the usual set operations.
(a) If $S\subseteq T$ are subsets of a vector space, is $[S]\subseteq[T]$? Always? Sometimes? Never?
(b) If S, T are subsets of a vector space, is $[S\cup T]=[S]\cup[T]$?
(c) If S, T are subsets of a vector space, is $[S\cap T]=[S]\cap[T]$?
(d) Is the span of the complement equal to the complement of the span?

2.48 Reprove Lemma 2.15 without doing the empty set separately.

2.49 Find a structure that is closed under linear combinations, and yet is not a vector space.

II Linear Independence

The prior section shows how to understand a vector space as a span, as an unrestricted linear combination of some of its elements. For example, the space of linear polynomials $\{a + bx \mid a, b \in \mathbb{R}\}$ is spanned by the set $\{1, x\}$. The prior section also showed that a space can have many sets that span it. Two more sets that span the space of linear polynomials are $\{1, 2x\}$ and $\{1, x, 2x\}$.

At the end of that section we described some spanning sets as 'minimal' but we never precisely defined that word. We could mean that a spanning set is minimal if it contains the smallest number of members of any set with the same span, so that $\{1, x, 2x\}$ is not minimal because it has three members while we can give two-element sets spanning the same space. Or we could mean that a spanning set is minimal when it has no elements that we can remove without changing the span. Under this meaning $\{1, x, 2x\}$ is not minimal because removing the $2x$ to get $\{1, x\}$ leaves the span unchanged.

The first sense of minimality appears to be a global requirement, in that to check if a spanning set is minimal we seemingly must look at all the sets that span and find one with the least number of elements. The second sense of minimality is local since we need to look only at the set and consider the span with and without various elements. For instance, using the second sense we could compare the span of $\{1, x, 2x\}$ with the span of $\{1, x\}$ and note that $2x$ is a "repeat" in that its removal doesn't shrink the span.

In this section we will use the second sense of 'minimal spanning set' because of this technical convenience. However, the most important result of this book is that the two senses coincide. We will prove that in the next section.

II.1 Definition and Examples

We saw "repeats" in the first chapter. There, Gauss's Method turned them into $0 = 0$ equations.

1.1 Example Recall the Statics example from Chapter One's opening. We got two balances with the pair of unknown-mass objects, one at 40 cm and 15 cm and another at -50 cm and 25 cm, and we then computed the value of those masses. Had we instead gotten the second balance at 20 cm and 7.5 cm then Gauss's Method on the resulting two-equations, two-unknowns system would not have yielded a solution, it would have yielded a $0 = 0$ equation along with an equation containing a free variable. Intuitively, the problem is that $(20\ \ 7.5)$ is half of $(40\ \ 15)$, that is, $(20\ \ 7.5)$ is in the span of the set $\{(40\ \ 15)\}$ and so is repeated data. We would have been trying to solve a two-unknowns problem with essentially only one piece of information.

We take $\vec{v}$ to be a "repeat" of the vectors in a set S if $\vec{v} \in [S]$ so that it depends on, that is, is expressible in terms of, elements of the set $\vec{v} = c_1\vec{s}_1 + \cdots + c_n\vec{s}_n$.

1.2 Lemma Where V is a vector space, S is a subset of that space, and $\vec{v}$ is an element of that space, $[S \cup \{\vec{v}\}] = [S]$ if and only if $\vec{v} \in [S]$.

Proof Half of the if and only if is immediate: if $\vec{v} \notin [S]$ then the sets are not equal because $\vec{v} \in [S \cup \{\vec{v}\}]$.

For the other half assume that $\vec{v} \in [S]$ so that $\vec{v} = c_1\vec{s}_1 + \cdots + c_n\vec{s}_n$ for some scalars c_i and vectors $\vec{s}_i \in S$. We will use mutual containment to show that the sets $[S \cup \{\vec{v}\}]$ and $[S]$ are equal. The containment $[S \cup \{\vec{v}\}] \supseteq [S]$ is clear.

To show containment in the other direction let $\vec{w}$ be an element of $[S \cup \{\vec{v}\}]$. Then $\vec{w}$ is a linear combination of elements of $S \cup \{\vec{v}\}$, which we can write as $\vec{w} = c_{n+1}\vec{s}_{n+1} + \cdots + c_{n+k}\vec{s}_{n+k} + c_{n+k+1}\vec{v}$. (Possibly some of the $\vec{s}_i$'s from $\vec{w}$'s equation are the same as some of those from $\vec{v}$'s equation but that does not matter.) Expand $\vec{v}$.

$$\vec{w} = c_{n+1}\vec{s}_{n+1} + \cdots + c_{n+k}\vec{s}_{n+k} + c_{n+k+1} \cdot (c_1\vec{s}_1 + \cdots + c_n\vec{s}_n)$$

Recognize the right hand side as a linear combination of linear combinations of vectors from S. Thus $\vec{w} \in [S]$. QED

The discussion at the section's opening involved removing vectors instead of adding them.

1.3 Corollary For $\vec{v} \in S$, omitting that vector does not shrink the span $[S] = [S - \{\vec{v}\}]$ if and only if it is dependent on other vectors in the set $\vec{v} \in [S]$.

The corollary says that to know whether removing a vector will decrease the span, we need to know whether the vector is a linear combination of others in the set.

1.4 Definition A multiset subset of a vector space is *linearly independent* if none of its elements is a linear combination of the others.* Otherwise it is *linearly dependent*.

That definition's use of the word 'others' means that writing $\vec{v}$ as a linear combination with $\vec{v} = 1 \cdot \vec{v}$ does not count.

*More information on multisets is in the appendix. Most of the time we won't need the set-multiset distinction and we will follow the standard terminology of referring to a linearly independent or dependent 'set'. Remark 1.12 explains why the definition requires a multiset.

Observe that, although this way of writing one vector as a combination of the others

$$\vec{s}_0 = c_1\vec{s}_1 + c_2\vec{s}_2 + \cdots + c_n\vec{s}_n$$

visually sets off $\vec{s}_0$, algebraically there is nothing special about that vector in that equation. For any $\vec{s}_i$ with a coefficient c_i that is non-0 we can rewrite to isolate $\vec{s}_i$.

$$\vec{s}_i = (1/c_i)\vec{s}_0 + \cdots + (-c_{i-1}/c_i)\vec{s}_{i-1} + (-c_{i+1}/c_i)\vec{s}_{i+1} + \cdots + (-c_n/c_i)\vec{s}_n$$

When we don't want to single out any vector we will instead say that $\vec{s}_0, \vec{s}_1, \ldots, \vec{s}_n$ are in a *linear relationship* and put all of the vectors on the same side. The next result rephrases the linear independence definition in this style. It is how we usually compute whether a finite set is dependent or independent.

1.5 Lemma A subset S of a vector space is linearly independent if and only if among its elements the only linear relationship $c_1\vec{s}_1 + \cdots + c_n\vec{s}_n = \vec{0}$ (with $\vec{s}_i \neq \vec{s}_j$ for all $i \neq j$) is the trivial one $c_1 = 0, \ldots, c_n = 0$.

PROOF If S is linearly independent then no vector $\vec{s}_i$ is a linear combination of other vectors from S so there is no linear relationship where some of the $\vec{s}$'s have nonzero coefficients.

If S is not linearly independent then some $\vec{s}_i$ is a linear combination $\vec{s}_i = c_1\vec{s}_1 + \cdots + c_{i-1}\vec{s}_{i-1} + c_{i+1}\vec{s}_{i+1} + \cdots + c_n\vec{s}_n$ of other vectors from S. Subtracting $\vec{s}_i$ from both sides gives a relationship involving a nonzero coefficient, the -1 in front of $\vec{s}_i$. QED

1.6 Example In the vector space of two-wide row vectors, the two-element set $\{(40\ \ 15), (-50\ \ 25)\}$ is linearly independent. To check this, take

$$c_1 \cdot (40\ \ 15) + c_2 \cdot (-50\ \ 25) = (0\ \ 0)$$

and solve the resulting system.

$$\begin{array}{rl} 40c_1 - 50c_2 = 0 \\ 15c_1 + 25c_2 = 0 \end{array} \xrightarrow{-(15/40)\rho_1+\rho_2} \begin{array}{rl} 40c_1 - \quad 50c_2 = 0 \\ (175/4)c_2 = 0 \end{array}$$

Both c_1 and c_2 are zero. So the only linear relationship between the two given row vectors is the trivial relationship.

In the same vector space, the set $\{(40\ \ 15), (20\ \ 7.5)\}$ is linearly dependent since we can satisfy $c_1 \cdot (40\ \ 15) + c_2 \cdot (20\ \ 7.5) = (0\ \ 0)$ with $c_1 = 1$ and $c_2 = -2$.

1.7 Example The set $\{1 + x, 1 - x\}$ is linearly independent in $\mathcal{P}_2$, the space of quadratic polynomials with real coefficients, because

$$0 + 0x + 0x^2 = c_1(1 + x) + c_2(1 - x) = (c_1 + c_2) + (c_1 - c_2)x + 0x^2$$

gives

$$\begin{array}{rcl} c_1 + c_2 = 0 & \xrightarrow{-\rho_1+\rho_2} & c_1 + c_2 = 0 \\ c_1 - c_2 = 0 & & 2c_2 = 0 \end{array}$$

since polynomials are equal only if their coefficients are equal. Thus, the only linear relationship between these two members of $\mathcal{P}_2$ is the trivial one.

1.8 Example The rows of this matrix

$$A = \begin{pmatrix} 2 & 3 & 1 & 0 \\ 0 & -1 & 0 & -2 \\ 0 & 0 & 0 & 1 \end{pmatrix}$$

form a linearly independent set. This is easy to check for this case but also recall that Lemma One.III.2.5 shows that the rows of any echelon form matrix form a linearly independent set.

1.9 Example In $\mathbb{R}^3$, where

$$\vec{v}_1 = \begin{pmatrix} 3 \\ 4 \\ 5 \end{pmatrix} \quad \vec{v}_2 = \begin{pmatrix} 2 \\ 9 \\ 2 \end{pmatrix} \quad \vec{v}_3 = \begin{pmatrix} 4 \\ 18 \\ 4 \end{pmatrix}$$

the set $S = \{\vec{v}_1, \vec{v}_2, \vec{v}_3\}$ is linearly dependent because this is a relationship

$$0 \cdot \vec{v}_1 + 2 \cdot \vec{v}_2 - 1 \cdot \vec{v}_3 = \vec{0}$$

where not all of the scalars are zero (the fact that some of the scalars are zero doesn't matter).

That example illustrates why, although Definition 1.4 is a clearer statement of what independence means, Lemma 1.5 is better for computations. Working straight from the definition, someone trying to compute whether S is linearly independent would start by setting $\vec{v}_1 = c_2\vec{v}_2 + c_3\vec{v}_3$ and concluding that there are no such c_2 and c_3. But knowing that the first vector is not dependent on the other two is not enough. This person would have to go on to try $\vec{v}_2 = c_1\vec{v}_1 + c_3\vec{v}_3$, in order to find the dependence $c_1 = 0$, $c_3 = 1/2$. Lemma 1.5 gets the same conclusion with only one computation.

1.10 Example The empty subset of a vector space is linearly independent. There is no nontrivial linear relationship among its members as it has no members.

1.11 Example In any vector space, any subset containing the zero vector is linearly dependent. One example is, in the space $\mathcal{P}_2$ of quadratic polynomials, the subset $\{1 + x, x + x^2, 0\}$. It is linearly dependent because $0 \cdot \vec{v}_1 + 0 \cdot \vec{v}_2 + 1 \cdot \vec{0} = \vec{0}$ is a nontrivial relationship, since not all of the coefficients are zero.

A subtler way to see that this subset is dependent is to remember that the zero vector is equal to the trivial sum, the sum of the empty set. So any set containing the zero vector has an element that is a combination of a subset of other vectors from the set, specifically, the zero vector is a combination of the empty subset.

1.12 Remark Definition 1.4 says that when we decide whether some S is linearly independent we must consider it as a multiset. Recall that in a set repeated elements collapse, so the set $\{0,1,0\}$ equals the set $\{0,1\}$. But in a multiset they do not collapse so the multiset $\{0,1,0\}$ contains the element 0 twice. Here is an example showing why the definition requires a multiset.

$$\begin{pmatrix} 1 & 1 & 1 \\ 2 & 2 & 2 \\ 1 & 2 & 3 \end{pmatrix} \xrightarrow{(1/2)\rho_2} \begin{pmatrix} 1 & 1 & 1 \\ 1 & 1 & 1 \\ 1 & 2 & 3 \end{pmatrix}$$

On the left the set of matrix rows $\{(1\ \ 1\ \ 1), (2\ \ 2\ \ 2), (1\ \ 2\ \ 3)\}$ is linearly dependent. But on the right the set $\{(1\ \ 1\ \ 1), (1\ \ 1\ \ 1), (1\ \ 2\ \ 3)\} = \{(1\ \ 1\ \ 1), (1\ \ 2\ \ 3)\}$ is linearly independent. That's not what we need; we rely on Gauss's Method to preserve dependence so we need that $(1\ \ 1\ \ 1)$ appears twice.

1.13 Corollary A set S is linearly independent if and only if for any $\vec{v} \in S$, its removal shrinks the span $[S - \{v\}] \subsetneq [S]$.

PROOF This follows from Corollary 1.3. If S is linearly independent then none of its vectors is dependent on the other elements, so removal of any vector will shrink the span. If S is not linearly independent then it contains a vector that is dependent on other elements of the set, and removal of that vector will not shrink the span. QED

So a spanning set is minimal if and only if it is linearly independent.

The prior result addresses removing elements from a linearly independent set. The next one adds elements.

1.14 Lemma Suppose that S is linearly independent and that $\vec{v} \notin S$. Then the set $S \cup \{\vec{v}\}$ is linearly independent if and only if $\vec{v} \notin [S]$.

PROOF We will show that $S \cup \{\vec{v}\}$ is not linearly independent if and only if $\vec{v} \in [S]$.

Suppose first that $v \in [S]$. Express $\vec{v}$ as a combination $\vec{v} = c_1\vec{s}_1 + \cdots + c_n\vec{s}_n$. Rewrite that $\vec{0} = c_1\vec{s}_1 + \cdots + c_n\vec{s}_n - 1 \cdot \vec{v}$. Since $v \notin S$, it does not equal any of the $\vec{s}_i$ so this is a nontrivial linear dependence among the elements of $S \cup \{\vec{v}\}$. Thus that set is not linearly independent.

Now suppose that $S \cup \{\vec{v}\}$ is not linearly independent and consider a nontrivial dependence among its members $\vec{0} = c_1\vec{s}_1 + \cdots + c_n\vec{s}_n + c_{n+1} \cdot \vec{v}$. If $c_{n+1} = 0$

then that is a dependence among the elements of S, but we are assuming that S is independent, so $c_{n+1} \neq 0$. Rewrite the equation as $\vec{v} = (c_1/c_{n+1})\vec{s}_1 + \cdots + (c_n/c_{n+1})\vec{s}_n$ to get $\vec{v} \in [S]$ QED

1.15 Example This subset of $\mathbb{R}^3$ is linearly independent.

$$S = \{ \begin{pmatrix} 1 \\ 0 \\ 0 \end{pmatrix} \}$$

The span of S is the x-axis. Here are two supersets, one that is linearly dependent and the other independent.

$$\text{dependent: } \{ \begin{pmatrix} 1 \\ 0 \\ 0 \end{pmatrix}, \begin{pmatrix} -3 \\ 0 \\ 0 \end{pmatrix} \} \qquad \text{independent: } \{ \begin{pmatrix} 1 \\ 0 \\ 0 \end{pmatrix}, \begin{pmatrix} 0 \\ 1 \\ 0 \end{pmatrix} \}$$

We got the dependent superset by adding a vector from the x-axis and so the span did not grow. We got the independent superset by adding a vector that isn't in [S], because it has a nonzero y component, causing the span to grow.

For the independent set

$$S = \{ \begin{pmatrix} 1 \\ 0 \\ 0 \end{pmatrix}, \begin{pmatrix} 0 \\ 1 \\ 0 \end{pmatrix} \}$$

the span [S] is the xy-plane. Here are two supersets.

$$\text{dependent: } \{ \begin{pmatrix} 1 \\ 0 \\ 0 \end{pmatrix}, \begin{pmatrix} 0 \\ 1 \\ 0 \end{pmatrix}, \begin{pmatrix} 3 \\ -2 \\ 0 \end{pmatrix} \} \qquad \text{independent: } \{ \begin{pmatrix} 1 \\ 0 \\ 0 \end{pmatrix}, \begin{pmatrix} 0 \\ 1 \\ 0 \end{pmatrix}, \begin{pmatrix} 0 \\ 0 \\ 1 \end{pmatrix} \}$$

As above, the additional member of the dependent superset comes from [S], the xy-plane, while the added member of the independent superset comes from outside of that span.

Finally, consider this independent set

$$S = \{ \begin{pmatrix} 1 \\ 0 \\ 0 \end{pmatrix}, \begin{pmatrix} 0 \\ 1 \\ 0 \end{pmatrix}, \begin{pmatrix} 0 \\ 0 \\ 1 \end{pmatrix} \}$$

with $[S] = \mathbb{R}^3$. We can get a linearly dependent superset.

$$\text{dependent: } \{ \begin{pmatrix} 1 \\ 0 \\ 0 \end{pmatrix}, \begin{pmatrix} 0 \\ 1 \\ 0 \end{pmatrix}, \begin{pmatrix} 0 \\ 0 \\ 1 \end{pmatrix}, \begin{pmatrix} 2 \\ -1 \\ 3 \end{pmatrix} \}$$

But there is no linearly independent superset os S. One way to see that is to note that for any vector that we would add to S, the equation

$$\begin{pmatrix} x \\ y \\ z \end{pmatrix} = c_1 \begin{pmatrix} 1 \\ 0 \\ 0 \end{pmatrix} + c_2 \begin{pmatrix} 0 \\ 1 \\ 0 \end{pmatrix} + c_3 \begin{pmatrix} 0 \\ 0 \\ 1 \end{pmatrix}$$

has a solution $c_1 = x$, $c_2 = y$, and $c_3 = z$. Another way to see it is that we cannot add any vectors from outside of the span $[S]$ because that span is $\mathbb{R}^3$.

1.16 Corollary In a vector space, any finite set has a linearly independent subset with the same span.

Proof If $S = \{\vec{s}_1, \ldots, \vec{s}_n\}$ is linearly independent then S itself satisfies the statement, so assume that it is linearly dependent.

By the definition of dependent, S contains a vector $\vec{v}_1$ that is a linear combination of the others. Define the set $S_1 = S - \{\vec{v}_1\}$. By Corollary 1.3 the span does not shrink $[S_1] = [S]$.

If S_1 is linearly independent then we are done. Otherwise iterate: take a vector $\vec{v}_2$ that is a linear combination of other members of S_1 and discard it to derive $S_2 = S_1 - \{\vec{v}_2\}$ such that $[S_2] = [S_1]$. Repeat this until a linearly independent set S_j appears; one must appear eventually because S is finite and the empty set is linearly independent. (Formally, this argument uses induction on the number of elements in S. Exercise 41 asks for the details.) QED

Thus if we have a set that is linearly dependent then we can, without changing the span, pare down by discarding what we have called "repeat" vectors.

1.17 Example This set spans $\mathbb{R}^3$ (the check is routine) but is not linearly independent.

$$S = \{ \begin{pmatrix} 1 \\ 0 \\ 0 \end{pmatrix}, \begin{pmatrix} 0 \\ 2 \\ 0 \end{pmatrix}, \begin{pmatrix} 1 \\ 2 \\ 0 \end{pmatrix}, \begin{pmatrix} 0 \\ -1 \\ 1 \end{pmatrix}, \begin{pmatrix} 3 \\ 3 \\ 0 \end{pmatrix} \}$$

We will calculate which vectors to drop in order to get a subset that is independent but has the same span. This linear relationship

$$c_1 \begin{pmatrix} 1 \\ 0 \\ 0 \end{pmatrix} + c_2 \begin{pmatrix} 0 \\ 2 \\ 0 \end{pmatrix} + c_3 \begin{pmatrix} 1 \\ 2 \\ 0 \end{pmatrix} + c_4 \begin{pmatrix} 0 \\ -1 \\ 1 \end{pmatrix} + c_5 \begin{pmatrix} 3 \\ 3 \\ 0 \end{pmatrix} = \begin{pmatrix} 0 \\ 0 \\ 0 \end{pmatrix} \qquad (*)$$

gives a system

$$\begin{aligned} c_1 \qquad + c_3 + \qquad + 3c_5 &= 0 \\ 2c_2 + 2c_3 - c_4 + 3c_5 &= 0 \\ c_4 \qquad &= 0 \end{aligned}$$

whose solution set has this parametrization.

$$\{\begin{pmatrix}c_1\\c_2\\c_3\\c_4\\c_5\end{pmatrix}=c_3\begin{pmatrix}-1\\-1\\1\\0\\0\end{pmatrix}+c_5\begin{pmatrix}-3\\-3/2\\0\\0\\1\end{pmatrix}\mid c_3,c_5\in\mathbb{R}\}$$

Set $c_5=1$ and $c_3=0$ to get an instance of $(*)$.

$$-3\cdot\begin{pmatrix}1\\0\\0\end{pmatrix}-\frac{3}{2}\cdot\begin{pmatrix}0\\2\\0\end{pmatrix}+0\cdot\begin{pmatrix}1\\2\\0\end{pmatrix}+0\cdot\begin{pmatrix}0\\-1\\1\end{pmatrix}+1\cdot\begin{pmatrix}3\\3\\0\end{pmatrix}=\begin{pmatrix}0\\0\\0\end{pmatrix}$$

This shows that the vector from S that we've associated with c_5 is in the span of the set of c_1's vector and c_2's vector. We can discard S's fifth vector without shrinking the span.

Similarly, set $c_3=1$, and $c_5=0$ to get an instance of $(*)$ that shows we can discard S's third vector without shrinking the span. Thus this set has the same span as S.

$$\{\begin{pmatrix}1\\0\\0\end{pmatrix},\begin{pmatrix}0\\2\\0\end{pmatrix},\begin{pmatrix}0\\-1\\1\end{pmatrix}\}$$

The check that it is linearly independent is routine.

1.18 Corollary A subset $S=\{\vec{s}_1,\ldots,\vec{s}_n\}$ of a vector space is linearly dependent if and only if some $\vec{s}_i$ is a linear combination of the vectors $\vec{s}_1$, ..., $\vec{s}_{i-1}$ listed before it.

Proof Consider $S_0=\{\}$, $S_1=\{\vec{s}_1\}$, $S_2=\{\vec{s}_1,\vec{s}_2\}$, etc. Some index $i\geqslant 1$ is the first one with $S_{i-1}\cup\{\vec{s}_i\}$ linearly dependent, and there $\vec{s}_i\in[S_{i-1}]$. QED

The proof of Corollary 1.16 describes producing a linearly independent set by shrinking, by taking subsets. And the proof of Corollary 1.18 describes finding a linearly dependent set by taking supersets. We finish this subsection by considering how linear independence and dependence interact with the subset relation between sets.

1.19 Lemma Any subset of a linearly independent set is also linearly independent. Any superset of a linearly dependent set is also linearly dependent.

Proof Both are clear. QED

Restated, subset preserves independence and superset preserves dependence.

Those are two of the four possible cases. The third case, whether subset preserves linear dependence, is covered by Example 1.17, which gives a linearly dependent set S with one subset that is linearly dependent and another that is independent. The fourth case, whether superset preserves linear independence, is covered by Example 1.15, which gives cases where a linearly independent set has both an independent and a dependent superset. This table summarizes.

	$\hat{S} \subset S$	$\hat{S} \supset S$
S *independent*	$\hat{S}$ must be independent	$\hat{S}$ may be either
S *dependent*	$\hat{S}$ may be either	$\hat{S}$ must be dependent

Example 1.15 has something else to say about the interaction between linear independence and superset. It names a linearly independent set that is maximal in that it has no supersets that are linearly independent. By Lemma 1.14 a linearly independent set is maximal if and only if it spans the entire space, because that is when all the vectors in the space are already in the span. This nicely complements Lemma 1.13, that a spanning set is minimal if and only if it is linearly independent.

Exercises

✓ **1.20** Decide whether each subset of $\mathbb{R}^3$ is linearly dependent or linearly independent.

(a) $\{\begin{pmatrix}1\\-3\\5\end{pmatrix}, \begin{pmatrix}2\\2\\4\end{pmatrix}, \begin{pmatrix}4\\-4\\14\end{pmatrix}\}$ (b) $\{\begin{pmatrix}1\\7\\7\end{pmatrix}, \begin{pmatrix}2\\7\\7\end{pmatrix}, \begin{pmatrix}3\\7\\7\end{pmatrix}\}$ (c) $\{\begin{pmatrix}0\\0\\-1\end{pmatrix}, \begin{pmatrix}1\\0\\4\end{pmatrix}\}$

(d) $\{\begin{pmatrix}9\\9\\0\end{pmatrix}, \begin{pmatrix}2\\0\\1\end{pmatrix}, \begin{pmatrix}3\\5\\-4\end{pmatrix}, \begin{pmatrix}12\\12\\-1\end{pmatrix}\}$

✓ **1.21** Which of these subsets of $\mathcal{P}_3$ are linearly dependent and which are independent?

(a) $\{3 - x + 9x^2, 5 - 6x + 3x^2, 1 + 1x - 5x^2\}$

(b) $\{-x^2, 1 + 4x^2\}$

(c) $\{2 + x + 7x^2, 3 - x + 2x^2, 4 - 3x^2\}$

(d) $\{8 + 3x + 3x^2, x + 2x^2, 2 + 2x + 2x^2, 8 - 2x + 5x^2\}$

1.22 Determine if each set is linearly independent in the natural space.

(a) $\{\begin{pmatrix}1\\2\\0\end{pmatrix}, \begin{pmatrix}-1\\1\\0\end{pmatrix}\}$ (b) $\{(1\ \ 3\ \ 1), (-1\ \ 4\ \ 3), (-1\ \ 11\ \ 7)\}$

(c) $\{\begin{pmatrix}5 & 4\\1 & 2\end{pmatrix}, \begin{pmatrix}0 & 0\\0 & 0\end{pmatrix}, \begin{pmatrix}1 & 0\\-1 & 4\end{pmatrix}\}$

✓ **1.23** Prove that each set $\{f, g\}$ is linearly independent in the vector space of all functions from $\mathbb{R}^+$ to $\mathbb{R}$.

(a) $f(x) = x$ and $g(x) = 1/x$

(b) $f(x) = \cos(x)$ and $g(x) = \sin(x)$
(c) $f(x) = e^x$ and $g(x) = \ln(x)$

✓ **1.24** Which of these subsets of the space of real-valued functions of one real variable is linearly dependent and which is linearly independent? (We have abbreviated some constant functions; e.g., in the first item, the '2' stands for the constant function $f(x) = 2$.)
(a) $\{2, 4\sin^2(x), \cos^2(x)\}$ (b) $\{1, \sin(x), \sin(2x)\}$ (c) $\{x, \cos(x)\}$
(d) $\{(1+x)^2, x^2+2x, 3\}$ (e) $\{\cos(2x), \sin^2(x), \cos^2(x)\}$ (f) $\{0, x, x^2\}$

1.25 Does the equation $\sin^2(x)/\cos^2(x) = \tan^2(x)$ show that this set of functions $\{\sin^2(x), \cos^2(x), \tan^2(x)\}$ is a linearly dependent subset of the set of all real-valued functions with domain the interval $(-\pi/2..\pi/2)$ of real numbers between $-\pi/2$ and $\pi/2$?

1.26 Is the xy-plane subset of the vector space $\mathbb{R}^3$ linearly independent?

✓ **1.27** Show that the nonzero rows of an echelon form matrix form a linearly independent set.

✓ **1.28** (a) Show that if the set $\{\vec{u}, \vec{v}, \vec{w}\}$ is linearly independent then so is the set $\{\vec{u}, \vec{u}+\vec{v}, \vec{u}+\vec{v}+\vec{w}\}$.
(b) What is the relationship between the linear independence or dependence of $\{\vec{u}, \vec{v}, \vec{w}\}$ and the independence or dependence of $\{\vec{u}-\vec{v}, \vec{v}-\vec{w}, \vec{w}-\vec{u}\}$?

1.29 Example 1.10 shows that the empty set is linearly independent.
(a) When is a one-element set linearly independent?
(b) How about a set with two elements?

1.30 In any vector space V, the empty set is linearly independent. What about all of V?

1.31 Show that if $\{\vec{x}, \vec{y}, \vec{z}\}$ is linearly independent then so are all of its proper subsets: $\{\vec{x}, \vec{y}\}$, $\{\vec{x}, \vec{z}\}$, $\{\vec{y}, \vec{z}\}$, $\{\vec{x}\}$,$\{\vec{y}\}$, $\{\vec{z}\}$, and $\{\}$. Is that 'only if' also?

1.32 (a) Show that this

$$S = \{\begin{pmatrix}1\\1\\0\end{pmatrix}, \begin{pmatrix}-1\\2\\0\end{pmatrix}\}$$

is a linearly independent subset of $\mathbb{R}^3$.
(b) Show that

$$\begin{pmatrix}3\\2\\0\end{pmatrix}$$

is in the span of S by finding c_1 and c_2 giving a linear relationship.

$$c_1\begin{pmatrix}1\\1\\0\end{pmatrix} + c_2\begin{pmatrix}-1\\2\\0\end{pmatrix} = \begin{pmatrix}3\\2\\0\end{pmatrix}$$

Show that the pair c_1, c_2 is unique.
(c) Assume that S is a subset of a vector space and that $\vec{v}$ is in $[S]$, so that $\vec{v}$ is a linear combination of vectors from S. Prove that if S is linearly independent then a linear combination of vectors from S adding to $\vec{v}$ is unique (that is, unique up to reordering and adding or taking away terms of the form $0 \cdot \vec{s}$). Thus S as a

spanning set is minimal in this strong sense: each vector in [S] is a combination of elements of S a minimum number of times — only once.

(d) Prove that it can happen when S is not linearly independent that distinct linear combinations sum to the same vector.

1.33 Prove that a polynomial gives rise to the zero function if and only if it is the zero polynomial. (*Comment.* This question is not a Linear Algebra matter but we often use the result. A polynomial gives rise to a function in the natural way: $x \mapsto c_n x^n + \cdots + c_1 x + c_0$.)

1.34 Return to Section 1.2 and redefine point, line, plane, and other linear surfaces to avoid degenerate cases.

1.35 (a) Show that any set of four vectors in $\mathbb{R}^2$ is linearly dependent.

(b) Is this true for any set of five? Any set of three?

(c) What is the most number of elements that a linearly independent subset of $\mathbb{R}^2$ can have?

✓ **1.36** Is there a set of four vectors in $\mathbb{R}^3$ such that any three form a linearly independent set?

1.37 Must every linearly dependent set have a subset that is dependent and a subset that is independent?

1.38 In $\mathbb{R}^4$ what is the biggest linearly independent set you can find? The smallest? The biggest linearly dependent set? The smallest? ('Biggest' and 'smallest' mean that there are no supersets or subsets with the same property.)

✓ **1.39** Linear independence and linear dependence are properties of sets. We can thus naturally ask how the properties of linear independence and dependence act with respect to the familiar elementary set relations and operations. In this body of this subsection we have covered the subset and superset relations. We can also consider the operations of intersection, complementation, and union.

(a) How does linear independence relate to intersection: can an intersection of linearly independent sets be independent? Must it be?

(b) How does linear independence relate to complementation?

(c) Show that the union of two linearly independent sets can be linearly independent.

(d) Show that the union of two linearly independent sets need not be linearly independent.

1.40 *Continued from prior exercise.* What is the interaction between the property of linear independence and the operation of union?

(a) We might conjecture that the union $S \cup T$ of linearly independent sets is linearly independent if and only if their spans have a trivial intersection $[S] \cap [T] = \{\vec{0}\}$. What is wrong with this argument for the 'if' direction of that conjecture? "If the union $S \cup T$ is linearly independent then the only solution to $c_1\vec{s}_1 + \cdots + c_n\vec{s}_n + d_1\vec{t}_1 + \cdots + d_m\vec{t}_m = \vec{0}$ is the trivial one $c_1 = 0, \ldots, d_m = 0$. So any member of the intersection of the spans must be the zero vector because in $c_1\vec{s}_1 + \cdots + c_n\vec{s}_n = d_1\vec{t}_1 + \cdots + d_m\vec{t}_m$ each scalar is zero."

(b) Give an example showing that the conjecture is false.

(c) Find linearly independent sets S and T so that the union of $S-(S\cap T)$ and $T-(S\cap T)$ is linearly independent, but the union $S\cup T$ is not linearly independent.

(d) Characterize when the union of two linearly independent sets is linearly independent, in terms of the intersection of spans.

✓ **1.41** For Corollary 1.16,

(a) fill in the induction for the proof;

(b) give an alternate proof that starts with the empty set and builds a sequence of linearly independent subsets of the given finite set until one appears with the same span as the given set.

1.42 With a some calculation we can get formulas to determine whether or not a set of vectors is linearly independent.

(a) Show that this subset of $\mathbb{R}^2$

$$\{\begin{pmatrix}a\\c\end{pmatrix},\begin{pmatrix}b\\d\end{pmatrix}\}$$

is linearly independent if and only if $ad-bc\neq 0$.

(b) Show that this subset of $\mathbb{R}^3$

$$\{\begin{pmatrix}a\\d\\g\end{pmatrix},\begin{pmatrix}b\\e\\h\end{pmatrix},\begin{pmatrix}c\\f\\i\end{pmatrix}\}$$

is linearly independent iff $aei+bfg+cdh-hfa-idb-gec\neq 0$.

(c) When is this subset of $\mathbb{R}^3$

$$\{\begin{pmatrix}a\\d\\g\end{pmatrix},\begin{pmatrix}b\\e\\h\end{pmatrix}\}$$

linearly independent?

(d) This is an opinion question: for a set of four vectors from $\mathbb{R}^4$, must there be a formula involving the sixteen entries that determines independence of the set? (You needn't produce such a formula, just decide if one exists.)

✓ **1.43** (a) Prove that a set of two perpendicular nonzero vectors from $\mathbb{R}^n$ is linearly independent when $n>1$.

(b) What if $n=1$? $n=0$?

(c) Generalize to more than two vectors.

1.44 Consider the set of functions from the interval $(-1\ldots 1)\subseteq\mathbb{R}$ to $\mathbb{R}$.

(a) Show that this set is a vector space under the usual operations.

(b) Recall the formula for the sum of an infinite geometric series: $1+x+x^2+\cdots=1/(1-x)$ for all $x\in(-1..1)$. Why does this not express a dependence inside of the set $\{g(x)=1/(1-x),f_0(x)=1,f_1(x)=x,f_2(x)=x^2,\ldots\}$ (in the vector space that we are considering)? (*Hint.* Review the definition of linear combination.)

(c) Show that the set in the prior item is linearly independent.

This shows that some vector spaces exist with linearly independent subsets that are infinite.

1.45 Show that, where S is a subspace of V, if a subset T of S is linearly independent in S then T is also linearly independent in V. Is that 'only if'?

III Basis and Dimension

The prior section ends with the observation that a spanning set is minimal when it is linearly independent and a linearly independent set is maximal when it spans the space. So the notions of minimal spanning set and maximal independent set coincide. In this section we will name this idea and study its properties.

III.1 Basis

1.1 Definition A *basis* for a vector space is a sequence of vectors that is linearly independent and that spans the space.

Because a basis is a sequence, meaning that bases are different if they contain the same elements but in different orders, we denote it with angle brackets $\langle\vec{\beta}_1,\vec{\beta}_2,\ldots\rangle$.* (A sequence is linearly independent if the multiset consisting of the elements of the sequence in is independent. Similarly, a sequence spans the space if the set of elements of the sequence spans the space.)

1.2 Example This is a basis for $\mathbb{R}^2$.

$$\langle\begin{pmatrix}2\\4\end{pmatrix},\begin{pmatrix}1\\1\end{pmatrix}\rangle$$

It is linearly independent

$$c_1\begin{pmatrix}2\\4\end{pmatrix}+c_2\begin{pmatrix}1\\1\end{pmatrix}=\begin{pmatrix}0\\0\end{pmatrix}\quad\Longrightarrow\quad\begin{aligned}2c_1+1c_2=0\\4c_1+1c_2=0\end{aligned}\quad\Longrightarrow\quad c_1=c_2=0$$

and it spans $\mathbb{R}^2$.

$$\begin{aligned}2c_1+1c_2=x\\4c_1+1c_2=y\end{aligned}\quad\Longrightarrow\quad c_2=2x-y\text{ and }c_1=(y-x)/2$$

1.3 Example This basis for $\mathbb{R}^2$ differs from the prior one

$$\langle\begin{pmatrix}1\\1\end{pmatrix},\begin{pmatrix}2\\4\end{pmatrix}\rangle$$

because it is in a different order. The verification that it is a basis is just as in the prior example.

* More information on sequences is in the appendix.

1.4 Example The space $\mathbb{R}^2$ has many bases. Another one is this.

$$\langle \begin{pmatrix} 1 \\ 0 \end{pmatrix}, \begin{pmatrix} 0 \\ 1 \end{pmatrix} \rangle$$

The verification is easy.

1.5 Definition For any $\mathbb{R}^n$

$$\mathcal{E}_n = \langle \begin{pmatrix} 1 \\ 0 \\ \vdots \\ 0 \end{pmatrix}, \begin{pmatrix} 0 \\ 1 \\ \vdots \\ 0 \end{pmatrix}, \dots, \begin{pmatrix} 0 \\ 0 \\ \vdots \\ 1 \end{pmatrix} \rangle$$

is the *standard* (or *natural*) basis. We denote these vectors $\vec{e}_1, \dots, \vec{e}_n$.

Calculus books denote $\mathbb{R}^2$'s standard basis vectors as $\vec{\imath}$ and $\vec{\jmath}$ instead of $\vec{e}_1$ and $\vec{e}_2$ and they denote to $\mathbb{R}^3$'s standard basis vectors as $\vec{\imath}$, $\vec{\jmath}$, and $\vec{k}$ instead of $\vec{e}_1$, $\vec{e}_2$, and $\vec{e}_3$. Note that $\vec{e}_1$ means something different in a discussion of $\mathbb{R}^3$ than it means in a discussion of $\mathbb{R}^2$.

1.6 Example Consider the space $\{ a \cdot \cos\theta + b \cdot \sin\theta \mid a, b \in \mathbb{R} \}$ of functions of the real variable θ. This is a natural basis $\langle \cos\theta, \sin\theta \rangle = \langle 1 \cdot \cos\theta + 0 \cdot \sin\theta, 0 \cdot \cos\theta + 1 \cdot \sin\theta \rangle$. A more generic basis for this space is $\langle \cos\theta - \sin\theta, 2\cos\theta + 3\sin\theta \rangle$. Verification that these two are bases is Exercise 27.

1.7 Example A natural basis for the vector space of cubic polynomials $\mathcal{P}_3$ is $\langle 1, x, x^2, x^3 \rangle$. Two other bases for this space are $\langle x^3, 3x^2, 6x, 6 \rangle$ and $\langle 1, 1 + x, 1 + x + x^2, 1 + x + x^2 + x^3 \rangle$. Checking that each is linearly independent and spans the space is easy.

1.8 Example The trivial space $\{\vec{0}\}$ has only one basis, the empty one $\langle \rangle$.

1.9 Example The space of finite-degree polynomials has a basis with infinitely many elements $\langle 1, x, x^2, \dots \rangle$.

1.10 Example We have seen bases before. In the first chapter we described the solution set of homogeneous systems such as this one

$$\begin{aligned} x + y \quad - w &= 0 \\ z + w &= 0 \end{aligned}$$

by parametrizing.

$$\{ \begin{pmatrix} -1 \\ 1 \\ 0 \\ 0 \end{pmatrix} y + \begin{pmatrix} 1 \\ 0 \\ -1 \\ 1 \end{pmatrix} w \mid y, w \in \mathbb{R} \}$$

Thus the vector space of solutions is the span of a two-element set. This two-vector set is also linearly independent, which is easy to check. Therefore the solution set is a subspace of $\mathbb{R}^4$ with a basis comprised of these two vectors.

1.11 Example Parametrization finds bases for other vector spaces, not just for solution sets of homogeneous systems. To find a basis for this subspace of $\mathcal{M}_{2\times2}$

$$\{\begin{pmatrix} a & b \\ c & 0 \end{pmatrix} \mid a+b-2c=0\}$$

we rewrite the condition as $a=-b+2c$.

$$\{\begin{pmatrix} -b+2c & b \\ c & 0 \end{pmatrix} \mid b,c\in\mathbb{R}\}=\{b\begin{pmatrix} -1 & 1 \\ 0 & 0 \end{pmatrix}+c\begin{pmatrix} 2 & 0 \\ 1 & 0 \end{pmatrix} \mid b,c\in\mathbb{R}\}$$

Thus, this is a natural candidate for a basis.

$$\langle\begin{pmatrix} -1 & 1 \\ 0 & 0 \end{pmatrix},\begin{pmatrix} 2 & 0 \\ 1 & 0 \end{pmatrix}\rangle$$

The above work shows that it spans the space. Linear independence is also easy.

Consider again Example 1.2. To verify that the set spans the space we looked at linear combinations that total to a member of the space $c_1\vec{\beta}_1+c_2\vec{\beta}_2=\binom{x}{y}$. We only noted in that example that such a combination exists, that for each x,y there exists a c_1,c_2, but in fact the calculation also shows that the combination is unique: c_1 must be $(y-x)/2$ and c_2 must be $2x-y$.

1.12 Theorem In any vector space, a subset is a basis if and only if each vector in the space can be expressed as a linear combination of elements of the subset in one and only one way.

We consider linear combinations to be the same if they have the same summands but in a different order, or if they differ only in the addition or deletion of terms of the form '$0\cdot\vec{\beta}$'.

PROOF A sequence is a basis if and only if its vectors form a set that spans and that is linearly independent. A subset is a spanning set if and only if each vector in the space is a linear combination of elements of that subset in at least one way. Thus we need only show that a spanning subset is linearly independent if and only if every vector in the space is a linear combination of elements from the subset in at most one way.

Consider two expressions of a vector as a linear combination of the members of the subset. Rearrange the two sums, and if necessary add some $0\cdot\vec{\beta}_i$

terms, so that the two sums combine the same $\vec{\beta}$'s in the same order: $\vec{v} = c_1\vec{\beta}_1 + c_2\vec{\beta}_2 + \cdots + c_n\vec{\beta}_n$ and $\vec{v} = d_1\vec{\beta}_1 + d_2\vec{\beta}_2 + \cdots + d_n\vec{\beta}_n$. Now

$$c_1\vec{\beta}_1 + c_2\vec{\beta}_2 + \cdots + c_n\vec{\beta}_n = d_1\vec{\beta}_1 + d_2\vec{\beta}_2 + \cdots + d_n\vec{\beta}_n$$

holds if and only if

$$(c_1 - d_1)\vec{\beta}_1 + \cdots + (c_n - d_n)\vec{\beta}_n = \vec{0}$$

holds. So, asserting that each coefficient in the lower equation is zero is the same thing as asserting that $c_i = d_i$ for each i, that is, that every vector is expressible as a linear combination of the $\vec{\beta}$'s in a unique way. QED

1.13 Definition In a vector space with basis B the *representation of $\vec{v}$ with respect to* B is the column vector of the coefficients used to express $\vec{v}$ as a linear combination of the basis vectors:

$$\mathrm{Rep}_B(\vec{v}) = \begin{pmatrix} c_1 \\ c_2 \\ \vdots \\ c_n \end{pmatrix}$$

where $B = \langle\vec{\beta}_1, \ldots, \vec{\beta}_n\rangle$ and $\vec{v} = c_1\vec{\beta}_1 + c_2\vec{\beta}_2 + \cdots + c_n\vec{\beta}_n$. The c's are the *coordinates of $\vec{v}$ with respect to* B.

1.14 Example In $\mathcal{P}_3$, with respect to the basis $B = \langle 1, 2x, 2x^2, 2x^3\rangle$, the representation of $x + x^2$ is

$$\mathrm{Rep}_B(x + x^2) = \begin{pmatrix} 0 \\ 1/2 \\ 1/2 \\ 0 \end{pmatrix}_B$$

because $x + x^2 = 0 \cdot 1 + (1/2) \cdot 2x + (1/2) \cdot 2x^2 + 0 \cdot 2x^3$. With respect to a different basis $D = \langle 1 + x, 1 - x, x + x^2, x + x^3\rangle$, the representation is different.

$$\mathrm{Rep}_D(x + x^2) = \begin{pmatrix} 0 \\ 0 \\ 1 \\ 0 \end{pmatrix}_D$$

(When there is more than one basis around, to help keep straight which representation is with respect to which basis we often write it as a subscript on the column vector.)

1.15 Remark Definition 1.1 requires that a basis be a sequence because without that we couldn't write these coordinates in a fixed order.

1.16 Example In $\mathbb{R}^2$, where $\vec{v} = \binom{3}{2}$, to find the coordinates of that vector with respect to the basis

$$B = \langle \begin{pmatrix} 1 \\ 1 \end{pmatrix}, \begin{pmatrix} 0 \\ 2 \end{pmatrix} \rangle$$

we solve

$$c_1 \begin{pmatrix} 1 \\ 1 \end{pmatrix} + c_2 \begin{pmatrix} 0 \\ 2 \end{pmatrix} = \begin{pmatrix} 3 \\ 2 \end{pmatrix}$$

and get that $c_1 = 3$ and $c_2 = -1/2$.

$$\text{Rep}_B(\vec{v}) = \begin{pmatrix} 3 \\ -1/2 \end{pmatrix}$$

1.17 Remark This use of column notation and the term 'coordinate' has both a disadvantage and an advantage. The disadvantage is that representations look like vectors from $\mathbb{R}^n$, which can be confusing when the vector space is $\mathbb{R}^n$, as in the prior example. We must infer the intent from the context. For example, the phrase 'in $\mathbb{R}^2$, where $\vec{v} = \binom{3}{2}$' refers to the plane vector that, when in canonical position, ends at $(3, 2)$. And in the end of that example, although we've omitted a subscript B from the column, that the right side is a representation is clear from the context.

The advantage of the notation and the term is that they generalize the familiar case: in $\mathbb{R}^n$ and with respect to the standard basis $\mathcal{E}_n$, the vector starting at the origin and ending at $(v_1, \ldots, v_n)$ has this representation.

$$\text{Rep}_{\mathcal{E}_n}(\begin{pmatrix} v_1 \\ \vdots \\ v_n \end{pmatrix}) = \begin{pmatrix} v_1 \\ \vdots \\ v_n \end{pmatrix}_{\mathcal{E}_n}$$

Our main use of representations will come later but the definition appears here because the fact that every vector is a linear combination of basis vectors in a unique way is a crucial property of bases, and also to help make a point. For calculation of coordinates among other things, we shall restrict our attention to spaces with bases having only finitely many elements. That will start in the next subsection.

Exercises

1.18 Decide if each is a basis for $\mathcal{P}_2$.
(a) $\langle x^2 - x + 1, 2x + 1, 2x - 1 \rangle$ (b) $\langle x + x^2, x - x^2 \rangle$

✓ 1.19 Decide if each is a basis for $\mathbb{R}^3$.

(a) $\langle\begin{pmatrix}1\\2\\3\end{pmatrix},\begin{pmatrix}3\\2\\1\end{pmatrix},\begin{pmatrix}0\\0\\1\end{pmatrix}\rangle$ (b) $\langle\begin{pmatrix}1\\2\\3\end{pmatrix},\begin{pmatrix}3\\2\\1\end{pmatrix}\rangle$ (c) $\langle\begin{pmatrix}0\\2\\-1\end{pmatrix},\begin{pmatrix}1\\1\\1\end{pmatrix},\begin{pmatrix}2\\5\\0\end{pmatrix}\rangle$

(d) $\langle\begin{pmatrix}0\\2\\-1\end{pmatrix},\begin{pmatrix}1\\1\\1\end{pmatrix},\begin{pmatrix}1\\3\\0\end{pmatrix}\rangle$

✓ **1.20** Represent the vector with respect to the basis.

(a) $\begin{pmatrix}1\\2\end{pmatrix}$, $B=\langle\begin{pmatrix}1\\1\end{pmatrix},\begin{pmatrix}-1\\1\end{pmatrix}\rangle\subseteq\mathbb{R}^2$

(b) x^2+x^3, $D=\langle 1,1+x,1+x+x^2,1+x+x^2+x^3\rangle\subseteq\mathcal{P}_3$

(c) $\begin{pmatrix}0\\-1\\0\\1\end{pmatrix}$, $\mathcal{E}_4\subseteq\mathbb{R}^4$

1.21 Represent the vector with respect to each of the two bases.

$$\vec{v}=\begin{pmatrix}3\\-1\end{pmatrix}\quad B_1=\langle\begin{pmatrix}1\\-1\end{pmatrix},\begin{pmatrix}1\\1\end{pmatrix}\rangle,\ B_2=\langle\begin{pmatrix}1\\2\end{pmatrix},\begin{pmatrix}1\\3\end{pmatrix}\rangle$$

1.22 Find a basis for $\mathcal{P}_2$, the space of all quadratic polynomials. Must any such basis contain a polynomial of each degree: degree zero, degree one, and degree two?

1.23 Find a basis for the solution set of this system.

$$\begin{aligned}x_1-4x_2+3x_3-\ x_4&=0\\2x_1-8x_2+6x_3-2x_4&=0\end{aligned}$$

✓ **1.24** Find a basis for $\mathcal{M}_{2\times2}$, the space of 2×2 matrices.

✓ **1.25** Find a basis for each.

(a) The subspace $\{a_2x^2+a_1x+a_0 \mid a_2-2a_1=a_0\}$ of $\mathcal{P}_2$

(b) The space of three-wide row vectors whose first and second components add to zero

(c) This subspace of the 2×2 matrices

$$\{\begin{pmatrix}a&b\\0&c\end{pmatrix} \mid c-2b=0\}$$

1.26 Find a basis for each space, and verify that it is a basis.

(a) The subspace $M=\{a+bx+cx^2+dx^3 \mid a-2b+c-d=0\}$ of $\mathcal{P}_3$.

(b) This subspace of $\mathcal{M}_{2\times2}$.

$$W=\{\begin{pmatrix}a&b\\c&d\end{pmatrix} \mid a-c=0\}$$

1.27 Check Example 1.6.

✓ **1.28** Find the span of each set and then find a basis for that span.

(a) $\{1+x,1+2x\}$ in $\mathcal{P}_2$ (b) $\{2-2x,3+4x^2\}$ in $\mathcal{P}_2$

✓ **1.29** Find a basis for each of these subspaces of the space $\mathcal{P}_3$ of cubic polynomials.

(a) The subspace of cubic polynomials $p(x)$ such that $p(7)=0$

(b) The subspace of polynomials $p(x)$ such that $p(7)=0$ and $p(5)=0$

(c) The subspace of polynomials $p(x)$ such that $p(7)=0$, $p(5)=0$, and $p(3)=0$

(d) The space of polynomials $p(x)$ such that $p(7) = 0$, $p(5) = 0$, $p(3) = 0$, and $p(1) = 0$

1.30 We've seen that the result of reordering a basis can be another basis. Must it be?

1.31 Can a basis contain a zero vector?

✓ 1.32 Let $\langle \vec{\beta}_1, \vec{\beta}_2, \vec{\beta}_3 \rangle$ be a basis for a vector space.

(a) Show that $\langle c_1\vec{\beta}_1, c_2\vec{\beta}_2, c_3\vec{\beta}_3 \rangle$ is a basis when $c_1, c_2, c_3 \neq 0$. What happens when at least one c_i is 0?

(b) Prove that $\langle \vec{\alpha}_1, \vec{\alpha}_2, \vec{\alpha}_3 \rangle$ is a basis where $\vec{\alpha}_i = \vec{\beta}_1 + \vec{\beta}_i$.

1.33 Find one vector $\vec{v}$ that will make each into a basis for the space.

(a) $\langle \begin{pmatrix} 1 \\ 1 \end{pmatrix}, \vec{v} \rangle$ in $\mathbb{R}^2$ (b) $\langle \begin{pmatrix} 1 \\ 1 \\ 0 \end{pmatrix}, \begin{pmatrix} 0 \\ 1 \\ 0 \end{pmatrix}, \vec{v} \rangle$ in $\mathbb{R}^3$ (c) $\langle x, 1 + x^2, \vec{v} \rangle$ in $\mathcal{P}_2$

✓ 1.34 Where $\langle \vec{\beta}_1, \ldots, \vec{\beta}_n \rangle$ is a basis, show that in this equation

$$c_1\vec{\beta}_1 + \cdots + c_k\vec{\beta}_k = c_{k+1}\vec{\beta}_{k+1} + \cdots + c_n\vec{\beta}_n$$

each of the c_i's is zero. Generalize.

1.35 A basis contains some of the vectors from a vector space; can it contain them all?

1.36 Theorem 1.12 shows that, with respect to a basis, every linear combination is unique. If a subset is not a basis, can linear combinations be not unique? If so, must they be?

✓ 1.37 A square matrix is *symmetric* if for all indices i and j, entry i, j equals entry j, i.

(a) Find a basis for the vector space of symmetric 2×2 matrices.

(b) Find a basis for the space of symmetric 3×3 matrices.

(c) Find a basis for the space of symmetric $n\times n$ matrices.

✓ 1.38 We can show that every basis for $\mathbb{R}^3$ contains the same number of vectors.

(a) Show that no linearly independent subset of $\mathbb{R}^3$ contains more than three vectors.

(b) Show that no spanning subset of $\mathbb{R}^3$ contains fewer than three vectors. *Hint:* recall how to calculate the span of a set and show that this method cannot yield all of $\mathbb{R}^3$ when we apply it to fewer than three vectors.

1.39 One of the exercises in the Subspaces subsection shows that the set

$$\{ \begin{pmatrix} x \\ y \\ z \end{pmatrix} \mid x + y + z = 1 \}$$

is a vector space under these operations.

$$\begin{pmatrix} x_1 \\ y_1 \\ z_1 \end{pmatrix} + \begin{pmatrix} x_2 \\ y_2 \\ z_2 \end{pmatrix} = \begin{pmatrix} x_1 + x_2 - 1 \\ y_1 + y_2 \\ z_1 + z_2 \end{pmatrix} \qquad r \begin{pmatrix} x \\ y \\ z \end{pmatrix} = \begin{pmatrix} rx - r + 1 \\ ry \\ rz \end{pmatrix}$$

Find a basis.

III.2 Dimension

The previous subsection defines a basis of a vector space and shows that a space can have many different bases. So we cannot talk about "the" basis for a vector space. True, some vector spaces have bases that strike us as more natural than others, for instance, $\mathbb{R}^2$'s basis $\mathcal{E}_2$ or $\mathcal{P}_2$'s basis $\langle 1, x, x^2 \rangle$. But for the vector space $\{a_2x^2 + a_1x + a_0 \mid 2a_2 - a_0 = a_1\}$, no particular basis leaps out at us as the natural one. We cannot, in general, associate with a space any single basis that best describes it.

We can however find something about the bases that is uniquely associated with the space. This subsection shows that any two bases for a space have the same number of elements. So with each space we can associate a number, the number of vectors in any of its bases.

Before we start, we first limit our attention to spaces where at least one basis has only finitely many members.

2.1 Definition A vector space is *finite-dimensional* if it has a basis with only finitely many vectors.

One space that is not finite-dimensional is the set of polynomials with real coefficients Example 1.11; this space is not spanned by any finite subset since that would contain a polynomial of largest degree but this space has polynomials of all degrees. Such spaces are interesting and important but we will focus in a different direction. From now on we will study only finite-dimensional vector spaces. In the rest of this book we shall take 'vector space' to mean 'finite-dimensional vector space'.

2.2 Remark One reason for sticking to finite-dimensional spaces is so that the representation of a vector with respect to a basis is a finitely-tall vector and we can easily write it. Another reason is that the statement 'any infinite-dimensional vector space has a basis' is equivalent to a statement called the Axiom of Choice [Blass 1984] and so covering this would move us far past this book's scope. (A discussion of the Axiom of Choice is in the Frequently Asked Questions list for `sci.math`, and another accessible one is [Rucker].)

To prove the main theorem we shall use a technical result, the Exchange Lemma. We first illustrate it with an example.

2.3 Example Here is a basis for $\mathbb{R}^3$ and a vector given as a linear combination of members of that basis.

$$B = \langle \begin{pmatrix} 1 \\ 0 \\ 0 \end{pmatrix}, \begin{pmatrix} 1 \\ 1 \\ 0 \end{pmatrix}, \begin{pmatrix} 0 \\ 0 \\ 2 \end{pmatrix} \rangle \qquad \begin{pmatrix} 1 \\ 2 \\ 0 \end{pmatrix} = (-1) \cdot \begin{pmatrix} 1 \\ 0 \\ 0 \end{pmatrix} + 2 \begin{pmatrix} 1 \\ 1 \\ 0 \end{pmatrix} + 0 \cdot \begin{pmatrix} 0 \\ 0 \\ 2 \end{pmatrix}$$

Two of the basis vectors have non-zero coefficients. Pick one, for instance the first. Replace it with the vector that we've expressed as the combination

$$\hat{B} = \langle \begin{pmatrix} 1 \\ 2 \\ 0 \end{pmatrix}, \begin{pmatrix} 1 \\ 1 \\ 0 \end{pmatrix}, \begin{pmatrix} 0 \\ 0 \\ 2 \end{pmatrix} \rangle$$

and the result is another basis for $\mathbb{R}^3$.

2.4 Lemma (Exchange Lemma) Assume that $B = \langle \vec{\beta}_1, \ldots, \vec{\beta}_n \rangle$ is a basis for a vector space, and that for the vector $\vec{v}$ the relationship $\vec{v} = c_1\vec{\beta}_1 + c_2\vec{\beta}_2 + \cdots + c_n\vec{\beta}_n$ has $c_i \neq 0$. Then exchanging $\vec{\beta}_i$ for $\vec{v}$ yields another basis for the space.

Proof Call the outcome of the exchange $\hat{B} = \langle \vec{\beta}_1, \ldots, \vec{\beta}_{i-1}, \vec{v}, \vec{\beta}_{i+1}, \ldots, \vec{\beta}_n \rangle$.

We first show that $\hat{B}$ is linearly independent. Any relationship $d_1\vec{\beta}_1 + \cdots + d_i\vec{v} + \cdots + d_n\vec{\beta}_n = \vec{0}$ among the members of $\hat{B}$, after substitution for $\vec{v}$,

$$d_1\vec{\beta}_1 + \cdots + d_i \cdot (c_1\vec{\beta}_1 + \cdots + c_i\vec{\beta}_i + \cdots + c_n\vec{\beta}_n) + \cdots + d_n\vec{\beta}_n = \vec{0} \qquad (*)$$

gives a linear relationship among the members of B. The basis B is linearly independent so the coefficient $d_i c_i$ of $\vec{\beta}_i$ is zero. Because we assumed that c_i is nonzero, $d_i = 0$. Using this in equation $(*)$ gives that all of the other d's are also zero. Therefore $\hat{B}$ is linearly independent.

We finish by showing that $\hat{B}$ has the same span as B. Half of this argument, that $[\hat{B}] \subseteq [B]$, is easy; we can write any member $d_1\vec{\beta}_1+\cdots+d_i\vec{v}+\cdots+d_n\vec{\beta}_n$ of $[\hat{B}]$ as $d_1\vec{\beta}_1+\cdots+d_i\cdot(c_1\vec{\beta}_1+\cdots+c_n\vec{\beta}_n)+\cdots+d_n\vec{\beta}_n$, which is a linear combination of linear combinations of members of B, and hence is in $[B]$. For the $[B] \subseteq [\hat{B}]$ half of the argument, recall that if $\vec{v} = c_1\vec{\beta}_1+\cdots+c_n\vec{\beta}_n$ with $c_i \neq 0$ then we can rearrange the equation to $\vec{\beta}_i = (-c_1/c_i)\vec{\beta}_1 + \cdots + (1/c_i)\vec{v} + \cdots + (-c_n/c_i)\vec{\beta}_n$. Now, consider any member $d_1\vec{\beta}_1 + \cdots + d_i\vec{\beta}_i + \cdots + d_n\vec{\beta}_n$ of $[B]$, substitute for $\vec{\beta}_i$ its expression as a linear combination of the members of $\hat{B}$, and recognize, as in the first half of this argument, that the result is a linear combination of linear combinations of members of $\hat{B}$, and hence is in $[\hat{B}]$. QED

2.5 Theorem In any finite-dimensional vector space, all bases have the same number of elements.

Proof Fix a vector space with at least one finite basis. Choose, from among all of this space's bases, one $B = \langle \vec{\beta}_1, \ldots, \vec{\beta}_n \rangle$ of minimal size. We will show that any other basis $D = \langle \vec{\delta}_1, \vec{\delta}_2, \ldots \rangle$ also has the same number of members, n. Because B has minimal size, D has no fewer than n vectors. We will argue that it cannot have more than n vectors.

The basis B spans the space and $\vec{\delta}_1$ is in the space, so $\vec{\delta}_1$ is a nontrivial linear combination of elements of B. By the Exchange Lemma, we can swap $\vec{\delta}_1$ for a vector from B, resulting in a basis B_1, where one element is $\vec{\delta}_1$ and all of the $n-1$ other elements are $\vec{\beta}$'s.

The prior paragraph forms the basis step for an induction argument. The inductive step starts with a basis B_k (for $1 \leqslant k < n$) containing k members of D and $n-k$ members of B. We know that D has at least n members so there is a $\vec{\delta}_{k+1}$. Represent it as a linear combination of elements of B_k. The key point: in that representation, at least one of the nonzero scalars must be associated with a $\vec{\beta}_i$ or else that representation would be a nontrivial linear relationship among elements of the linearly independent set D. Exchange $\vec{\delta}_{k+1}$ for $\vec{\beta}_i$ to get a new basis B_{k+1} with one $\vec{\delta}$ more and one $\vec{\beta}$ fewer than the previous basis B_k.

Repeat that until no $\vec{\beta}$'s remain, so that B_n contains $\vec{\delta}_1, \ldots, \vec{\delta}_n$. Now, D cannot have more than these n vectors because any $\vec{\delta}_{n+1}$ that remains would be in the span of B_n (since it is a basis) and hence would be a linear combination of the other $\vec{\delta}$'s, contradicting that D is linearly independent. QED

2.6 Definition The *dimension* of a vector space is the number of vectors in any of its bases.

2.7 Example Any basis for $\mathbb{R}^n$ has n vectors since the standard basis $\mathcal{E}_n$ has n vectors. Thus, this definition of 'dimension' generalizes the most familiar use of term, that $\mathbb{R}^n$ is n-dimensional.

2.8 Example The space $\mathcal{P}_n$ of polynomials of degree at most n has dimension $n+1$. We can show this by exhibiting any basis — $\langle 1, x, \ldots, x^n \rangle$ comes to mind — and counting its members.

2.9 Example The space of functions $\{a \cdot \cos\theta + b \cdot \sin\theta \mid a, b \in \mathbb{R}\}$ of the real variable θ has dimension 2 since this space has the basis $\langle \cos\theta, \sin\theta \rangle$.

2.10 Example A trivial space is zero-dimensional since its basis is empty.

Again, although we sometimes say 'finite-dimensional' for emphasis, from now on we take all vector spaces to be finite-dimensional. So in the next result the word 'space' means 'finite-dimensional vector space'.

2.11 Corollary No linearly independent set can have a size greater than the dimension of the enclosing space.

PROOF The proof of Theorem 2.5 never uses that D spans the space, only that it is linearly independent. QED

2.12 Example Recall the diagram from Example I.2.19 showing the subspaces of $\mathbb{R}^3$. Each subspace is described with a minimal spanning set, a basis. The whole space has a basis with three members, the plane subspaces have bases with two members, the line subspaces have bases with one member, and the trivial subspace has a basis with zero members. We could not in that section show that these are $\mathbb{R}^3$'s only subspaces. We can show it now. The prior corollary proves that the only subspaces of $\mathbb{R}^3$ are either three-, two-, one-, or zero-dimensional. There are no subspaces somehow, say, between lines and planes.

2.13 Corollary Any linearly independent set can be expanded to make a basis.

Proof If a linearly independent set is not already a basis then it must not span the space. Adding to the set a vector that is not in the span will preserve linear independence by Lemma II.1.14. Keep adding until the resulting set does span the space, which the prior corollary shows will happen after only a finite number of steps. QED

2.14 Corollary Any spanning set can be shrunk to a basis.

Proof Call the spanning set S. If S is empty then it is already a basis (the space must be a trivial space). If $S = \{\vec{0}\}$ then it can be shrunk to the empty basis, thereby making it linearly independent, without changing its span.

Otherwise, S contains a vector $\vec{s}_1$ with $\vec{s}_1 \neq \vec{0}$ and we can form a basis $B_1 = \langle \vec{s}_1 \rangle$. If $[B_1] = [S]$ then we are done. If not then there is a $\vec{s}_2 \in [S]$ such that $\vec{s}_2 \notin [B_1]$. Let $B_2 = \langle \vec{s}_1, \vec{s}_2 \rangle$; by Lemma II.1.14 this is linearly independent so if $[B_2] = [S]$ then we are done.

We can repeat this process until the spans are equal, which must happen in at most finitely many steps. QED

2.15 Corollary In an n-dimensional space, a set composed of n vectors is linearly independent if and only if it spans the space.

Proof First we will show that a subset with n vectors is linearly independent if and only if it is a basis. The 'if' is trivially true — bases are linearly independent. 'Only if' holds because a linearly independent set can be expanded to a basis, but a basis has n elements, so this expansion is actually the set that we began with.

To finish, we will show that any subset with n vectors spans the space if and only if it is a basis. Again, 'if' is trivial. 'Only if' holds because any spanning set can be shrunk to a basis, but a basis has n elements and so this shrunken set is just the one we started with. QED

The main result of this subsection, that all of the bases in a finite-dimensional vector space have the same number of elements, is the single most important result in this book. As Example 2.12 shows, it describes what vector spaces and subspaces there can be.

One immediate consequence brings us back to when we considered the two things that could be meant by the term 'minimal spanning set'. At that point we defined 'minimal' as linearly independent but we noted that another reasonable interpretation of the term is that a spanning set is 'minimal' when it has the fewest number of elements of any set with the same span. Now that we have shown that all bases have the same number of elements, we know that the two senses of 'minimal' are equivalent.

Exercises

Assume that all spaces are finite-dimensional unless otherwise stated.

✓ **2.16** Find a basis for, and the dimension of, $\mathcal{P}_2$.

2.17 Find a basis for, and the dimension of, the solution set of this system.

$$\begin{aligned} x_1 - 4x_2 + 3x_3 - x_4 &= 0 \\ 2x_1 - 8x_2 + 6x_3 - 2x_4 &= 0 \end{aligned}$$

✓ **2.18** Find a basis for, and the dimension of, each space.

(a) $\{\begin{pmatrix} x \\ y \\ z \\ w \end{pmatrix} \in \mathbb{R}^4 \mid x - w + z = 0\}$

(b) the set of 5×5 matrices whose only nonzero entries are on the diagonal (e.g., in entry $1,1$ and $2,2$, etc.)

(c) $\{a_0 + a_1x + a_2x^2 + a_3x^3 \mid a_0 + a_1 = 0 \text{ and } a_2 - 2a_3 = 0\} \subseteq \mathcal{P}_3$

2.19 Find a basis for, and the dimension of, $\mathcal{M}_{2\times2}$, the vector space of 2×2 matrices.

2.20 Find the dimension of the vector space of matrices

$$\begin{pmatrix} a & b \\ c & d \end{pmatrix}$$

subject to each condition.

(a) $a, b, c, d \in \mathbb{R}$

(b) $a - b + 2c = 0$ and $d \in \mathbb{R}$

(c) $a + b + c = 0$, $a + b - c = 0$, and $d \in \mathbb{R}$

✓ **2.21** Find the dimension of this subspace of $\mathbb{R}^2$.

$$S = \{\begin{pmatrix} a+b \\ a+c \end{pmatrix} \mid a, b, c \in \mathbb{R}\}$$

✓ **2.22** Find the dimension of each.

(a) The space of cubic polynomials $p(x)$ such that $p(7) = 0$

(b) The space of cubic polynomials $p(x)$ such that $p(7) = 0$ and $p(5) = 0$

(c) The space of cubic polynomials $p(x)$ such that $p(7) = 0$, $p(5) = 0$, and $p(3) = 0$

(d) The space of cubic polynomials $p(x)$ such that $p(7) = 0$, $p(5) = 0$, $p(3) = 0$, and $p(1) = 0$

2.23 What is the dimension of the span of the set $\{\cos^2\theta, \sin^2\theta, \cos 2\theta, \sin 2\theta\}$? This span is a subspace of the space of all real-valued functions of one real variable.

2.24 Find the dimension of $\mathbb{C}^{47}$, the vector space of 47-tuples of complex numbers.

2.25 What is the dimension of the vector space $\mathcal{M}_{3\times5}$ of 3×5 matrices?

✓ **2.26** Show that this is a basis for $\mathbb{R}^4$.

$$\langle \begin{pmatrix}1\\0\\0\\0\end{pmatrix}, \begin{pmatrix}1\\1\\0\\0\end{pmatrix}, \begin{pmatrix}1\\1\\1\\0\end{pmatrix}, \begin{pmatrix}1\\1\\1\\1\end{pmatrix} \rangle$$

(We can use the results of this subsection to simplify this job.)

2.27 Refer to Example 2.12.

(a) Sketch a similar subspace diagram for $\mathcal{P}_2$.

(b) Sketch one for $\mathcal{M}_{2\times2}$.

✓ **2.28** Where S is a set, the functions $f\colon S \to \mathbb{R}$ form a vector space under the natural operations: the sum $f + g$ is the function given by $f + g\,(s) = f(s) + g(s)$ and the scalar product is $r \cdot f\,(s) = r \cdot f(s)$. What is the dimension of the space resulting for each domain?

(a) $S = \{1\}$ **(b)** $S = \{1, 2\}$ **(c)** $S = \{1, \ldots, n\}$

2.29 (See Exercise 28.) Prove that this is an infinite-dimensional space: the set of all functions $f\colon \mathbb{R} \to \mathbb{R}$ under the natural operations.

2.30 (See Exercise 28.) What is the dimension of the vector space of functions $f\colon S \to \mathbb{R}$, under the natural operations, where the domain S is the empty set?

2.31 Show that any set of four vectors in $\mathbb{R}^2$ is linearly dependent.

2.32 Show that $\langle \vec{\alpha}_1, \vec{\alpha}_2, \vec{\alpha}_3 \rangle \subset \mathbb{R}^3$ is a basis if and only if there is no plane through the origin containing all three vectors.

2.33 Prove that any subspace of a finite dimensional space is finite dimensional.

2.34 Where is the finiteness of B used in Theorem 2.5?

2.35 Prove that if U and W are both three-dimensional subspaces of $\mathbb{R}^5$ then $U \cap W$ is non-trivial. Generalize.

2.36 A basis for a space consists of elements of that space. So we are naturally led to how the property 'is a basis' interacts with operations $\subseteq$ and $\cap$ and $\cup$. (Of course, a basis is actually a sequence in that it is ordered, but there is a natural extension of these operations.)

(a) Consider first how bases might be related by $\subseteq$. Assume that U, W are subspaces of some vector space and that $U \subseteq W$. Can there exist bases B_U for U and B_W for W such that $B_U \subseteq B_W$? Must such bases exist?

For any basis B_U for U, must there be a basis B_W for W such that $B_U \subseteq B_W$?

For any basis B_W for W, must there be a basis B_U for U such that $B_U \subseteq B_W$?

For any bases B_U, B_W for U and W, must B_U be a subset of B_W?

(b) Is the $\cap$ of bases a basis? For what space?

(c) Is the $\cup$ of bases a basis? For what space?

(d) What about the complement operation?

(*Hint.* Test any conjectures against some subspaces of $\mathbb{R}^3$.)

✓ **2.37** Consider how 'dimension' interacts with 'subset'. Assume U and W are both subspaces of some vector space, and that $U \subseteq W$.

(a) Prove that $\dim(U) \leqslant \dim(W)$.

(b) Prove that equality of dimension holds if and only if $U = W$.

(c) Show that the prior item does not hold if they are infinite-dimensional.

? **2.38** [Wohascum no. 47] For any vector $\vec{v}$ in $\mathbb{R}^n$ and any permutation σ of the numbers $1, 2, \ldots, n$ (that is, σ is a rearrangement of those numbers into a new order), define $\sigma(\vec{v})$ to be the vector whose components are $v_{\sigma(1)}, v_{\sigma(2)}, \ldots,$ and $v_{\sigma(n)}$ (where $\sigma(1)$ is the first number in the rearrangement, etc.). Now fix $\vec{v}$ and let V be the span of $\{\sigma(\vec{v}) \mid \sigma \text{ permutes } 1, \ldots, n\}$. What are the possibilities for the dimension of V?

III.3 Vector Spaces and Linear Systems

We will now reconsider linear systems and Gauss's Method, aided by the tools and terms of this chapter. We will make three points.

For the first, recall the insight from the Chapter One that Gauss's Method works by taking linear combinations of rows—if two matrices are related by row operations $A \longrightarrow \cdots \longrightarrow B$ then each row of B is a linear combination of the rows of A. Therefore, the right setting in which to study row operations in general, and Gauss's Method in particular, is the following vector space.

3.1 Definition The *row space* of a matrix is the span of the set of its rows. The *row rank* is the dimension of this space, the number of linearly independent rows.

3.2 Example If

$$A = \begin{pmatrix} 2 & 3 \\ 4 & 6 \end{pmatrix}$$

then $\text{Rowspace}(A)$ is this subspace of the space of two-component row vectors.

$$\{c_1 \cdot (2 \;\; 3) + c_2 \cdot (4 \;\; 6) \mid c_1, c_2 \in \mathbb{R}\}$$

The second row vector is linearly dependent on the first and so we can simplify the above description to $\{c \cdot (2 \;\; 3) \mid c \in \mathbb{R}\}$.

3.3 Lemma If two matrices A and B are related by a row operation

$$A \xrightarrow{\rho_i \leftrightarrow \rho_j} B \quad \text{or} \quad A \xrightarrow{k\rho_i} B \quad \text{or} \quad A \xrightarrow{k\rho_i + \rho_j} B$$

(for $i \neq j$ and $k \neq 0$) then their row spaces are equal. Hence, row-equivalent matrices have the same row space and therefore the same row rank.

Proof Corollary One.III.2.4 shows that when $A \longrightarrow B$ then each row of B is a linear combination of the rows of A. That is, in the above terminology, each row of B is an element of the row space of A. Then $\text{Rowspace}(B) \subseteq \text{Rowspace}(A)$ follows because a member of the set $\text{Rowspace}(B)$ is a linear combination of the rows of B, so it is a combination of combinations of the rows of A, and by the Linear Combination Lemma is also a member of $\text{Rowspace}(A)$.

For the other set containment, recall Lemma One.III.1.5, that row operations are reversible so $A \longrightarrow B$ if and only if $B \longrightarrow A$. Then $\text{Rowspace}(A) \subseteq \text{Rowspace}(B)$ follows as in the previous paragraph. QED

Of course, Gauss's Method performs the row operations systematically, with the goal of echelon form.

3.4 Lemma The nonzero rows of an echelon form matrix make up a linearly independent set.

Proof Lemma One.III.2.5 says that no nonzero row of an echelon form matrix is a linear combination of the other rows. This result just restates that in this chapter's terminology. QED

Thus, in the language of this chapter, Gaussian reduction works by eliminating linear dependences among rows, leaving the span unchanged, until no nontrivial linear relationships remain among the nonzero rows. In short, Gauss's Method produces a basis for the row space.

3.5 Example From any matrix, we can produce a basis for the row space by performing Gauss's Method and taking the nonzero rows of the resulting echelon form matrix. For instance,

$$\begin{pmatrix} 1 & 3 & 1 \\ 1 & 4 & 1 \\ 2 & 0 & 5 \end{pmatrix} \xrightarrow[-2\rho_1+\rho_3]{-\rho_1+\rho_2} \xrightarrow{6\rho_2+\rho_3} \begin{pmatrix} 1 & 3 & 1 \\ 0 & 1 & 0 \\ 0 & 0 & 3 \end{pmatrix}$$

produces the basis $\langle (1\ \ 3\ \ 1), (0\ \ 1\ \ 0), (0\ \ 0\ \ 3) \rangle$ for the row space. This is a basis for the row space of both the starting and ending matrices, since the two row spaces are equal.

Using this technique, we can also find bases for spans not directly involving row vectors.

3.6 Definition The *column space* of a matrix is the span of the set of its columns. The *column rank* is the dimension of the column space, the number of linearly independent columns.

Our interest in column spaces stems from our study of linear systems. An example is that this system

$$\begin{aligned} c_1 + 3c_2 + 7c_3 &= d_1 \\ 2c_1 + 3c_2 + 8c_3 &= d_2 \\ c_2 + 2c_3 &= d_3 \\ 4c_1 \qquad + 4c_3 &= d_4 \end{aligned}$$

has a solution if and only if the vector of d's is a linear combination of the other column vectors,

$$c_1\begin{pmatrix}1\\2\\0\\4\end{pmatrix} + c_2\begin{pmatrix}3\\3\\1\\0\end{pmatrix} + c_3\begin{pmatrix}7\\8\\2\\4\end{pmatrix} = \begin{pmatrix}d_1\\d_2\\d_3\\d_4\end{pmatrix}$$

meaning that the vector of d's is in the column space of the matrix of coefficients.

3.7 Example Given this matrix,

$$\begin{pmatrix}1&3&7\\2&3&8\\0&1&2\\4&0&4\end{pmatrix}$$

to get a basis for the column space, temporarily turn the columns into rows and reduce.

$$\begin{pmatrix}1&2&0&4\\3&3&1&0\\7&8&2&4\end{pmatrix} \xrightarrow[-7\rho_1+\rho_3]{-3\rho_1+\rho_2} \xrightarrow{-2\rho_2+\rho_3} \begin{pmatrix}1&2&0&4\\0&-3&1&-12\\0&0&0&0\end{pmatrix}$$

Now turn the rows back to columns.

$$\langle \begin{pmatrix}1\\2\\0\\4\end{pmatrix}, \begin{pmatrix}0\\-3\\1\\-12\end{pmatrix} \rangle$$

The result is a basis for the column space of the given matrix.

3.8 Definition The *transpose* of a matrix is the result of interchanging its rows and columns, so that column j of the matrix A is row j of A^T and vice versa.

So we can summarize the prior example as "transpose, reduce, and transpose back."

We can even, at the price of tolerating the as-yet-vague idea of vector spaces being "the same," use Gauss's Method to find bases for spans in other types of vector spaces.

3.9 Example To get a basis for the span of $\{x^2+x^4, 2x^2+3x^4, -x^2-3x^4\}$ in the space $\mathcal{P}_4$, think of these three polynomials as "the same" as the row vectors $(0\ 0\ 1\ 0\ 1)$, $(0\ 0\ 2\ 0\ 3)$, and $(0\ 0\ -1\ 0\ -3)$, apply Gauss's Method

$$\begin{pmatrix} 0 & 0 & 1 & 0 & 1 \\ 0 & 0 & 2 & 0 & 3 \\ 0 & 0 & -1 & 0 & -3 \end{pmatrix} \xrightarrow[\rho_1+\rho_3]{-2\rho_1+\rho_2} \xrightarrow{2\rho_2+\rho_3} \begin{pmatrix} 0 & 0 & 1 & 0 & 1 \\ 0 & 0 & 0 & 0 & 1 \\ 0 & 0 & 0 & 0 & 0 \end{pmatrix}$$

and translate back to get the basis $\langle x^2+x^4, x^4\rangle$. (As mentioned earlier, we will make the phrase "the same" precise at the start of the next chapter.)

Thus, the first point for this subsection is that the tools of this chapter give us a more conceptual understanding of Gaussian reduction.

For the second point observe that row operations on a matrix can change its column space.

$$\begin{pmatrix} 1 & 2 \\ 2 & 4 \end{pmatrix} \xrightarrow{-2\rho_1+\rho_2} \begin{pmatrix} 1 & 2 \\ 0 & 0 \end{pmatrix}$$

The column space of the left-hand matrix contains vectors with a second component that is nonzero but the column space of the right-hand matrix contains only vectors whose second component is zero, so the two spaces are different. This observation makes next result surprising.

3.10 Lemma Row operations do not change the column rank.

PROOF Restated, if A reduces to B then the column rank of B equals the column rank of A.

This proof will be finished if we show that row operations do not affect linear relationships among columns, because the column rank is the size of the largest set of unrelated columns. That is, we will show that a relationship exists among columns (such as that the fifth column is twice the second plus the fourth) if and only if that relationship exists after the row operation. But this is exactly the first theorem of this book, Theorem One.I.1.5: in a relationship among columns,

$$c_1 \cdot \begin{pmatrix} a_{1,1} \\ a_{2,1} \\ \vdots \\ a_{m,1} \end{pmatrix} + \cdots + c_n \cdot \begin{pmatrix} a_{1,n} \\ a_{2,n} \\ \vdots \\ a_{m,n} \end{pmatrix} = \begin{pmatrix} 0 \\ 0 \\ \vdots \\ 0 \end{pmatrix}$$

row operations leave unchanged the set of solutions $(c_1, \ldots, c_n)$. QED

Another way to make the point that Gauss's Method has something to say about the column space as well as about the row space is with Gauss-Jordan reduction. It ends with the reduced echelon form of a matrix, as here.

$$\begin{pmatrix} 1 & 3 & 1 & 6 \\ 2 & 6 & 3 & 16 \\ 1 & 3 & 1 & 6 \end{pmatrix} \longrightarrow \cdots \longrightarrow \begin{pmatrix} 1 & 3 & 0 & 2 \\ 0 & 0 & 1 & 4 \\ 0 & 0 & 0 & 0 \end{pmatrix}$$

Consider the row space and the column space of this result.

The first point made earlier in this subsection says that to get a basis for the row space we can just collect the rows with leading entries. However, because this is in reduced echelon form, a basis for the column space is just as easy: collect the columns containing the leading entries, $\langle \vec{e}_1, \vec{e}_2 \rangle$. Thus, for a reduced echelon form matrix we can find bases for the row and column spaces in essentially the same way, by taking the parts of the matrix, the rows or columns, containing the leading entries.

3.11 Theorem For any matrix, the row rank and column rank are equal.

Proof Bring the matrix to reduced echelon form. Then the row rank equals the number of leading entries since that equals the number of nonzero rows. Then also, the number of leading entries equals the column rank because the set of columns containing leading entries consists of some of the $\vec{e}_i$'s from a standard basis, and that set is linearly independent and spans the set of columns. Hence, in the reduced echelon form matrix, the row rank equals the column rank, because each equals the number of leading entries.

But Lemma 3.3 and Lemma 3.10 show that the row rank and column rank are not changed by using row operations to get to reduced echelon form. Thus the row rank and the column rank of the original matrix are also equal. QED

3.12 Definition The *rank* of a matrix is its row rank or column rank.

So the second point that we have made in this subsection is that the column space and row space of a matrix have the same dimension.

Our final point is that the concepts that we've seen arising naturally in the study of vector spaces are exactly the ones that we have studied with linear systems.

3.13 Theorem For linear systems with n unknowns and with matrix of coefficients A, the statements

(1) the rank of A is r

(2) the vector space of solutions of the associated homogeneous system has dimension $n - r$

are equivalent.

So if the system has at least one particular solution then for the set of solutions, the number of parameters equals $n - r$, the number of variables minus the rank of the matrix of coefficients.

Proof The rank of A is r if and only if Gaussian reduction on A ends with r nonzero rows. That's true if and only if echelon form matrices row equivalent to A have r-many leading variables. That in turn holds if and only if there are $n - r$ free variables. QED

3.14 Corollary Where the matrix A is $n \times n$, these statements

(1) the rank of A is n

(2) A is nonsingular

(3) the rows of A form a linearly independent set

(4) the columns of A form a linearly independent set

(5) any linear system whose matrix of coefficients is A has one and only one solution

are equivalent.

Proof Clearly (1) $\iff$ (2) $\iff$ (3) $\iff$ (4). The last, (4) $\iff$ (5), holds because a set of n column vectors is linearly independent if and only if it is a basis for $\mathbb{R}^n$, but the system

$$c_1 \begin{pmatrix} a_{1,1} \\ a_{2,1} \\ \vdots \\ a_{m,1} \end{pmatrix} + \cdots + c_n \begin{pmatrix} a_{1,n} \\ a_{2,n} \\ \vdots \\ a_{m,n} \end{pmatrix} = \begin{pmatrix} d_1 \\ d_2 \\ \vdots \\ d_m \end{pmatrix}$$

has a unique solution for all choices of $d_1, \ldots, d_n \in \mathbb{R}$ if and only if the vectors of a's on the left form a basis. QED

3.15 Remark [Munkres] Sometimes the results of this subsection are mistakenly remembered to say that the general solution of an n unknowns system of

m equations uses $n-m$ parameters. The number of equations is not the relevant figure, rather, what matters is the number of independent equations, the number of equations in a maximal independent set. Where there are r independent equations, the general solution involves $n-r$ parameters.

Exercises

3.16 Transpose each.

(a) $\begin{pmatrix} 2 & 1 \\ 3 & 1 \end{pmatrix}$ (b) $\begin{pmatrix} 2 & 1 \\ 1 & 3 \end{pmatrix}$ (c) $\begin{pmatrix} 1 & 4 & 3 \\ 6 & 7 & 8 \end{pmatrix}$ (d) $\begin{pmatrix} 0 \\ 0 \\ 0 \end{pmatrix}$

(e) $\begin{pmatrix} -1 & -2 \end{pmatrix}$

✓ **3.17** Decide if the vector is in the row space of the matrix.

(a) $\begin{pmatrix} 2 & 1 \\ 3 & 1 \end{pmatrix}$, $\begin{pmatrix} 1 & 0 \end{pmatrix}$ (b) $\begin{pmatrix} 0 & 1 & 3 \\ -1 & 0 & 1 \\ -1 & 2 & 7 \end{pmatrix}$, $\begin{pmatrix} 1 & 1 & 1 \end{pmatrix}$

✓ **3.18** Decide if the vector is in the column space.

(a) $\begin{pmatrix} 1 & 1 \\ 1 & 1 \end{pmatrix}$, $\begin{pmatrix} 1 \\ 3 \end{pmatrix}$ (b) $\begin{pmatrix} 1 & 3 & 1 \\ 2 & 0 & 4 \\ 1 & -3 & -3 \end{pmatrix}$, $\begin{pmatrix} 1 \\ 0 \\ 0 \end{pmatrix}$

✓ **3.19** Decide if the vector is in the column space of the matrix.

(a) $\begin{pmatrix} 2 & 1 \\ 2 & 5 \end{pmatrix}$, $\begin{pmatrix} 1 \\ -3 \end{pmatrix}$ (b) $\begin{pmatrix} 4 & -8 \\ 2 & -4 \end{pmatrix}$, $\begin{pmatrix} 0 \\ 1 \end{pmatrix}$ (c) $\begin{pmatrix} 1 & -1 & 1 \\ 1 & 1 & -1 \\ -1 & -1 & 1 \end{pmatrix}$, $\begin{pmatrix} 2 \\ 0 \\ 0 \end{pmatrix}$

✓ **3.20** Find a basis for the row space of this matrix.

$$\begin{pmatrix} 2 & 0 & 3 & 4 \\ 0 & 1 & 1 & -1 \\ 3 & 1 & 0 & 2 \\ 1 & 0 & -4 & -1 \end{pmatrix}$$

✓ **3.21** Find the rank of each matrix.

(a) $\begin{pmatrix} 2 & 1 & 3 \\ 1 & -1 & 2 \\ 1 & 0 & 3 \end{pmatrix}$ (b) $\begin{pmatrix} 1 & -1 & 2 \\ 3 & -3 & 6 \\ -2 & 2 & -4 \end{pmatrix}$ (c) $\begin{pmatrix} 1 & 3 & 2 \\ 5 & 1 & 1 \\ 6 & 4 & 3 \end{pmatrix}$

(d) $\begin{pmatrix} 0 & 0 & 0 \\ 0 & 0 & 0 \\ 0 & 0 & 0 \end{pmatrix}$

3.22 Give a basis for the column space of this matrix. Give the matrix's rank.

$$\begin{pmatrix} 1 & 3 & -1 & 2 \\ 2 & 1 & 1 & 0 \\ 0 & 1 & 1 & 4 \end{pmatrix}$$

✓ **3.23** Find a basis for the span of each set.

(a) $\{\begin{pmatrix} 1 & 3 \end{pmatrix}, \begin{pmatrix} -1 & 3 \end{pmatrix}, \begin{pmatrix} 1 & 4 \end{pmatrix}, \begin{pmatrix} 2 & 1 \end{pmatrix}\} \subseteq \mathcal{M}_{1\times 2}$

(b) $\{\begin{pmatrix} 1 \\ 2 \\ 1 \end{pmatrix}, \begin{pmatrix} 3 \\ 1 \\ -1 \end{pmatrix}, \begin{pmatrix} 1 \\ -3 \\ -3 \end{pmatrix}\} \subseteq \mathbb{R}^3$

(c) $\{1+x, 1-x^2, 3+2x-x^2\} \subseteq \mathcal{P}_3$

(d) $\{\begin{pmatrix}1&0&1\\3&1&-1\end{pmatrix}, \begin{pmatrix}1&0&3\\2&1&4\end{pmatrix}, \begin{pmatrix}-1&0&-5\\-1&-1&-9\end{pmatrix}\} \subseteq \mathcal{M}_{2\times 3}$

3.24 Give a basis for the span of each set, in the natural vector space.

(a) $\{\begin{pmatrix}1\\1\\3\end{pmatrix}, \begin{pmatrix}-1\\2\\0\end{pmatrix}, \begin{pmatrix}0\\12\\6\end{pmatrix}\}$

(b) $\{x+x^2, 2-2x, 7, 4+3x+2x^2\}$

3.25 Which matrices have rank zero? Rank one?

✓ **3.26** Given $a, b, c \in \mathbb{R}$, what choice of d will cause this matrix to have the rank of one?

$$\begin{pmatrix}a&b\\c&d\end{pmatrix}$$

3.27 Find the column rank of this matrix.

$$\begin{pmatrix}1&3&-1&5&0&4\\2&0&1&0&4&1\end{pmatrix}$$

3.28 Show that a linear system with at least one solution has at most one solution if and only if the matrix of coefficients has rank equal to the number of its columns.

✓ **3.29** If a matrix is 5×9, which set must be dependent, its set of rows or its set of columns?

3.30 Give an example to show that, despite that they have the same dimension, the row space and column space of a matrix need not be equal. Are they ever equal?

3.31 Show that the set $\{(1,-1,2,-3), (1,1,2,0), (3,-1,6,-6)\}$ does not have the same span as $\{(1,0,1,0), (0,2,0,3)\}$. What, by the way, is the vector space?

✓ **3.32** Show that this set of column vectors

$$\{\begin{pmatrix}d_1\\d_2\\d_3\end{pmatrix} \mid \text{there are } x,\ y, \text{ and } z \text{ such that: } \begin{aligned}3x+2y+4z&=d_1\\ x\qquad\ \ -z&=d_2\\ 2x+2y+5z&=d_3\end{aligned}\}$$

is a subspace of $\mathbb{R}^3$. Find a basis.

3.33 Show that the transpose operation is linear:

$$(rA+sB)^{\mathsf{T}} = rA^{\mathsf{T}} + sB^{\mathsf{T}}$$

for $r, s \in \mathbb{R}$ and $A, B \in \mathcal{M}_{m\times n}$.

✓ **3.34** In this subsection we have shown that Gaussian reduction finds a basis for the row space.

(a) Show that this basis is not unique—different reductions may yield different bases.

(b) Produce matrices with equal row spaces but unequal numbers of rows.

(c) Prove that two matrices have equal row spaces if and only if after Gauss-Jordan reduction they have the same nonzero rows.

3.35 Why is there not a problem with Remark 3.15 in the case that r is bigger than n?

3.36 Show that the row rank of an $m\times n$ matrix is at most m. Is there a better bound?

✓ **3.37** Show that the rank of a matrix equals the rank of its transpose.

3.38 True or false: the column space of a matrix equals the row space of its transpose.

✓ **3.39** We have seen that a row operation may change the column space. Must it?

3.40 Prove that a linear system has a solution if and only if that system's matrix of coefficients has the same rank as its augmented matrix.

3.41 An $m \times n$ matrix has *full row rank* if its row rank is m, and it has *full column rank* if its column rank is n.

(a) Show that a matrix can have both full row rank and full column rank only if it is square.

(b) Prove that the linear system with matrix of coefficients A has a solution for any $d_1, \ldots, d_n$'s on the right side if and only if A has full row rank.

(c) Prove that a homogeneous system has a unique solution if and only if its matrix of coefficients A has full column rank.

(d) Prove that the statement "if a system with matrix of coefficients A has any solution then it has a unique solution" holds if and only if A has full column rank.

3.42 How would the conclusion of Lemma 3.3 change if Gauss's Method were changed to allow multiplying a row by zero?

✓ **3.43** What is the relationship between $\text{rank}(A)$ and $\text{rank}(-A)$? Between $\text{rank}(A)$ and $\text{rank}(kA)$? What, if any, is the relationship between $\text{rank}(A)$, $\text{rank}(B)$, and $\text{rank}(A+B)$?

III.4 Combining Subspaces

This subsection is optional. It is required only for the last sections of Chapter Three and Chapter Five and for occasional exercises. You can pass it over without loss of continuity.

One way to understand something is to see how to build it from component parts. For instance, we sometimes think of $\mathbb{R}^3$ put together from the x-axis, the y-axis, and z-axis. In this subsection we will describe how to decompose a vector space into a combination of some of its subspaces. In developing this idea of subspace combination, we will keep the $\mathbb{R}^3$ example in mind as a prototype.

Subspaces are subsets and sets combine via union. But taking the combination operation for subspaces to be the simple set union operation isn't what we want. For instance, the union of the x-axis, the y-axis, and z-axis is not all of $\mathbb{R}^3$. In fact this union is not a subspace because it is not closed under addition: this vector

$$\begin{pmatrix} 1 \\ 0 \\ 0 \end{pmatrix} + \begin{pmatrix} 0 \\ 1 \\ 0 \end{pmatrix} + \begin{pmatrix} 0 \\ 0 \\ 1 \end{pmatrix} = \begin{pmatrix} 1 \\ 1 \\ 1 \end{pmatrix}$$

is in none of the three axes and hence is not in the union. Therefore to combine subspaces, in addition to the members of those subspaces, we must at least also include all of their linear combinations.

4.1 Definition Where $W_1, \ldots, W_k$ are subspaces of a vector space, their *sum* is the span of their union $W_1 + W_2 + \cdots + W_k = [W_1 \cup W_2 \cup \cdots W_k]$.

Writing '+' fits with the conventional practice of using this symbol for a natural accumulation operation.

4.2 Example Our $\mathbb{R}^3$ prototype works with this. Any vector $\vec{w} \in \mathbb{R}^3$ is a linear combination $c_1\vec{v}_1 + c_2\vec{v}_2 + c_3\vec{v}_3$ where $\vec{v}_1$ is a member of the x-axis, etc., in this way

$$\begin{pmatrix} w_1 \\ w_2 \\ w_3 \end{pmatrix} = 1 \cdot \begin{pmatrix} w_1 \\ 0 \\ 0 \end{pmatrix} + 1 \cdot \begin{pmatrix} 0 \\ w_2 \\ 0 \end{pmatrix} + 1 \cdot \begin{pmatrix} 0 \\ 0 \\ w_3 \end{pmatrix}$$

and so $x\text{-axis} + y\text{-axis} + z\text{-axis} = \mathbb{R}^3$.

4.3 Example A sum of subspaces can be less than the entire space. Inside of $\mathcal{P}_4$, let L be the subspace of linear polynomials $\{a + bx \mid a, b \in \mathbb{R}\}$ and let C be the subspace of purely-cubic polynomials $\{cx^3 \mid c \in \mathbb{R}\}$. Then $L + C$ is not all of $\mathcal{P}_4$. Instead, $L + C = \{a + bx + cx^3 \mid a, b, c \in \mathbb{R}\}$.

4.4 Example A space can be described as a combination of subspaces in more than one way. Besides the decomposition $\mathbb{R}^3 = x\text{-axis} + y\text{-axis} + z\text{-axis}$, we can also write $\mathbb{R}^3 = xy\text{-plane} + yz\text{-plane}$. To check this, note that any $\vec{w} \in \mathbb{R}^3$ can be written as a linear combination of a member of the xy-plane and a member of the yz-plane; here are two such combinations.

$$\begin{pmatrix} w_1 \\ w_2 \\ w_3 \end{pmatrix} = 1 \cdot \begin{pmatrix} w_1 \\ w_2 \\ 0 \end{pmatrix} + 1 \cdot \begin{pmatrix} 0 \\ 0 \\ w_3 \end{pmatrix} \qquad \begin{pmatrix} w_1 \\ w_2 \\ w_3 \end{pmatrix} = 1 \cdot \begin{pmatrix} w_1 \\ w_2/2 \\ 0 \end{pmatrix} + 1 \cdot \begin{pmatrix} 0 \\ w_2/2 \\ w_3 \end{pmatrix}$$

The above definition gives one way in which we can think of a space as a combination of some of its parts. However, the prior example shows that there is at least one interesting property of our benchmark model that is not captured by the definition of the sum of subspaces. In the familiar decomposition of $\mathbb{R}^3$, we often speak of a vector's 'x part' or 'y part' or 'z part'. That is, in our prototype each vector has a unique decomposition into pieces from the parts making up the whole space. But in the decomposition used in Example 4.4, we cannot refer to the "xy part" of a vector — these three sums

$$\begin{pmatrix} 1 \\ 2 \\ 3 \end{pmatrix} = \begin{pmatrix} 1 \\ 2 \\ 0 \end{pmatrix} + \begin{pmatrix} 0 \\ 0 \\ 3 \end{pmatrix} = \begin{pmatrix} 1 \\ 0 \\ 0 \end{pmatrix} + \begin{pmatrix} 0 \\ 2 \\ 3 \end{pmatrix} = \begin{pmatrix} 1 \\ 1 \\ 0 \end{pmatrix} + \begin{pmatrix} 0 \\ 1 \\ 3 \end{pmatrix}$$

all describe the vector as comprised of something from the first plane plus something from the second plane, but the "xy part" is different in each.

That is, when we consider how $\mathbb{R}^3$ is put together from the three axes we might mean "in such a way that every vector has at least one decomposition," which gives the definition above. But if we take it to mean "in such a way that every vector has one and only one decomposition" then we need another condition on combinations. To see what this condition is, recall that vectors are uniquely represented in terms of a basis. We can use this to break a space into a sum of subspaces such that any vector in the space breaks uniquely into a sum of members of those subspaces.

4.5 Example Consider $\mathbb{R}^3$ with its standard basis $\mathcal{E}_3 = \langle \vec{e}_1, \vec{e}_2, \vec{e}_3 \rangle$. The subspace with the basis $B_1 = \langle \vec{e}_1 \rangle$ is the x-axis, the subspace with the basis $B_2 = \langle \vec{e}_2 \rangle$ is the y-axis, and the subspace with the basis $B_3 = \langle \vec{e}_3 \rangle$ is the z-axis. The fact that any member of $\mathbb{R}^3$ is expressible as a sum of vectors from these subspaces

$$\begin{pmatrix} x \\ y \\ z \end{pmatrix} = \begin{pmatrix} x \\ 0 \\ 0 \end{pmatrix} + \begin{pmatrix} 0 \\ y \\ 0 \end{pmatrix} + \begin{pmatrix} 0 \\ 0 \\ z \end{pmatrix}$$

reflects the fact that $\mathcal{E}_3$ spans the space—this equation

$$\begin{pmatrix} x \\ y \\ z \end{pmatrix} = c_1 \begin{pmatrix} 1 \\ 0 \\ 0 \end{pmatrix} + c_2 \begin{pmatrix} 0 \\ 1 \\ 0 \end{pmatrix} + c_3 \begin{pmatrix} 0 \\ 0 \\ 1 \end{pmatrix}$$

has a solution for any $x, y, z \in \mathbb{R}$. And the fact that each such expression is unique reflects that fact that $\mathcal{E}_3$ is linearly independent, so any equation like the one above has a unique solution.

4.6 Example We don't have to take the basis vectors one at a time, we can conglomerate them into larger sequences. Consider again the space $\mathbb{R}^3$ and the vectors from the standard basis $\mathcal{E}_3$. The subspace with the basis $B_1 = \langle \vec{e}_1, \vec{e}_3 \rangle$ is the xz-plane. The subspace with the basis $B_2 = \langle \vec{e}_2 \rangle$ is the y-axis. As in the prior example, the fact that any member of the space is a sum of members of the two subspaces in one and only one way

$$\begin{pmatrix} x \\ y \\ z \end{pmatrix} = \begin{pmatrix} x \\ 0 \\ z \end{pmatrix} + \begin{pmatrix} 0 \\ y \\ 0 \end{pmatrix}$$

is a reflection of the fact that these vectors form a basis—this equation

$$\begin{pmatrix} x \\ y \\ z \end{pmatrix} = (c_1 \begin{pmatrix} 1 \\ 0 \\ 0 \end{pmatrix} + c_3 \begin{pmatrix} 0 \\ 0 \\ 1 \end{pmatrix}) + c_2 \begin{pmatrix} 0 \\ 1 \\ 0 \end{pmatrix}$$

has one and only one solution for any $x, y, z \in \mathbb{R}$.

4.7 Definition The *concatenation* of the sequences $B_1 = \langle \vec{\beta}_{1,1}, \ldots, \vec{\beta}_{1,n_1} \rangle$, ..., $B_k = \langle \vec{\beta}_{k,1}, \ldots, \vec{\beta}_{k,n_k} \rangle$ adjoins them into a single sequence.

$$B_1 {}^\frown B_2 {}^\frown \cdots {}^\frown B_k = \langle \vec{\beta}_{1,1}, \ldots, \vec{\beta}_{1,n_1}, \vec{\beta}_{2,1}, \ldots, \vec{\beta}_{k,n_k} \rangle$$

4.8 Lemma Let V be a vector space that is the sum of some of its subspaces $V = W_1 + \cdots + W_k$. Let B_1, ..., B_k be bases for these subspaces. The following are equivalent.

(1) The expression of any $\vec{v} \in V$ as a combination $\vec{v} = \vec{w}_1 + \cdots + \vec{w}_k$ with $\vec{w}_i \in W_i$ is unique.

(2) The concatenation $B_1 {}^\frown \cdots {}^\frown B_k$ is a basis for V.

(3) Among nonzero vectors from different W_i's every linear relationship is trivial.

Proof We will show that (1) $\Longrightarrow$ (2), that (2) $\Longrightarrow$ (3), and finally that (3) $\Longrightarrow$ (1). For these arguments, observe that we can pass from a combination of $\vec{w}$'s to a combination of $\vec{\beta}$'s

$$\begin{aligned} d_1\vec{w}_1 + \cdots + d_k\vec{w}_k &= d_1(c_{1,1}\vec{\beta}_{1,1} + \cdots + c_{1,n_1}\vec{\beta}_{1,n_1}) \\ &\quad + \cdots + d_k(c_{k,1}\vec{\beta}_{k,1} + \cdots + c_{k,n_k}\vec{\beta}_{k,n_k}) \\ &= d_1c_{1,1} \cdot \vec{\beta}_{1,1} + \cdots + d_kc_{k,n_k} \cdot \vec{\beta}_{k,n_k} \end{aligned} \qquad (*)$$

and vice versa (we can move from the bottom to the top by taking each d_i to be 1).

For (1) $\Longrightarrow$ (2), assume that all decompositions are unique. We will show that $B_1 {}^\frown \cdots {}^\frown B_k$ spans the space and is linearly independent. It spans the space because the assumption that $V = W_1 + \cdots + W_k$ means that every $\vec{v}$ can be expressed as $\vec{v} = \vec{w}_1 + \cdots + \vec{w}_k$, which translates by equation $(*)$ to an expression of $\vec{v}$ as a linear combination of the $\vec{\beta}$'s from the concatenation. For linear independence, consider this linear relationship.

$$\vec{0} = c_{1,1}\vec{\beta}_{1,1} + \cdots + c_{k,n_k}\vec{\beta}_{k,n_k}$$

Regroup as in $(*)$ (that is, move from bottom to top) to get the decomposition $\vec{0} = \vec{w}_1 + \cdots + \vec{w}_k$. Because the zero vector obviously has the decomposition $\vec{0} = \vec{0} + \cdots + \vec{0}$, the assumption that decompositions are unique shows that each $\vec{w}_i$ is the zero vector. This means that $c_{i,1}\vec{\beta}_{i,1} + \cdots + c_{i,n_i}\vec{\beta}_{i,n_i} = \vec{0}$, and since each B_i is a basis we have the desired conclusion that all of the c's are zero.

For (2) $\Longrightarrow$ (3) assume that the concatenation of the bases is a basis for the entire space. Consider a linear relationship among nonzero vectors from different

W_i's. This might or might not involve a vector from W_1, or one from W_2, etc., so we write it $\vec{0} = \cdots + d_i\vec{w}_i + \cdots$. As in equation (*) expand the vector.

$$\begin{aligned}\vec{0} &= \cdots + d_i(c_{i,1}\vec{\beta}_{i,1} + \cdots + c_{i,n_i}\vec{\beta}_{i,n_i}) + \cdots \\ &= \cdots + d_i c_{i,1} \cdot \vec{\beta}_{i,1} + \cdots + d_i c_{i,n_i} \cdot \vec{\beta}_{i,n_i} + \cdots\end{aligned}$$

The linear independence of $B_1 {}^\frown \cdots {}^\frown B_k$ gives that each coefficient $d_i c_{i,j}$ is zero. Since $\vec{w}_i$ is nonzero vector, at least one of the $c_{i,j}$'s is not zero, and thus d_i is zero. This holds for each d_i, and therefore the linear relationship is trivial.

Finally, for (3) $\implies$ (1), assume that among nonzero vectors from different W_i's any linear relationship is trivial. Consider two decompositions of a vector $\vec{v} = \cdots + \vec{w}_i + \cdots$ and $\vec{v} = \cdots + \vec{u}_j + \cdots$ where $\vec{w}_i \in W_i$ and $\vec{u}_j \in W_j$. Subtract one from the other to get a linear relationship, something like this (if there is no $\vec{u}_i$ or $\vec{w}_j$ then leave those out).

$$\vec{0} = \cdots + (\vec{w}_i - \vec{u}_i) + \cdots + (\vec{w}_j - \vec{u}_j) + \cdots$$

The case assumption that statement (3) holds implies that the terms each equal the zero vector $\vec{w}_i - \vec{u}_i = \vec{0}$. Hence decompositions are unique. QED

4.9 Definition A collection of subspaces $\{W_1, \ldots, W_k\}$ is *independent* if no nonzero vector from any W_i is a linear combination of vectors from the other subspaces $W_1, \ldots, W_{i-1}, W_{i+1}, \ldots, W_k$.

4.10 Definition A vector space V is the *direct sum* (or *internal direct sum*) of its subspaces $W_1, \ldots, W_k$ if $V = W_1 + W_2 + \cdots + W_k$ and the collection $\{W_1, \ldots, W_k\}$ is independent. We write $V = W_1 \oplus W_2 \oplus \cdots \oplus W_k$.

4.11 Example Our prototype works: $\mathbb{R}^3 = x\text{-axis} \oplus y\text{-axis} \oplus z\text{-axis}$.

4.12 Example The space of 2×2 matrices is this direct sum.

$$\{\begin{pmatrix} a & 0 \\ 0 & d \end{pmatrix} \mid a, d \in \mathbb{R}\} \oplus \{\begin{pmatrix} 0 & b \\ 0 & 0 \end{pmatrix} \mid b \in \mathbb{R}\} \oplus \{\begin{pmatrix} 0 & 0 \\ c & 0 \end{pmatrix} \mid c \in \mathbb{R}\}$$

It is the direct sum of subspaces in many other ways as well; direct sum decompositions are not unique.

4.13 Corollary The dimension of a direct sum is the sum of the dimensions of its summands.

PROOF In Lemma 4.8, the number of basis vectors in the concatenation equals the sum of the number of vectors in the sub-bases. QED

The special case of two subspaces is worth its own mention.

4.14 Definition When a vector space is the direct sum of two of its subspaces then they are *complements*.

4.15 Lemma A vector space V is the direct sum of two of its subspaces W_1 and W_2 if and only if it is the sum of the two $V = W_1 + W_2$ and their intersection is trivial $W_1 \cap W_2 = \{\vec{0}\}$.

PROOF Suppose first that $V = W_1 \oplus W_2$. By definition, V is the sum of the two $V = W_1 + W_2$. To show that their intersection is trivial let $\vec{v}$ be a vector from $W_1 \cap W_2$ and consider the equation $\vec{v} = \vec{v}$. On that equation's left side is a member of W_1 and on the right is a member of W_2, which we can think of as a linear combination of members of W_2. But the two spaces are independent so the only way that a member of W_1 can be a linear combination of vectors from W_2 is if that member is the zero vector $\vec{v} = \vec{0}$.

For the other direction, suppose that V is the sum of two spaces with a trivial intersection. To show that V is a direct sum of the two we need only show that the spaces are independent — that no nonzero member of the first is expressible as a linear combination of members of the second, and vice versa. This holds because any relationship $\vec{w}_1 = c_1\vec{w}_{2,1} + \cdots + c_k\vec{w}_{2,k}$ (with $\vec{w}_1 \in W_1$ and $\vec{w}_{2,j} \in W_2$ for all j) shows that the vector on the left is also in W_2, since the right side is a combination of members of W_2. The intersection of these two spaces is trivial, so $\vec{w}_1 = \vec{0}$. The same argument works for any $\vec{w}_2$. QED

4.16 Example In $\mathbb{R}^2$ the x-axis and the y-axis are complements, that is, $\mathbb{R}^2 = x\text{-axis} \oplus y\text{-axis}$. A space can have more than one pair of complementary subspaces; another pair for $\mathbb{R}^2$ are the subspaces consisting of the lines $y = x$ and $y = 2x$.

4.17 Example In the space $F = \{a\cos\theta + b\sin\theta \mid a, b \in \mathbb{R}\}$, the subspaces $W_1 = \{a\cos\theta \mid a \in \mathbb{R}\}$ and $W_2 = \{b\sin\theta \mid b \in \mathbb{R}\}$ are complements. The prior example noted that a space can be decomposed into more than one pair of complements. In addition note that F can has more than one pair of complementary subspaces where the first in the pair is W_1 — another complement of W_1 is $W_3 = \{b\sin\theta + b\cos\theta \mid b \in \mathbb{R}\}$.

4.18 Example In $\mathbb{R}^3$, the xy-plane and the yz-planes are not complements, which is the point of the discussion following Example 4.4. One complement of the xy-plane is the z-axis.

Here is a natural question that arises from Lemma 4.15: for $k > 2$ is the simple sum $V = W_1 + \cdots + W_k$ also a direct sum if and only if the intersection of the subspaces is trivial?

4.19 Example If there are more than two subspaces then having a trivial intersection is not enough to guarantee unique decomposition (i.e., is not enough to ensure that the spaces are independent). In $\mathbb{R}^3$, let W_1 be the x-axis, let W_2 be the y-axis, and let W_3 be this.

$$W_3 = \{\begin{pmatrix} q \\ q \\ r \end{pmatrix} \mid q, r \in \mathbb{R}\}$$

The check that $\mathbb{R}^3 = W_1 + W_2 + W_3$ is easy. The intersection $W_1 \cap W_2 \cap W_3$ is trivial, but decompositions aren't unique.

$$\begin{pmatrix} x \\ y \\ z \end{pmatrix} = \begin{pmatrix} 0 \\ 0 \\ 0 \end{pmatrix} + \begin{pmatrix} 0 \\ y-x \\ 0 \end{pmatrix} + \begin{pmatrix} x \\ x \\ z \end{pmatrix} = \begin{pmatrix} x-y \\ 0 \\ 0 \end{pmatrix} + \begin{pmatrix} 0 \\ 0 \\ 0 \end{pmatrix} + \begin{pmatrix} y \\ y \\ z \end{pmatrix}$$

(This example also shows that this requirement is also not enough: that all pairwise intersections of the subspaces be trivial. See Exercise 30.)

In this subsection we have seen two ways to regard a space as built up from component parts. Both are useful; in particular we will use the direct sum definition at the end of the Chapter Five.

Exercises

✓ **4.20** Decide if $\mathbb{R}^2$ is the direct sum of each W_1 and W_2.

(a) $W_1 = \{\begin{pmatrix} x \\ 0 \end{pmatrix} \mid x \in \mathbb{R}\}$, $W_2 = \{\begin{pmatrix} x \\ x \end{pmatrix} \mid x \in \mathbb{R}\}$

(b) $W_1 = \{\begin{pmatrix} s \\ s \end{pmatrix} \mid s \in \mathbb{R}\}$, $W_2 = \{\begin{pmatrix} s \\ 1.1s \end{pmatrix} \mid s \in \mathbb{R}\}$

(c) $W_1 = \mathbb{R}^2$, $W_2 = \{\vec{0}\}$

(d) $W_1 = W_2 = \{\begin{pmatrix} t \\ t \end{pmatrix} \mid t \in \mathbb{R}\}$

(e) $W_1 = \{\begin{pmatrix} 1 \\ 0 \end{pmatrix} + \begin{pmatrix} x \\ 0 \end{pmatrix} \mid x \in \mathbb{R}\}$, $W_2 = \{\begin{pmatrix} -1 \\ 0 \end{pmatrix} + \begin{pmatrix} 0 \\ y \end{pmatrix} \mid y \in \mathbb{R}\}$

✓ **4.21** Show that $\mathbb{R}^3$ is the direct sum of the xy-plane with each of these.

(a) the z-axis

(b) the line

$$\{\begin{pmatrix} z \\ z \\ z \end{pmatrix} \mid z \in \mathbb{R}\}$$

4.22 Is $\mathcal{P}_2$ the direct sum of $\{a + bx^2 \mid a, b \in \mathbb{R}\}$ and $\{cx \mid c \in \mathbb{R}\}$?

✓ **4.23** In $\mathcal{P}_n$, the *even* polynomials are the members of this set

$$\mathcal{E} = \{p \in \mathcal{P}_n \mid p(-x) = p(x) \text{ for all } x\}$$

and the *odd* polynomials are the members of this set.

$$\mathcal{O} = \{p \in \mathcal{P}_n \mid p(-x) = -p(x) \text{ for all } x\}$$

Show that these are complementary subspaces.

4.24 Which of these subspaces of $\mathbb{R}^3$

$$W_1\colon \text{the } x\text{-axis},\quad W_2\colon \text{the } y\text{-axis},\quad W_3\colon \text{the } z\text{-axis},$$
$$W_4\colon \text{the plane } x+y+z=0,\quad W_5\colon \text{the } yz\text{-plane}$$

can be combined to

(a) sum to $\mathbb{R}^3$? (b) direct sum to $\mathbb{R}^3$?

✓ **4.25** Show that $\mathcal{P}_n = \{a_0 \mid a_0 \in \mathbb{R}\} \oplus \ldots \oplus \{a_n x^n \mid a_n \in \mathbb{R}\}$.

4.26 What is $W_1 + W_2$ if $W_1 \subseteq W_2$?

4.27 Does Example 4.5 generalize? That is, is this true or false: if a vector space V has a basis $\langle \vec{\beta}_1, \ldots, \vec{\beta}_n \rangle$ then it is the direct sum of the spans of the one-dimensional subspaces $V = [\{\vec{\beta}_1\}] \oplus \ldots \oplus [\{\vec{\beta}_n\}]$?

4.28 Can $\mathbb{R}^4$ be decomposed as a direct sum in two different ways? Can $\mathbb{R}^1$?

4.29 This exercise makes the notation of writing '+' between sets more natural. Prove that, where $W_1, \ldots, W_k$ are subspaces of a vector space,

$$W_1 + \cdots + W_k = \{\vec{w}_1 + \vec{w}_2 + \cdots + \vec{w}_k \mid \vec{w}_1 \in W_1, \ldots, \vec{w}_k \in W_k\},$$

and so the sum of subspaces is the subspace of all sums.

4.30 (Refer to Example 4.19. This exercise shows that the requirement that pairwise intersections be trivial is genuinely stronger than the requirement only that the intersection of all of the subspaces be trivial.) Give a vector space and three subspaces W_1, W_2, and W_3 such that the space is the sum of the subspaces, the intersection of all three subspaces $W_1 \cap W_2 \cap W_3$ is trivial, but the pairwise intersections $W_1 \cap W_2$, $W_1 \cap W_3$, and $W_2 \cap W_3$ are nontrivial.

✓ **4.31** Prove that if $V = W_1 \oplus \ldots \oplus W_k$ then $W_i \cap W_j$ is trivial whenever $i \neq j$. This shows that the first half of the proof of Lemma 4.15 extends to the case of more than two subspaces. (Example 4.19 shows that this implication does not reverse; the other half does not extend.)

4.32 Recall that no linearly independent set contains the zero vector. Can an independent set of subspaces contain the trivial subspace?

✓ **4.33** Does every subspace have a complement?

✓ **4.34** Let W_1, W_2 be subspaces of a vector space.

(a) Assume that the set S_1 spans W_1, and that the set S_2 spans W_2. Can $S_1 \cup S_2$ span $W_1 + W_2$? Must it?

(b) Assume that S_1 is a linearly independent subset of W_1 and that S_2 is a linearly independent subset of W_2. Can $S_1 \cup S_2$ be a linearly independent subset of $W_1 + W_2$? Must it?

4.35 When we decompose a vector space as a direct sum, the dimensions of the subspaces add to the dimension of the space. The situation with a space that is given as the sum of its subspaces is not as simple. This exercise considers the two-subspace special case.

(a) For these subspaces of $\mathcal{M}_{2\times 2}$ find $W_1 \cap W_2$, $\dim(W_1 \cap W_2)$, $W_1 + W_2$, and $\dim(W_1 + W_2)$.

$$W_1 = \{\begin{pmatrix} 0 & 0 \\ c & d \end{pmatrix} \mid c, d \in \mathbb{R}\} \qquad W_2 = \{\begin{pmatrix} 0 & b \\ c & 0 \end{pmatrix} \mid b, c \in \mathbb{R}\}$$

(b) Suppose that U and W are subspaces of a vector space. Suppose that the sequence $\langle\vec{\beta}_1,\ldots,\vec{\beta}_k\rangle$ is a basis for $U\cap W$. Finally, suppose that the prior sequence has been expanded to give a sequence $\langle\vec{\mu}_1,\ldots,\vec{\mu}_j,\vec{\beta}_1,\ldots,\vec{\beta}_k\rangle$ that is a basis for U, and a sequence $\langle\vec{\beta}_1,\ldots,\vec{\beta}_k,\vec{\omega}_1,\ldots,\vec{\omega}_p\rangle$ that is a basis for W. Prove that this sequence

$$\langle\vec{\mu}_1,\ldots,\vec{\mu}_j,\vec{\beta}_1,\ldots,\vec{\beta}_k,\vec{\omega}_1,\ldots,\vec{\omega}_p\rangle$$

is a basis for the sum $U+W$.

(c) Conclude that $\dim(U+W)=\dim(U)+\dim(W)-\dim(U\cap W)$.

(d) Let W_1 and W_2 be eight-dimensional subspaces of a ten-dimensional space. List all values possible for $\dim(W_1\cap W_2)$.

4.36 Let $V=W_1\oplus\cdots\oplus W_k$ and for each index i suppose that S_i is a linearly independent subset of W_i. Prove that the union of the S_i's is linearly independent.

4.37 A matrix is *symmetric* if for each pair of indices i and j, the i,j entry equals the j,i entry. A matrix is *antisymmetric* if each i,j entry is the negative of the j,i entry.

(a) Give a symmetric 2×2 matrix and an antisymmetric 2×2 matrix. (*Remark.* For the second one, be careful about the entries on the diagonal.)

(b) What is the relationship between a square symmetric matrix and its transpose? Between a square antisymmetric matrix and its transpose?

(c) Show that $\mathcal{M}_{n\times n}$ is the direct sum of the space of symmetric matrices and the space of antisymmetric matrices.

4.38 Let W_1,W_2,W_3 be subspaces of a vector space. Prove that $(W_1\cap W_2)+(W_1\cap W_3)\subseteq W_1\cap(W_2+W_3)$. Does the inclusion reverse?

4.39 The example of the x-axis and the y-axis in $\mathbb{R}^2$ shows that $W_1\oplus W_2=V$ does not imply that $W_1\cup W_2=V$. Can $W_1\oplus W_2=V$ and $W_1\cup W_2=V$ happen?

✓ **4.40** Consider Corollary 4.13. Does it work both ways—that is, supposing that $V=W_1+\cdots+W_k$, is $V=W_1\oplus\cdots\oplus W_k$ if and only if $\dim(V)=\dim(W_1)+\cdots+\dim(W_k)$?

4.41 We know that if $V=W_1\oplus W_2$ then there is a basis for V that splits into a basis for W_1 and a basis for W_2. Can we make the stronger statement that every basis for V splits into a basis for W_1 and a basis for W_2?

4.42 We can ask about the algebra of the '+' operation.

(a) Is it commutative; is $W_1+W_2=W_2+W_1$?

(b) Is it associative; is $(W_1+W_2)+W_3=W_1+(W_2+W_3)$?

(c) Let W be a subspace of some vector space. Show that $W+W=W$.

(d) Must there be an identity element, a subspace I such that $I+W=W+I=W$ for all subspaces W?

(e) Does left-cancellation hold: if $W_1+W_2=W_1+W_3$ then $W_2=W_3$? Right cancellation?

Topic

Fields

Computations involving only integers or only rational numbers are much easier than those with real numbers. Could other algebraic structures, such as the integers or the rationals, work in the place of $\mathbb{R}$ in the definition of a vector space?

If we take "work" to mean that the results of this chapter remain true then there is a natural list of conditions that a structure (that is, number system) must have in order to work in the place of $\mathbb{R}$. A *field* is a set $\mathcal{F}$ with operations '+' and '·' such that

(1) for any $a, b \in \mathcal{F}$ the result of $a + b$ is in $\mathcal{F}$, and $a + b = b + a$, and if $c \in \mathcal{F}$ then $a + (b + c) = (a + b) + c$

(2) for any $a, b \in \mathcal{F}$ the result of $a \cdot b$ is in $\mathcal{F}$, and $a \cdot b = b \cdot a$, and if $c \in \mathcal{F}$ then $a \cdot (b \cdot c) = (a \cdot b) \cdot c$

(3) if $a, b, c \in \mathcal{F}$ then $a \cdot (b + c) = a \cdot b + a \cdot c$

(4) there is an element $0 \in \mathcal{F}$ such that if $a \in \mathcal{F}$ then $a + 0 = a$, and for each $a \in \mathcal{F}$ there is an element $-a \in \mathcal{F}$ such that $(-a) + a = 0$

(5) there is an element $1 \in \mathcal{F}$ such that if $a \in \mathcal{F}$ then $a \cdot 1 = a$, and for each element $a \neq 0$ of $\mathcal{F}$ there is an element $a^{-1} \in \mathcal{F}$ such that $a^{-1} \cdot a = 1$.

For example, the algebraic structure consisting of the set of real numbers along with its usual addition and multiplication operation is a field. Another field is the set of rational numbers with its usual addition and multiplication operations. An example of an algebraic structure that is not a field is the integers, because it fails the final condition.

Some examples are more surprising. The set $\mathbb{B} = \{0, 1\}$ under these operations:

+	0	1
0	0	1
1	1	0

·	0	1
0	0	0
1	0	1

is a field; see Exercise 4.

We could in this book develop Linear Algebra as the theory of vector spaces with scalars from an arbitrary field. In that case, almost all of the statements here would carry over by replacing '$\mathbb{R}$' with '$\mathcal{F}$', that is, by taking coefficients, vector entries, and matrix entries to be elements of $\mathcal{F}$ (the exceptions are statements involving distances or angles, which would need additional development). Here are some examples; each applies to a vector space V over a field $\mathcal{F}$.

* For any $\vec{v} \in V$ and $a \in \mathcal{F}$, (i) $0 \cdot \vec{v} = \vec{0}$, (ii) $-1 \cdot \vec{v} + \vec{v} = \vec{0}$, and (iii) $a \cdot \vec{0} = \vec{0}$.
* The span, the set of linear combinations, of a subset of V is a subspace of V.
* Any subset of a linearly independent set is also linearly independent.
* In a finite-dimensional vector space, any two bases have the same number of elements.

(Even statements that don't explicitly mention $\mathcal{F}$ use field properties in their proof.)

We will not develop vector spaces in this more general setting because the additional abstraction can be a distraction. The ideas we want to bring out already appear when we stick to the reals.

The exception is Chapter Five. There we must factor polynomials, so we will switch to considering vector spaces over the field of complex numbers.

Exercises

1 Check that the real numbers form a field.

2 Prove that these are fields.
(a) The rational numbers $\mathbb{Q}$ (b) The complex numbers $\mathbb{C}$

3 Give an example that shows that the integer number system is not a field.

4 Check that the set $\mathbb{B} = \{0, 1\}$ is a field under the operations listed above,

5 Give suitable operations to make the set $\{0, 1, 2\}$ a field.

Topic

Crystals

Everyone has noticed that table salt comes in little cubes.

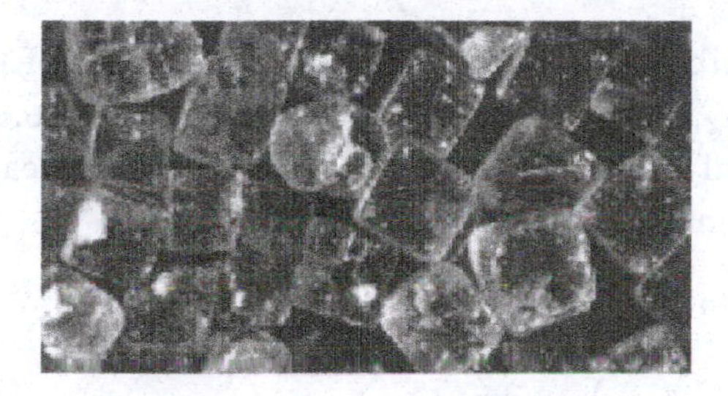

This orderly outside arises from an orderly inside — the way the atoms lie is also cubical, these cubes stack in neat rows and columns, and the salt faces tend to be just an outer layer of cubes. One cube of atoms is shown below. Salt is sodium chloride and the small spheres shown are sodium while the big ones are chloride. To simplify the view, it only shows the sodiums and chlorides on the front, top, and right.

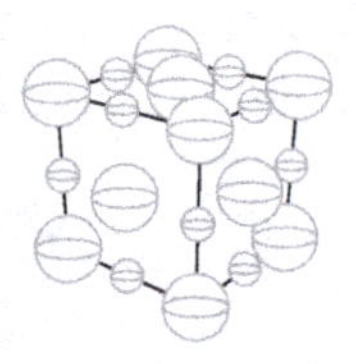

The specks of salt that we see above have many repetitions of this fundamental unit. A solid, such as table salt, with a regular internal structure is a *crystal*.

We can restrict our attention to the front face. There we have a square repeated many times giving a lattice of atoms.

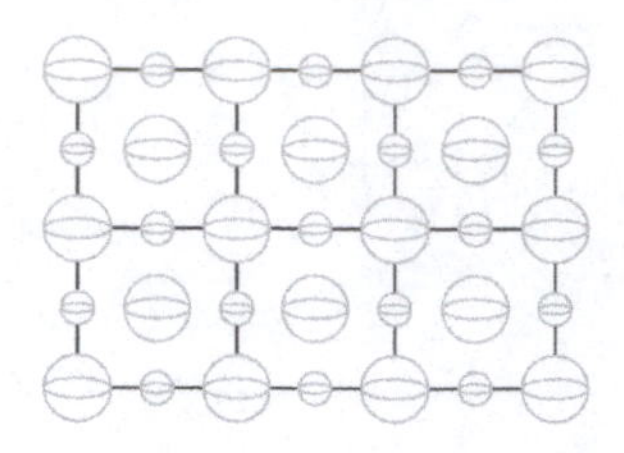

The distance along the sides of each square cell is about 3.34 Ångstroms (an Ångstrom is 10^{-10} meters). When we want to refer to atoms in the lattice that number is unwieldy, and so we take the square's side length as a unit. That is, we naturally adopt this basis.

$$\langle \begin{pmatrix} 3.34 \\ 0 \end{pmatrix}, \begin{pmatrix} 0 \\ 3.34 \end{pmatrix} \rangle$$

Now we can describe, say, the atom in the upper right of the lattice picture above as $3\vec{\beta}_1 + 2\vec{\beta}_2$, instead of 10.02 Ångstroms over and 6.68 up.

Another crystal from everyday experience is pencil lead. It is *graphite*, formed from carbon atoms arranged in this shape.

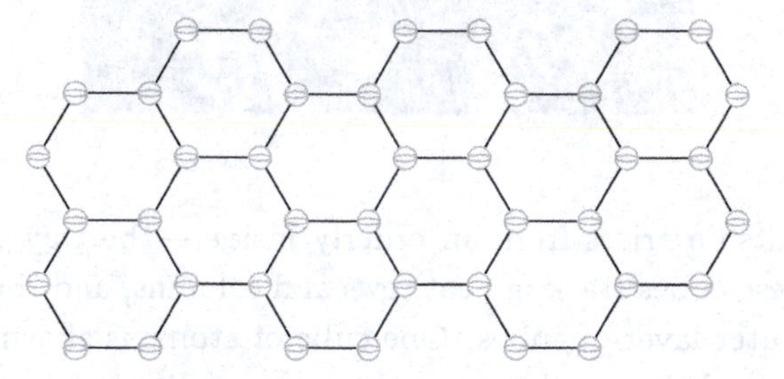

This is a single plane of graphite, called *graphene*. A piece of graphite consists of many of these planes, layered. The chemical bonds between the planes are much weaker than the bonds inside the planes, which explains why pencils write — the graphite can be sheared so that the planes slide off and are left on the paper.

We can get a convenient unit of length by decomposing the hexagonal ring into three regions that are rotations of this *unit cell*.

The vectors that form the sides of that unit cell make a convenient basis. The distance along the bottom and slant is 1.42 Ångstroms, so this

$$\langle \begin{pmatrix} 1.42 \\ 0 \end{pmatrix}, \begin{pmatrix} 0.71 \\ 1.23 \end{pmatrix} \rangle$$

is a good basis.

Another familiar crystal formed from carbon is diamond. Like table salt it is built from cubes but the structure inside each cube is more complicated. In addition to carbons at each corner,

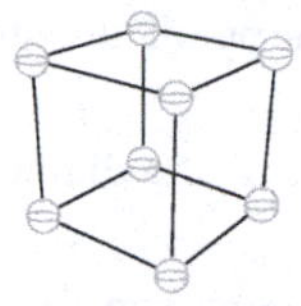

there are carbons in the middle of each face.

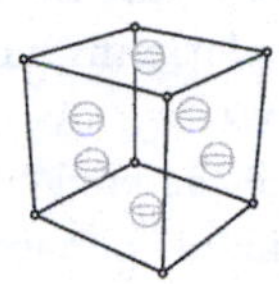

(To show the new face carbons clearly, the corner carbons are reduced to dots.) There are also four more carbons inside the cube, two that are a quarter of the way up from the bottom and two that are a quarter of the way down from the top.

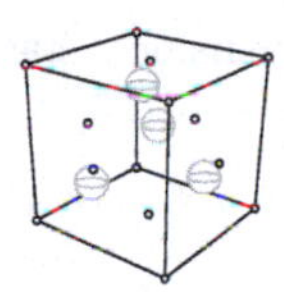

(As before, carbons shown earlier are reduced here to dots.) The distance along any edge of the cube is 2.18 Ångstroms. Thus, a natural basis for describing the locations of the carbons and the bonds between them, is this.

$$\langle \begin{pmatrix} 2.18 \\ 0 \\ 0 \end{pmatrix}, \begin{pmatrix} 0 \\ 2.18 \\ 0 \end{pmatrix}, \begin{pmatrix} 0 \\ 0 \\ 2.18 \end{pmatrix} \rangle$$

The examples here show that the structures of crystals is complicated enough to need some organized system to give the locations of the atoms and how they are chemically bound. One tool for that organization is a convenient basis. This application of bases is simple but it shows a science context where the idea arises naturally.

Exercises

1 How many fundamental regions are there in one face of a speck of salt? (With a ruler, we can estimate that face is a square that is 0.1 cm on a side.)

2 In the graphite picture, imagine that we are interested in a point 5.67 Ångstroms over and 3.14 Ångstroms up from the origin.

(a) Express that point in terms of the basis given for graphite.

(b) How many hexagonal shapes away is this point from the origin?

(c) Express that point in terms of a second basis, where the first basis vector is the same, but the second is perpendicular to the first (going up the plane) and of the same length.

3 Give the locations of the atoms in the diamond cube both in terms of the basis, and in Ångstroms.

4 This illustrates how we could compute the dimensions of a unit cell from the shape in which a substance crystallizes ([Ebbing], p. 462).

(a) Recall that there are 6.022×10^{23} atoms in a mole (this is Avogadro's number). From that, and the fact that platinum has a mass of 195.08 grams per mole, calculate the mass of each atom.

(b) Platinum crystallizes in a face-centered cubic lattice with atoms at each lattice point, that is, it looks like the middle picture given above for the diamond crystal. Find the number of platinum's per unit cell (hint: sum the fractions of platinum's that are inside of a single cell).

(c) From that, find the mass of a unit cell.

(d) Platinum crystal has a density of 21.45 grams per cubic centimeter. From this, and the mass of a unit cell, calculate the volume of a unit cell.

(e) Find the length of each edge.

(f) Describe a natural three-dimensional basis.

Topic

Voting Paradoxes

Imagine that a Political Science class studying the American presidential process holds a mock election. The 29 class members rank the Democratic Party, Republican Party, and Third Party nominees, from most preferred to least preferred (> means 'is preferred to').

preference order	*number with that preference*
Democrat > Republican > Third	5
Democrat > Third > Republican	4
Republican > Democrat > Third	2
Republican > Third > Democrat	8
Third > Democrat > Republican	8
Third > Republican > Democrat	2

What is the preference of the group as a whole?

Overall, the group prefers the Democrat to the Republican by five votes; seventeen voters ranked the Democrat above the Republican versus twelve the other way. And the group prefers the Republican to the Third's nominee, fifteen to fourteen. But, strangely enough, the group also prefers the Third to the Democrat, eighteen to eleven.

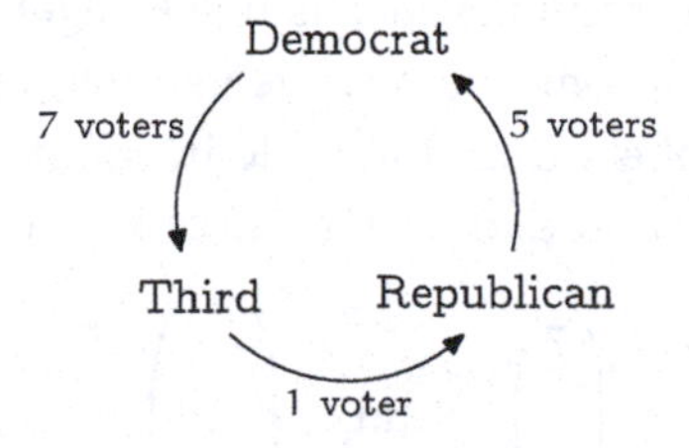

This is a *voting paradox*, specifically, a *majority cycle*.

Mathematicians study voting paradoxes in part because of their implications for practical politics. For instance, the instructor can manipulate this class into choosing the Democrat as the overall winner by first asking for a vote between the Republican and the Third, and then asking for a vote between the winner of that contest, who will be the Republican, and the Democrat. By similar manipulations the instructor can make any of the other two candidates come out as the winner. (We will stick to three-candidate elections but the same thing happens in larger elections.)

Mathematicians also study voting paradoxes simply because they are interesting. One interesting aspect is that the group's overall majority cycle occurs despite that each single voter's preference list is *rational*, in a straight-line order. That is, the majority cycle seems to arise in the aggregate without being present in the components of that aggregate, the preference lists. However we can use linear algebra to argue that a tendency toward cyclic preference is actually present in each voter's list and that it surfaces when there is more adding of the tendency than canceling.

For this, abbreviating the choices as D, R, and T, we can describe how a voter with preference order D > R > T contributes to the above cycle.

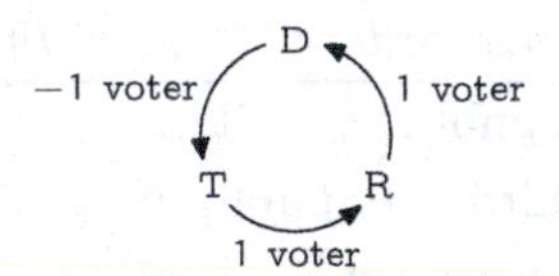

(The negative sign is here because the arrow describes T as preferred to D, but this voter likes them the other way.) The descriptions for the other preference lists are in the table on page 152.

Now, to conduct the election we linearly combine these descriptions; for instance, the Political Science mock election

5 · (D, T, R cycle: −1, 1, 1) + 4 · (D, T, R cycle: −1, −1, 1) + ··· + 2 · (D, T, R cycle: 1, −1, −1)

yields the circular group preference shown earlier.

Of course, taking linear combinations is linear algebra. The graphical cycle notation is suggestive but inconvenient so we use column vectors by starting at the D and taking the numbers from the cycle in counterclockwise order. Thus, we represent the mock election and a single D > R > T vote in this way.

$$\begin{pmatrix}7\\1\\5\end{pmatrix} \quad\text{and}\quad \begin{pmatrix}-1\\1\\1\end{pmatrix}$$

We will decompose vote vectors into two parts, one cyclic and the other acyclic. For the first part, we say that a vector is *purely cyclic* if it is in this subspace of $\mathbb{R}^3$.

$$C = \{\begin{pmatrix} k \\ k \\ k \end{pmatrix} \mid k \in \mathbb{R}\} = \{k \cdot \begin{pmatrix} 1 \\ 1 \\ 1 \end{pmatrix} \mid k \in \mathbb{R}\}$$

For the second part, consider the set of vectors that are perpendicular to all of the vectors in C. Exercise 6 shows that this is a subspace

$$\begin{aligned} C^{\perp} &= \{\begin{pmatrix} c_1 \\ c_2 \\ c_3 \end{pmatrix} \mid \begin{pmatrix} c_1 \\ c_2 \\ c_3 \end{pmatrix} \cdot \begin{pmatrix} k \\ k \\ k \end{pmatrix} = 0 \text{ for all } k \in \mathbb{R}\} \\ &= \{\begin{pmatrix} c_1 \\ c_2 \\ c_3 \end{pmatrix} \mid c_1 + c_2 + c_3 = 0\} = \{c_2 \begin{pmatrix} -1 \\ 1 \\ 0 \end{pmatrix} + c_3 \begin{pmatrix} -1 \\ 0 \\ 1 \end{pmatrix} \mid c_2, c_3 \in \mathbb{R}\} \end{aligned}$$

(read that aloud as "C perp"). So we are led to this basis for $\mathbb{R}^3$.

$$\langle \begin{pmatrix} 1 \\ 1 \\ 1 \end{pmatrix}, \begin{pmatrix} -1 \\ 1 \\ 0 \end{pmatrix}, \begin{pmatrix} -1 \\ 0 \\ 1 \end{pmatrix} \rangle$$

We can represent votes with respect to this basis, and thereby decompose them into a cyclic part and an acyclic part. *(Note for readers who have covered the optional section in this chapter: that is, the space is the direct sum of C and $C^{\perp}$.)*

For example, consider the $D > R > T$ voter discussed above. We represent it with respect to the basis

$$\begin{array}{rcl} c_1 - c_2 - c_3 &=& -1 \\ c_1 + c_2 \quad\;\; &=& 1 \\ c_1 \quad\;\; + c_3 &=& 1 \end{array} \quad \xrightarrow[-\rho_1+\rho_3]{-\rho_1+\rho_2} \quad \xrightarrow{(-1/2)\rho_2+\rho_3} \quad \begin{array}{rcl} c_1 - c_2 - c_3 &=& -1 \\ 2c_2 + c_3 &=& 2 \\ (3/2)c_3 &=& 1 \end{array}$$

using the coordinates $c_1 = 1/3$, $c_2 = 2/3$, and $c_3 = 2/3$. Then

$$\begin{pmatrix} -1 \\ 1 \\ 1 \end{pmatrix} = \frac{1}{3} \cdot \begin{pmatrix} 1 \\ 1 \\ 1 \end{pmatrix} + \frac{2}{3} \cdot \begin{pmatrix} -1 \\ 1 \\ 0 \end{pmatrix} + \frac{2}{3} \cdot \begin{pmatrix} -1 \\ 0 \\ 1 \end{pmatrix} = \begin{pmatrix} 1/3 \\ 1/3 \\ 1/3 \end{pmatrix} + \begin{pmatrix} -4/3 \\ 2/3 \\ 2/3 \end{pmatrix}$$

gives the desired decomposition into a cyclic part and an acyclic part.

D, T, R: −1, 1, 1 = D, T, R: 1/3, 1/3, 1/3 + D, T, R: −4/3, 2/3, 2/3

Thus we can see that this D > R > T voter's rational preference list does have a cyclic part.

The T > R > D voter is opposite to the one just considered in that the '>' symbols are reversed. This voter's decomposition

shows that these opposite preferences have decompositions that are opposite. We say that the first voter has positive *spin* since the cycle part is with the direction that we have chosen for the arrows, while the second voter's spin is negative.

The fact that these opposite voters cancel each other is reflected in the fact that their vote vectors add to zero. This suggests an alternate way to tally an election. We could first cancel as many opposite preference lists as possible, and then determine the outcome by adding the remaining lists.

The table below contains the three pairs of opposite preference lists. For instance, the top line contains the voters discussed above.

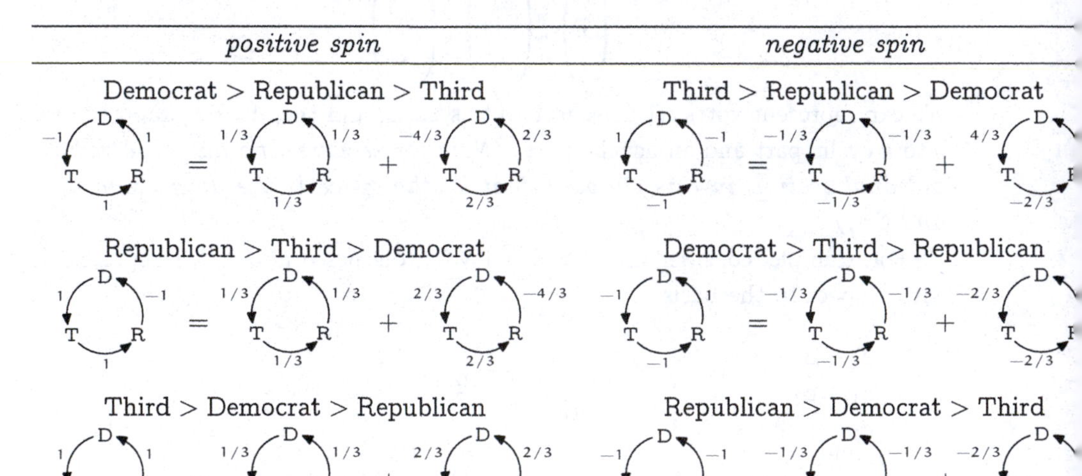

If we conduct the election as just described then after the cancellation of as many opposite pairs of voters as possible there will remain three sets of preference lists: one set from the first row, one from the second row, and one from the third row. We will finish by proving that a voting paradox can happen only if the spins of these three sets are in the same direction. That is, for a voting paradox to occur the three remaining sets must all come from the left of the table or all come from the right (see Exercise 3). This shows that there is some connection between the majority cycle and the decomposition that we are using — a voting paradox can happen only when the tendencies toward cyclic preference reinforce each other.

For the proof, assume that we have canceled opposite preference orders and we are left with one set of preference lists from each of the three rows. Consider the sum of these three (here, the numbers a, b, and c could be positive, negative, or zero).

D, T, R cycle with $-a$, a, a + D, T, R cycle with b, $-b$, b + D, T, R cycle with c, c, $-c$ = D, T, R cycle with $-a+b+c$, $a-b+c$, $a+b-c$

A voting paradox occurs when the three numbers on the right, $a-b+c$ and $a+b-c$ and $-a+b+c$, are all nonnegative or all nonpositive. On the left, at least two of the three numbers a and b and c are both nonnegative or both nonpositive. We can assume that they are a and b. That makes four cases: the cycle is nonnegative and a and b are nonnegative, the cycle is nonpositive and a and b are nonpositive, etc. We will do only the first case, since the second is similar and the other two are also easy.

So assume that the cycle is nonnegative and that a and b are nonnegative. The conditions $0 \leqslant a-b+c$ and $0 \leqslant -a+b+c$ add to give that $0 \leqslant 2c$, which implies that c is also nonnegative, as desired. That ends the proof.

This result says only that having all three spin in the same direction is a necessary condition for a majority cycle. It is not sufficient; see Exercise 4.

Voting theory and associated topics are the subject of current research. There are many intriguing results, notably the one produced by K Arrow [Arrow] who won the Nobel Prize in part for this work, showing that no voting system is entirely fair (for a reasonable definition of "fair"). Some good introductory articles are [Gardner, 1970], [Gardner, 1974], [Gardner, 1980], and [Neimi & Riker]. [Taylor] is a readable recent book. The long list of cases from recent American political history in [Poundstone] shows these paradoxes are routinely manipulated in practice.

This Topic is largely drawn from [Zwicker]. *(Author's Note: I would like to thank Professor Zwicker for his kind and illuminating discussions.)*

Exercises

1 Here is a reasonable way in which a voter could have a cyclic preference. Suppose that this voter ranks each candidate on each of three criteria.

(a) Draw up a table with the rows labeled 'Democrat', 'Republican', and 'Third', and the columns labeled 'character', 'experience', and 'policies'. Inside each column, rank some candidate as most preferred, rank another as in the middle, and rank the remaining one as least preferred.

(b) In this ranking, is the Democrat preferred to the Republican in (at least) two out of three criteria, or vice versa? Is the Republican preferred to the Third?

(c) Does the table that was just constructed have a cyclic preference order? If not, make one that does.

So it is possible for a voter to have a cyclic preference among candidates. The paradox described above, however, is that even if each voter has a straight-line preference list, a cyclic preference can still arise for the entire group.

2 Compute the values in the table of decompositions.

3 Do the cancellations of opposite preference orders for the Political Science class's mock election. Are all the remaining preferences from the left three rows of the table or from the right?

4 The necessary condition that is proved above — a voting paradox can happen only if all three preference lists remaining after cancellation have the same spin—is not also sufficient.

(a) Continuing the positive cycle case considered in the proof, use the two inequalities $0 \leqslant a - b + c$ and $0 \leqslant -a + b + c$ to show that $|a - b| \leqslant c$.

(b) Also show that $c \leqslant a + b$, and hence that $|a - b| \leqslant c \leqslant a + b$.

(c) Give an example of a vote where there is a majority cycle, and addition of one more voter with the same spin causes the cycle to go away.

(d) Can the opposite happen; can addition of one voter with a "wrong" spin cause a cycle to appear?

(e) Give a condition that is both necessary and sufficient to get a majority cycle.

5 A one-voter election cannot have a majority cycle because of the requirement that we've imposed that the voter's list must be rational.

(a) Show that a two-voter election may have a majority cycle. (We consider the group preference a majority cycle if all three group totals are nonnegative or if all three are nonpositive—that is, we allow some zero's in the group preference.)

(b) Show that for any number of voters greater than one, there is an election involving that many voters that results in a majority cycle.

6 Let U be a subspace of $\mathbb{R}^3$. Prove that the set $U^{\perp} = \{\vec{v} \mid \vec{v} \cdot \vec{u} = 0 \text{ for all } \vec{u} \in U\}$ of vectors that are perpendicular to each vector in U is also subspace of $\mathbb{R}^3$. Does this hold if U is not a subspace?

Topic

Dimensional Analysis

"You can't add apples and oranges," the old saying goes. It reflects our experience that in applications the quantities have units and keeping track of those units can help. Everyone has done calculations such as this one that use the units as a check.

$$60\,\frac{\text{sec}}{\text{min}}\cdot 60\,\frac{\text{min}}{\text{hr}}\cdot 24\,\frac{\text{hr}}{\text{day}}\cdot 365\,\frac{\text{day}}{\text{year}}=31\,536\,000\,\frac{\text{sec}}{\text{year}}$$

We can take the idea of including the units beyond bookkeeping. We can use units to draw conclusions about what relationships are possible among the physical quantities.

To start, consider the falling body equation distance $= 16\cdot(\text{time})^2$. If the distance is in feet and the time is in seconds then this is a true statement. However it is not correct in other unit systems, such as meters and seconds, because 16 isn't the right constant in those systems. We can fix that by attaching units to the 16, making it a *dimensional constant.*

$$\text{dist}=16\,\frac{\text{ft}}{\text{sec}^2}\cdot(\text{time})^2$$

Now the equation holds also in the meter-second system because when we align the units (a foot is approximately 0.30 meters),

$$\text{distance in meters}=16\,\frac{0.30\,\text{m}}{\text{sec}^2}\cdot(\text{time in sec})^2=4.8\,\frac{\text{m}}{\text{sec}^2}\cdot(\text{time in sec})^2$$

the constant gets adjusted. So in order to have equations that are correct across unit systems, we restrict our attention to those that use dimensional constants. Such an equation is *complete.*

Moving away from a particular unit system allows us to just measure quantities in combinations of some units of length L, mass M, and time T. These three are our *physical dimensions*. For instance, we could measure velocity in feet/second or fathoms/hour but at all events it involves a unit of length divided by a unit of time so the *dimensional formula* of velocity is L/T. Similarly, density's dimensional formula is M/L^3.

To write the dimensional formula we shall use negative exponents instead of fractions and we shall include the dimensions with a zero exponent. Thus we will write the dimensional formula of velocity as $L^1M^0T^{-1}$ and that of density as $L^{-3}M^1T^0$.

With that, "you can't add apples and oranges" becomes the advice to check that all of an equation's terms have the same dimensional formula. An example is this version of the falling body equation $d - gt^2 = 0$. The dimensional formula of the d term is $L^1M^0T^0$. For the other term, the dimensional formula of g is $L^1M^0T^{-2}$ (g is given above as $16\,\text{ft}/\text{sec}^2$) and the dimensional formula of t is $L^0M^0T^1$ so that of the entire gt^2 term is $L^1M^0T^{-2}(L^0M^0T^1)^2 = L^1M^0T^0$. Thus the two terms have the same dimensional formula. An equation with this property is *dimensionally homogeneous*.

Quantities with dimensional formula $L^0M^0T^0$ are *dimensionless*. For example, we measure an angle by taking the ratio of the subtended arc to the radius

arc

r

which is the ratio of a length to a length $(L^1M^0T^0)(L^1M^0T^0)^{-1}$ and thus angles have the dimensional formula $L^0M^0T^0$.

The classic example of using the units for more than bookkeeping, using them to draw conclusions, considers the formula for the period of a pendulum.

$$p = \text{–some expression involving the length of the string, etc.–}$$

The period is in units of time $L^0M^0T^1$. So the quantities on the other side of the equation must have dimensional formulas that combine in such a way that their L's and M's cancel and only a single T remains. The table on page 157 has the quantities that an experienced investigator would consider possibly relevant to the period of a pendulum. The only dimensional formulas involving L are for the length of the string and the acceleration due to gravity. For the L's of these two to cancel when they appear in the equation they must be in ratio, e.g., as $(\ell/g)^2$, or as $\cos(\ell/g)$, or as $(\ell/g)^{-1}$. Therefore the period is a function of ℓ/g.

This is a remarkable result: with a pencil and paper analysis, before we ever took out the pendulum and made measurements, we have determined something about what makes up its period.

To do dimensional analysis systematically, we need two facts (arguments for these are in [Bridgman], Chapter II and IV). The first is that each equation relating physical quantities that we shall see involves a sum of terms, where each term has the form

$$m_1^{p_1} m_2^{p_2} \cdots m_k^{p_k}$$

for numbers $m_1, \ldots, m_k$ that measure the quantities.

For the second fact, observe that an easy way to construct a dimensionally homogeneous expression is by taking a product of dimensionless quantities or by adding such dimensionless terms. Buckingham's Theorem states that any complete relationship among quantities with dimensional formulas can be algebraically manipulated into a form where there is some function f such that

$$f(\Pi_1, \ldots, \Pi_n) = 0$$

for a complete set $\{\Pi_1, \ldots, \Pi_n\}$ of dimensionless products. (The first example below describes what makes a set of dimensionless products 'complete'.) We usually want to express one of the quantities, m_1 for instance, in terms of the others. For that we will assume that the above equality can be rewritten

$$m_1 = m_2^{-p_2} \cdots m_k^{-p_k} \cdot \hat{f}(\Pi_2, \ldots, \Pi_n)$$

where $\Pi_1 = m_1 m_2^{p_2} \cdots m_k^{p_k}$ is dimensionless and the products $\Pi_2, \ldots, \Pi_n$ don't involve m_1 (as with f, here $\hat{f}$ is an arbitrary function, this time of $n-1$ arguments). Thus, to do dimensional analysis we should find which dimensionless products are possible.

For example, again consider the formula for a pendulum's period.

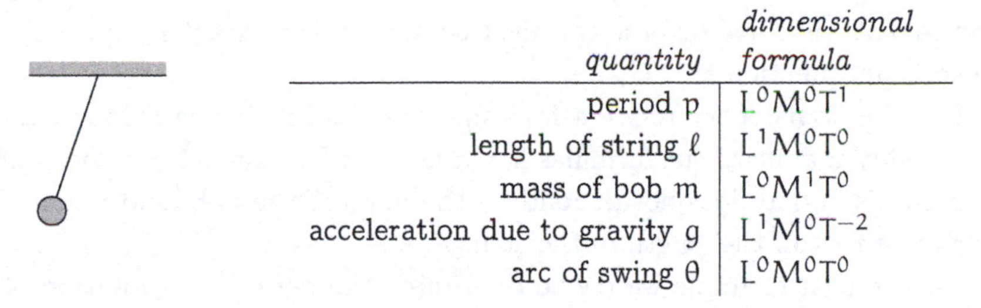

quantity	*dimensional formula*
period p	$L^0M^0T^1$
length of string ℓ	$L^1M^0T^0$
mass of bob m	$L^0M^1T^0$
acceleration due to gravity g	$L^1M^0T^{-2}$
arc of swing θ	$L^0M^0T^0$

By the first fact cited above, we expect the formula to have (possibly sums of terms of) the form $p^{p_1}\ell^{p_2}m^{p_3}g^{p_4}\theta^{p_5}$. To use the second fact, to find which combinations of the powers $p_1, \ldots, p_5$ yield dimensionless products, consider this equation.

$$(L^0M^0T^1)^{p_1}(L^1M^0T^0)^{p_2}(L^0M^1T^0)^{p_3}(L^1M^0T^{-2})^{p_4}(L^0M^0T^0)^{p_5} = L^0M^0T^0$$

It gives three conditions on the powers.

$$\begin{aligned} p_2 \quad + p_4 &= 0 \\ p_3 \qquad &= 0 \\ p_1 \qquad - 2p_4 &= 0 \end{aligned}$$

Note that $p_3 = 0$ so the mass of the bob does not affect the period. Gaussian reduction and parametrization of that system gives this

$$\{\begin{pmatrix}p_1\\p_2\\p_3\\p_4\\p_5\end{pmatrix} = \begin{pmatrix}1\\-1/2\\0\\1/2\\0\end{pmatrix} p_1 + \begin{pmatrix}0\\0\\0\\0\\1\end{pmatrix} p_5 \mid p_1, p_5 \in \mathbb{R}\}$$

(we've taken p_1 as one of the parameters in order to express the period in terms of the other quantities).

The set of dimensionless products contains all terms $p^{p_1} \ell^{p_2} m^{p_3} a^{p_4} \theta^{p_5}$ subject to the conditions above. This set forms a vector space under the '+' operation of multiplying two such products and the '·' operation of raising such a product to the power of the scalar (see Exercise 5). The term 'complete set of dimensionless products' in Buckingham's Theorem means a basis for this vector space.

We can get a basis by first taking $p_1 = 1$, $p_5 = 0$, and then taking $p_1 = 0$, $p_5 = 1$. The associated dimensionless products are $\Pi_1 = p\ell^{-1/2}g^{1/2}$ and $\Pi_2 = \theta$. Because the set $\{\Pi_1, \Pi_2\}$ is complete, Buckingham's Theorem says that

$$p = \ell^{1/2}g^{-1/2} \cdot \hat{f}(\theta) = \sqrt{\ell/g} \cdot \hat{f}(\theta)$$

where $\hat{f}$ is a function that we cannot determine from this analysis (a first year physics text will show by other means that for small angles it is approximately the constant function $\hat{f}(\theta) = 2\pi$).

Thus, analysis of the relationships that are possible between the quantities with the given dimensional formulas has given us a fair amount of information: a pendulum's period does not depend on the mass of the bob, and it rises with the square root of the length of the string.

For the next example we try to determine the period of revolution of two bodies in space orbiting each other under mutual gravitational attraction. An experienced investigator could expect that these are the relevant quantities.

quantity	*dimensional formula*
period p	$L^0M^0T^1$
mean separation r	$L^1M^0T^0$
first mass m_1	$L^0M^1T^0$
second mass m_2	$L^0M^1T^0$
gravitational constant G	$L^3M^{-1}T^{-2}$

To get the complete set of dimensionless products we consider the equation

$$(L^0M^0T^1)^{p_1}(L^1M^0T^0)^{p_2}(L^0M^1T^0)^{p_3}(L^0M^1T^0)^{p_4}(L^3M^{-1}T^{-2})^{p_5} = L^0M^0T^0$$

which results in a system

$$\begin{aligned} p_2 \qquad\qquad + 3p_5 &= 0 \\ p_3 + p_4 - p_5 &= 0 \\ p_1 \qquad\qquad\qquad - 2p_5 &= 0 \end{aligned}$$

with this solution.

$$\{ \begin{pmatrix} 1 \\ -3/2 \\ 1/2 \\ 0 \\ 1/2 \end{pmatrix} p_1 + \begin{pmatrix} 0 \\ 0 \\ -1 \\ 1 \\ 0 \end{pmatrix} p_4 \mid p_1, p_4 \in \mathbb{R} \}$$

As earlier, the set of dimensionless products of these quantities forms a vector space and we want to produce a basis for that space, a 'complete' set of dimensionless products. One such set, gotten from setting $p_1 = 1$ and $p_4 = 0$ and also setting $p_1 = 0$ and $p_4 = 1$ is $\{\Pi_1 = pr^{-3/2}m_1^{1/2}G^{1/2}, \Pi_2 = m_1^{-1}m_2\}$. With that, Buckingham's Theorem says that any complete relationship among these quantities is stateable this form.

$$p = r^{3/2}m_1^{-1/2}G^{-1/2} \cdot \hat{f}(m_1^{-1}m_2) = \frac{r^{3/2}}{\sqrt{Gm_1}} \cdot \hat{f}(m_2/m_1)$$

Remark. An important application of the prior formula is when m_1 is the mass of the sun and m_2 is the mass of a planet. Because m_1 is very much greater than m_2, the argument to $\hat{f}$ is approximately 0, and we can wonder whether this part of the formula remains approximately constant as m_2 varies. One way to see that it does is this. The sun is so much larger than the planet that the mutual rotation is approximately about the sun's center. If we vary the planet's mass m_2 by a factor of x (e.g., Venus's mass is $x = 0.815$ times Earth's mass), then the force of attraction is multiplied by x, and x times the force acting on x times the mass gives, since $F = ma$, the same acceleration, about the same center (approximately). Hence, the orbit will be the same and so its period will be the same, and thus the right side of the above equation also remains unchanged (approximately). Therefore, $\hat{f}(m_2/m_1)$ is approximately constant as m_2 varies. This is Kepler's Third Law: the square of the period of a planet is proportional to the cube of the mean radius of its orbit about the sun.

The final example was one of the first explicit applications of dimensional analysis. Lord Raleigh considered the speed of a wave in deep water and suggested these as the relevant quantities.

quantity	*dimensional formula*
velocity of the wave v	$L^1M^0T^{-1}$
density of the water d	$L^{-3}M^1T^0$
acceleration due to gravity g	$L^1M^0T^{-2}$
wavelength λ	$L^1M^0T^0$

The equation

$$(L^1M^0T^{-1})^{p_1}(L^{-3}M^1T^0)^{p_2}(L^1M^0T^{-2})^{p_3}(L^1M^0T^0)^{p_4} = L^0M^0T^0$$

gives this system

$$\begin{aligned} p_1 - 3p_2 + p_3 + p_4 &= 0 \\ p_2 &= 0 \\ -p_1 - 2p_3 &= 0 \end{aligned}$$

with this solution space.

$$\{\begin{pmatrix} 1 \\ 0 \\ -1/2 \\ -1/2 \end{pmatrix} p_1 \mid p_1 \in \mathbb{R}\}$$

There is one dimensionless product, $\Pi_1 = vg^{-1/2}\lambda^{-1/2}$, and so v is $\sqrt{\lambda g}$ times a constant; $\hat{f}$ is constant since it is a function of no arguments. The quantity d is not involved in the relationship.

The three examples above show that dimensional analysis can bring us far toward expressing the relationship among the quantities. For further reading, the classic reference is [Bridgman]—this brief book is delightful. Another source is [Giordano, Wells, Wilde]. A description of dimensional analysis's place in modeling is in [Giordano, Jaye, Weir].

Exercises

1 [de Mestre] Consider a projectile, launched with initial velocity v_0, at an angle θ. To study its motion we may guess that these are the relevant quantities.

quantity	*dimensional formula*
horizontal position x	$L^1M^0T^0$
vertical position y	$L^1M^0T^0$
initial speed v_0	$L^1M^0T^{-1}$
angle of launch θ	$L^0M^0T^0$
acceleration due to gravity g	$L^1M^0T^{-2}$
time t	$L^0M^0T^1$

(a) Show that $\{gt/v_0, gx/v_0^2, gy/v_0^2, \theta\}$ is a complete set of dimensionless products. (*Hint.* One way to go is to find the appropriate free variables in the linear system that arises but there is a shortcut that uses the properties of a basis.)

(b) These two equations of motion for projectiles are familiar: $x = v_0 \cos(\theta)t$ and $y = v_0 \sin(\theta)t - (g/2)t^2$. Manipulate each to rewrite it as a relationship among the dimensionless products of the prior item.

2 [Einstein] conjectured that the infrared characteristic frequencies of a solid might be determined by the same forces between atoms as determine the solid's ordinary elastic behavior. The relevant quantities are these.

quantity	*dimensional formula*
characteristic frequency ν	$L^0M^0T^{-1}$
compressibility k	$L^1M^{-1}T^2$
number of atoms per cubic cm N	$L^{-3}M^0T^0$
mass of an atom m	$L^0M^1T^0$

Show that there is one dimensionless product. Conclude that, in any complete relationship among quantities with these dimensional formulas, k is a constant times $\nu^{-2}N^{-1/3}m^{-1}$. This conclusion played an important role in the early study of quantum phenomena.

3 [Giordano, Wells, Wilde] The torque produced by an engine has dimensional formula $L^2M^1T^{-2}$. We may first guess that it depends on the engine's rotation rate (with dimensional formula $L^0M^0T^{-1}$), and the volume of air displaced (with dimensional formula $L^3M^0T^0$).

(a) Try to find a complete set of dimensionless products. What goes wrong?

(b) Adjust the guess by adding the density of the air (with dimensional formula $L^{-3}M^1T^0$). Now find a complete set of dimensionless products.

4 [Tilley] Dominoes falling make a wave. We may conjecture that the wave speed v depends on the spacing d between the dominoes, the height h of each domino, and the acceleration due to gravity g.

(a) Find the dimensional formula for each of the four quantities.

(b) Show that $\{\Pi_1 = h/d, \Pi_2 = dg/v^2\}$ is a complete set of dimensionless products.

(c) Show that if h/d is fixed then the propagation speed is proportional to the square root of d.

5 Prove that the dimensionless products form a vector space under the $\vec{+}$ operation of multiplying two such products and the $\vec{\cdot}$ operation of raising such the product to the power of the scalar. (The vector arrows are a precaution against confusion.) That is, prove that, for any particular homogeneous system, this set of products of powers of $m_1, \ldots, m_k$

$$\{m_1^{p_1} \ldots m_k^{p_k} \mid p_1, \ldots, p_k \text{ satisfy the system}\}$$

is a vector space under:

$$m_1^{p_1} \ldots m_k^{p_k} \vec{+} m_1^{q_1} \ldots m_k^{q_k} = m_1^{p_1+q_1} \ldots m_k^{p_k+q_k}$$

and

$$r \vec{\cdot} (m_1^{p_1} \ldots m_k^{p_k}) = m_1^{rp_1} \ldots m_k^{rp_k}$$

(assume that all variables represent real numbers).

6 The advice about apples and oranges is not right. Consider the familiar equations for a circle $C = 2\pi r$ and $A = \pi r^2$.

(a) Check that C and A have different dimensional formulas.

(b) Produce an equation that is not dimensionally homogeneous (i.e., it adds apples and oranges) but is nonetheless true of any circle.

(c) The prior item asks for an equation that is complete but not dimensionally homogeneous. Produce an equation that is dimensionally homogeneous but not complete.

(Just because the old saying isn't strictly right, doesn't keep it from being a useful strategy. Dimensional homogeneity is often used to check the plausibility of equations used in models. For an argument that any complete equation can easily be made dimensionally homogeneous, see [Bridgman], Chapter I, especially page 15.)

Chapter Three

Maps Between Spaces

I Isomorphisms

In the examples following the definition of a vector space we expressed the intuition that some spaces are "the same" as others. For instance, the space of two-tall column vectors and the space of two-wide row vectors are not equal because their elements — column vectors and row vectors — are not equal, but we feel that these spaces differ only in how their elements appear. We will now make this precise.

This section illustrates a common phase of a mathematical investigation. With the help of some examples we've gotten an idea. We will next give a formal definition and then we will produce some results backing our contention that the definition captures the idea. We've seen this happen already, for instance in the first section of the Vector Space chapter. There, the study of linear systems led us to consider collections closed under linear combinations. We defined such a collection as a vector space and we followed it with some supporting results.

That wasn't an end point, instead it led to new insights such as the idea of a basis. Here also, after producing a definition and supporting it, we will get two surprises (pleasant ones). First, we will find that the definition applies to some unforeseen, and interesting, cases. Second, the study of the definition will lead to new ideas. In this way, our investigation will build momentum.

I.1 Definition and Examples

We start with two examples that suggest the right definition.

1.1 Example The space of two-wide row vectors and the space of two-tall column vectors are "the same" in that if we associate the vectors that have the same components, e.g.,

$$\begin{pmatrix}1 & 2\end{pmatrix} \quad\longleftrightarrow\quad \begin{pmatrix}1\\2\end{pmatrix}$$

(read the double arrow as "corresponds to") then this association respects the operations. For instance these corresponding vectors add to corresponding totals

$$\begin{pmatrix}1 & 2\end{pmatrix}+\begin{pmatrix}3 & 4\end{pmatrix}=\begin{pmatrix}4 & 6\end{pmatrix} \quad\longleftrightarrow\quad \begin{pmatrix}1\\2\end{pmatrix}+\begin{pmatrix}3\\4\end{pmatrix}=\begin{pmatrix}4\\6\end{pmatrix}$$

and here is an example of the correspondence respecting scalar multiplication.

$$5\cdot\begin{pmatrix}1 & 2\end{pmatrix}=\begin{pmatrix}5 & 10\end{pmatrix} \quad\longleftrightarrow\quad 5\cdot\begin{pmatrix}1\\2\end{pmatrix}=\begin{pmatrix}5\\10\end{pmatrix}$$

Stated generally, under the correspondence

$$\begin{pmatrix}a_0 & a_1\end{pmatrix} \quad\longleftrightarrow\quad \begin{pmatrix}a_0\\a_1\end{pmatrix}$$

both operations are preserved:

$$\begin{pmatrix}a_0 & a_1\end{pmatrix}+\begin{pmatrix}b_0 & b_1\end{pmatrix}=\begin{pmatrix}a_0+b_0 & a_1+b_1\end{pmatrix} \longleftrightarrow \begin{pmatrix}a_0\\a_1\end{pmatrix}+\begin{pmatrix}b_0\\b_1\end{pmatrix}=\begin{pmatrix}a_0+b_0\\a_1+b_1\end{pmatrix}$$

and

$$r\cdot\begin{pmatrix}a_0 & a_1\end{pmatrix}=\begin{pmatrix}ra_0 & ra_1\end{pmatrix} \quad\longleftrightarrow\quad r\cdot\begin{pmatrix}a_0\\a_1\end{pmatrix}=\begin{pmatrix}ra_0\\ra_1\end{pmatrix}$$

(all of the variables are scalars).

1.2 Example Another two spaces that we can think of as "the same" are $\mathcal{P}_2$, the space of quadratic polynomials, and $\mathbb{R}^3$. A natural correspondence is this.

$$a_0+a_1x+a_2x^2 \quad\longleftrightarrow\quad \begin{pmatrix}a_0\\a_1\\a_2\end{pmatrix} \qquad (\text{e.g., } 1+2x+3x^2 \longleftrightarrow \begin{pmatrix}1\\2\\3\end{pmatrix})$$

This preserves structure: corresponding elements add in a corresponding way

$$\begin{array}{r} a_0+a_1x+a_2x^2\\ +\ b_0+b_1x+b_2x^2\\ \hline (a_0+b_0)+(a_1+b_1)x+(a_2+b_2)x^2 \end{array} \quad\longleftrightarrow\quad \begin{pmatrix}a_0\\a_1\\a_2\end{pmatrix}+\begin{pmatrix}b_0\\b_1\\b_2\end{pmatrix}=\begin{pmatrix}a_0+b_0\\a_1+b_1\\a_2+b_2\end{pmatrix}$$

and scalar multiplication corresponds also.

$$r \cdot (a_0 + a_1 x + a_2 x^2) = (ra_0) + (ra_1)x + (ra_2)x^2 \quad \longleftrightarrow \quad r \cdot \begin{pmatrix} a_0 \\ a_1 \\ a_2 \end{pmatrix} = \begin{pmatrix} ra_0 \\ ra_1 \\ ra_2 \end{pmatrix}$$

1.3 Definition An *isomorphism* between two vector spaces V and W is a map $f\colon V \to W$ that

(1) is a correspondence: f is one-to-one and onto;*

(2) *preserves structure:* if $\vec{v}_1, \vec{v}_2 \in V$ then

$$f(\vec{v}_1 + \vec{v}_2) = f(\vec{v}_1) + f(\vec{v}_2)$$

and if $\vec{v} \in V$ and $r \in \mathbb{R}$ then

$$f(r\vec{v}) = rf(\vec{v})$$

(we write $V \cong W$, read "V is isomorphic to W", when such a map exists).

"Morphism" means map, so "isomorphism" means a map expressing sameness.

1.4 Example The vector space $G = \{c_1 \cos\theta + c_2 \sin\theta \mid c_1, c_2 \in \mathbb{R}\}$ of functions of θ is isomorphic to $\mathbb{R}^2$ under this map.

$$c_1 \cos\theta + c_2 \sin\theta \stackrel{f}{\longmapsto} \begin{pmatrix} c_1 \\ c_2 \end{pmatrix}$$

We will check this by going through the conditions in the definition. We will first verify condition (1), that the map is a correspondence between the sets underlying the spaces.

To establish that f is one-to-one we must prove that $f(\vec{a}) = f(\vec{b})$ only when $\vec{a} = \vec{b}$. If

$$f(a_1 \cos\theta + a_2 \sin\theta) = f(b_1 \cos\theta + b_2 \sin\theta)$$

then by the definition of f

$$\begin{pmatrix} a_1 \\ a_2 \end{pmatrix} = \begin{pmatrix} b_1 \\ b_2 \end{pmatrix}$$

from which we conclude that $a_1 = b_1$ and $a_2 = b_2$, because column vectors are equal only when they have equal components. Thus $a_1 \cos\theta + a_2 \sin\theta = b_1 \cos\theta + b_2 \sin\theta$, and as required we've verified that $f(\vec{a}) = f(\vec{b})$ implies that $\vec{a} = \vec{b}$.

*More information on correspondences is in the appendix.

To prove that f is onto we must check that any member of the codomain $\mathbb{R}^2$ is the image of some member of the domain G. So, consider a member of the codomain

$$\begin{pmatrix} x \\ y \end{pmatrix}$$

and note that it is the image under f of $x\cos\theta + y\sin\theta$.

Next we will verify condition (2), that f preserves structure. This computation shows that f preserves addition.

$$\begin{aligned}
f\big(\,(a_1\cos\theta + a_2\sin\theta) + (b_1\cos\theta + b_2\sin\theta)\,\big) \\
&= f\big(\,(a_1+b_1)\cos\theta + (a_2+b_2)\sin\theta\,\big) \\
&= \begin{pmatrix} a_1+b_1 \\ a_2+b_2 \end{pmatrix} \\
&= \begin{pmatrix} a_1 \\ a_2 \end{pmatrix} + \begin{pmatrix} b_1 \\ b_2 \end{pmatrix} \\
&= f(a_1\cos\theta + a_2\sin\theta) + f(b_1\cos\theta + b_2\sin\theta)
\end{aligned}$$

The computation showing that f preserves scalar multiplication is similar.

$$\begin{aligned}
f\big(\,r\cdot(a_1\cos\theta + a_2\sin\theta)\,\big) &= f(\,ra_1\cos\theta + ra_2\sin\theta\,) \\
&= \begin{pmatrix} ra_1 \\ ra_2 \end{pmatrix} \\
&= r\cdot\begin{pmatrix} a_1 \\ a_2 \end{pmatrix} \\
&= r\cdot f(a_1\cos\theta + a_2\sin\theta)
\end{aligned}$$

With both (1) and (2) verified, we know that f is an isomorphism and we can say that the spaces are isomorphic $G \cong \mathbb{R}^2$.

1.5 Example Let V be the space $\{c_1x + c_2y + c_3z \mid c_1, c_2, c_3 \in \mathbb{R}\}$ of linear combinations of the three variables under the natural addition and scalar multiplication operations. Then V is isomorphic to $\mathcal{P}_2$, the space of quadratic polynomials.

To show this we must produce an isomorphism map. There is more than one possibility; for instance, here are four to choose among.

$$c_1x + c_2y + c_3z \quad \begin{array}{ll}
\stackrel{f_1}{\longmapsto} & c_1 + c_2x + c_3x^2 \\
\stackrel{f_2}{\longmapsto} & c_2 + c_3x + c_1x^2 \\
\stackrel{f_3}{\longmapsto} & -c_1 - c_2x - c_3x^2 \\
\stackrel{f_4}{\longmapsto} & c_1 + (c_1+c_2)x + (c_1+c_3)x^2
\end{array}$$

The first map is the more natural correspondence in that it just carries the coefficients over. However we shall do f_2 to underline that there are isomorphisms other than the obvious one. (Checking that f_1 is an isomorphism is Exercise 14.)

To show that f_2 is one-to-one we will prove that if $f_2(c_1x + c_2y + c_3z) = f_2(d_1x + d_2y + d_3z)$ then $c_1x + c_2y + c_3z = d_1x + d_2y + d_3z$. The assumption that $f_2(c_1x+c_2y+c_3z) = f_2(d_1x+d_2y+d_3z)$ gives, by the definition of f_2, that $c_2+c_3x+c_1x^2 = d_2+d_3x+d_1x^2$. Equal polynomials have equal coefficients so $c_2 = d_2$, $c_3 = d_3$, and $c_1 = d_1$. Hence $f_2(c_1x+c_2y+c_3z) = f_2(d_1x+d_2y+d_3z)$ implies that $c_1x + c_2y + c_3z = d_1x + d_2y + d_3z$, and f_2 is one-to-one.

The map f_2 is onto because a member $a + bx + cx^2$ of the codomain is the image of a member of the domain, namely it is $f_2(cx + ay + bz)$. For instance, $2 + 3x - 4x^2$ is $f_2(-4x + 2y + 3z)$.

The computations for structure preservation are like those in the prior example. The map f_2 preserves addition

$$\begin{aligned}
f_2\bigl((c_1x + c_2y + c_3z) + (d_1x + d_2y + d_3z)\bigr)& \\
&= f_2\bigl((c_1 + d_1)x + (c_2 + d_2)y + (c_3 + d_3)z\bigr) \\
&= (c_2 + d_2) + (c_3 + d_3)x + (c_1 + d_1)x^2 \\
&= (c_2 + c_3x + c_1x^2) + (d_2 + d_3x + d_1x^2) \\
&= f_2(c_1x + c_2y + c_3z) + f_2(d_1x + d_2y + d_3z)
\end{aligned}$$

and scalar multiplication.

$$\begin{aligned}
f_2\bigl(r \cdot (c_1x + c_2y + c_3z)\bigr) &= f_2(rc_1x + rc_2y + rc_3z) \\
&= rc_2 + rc_3x + rc_1x^2 \\
&= r \cdot (c_2 + c_3x + c_1x^2) \\
&= r \cdot f_2(c_1x + c_2y + c_3z)
\end{aligned}$$

Thus f_2 is an isomorphism. We write $V \cong \mathcal{P}_2$.

1.6 Example Every space is isomorphic to itself under the identity map. The check is easy.

1.7 Definition An *automorphism* is an isomorphism of a space with itself.

1.8 Example A *dilation* map $d_s\colon \mathbb{R}^2 \to \mathbb{R}^2$ that multiplies all vectors by a nonzero scalar s is an automorphism of $\mathbb{R}^2$.

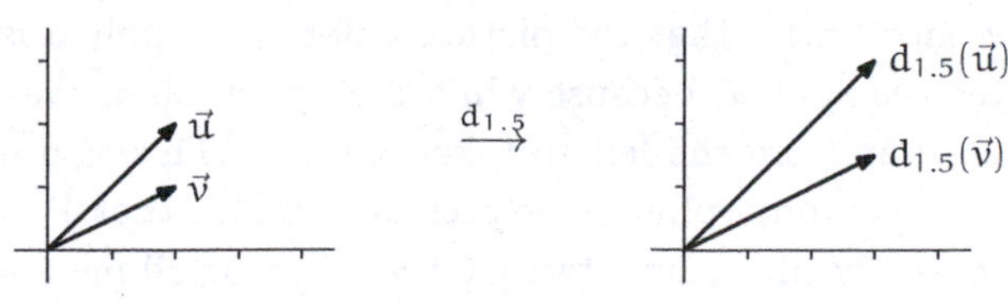

Another automorphism is a *rotation* or *turning map*, $t_\theta\colon \mathbb{R}^2 \to \mathbb{R}^2$ that rotates all vectors through an angle θ.

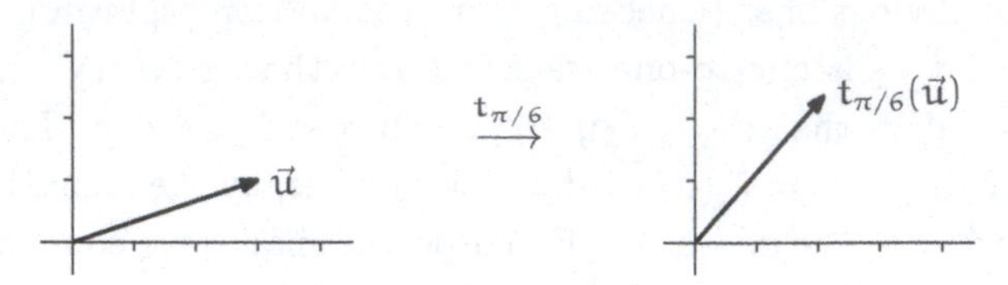

A third type of automorphism of $\mathbb{R}^2$ is a map $f_\ell\colon \mathbb{R}^2 \to \mathbb{R}^2$ that *flips* or *reflects* all vectors over a line ℓ through the origin.

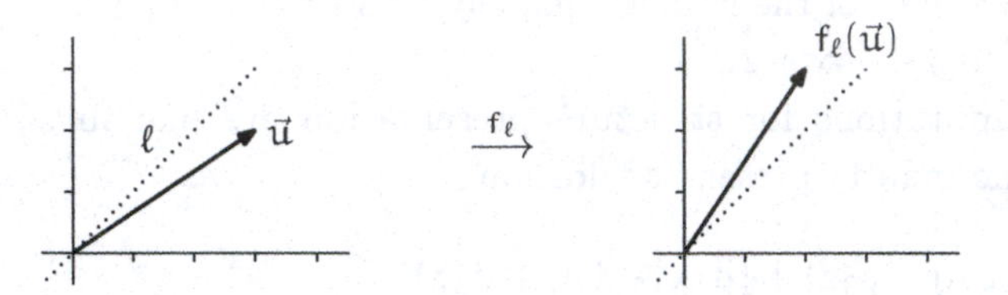

Checking that these are automorphisms is Exercise 33.

1.9 Example Consider the space $\mathcal{P}_5$ of polynomials of degree 5 or less and the map f that sends a polynomial $p(x)$ to $p(x-1)$. For instance, under this map $x^2 \mapsto (x-1)^2 = x^2-2x+1$ and $x^3+2x \mapsto (x-1)^3+2(x-1) = x^3-3x^2+5x-3$. This map is an automorphism of this space; the check is Exercise 25.

This isomorphism of $\mathcal{P}_5$ with itself does more than just tell us that the space is "the same" as itself. It gives us some insight into the space's structure. Below is a family of parabolas, graphs of members of $\mathcal{P}_5$. Each has a vertex at $y=-1$, and the left-most one has zeroes at -2.25 and -1.75, the next one has zeroes at -1.25 and -0.75, etc.

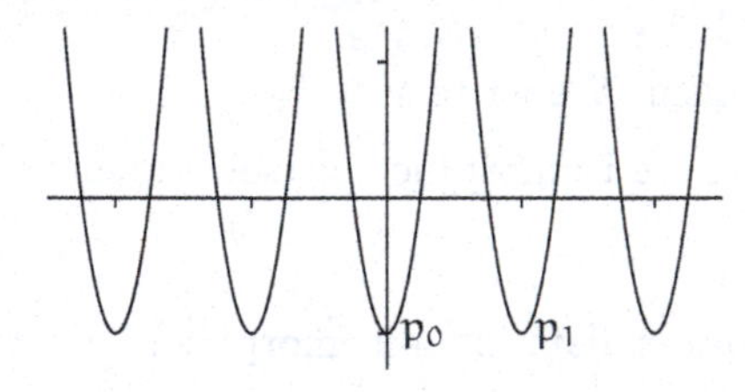

Substitution of $x-1$ for x in any function's argument shifts its graph to the right by one. Thus, $f(p_0) = p_1$, and f's action is to shift all of the parabolas to the right by one. Notice that the picture before f is applied is the same as the picture after f is applied because while each parabola moves to the right, another one comes in from the left to take its place. This also holds true for cubics, etc. So the automorphism f expresses the idea that P_5 has a certain horizontal-homogeneity: if we draw two pictures showing all members of $\mathcal{P}_5$, one picture centered at $x = 0$ and the other centered at $x = 1$, then the two pictures would be indistinguishable.

As described in the opening to this section, having given the definition of isomorphism, we next look to support the thesis that it captures our intuition of vector spaces being the same. First, the definition itself is persuasive: a vector space consists of a set and some structure and the definition simply requires that the sets correspond and that the structures correspond also. Also persuasive are the examples above, such as Example 1.1, which dramatize that isomorphic spaces are the same in all relevant respects. Sometimes people say, where $V \cong W$, that "W is just V painted green" — differences are merely cosmetic.

The results below further support our contention that under an isomorphism all the things of interest in the two vector spaces correspond. Because we introduced vector spaces to study linear combinations, "of interest" means "pertaining to linear combinations." Not of interest is the way that the vectors are presented typographically (or their color!).

1.10 Lemma An isomorphism maps a zero vector to a zero vector.

PROOF Where $f\colon V \to W$ is an isomorphism, fix some $\vec{v} \in V$. Then $f(\vec{0}_V) = f(0 \cdot \vec{v}) = 0 \cdot f(\vec{v}) = \vec{0}_W$. QED

1.11 Lemma For any map $f\colon V \to W$ between vector spaces these statements are equivalent.

(1) f preserves structure

$$f(\vec{v}_1 + \vec{v}_2) = f(\vec{v}_1) + f(\vec{v}_2) \quad \text{and} \quad f(c\vec{v}) = c\,f(\vec{v})$$

(2) f preserves linear combinations of two vectors

$$f(c_1\vec{v}_1 + c_2\vec{v}_2) = c_1 f(\vec{v}_1) + c_2 f(\vec{v}_2)$$

(3) f preserves linear combinations of any finite number of vectors

$$f(c_1\vec{v}_1 + \cdots + c_n\vec{v}_n) = c_1 f(\vec{v}_1) + \cdots + c_n f(\vec{v}_n)$$

PROOF Since the implications $(3) \implies (2)$ and $(2) \implies (1)$ are clear, we need only show that $(1) \implies (3)$. So assume statement (1). We will prove (3) by induction on the number of summands n.

The one-summand base case, that $f(c\vec{v}_1) = c\,f(\vec{v}_1)$, is covered by the second clause of statement (1).

For the inductive step assume that statement (3) holds whenever there are k or fewer summands. Consider the $k+1$-summand case. Use the first half of (1) to break the sum along the final '+'.

$$f(c_1\vec{v}_1 + \cdots + c_k\vec{v}_k + c_{k+1}\vec{v}_{k+1}) = f(c_1\vec{v}_1 + \cdots + c_k\vec{v}_k) + f(c_{k+1}\vec{v}_{k+1})$$

Use the inductive hypothesis to break up the k-term sum on the left.

$$= f(c_1\vec{v}_1) + \cdots + f(c_k\vec{v}_k) + f(c_{k+1}\vec{v}_{k+1})$$

Now the second half of (1) gives

$$= c_1\,f(\vec{v}_1) + \cdots + c_k\,f(\vec{v}_k) + c_{k+1}\,f(\vec{v}_{k+1})$$

when applied $k+1$ times. QED

We often use item (2) to simplify the verification that a map preserves structure.

Finally, a summary. In the prior chapter, after giving the definition of a vector space, we looked at examples and noted that some spaces seemed to be essentially the same as others. Here we have defined the relation '$\cong$' and have argued that it is the right way to precisely say what we mean by "the same" because it preserves the features of interest in a vector space — in particular, it preserves linear combinations. In the next section we will show that isomorphism is an equivalence relation and so partitions the collection of vector spaces.

Exercises

✓ **1.12** Verify, using Example 1.4 as a model, that the two correspondences given before the definition are isomorphisms.
(a) Example 1.1 (b) Example 1.2

✓ **1.13** For the map $f\colon \mathcal{P}_1 \to \mathbb{R}^2$ given by

$$a + bx \stackrel{f}{\longmapsto} \begin{pmatrix} a-b \\ b \end{pmatrix}$$

Find the image of each of these elements of the domain.
(a) $3-2x$ (b) $2+2x$ (c) x
Show that this map is an isomorphism.

1.14 Show that the natural map f_1 from Example 1.5 is an isomorphism.

1.15 Show that the map $t\colon \mathcal{P}_2 \to \mathcal{P}_2$ given by $t(ax^2+bx+c) = bx^2-(a+c)x+a$ is an isomorphism.

1.16 Verify that this map is an isomorphism: $h\colon \mathbb{R}^4 \to \mathcal{M}_{2\times2}$ given by

$$\begin{pmatrix} a \\ b \\ c \\ d \end{pmatrix} \mapsto \begin{pmatrix} c & a+d \\ b & d \end{pmatrix}$$

✓ **1.17** Decide whether each map is an isomorphism (if it is an isomorphism then prove it and if it isn't then state a condition that it fails to satisfy).
(a) $f\colon \mathcal{M}_{2\times2} \to \mathbb{R}$ given by

$$\begin{pmatrix} a & b \\ c & d \end{pmatrix} \mapsto ad-bc$$

(b) $f\colon \mathcal{M}_{2\times2} \to \mathbb{R}^4$ given by

$$\begin{pmatrix} a & b \\ c & d \end{pmatrix} \mapsto \begin{pmatrix} a+b+c+d \\ a+b+c \\ a+b \\ a \end{pmatrix}$$

(c) $f\colon \mathcal{M}_{2\times2} \to \mathcal{P}_3$ given by

$$\begin{pmatrix} a & b \\ c & d \end{pmatrix} \mapsto c + (d+c)x + (b+a)x^2 + ax^3$$

(d) $f\colon \mathcal{M}_{2\times2} \to \mathcal{P}_3$ given by

$$\begin{pmatrix} a & b \\ c & d \end{pmatrix} \mapsto c + (d+c)x + (b+a+1)x^2 + ax^3$$

1.18 Show that the map $f\colon \mathbb{R}^1 \to \mathbb{R}^1$ given by $f(x) = x^3$ is one-to-one and onto. Is it an isomorphism?

✓ **1.19** Refer to Example 1.1. Produce two more isomorphisms (of course, you must also verify that they satisfy the conditions in the definition of isomorphism).

1.20 Refer to Example 1.2. Produce two more isomorphisms (and verify that they satisfy the conditions).

✓ **1.21** Show that, although $\mathbb{R}^2$ is not itself a subspace of $\mathbb{R}^3$, it is isomorphic to the xy-plane subspace of $\mathbb{R}^3$.

1.22 Find two isomorphisms between $\mathbb{R}^{16}$ and $\mathcal{M}_{4\times4}$.

✓ **1.23** For what k is $\mathcal{M}_{m\times n}$ isomorphic to $\mathbb{R}^k$?

1.24 For what k is $\mathcal{P}_k$ isomorphic to $\mathbb{R}^n$?

1.25 Prove that the map in Example 1.9, from $\mathcal{P}_5$ to $\mathcal{P}_5$ given by $p(x) \mapsto p(x-1)$, is a vector space isomorphism.

1.26 Why, in Lemma 1.10, must there be a $\vec{v} \in V$? That is, why must V be nonempty?

1.27 Are any two trivial spaces isomorphic?

1.28 In the proof of Lemma 1.11, what about the zero-summands case (that is, if n is zero)?

1.29 Show that any isomorphism $f\colon \mathcal{P}_0 \to \mathbb{R}^1$ has the form $a \mapsto ka$ for some nonzero real number k.

✓ **1.30** These prove that isomorphism is an equivalence relation.

(a) Show that the identity map $\mathrm{id}\colon V \to V$ is an isomorphism. Thus, any vector space is isomorphic to itself.

(b) Show that if $f\colon V \to W$ is an isomorphism then so is its inverse $f^{-1}\colon W \to V$. Thus, if V is isomorphic to W then also W is isomorphic to V.

(c) Show that a composition of isomorphisms is an isomorphism: if $f\colon V \to W$ is an isomorphism and $g\colon W \to U$ is an isomorphism then so also is $g \circ f\colon V \to U$. Thus, if V is isomorphic to W and W is isomorphic to U, then also V is isomorphic to U.

1.31 Suppose that $f\colon V \to W$ preserves structure. Show that f is one-to-one if and only if the unique member of V mapped by f to $\vec{0}_W$ is $\vec{0}_V$.

1.32 Suppose that $f\colon V \to W$ is an isomorphism. Prove that the set $\{\vec{v}_1,\ldots,\vec{v}_k\} \subseteq V$ is linearly dependent if and only if the set of images $\{f(\vec{v}_1),\ldots,f(\vec{v}_k)\} \subseteq W$ is linearly dependent.

✓ 1.33 Show that each type of map from Example 1.8 is an automorphism.

(a) Dilation d_s by a nonzero scalar s.

(b) Rotation t_θ through an angle θ.

(c) Reflection f_ℓ over a line through the origin.

Hint. For the second and third items, polar coordinates are useful.

1.34 Produce an automorphism of $\mathcal{P}_2$ other than the identity map, and other than a shift map $p(x) \mapsto p(x-k)$.

1.35 (a) Show that a function $f\colon \mathbb{R}^1 \to \mathbb{R}^1$ is an automorphism if and only if it has the form $x \mapsto kx$ for some $k \neq 0$.

(b) Let f be an automorphism of $\mathbb{R}^1$ such that $f(3)=7$. Find $f(-2)$.

(c) Show that a function $f\colon \mathbb{R}^2 \to \mathbb{R}^2$ is an automorphism if and only if it has the form

$$\begin{pmatrix} x \\ y \end{pmatrix} \mapsto \begin{pmatrix} ax+by \\ cx+dy \end{pmatrix}$$

for some $a,b,c,d \in \mathbb{R}$ with $ad-bc \neq 0$. *Hint.* Exercises in prior subsections have shown that

$$\begin{pmatrix} b \\ d \end{pmatrix} \text{ is not a multiple of } \begin{pmatrix} a \\ c \end{pmatrix}$$

if and only if $ad-bc \neq 0$.

(d) Let f be an automorphism of $\mathbb{R}^2$ with

$$f(\begin{pmatrix} 1 \\ 3 \end{pmatrix}) = \begin{pmatrix} 2 \\ -1 \end{pmatrix} \quad\text{and}\quad f(\begin{pmatrix} 1 \\ 4 \end{pmatrix}) = \begin{pmatrix} 0 \\ 1 \end{pmatrix}.$$

Find

$$f(\begin{pmatrix} 0 \\ -1 \end{pmatrix}).$$

1.36 Refer to Lemma 1.10 and Lemma 1.11. Find two more things preserved by isomorphism.

1.37 We show that isomorphisms can be tailored to fit in that, sometimes, given vectors in the domain and in the range we can produce an isomorphism associating those vectors.

(a) Let $B = \langle \vec{\beta}_1, \vec{\beta}_2, \vec{\beta}_3 \rangle$ be a basis for $\mathcal{P}_2$ so that any $\vec{p} \in \mathcal{P}_2$ has a unique representation as $\vec{p} = c_1\vec{\beta}_1 + c_2\vec{\beta}_2 + c_3\vec{\beta}_3$, which we denote in this way.

$$\text{Rep}_B(\vec{p}) = \begin{pmatrix} c_1 \\ c_2 \\ c_3 \end{pmatrix}$$

Show that the $\text{Rep}_B(\cdot)$ operation is a function from $\mathcal{P}_2$ to $\mathbb{R}^3$ (this entails showing that with every domain vector $\vec{v} \in \mathcal{P}_2$ there is an associated image vector in $\mathbb{R}^3$, and further, that with every domain vector $\vec{v} \in \mathcal{P}_2$ there is at most one associated image vector).

(b) Show that this $\text{Rep}_B(\cdot)$ function is one-to-one and onto.

(c) Show that it preserves structure.

(d) Produce an isomorphism from $\mathcal{P}_2$ to $\mathbb{R}^3$ that fits these specifications.

$$x + x^2 \mapsto \begin{pmatrix}1\\0\\0\end{pmatrix} \quad\text{and}\quad 1 - x \mapsto \begin{pmatrix}0\\1\\0\end{pmatrix}$$

1.38 Prove that a space is n-dimensional if and only if it is isomorphic to $\mathbb{R}^n$. *Hint.* Fix a basis B for the space and consider the map sending a vector over to its representation with respect to B.

1.39 *(Requires the subsection on Combining Subspaces, which is optional.)* Let U and W be vector spaces. Define a new vector space, consisting of the set $U \times W = \{(\vec{u}, \vec{w}) \mid \vec{u} \in U \text{ and } \vec{w} \in W\}$ along with these operations.

$$(\vec{u}_1, \vec{w}_1) + (\vec{u}_2, \vec{w}_2) = (\vec{u}_1 + \vec{u}_2, \vec{w}_1 + \vec{w}_2) \quad\text{and}\quad r \cdot (\vec{u}, \vec{w}) = (r\vec{u}, r\vec{w})$$

This is a vector space, the *external direct sum* of U and W.

(a) Check that it is a vector space.

(b) Find a basis for, and the dimension of, the external direct sum $\mathcal{P}_2 \times \mathbb{R}^2$.

(c) What is the relationship among $\dim(U)$, $\dim(W)$, and $\dim(U \times W)$?

(d) Suppose that U and W are subspaces of a vector space V such that $V = U \oplus W$ (in this case we say that V is the *internal direct sum* of U and W). Show that the map $f\colon U \times W \to V$ given by

$$(\vec{u}, \vec{w}) \stackrel{f}{\longmapsto} \vec{u} + \vec{w}$$

is an isomorphism. Thus if the internal direct sum is defined then the internal and external direct sums are isomorphic.

I.2 Dimension Characterizes Isomorphism

In the prior subsection, after stating the definition of isomorphism, we gave some results supporting our sense that such a map describes spaces as "the same." Here we will develop this intuition. When two (unequal) spaces are isomorphic we think of them as almost equal, as equivalent. We shall make that precise by proving that the relationship 'is isomorphic to' is an equivalence relation.

2.1 Lemma The inverse of an isomorphism is also an isomorphism.

Proof Suppose that V is isomorphic to W via $f\colon V \to W$. An isomorphism is a correspondence between the sets so f has an inverse function $f^{-1}\colon W \to V$ that is also a correspondence.*

We will show that because f preserves linear combinations, so also does f^{-1}. Suppose that $\vec{w}_1, \vec{w}_2 \in W$. Because it is an isomorphism, f is onto and there

* More information on inverse functions is in the appendix.

are $\vec{v}_1, \vec{v}_2 \in V$ such that $\vec{w}_1 = f(\vec{v}_1)$ and $\vec{w}_2 = f(\vec{v}_2)$. Then

$$\begin{aligned} f^{-1}(c_1 \cdot \vec{w}_1 + c_2 \cdot \vec{w}_2) &= f^{-1}\big(c_1 \cdot f(\vec{v}_1) + c_2 \cdot f(\vec{v}_2)\big) \\ &= f^{-1}\big(f(c_1\vec{v}_1 + c_2\vec{v}_2)\big) = c_1\vec{v}_1 + c_2\vec{v}_2 = c_1 \cdot f^{-1}(\vec{w}_1) + c_2 \cdot f^{-1}(\vec{w}_2) \end{aligned}$$

since $f^{-1}(\vec{w}_1) = \vec{v}_1$ and $f^{-1}(\vec{w}_2) = \vec{v}_2$. With that, by Lemma 1.11's second statement, this map preserves structure. QED

2.2 Theorem Isomorphism is an equivalence relation between vector spaces.

Proof We must prove that the relation is symmetric, reflexive, and transitive.

To check reflexivity, that any space is isomorphic to itself, consider the identity map. It is clearly one-to-one and onto. This shows that it preserves linear combinations.

$$\mathrm{id}(c_1 \cdot \vec{v}_1 + c_2 \cdot \vec{v}_2) = c_1\vec{v}_1 + c_2\vec{v}_2 = c_1 \cdot \mathrm{id}(\vec{v}_1) + c_2 \cdot \mathrm{id}(\vec{v}_2)$$

Symmetry, that if V is isomorphic to W then also W is isomorphic to V, holds by Lemma 2.1 since each isomorphism map from V to W is paired with an isomorphism from W to V.

To finish we must check transitivity, that if V is isomorphic to W and W is isomorphic to U then V is isomorphic to U. Let $f\colon V \to W$ and $g\colon W \to U$ be isomorphisms. Consider their composition $g \circ f\colon V \to U$. Because the composition of correspondences is a correspondence, we need only check that the composition preserves linear combinations.

$$\begin{aligned} g \circ f\,(c_1 \cdot \vec{v}_1 + c_2 \cdot \vec{v}_2) &= g\big(f(c_1 \cdot \vec{v}_1 + c_2 \cdot \vec{v}_2)\big) \\ &= g\big(c_1 \cdot f(\vec{v}_1) + c_2 \cdot f(\vec{v}_2)\big) \\ &= c_1 \cdot g(f(\vec{v}_1)) + c_2 \cdot g(f(\vec{v}_2)) \\ &= c_1 \cdot (g \circ f)\,(\vec{v}_1) + c_2 \cdot (g \circ f)\,(\vec{v}_2) \end{aligned}$$

Thus the composition is an isomorphism. QED

Since it is an equivalence, isomorphism partitions the universe of vector spaces into classes: each space is in one and only one isomorphism class.

All finite dimensional vector spaces:

$V \cong W$

The next result characterizes these classes by dimension. That is, we can describe each class simply by giving the number that is the dimension of all of the spaces in that class.

2.3 Theorem Vector spaces are isomorphic if and only if they have the same dimension.

In this double implication statement the proof of each half involves a significant idea so we will do the two separately.

2.4 Lemma If spaces are isomorphic then they have the same dimension.

Proof We shall show that an isomorphism of two spaces gives a correspondence between their bases. That is, we shall show that if $f\colon V \to W$ is an isomorphism and a basis for the domain V is $B = \langle \vec{\beta}_1, \dots, \vec{\beta}_n \rangle$ then its image $D = \langle f(\vec{\beta}_1), \dots, f(\vec{\beta}_n) \rangle$ is a basis for the codomain W. (The other half of the correspondence, that for any basis of W the inverse image is a basis for V, follows from the fact that f^{-1} is also an isomorphism and so we can apply the prior sentence to f^{-1}.)

To see that D spans W, fix any $\vec{w} \in W$. Because f is an isomorphism it is onto and so there is a $\vec{v} \in V$ with $\vec{w} = f(\vec{v})$. Expand $\vec{v}$ as a combination of basis vectors.

$$\vec{w} = f(\vec{v}) = f(v_1\vec{\beta}_1 + \cdots + v_n\vec{\beta}_n) = v_1 \cdot f(\vec{\beta}_1) + \cdots + v_n \cdot f(\vec{\beta}_n)$$

For linear independence of D, if

$$\vec{0}_W = c_1 f(\vec{\beta}_1) + \cdots + c_n f(\vec{\beta}_n) = f(c_1\vec{\beta}_1 + \cdots + c_n\vec{\beta}_n)$$

then, since f is one-to-one and so the only vector sent to $\vec{0}_W$ is $\vec{0}_V$, we have that $\vec{0}_V = c_1\vec{\beta}_1 + \cdots + c_n\vec{\beta}_n$, which implies that all of the c's are zero. QED

2.5 Lemma If spaces have the same dimension then they are isomorphic.

Proof We will prove that any space of dimension n is isomorphic to $\mathbb{R}^n$. Then we will have that all such spaces are isomorphic to each other by transitivity, which was shown in Theorem 2.2.

Let V be n-dimensional. Fix a basis $B = \langle \vec{\beta}_1, \dots, \vec{\beta}_n \rangle$ for the domain V. Consider the operation of representing the members of V with respect to B as a function from V to $\mathbb{R}^n$.

$$\vec{v} = v_1\vec{\beta}_1 + \cdots + v_n\vec{\beta}_n \overset{\text{Rep}_B}{\longmapsto} \begin{pmatrix} v_1 \\ \vdots \\ v_n \end{pmatrix}$$

It is well-defined* since every $\vec{v}$ has one and only one such representation (see Remark 2.6 following this proof).

* More information on well-defined is in the appendix.

This function is one-to-one because if

$$\mathrm{Rep}_B(u_1\vec{\beta}_1+\cdots+u_n\vec{\beta}_n)=\mathrm{Rep}_B(v_1\vec{\beta}_1+\cdots+v_n\vec{\beta}_n)$$

then

$$\begin{pmatrix}u_1\\ \vdots\\ u_n\end{pmatrix}=\begin{pmatrix}v_1\\ \vdots\\ v_n\end{pmatrix}$$

and so $u_1=v_1, \ldots, u_n=v_n$, implying that the original arguments $u_1\vec{\beta}_1+\cdots+u_n\vec{\beta}_n$ and $v_1\vec{\beta}_1+\cdots+v_n\vec{\beta}_n$ are equal.

This function is onto; any member of $\mathbb{R}^n$

$$\vec{w}=\begin{pmatrix}w_1\\ \vdots\\ w_n\end{pmatrix}$$

is the image of some $\vec{v}\in V$, namely $\vec{w}=\mathrm{Rep}_B(w_1\vec{\beta}_1+\cdots+w_n\vec{\beta}_n)$.

Finally, this function preserves structure.

$$\begin{aligned}\mathrm{Rep}_B(r\cdot\vec{u}+s\cdot\vec{v})&=\mathrm{Rep}_B(\,(ru_1+sv_1)\vec{\beta}_1+\cdots+(ru_n+sv_n)\vec{\beta}_n\,)\\ &=\begin{pmatrix}ru_1+sv_1\\ \vdots\\ ru_n+sv_n\end{pmatrix}\\ &=r\cdot\begin{pmatrix}u_1\\ \vdots\\ u_n\end{pmatrix}+s\cdot\begin{pmatrix}v_1\\ \vdots\\ v_n\end{pmatrix}\\ &=r\cdot\mathrm{Rep}_B(\vec{u})+s\cdot\mathrm{Rep}_B(\vec{v})\end{aligned}$$

Therefore Rep_B is an isomorphism. Consequently any n-dimensional space is isomorphic to $\mathbb{R}^n$. QED

2.6 Remark The proof has a sentence about 'well-defined.' Its point is that to be an isomorphism Rep_B must be a function, and the definition of function requires that for all inputs the associated output exists and is determined by the input. So we must check that every $\vec{v}$ is associated with at least one $\mathrm{Rep}_B(\vec{v})$, and no more than one.

In the proof we express elements $\vec{v}$ of the domain space as combinations of members of the basis B and then associate $\vec{v}$ with the column vector of coefficients. That there is at least one expansion of each $\vec{v}$ holds because B is a basis and so spans the space.

The worry that there is no more than one associated member of the codomain is subtler. A contrasting example, where an association fails this unique output requirement, illuminates the issue. Let the domain be $\mathcal{P}_2$ and consider a set that is not a basis (it is not linearly independent, although it does span the space).

$$A = \{1 + 0x + 0x^2, 0 + 1x + 0x^2, 0 + 0x + 1x^2, 1 + 1x + 2x^2\}$$

Call those polynomials $\vec{\alpha}_1, \ldots, \vec{\alpha}_4$. In contrast to the situation when the set is a basis, here there can be more than one expression of a domain vector in terms of members of the set. For instance, consider $\vec{v} = 1 + x + x^2$. Here are two different expansions.

$$\vec{v} = 1\vec{\alpha}_1 + 1\vec{\alpha}_2 + 1\vec{\alpha}_3 + 0\vec{\alpha}_4 \qquad \vec{v} = 0\vec{\alpha}_1 + 0\vec{\alpha}_2 - 1\vec{\alpha}_3 + 1\vec{\alpha}_4$$

So this input vector $\vec{v}$ is associated with more than one column.

$$\begin{pmatrix} 1 \\ 1 \\ 1 \\ 0 \end{pmatrix} \quad \begin{pmatrix} 0 \\ 0 \\ -1 \\ 1 \end{pmatrix}$$

Thus, with A the association is not well-defined. (The issue is that A is not linearly independent; to show uniqueness Theorem Two.III.1.12's proof uses only linear independence.)

In general, any time that we define a function we must check that output values are well-defined. Most of the time that condition is perfectly obvious but in the above proof it needs verification. See Exercise 19.

2.7 Corollary A finite-dimensional vector space is isomorphic to one and only one of the $\mathbb{R}^n$.

This gives us a collection of representatives of the isomorphism classes.

All finite dimensional vector spaces:

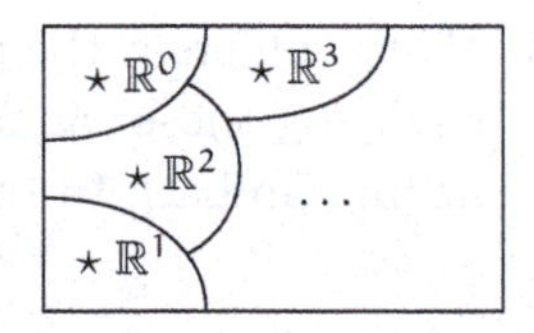

One representative per class

The proofs above pack many ideas into a small space. Through the rest of this chapter we'll consider these ideas again, and fill them out. As a taste of this we will expand here on the proof of Lemma 2.5.

2.8 Example The space $\mathcal{M}_{2\times 2}$ of 2×2 matrices is isomorphic to $\mathbb{R}^4$. With this basis for the domain

$$B = \langle \begin{pmatrix} 1 & 0 \\ 0 & 0 \end{pmatrix}, \begin{pmatrix} 0 & 1 \\ 0 & 0 \end{pmatrix}, \begin{pmatrix} 0 & 0 \\ 1 & 0 \end{pmatrix}, \begin{pmatrix} 0 & 0 \\ 0 & 1 \end{pmatrix} \rangle$$

the isomorphism given in the lemma, the representation map $f_1 = \mathrm{Rep}_B$, carries the entries over.

$$\begin{pmatrix} a & b \\ c & d \end{pmatrix} \overset{f_1}{\longmapsto} \begin{pmatrix} a \\ b \\ c \\ d \end{pmatrix}$$

One way to think of the map f_1 is: fix the basis B for the domain, use the standard basis $\mathcal{E}_4$ for the codomain, and associate $\vec{\beta}_1$ with $\vec{e}_1$, $\vec{\beta}_2$ with $\vec{e}_2$, etc. Then extend this association to all of the members of two spaces.

$$\begin{pmatrix} a & b \\ c & d \end{pmatrix} = a\vec{\beta}_1 + b\vec{\beta}_2 + c\vec{\beta}_3 + d\vec{\beta}_4 \overset{f_1}{\longmapsto} a\vec{e}_1 + b\vec{e}_2 + c\vec{e}_3 + d\vec{e}_4 = \begin{pmatrix} a \\ b \\ c \\ d \end{pmatrix}$$

We can do the same thing with different bases, for instance, taking this basis for the domain.

$$A = \langle \begin{pmatrix} 2 & 0 \\ 0 & 0 \end{pmatrix}, \begin{pmatrix} 0 & 2 \\ 0 & 0 \end{pmatrix}, \begin{pmatrix} 0 & 0 \\ 2 & 0 \end{pmatrix}, \begin{pmatrix} 0 & 0 \\ 0 & 2 \end{pmatrix} \rangle$$

Associating corresponding members of A and $\mathcal{E}_4$ gives this.

$$\begin{pmatrix} a & b \\ c & d \end{pmatrix} = (a/2)\vec{\alpha}_1 + (b/2)\vec{\alpha}_2 + (c/2)\vec{\alpha}_3 + (d/2)\vec{\alpha}_4$$

$$\overset{f_2}{\longmapsto} (a/2)\vec{e}_1 + (b/2)\vec{e}_2 + (c/2)\vec{e}_3 + (d/2)\vec{e}_4 = \begin{pmatrix} a/2 \\ b/2 \\ c/2 \\ d/2 \end{pmatrix}$$

gives rise to an isomorphism that is different than f_1.

The prior map arose by changing the basis for the domain. We can also change the basis for the codomain. Go back to the basis B above and use this basis for the codomain.

$$D = \langle \begin{pmatrix} 1 \\ 0 \\ 0 \\ 0 \end{pmatrix}, \begin{pmatrix} 0 \\ 1 \\ 0 \\ 0 \end{pmatrix}, \begin{pmatrix} 0 \\ 0 \\ 0 \\ 1 \end{pmatrix}, \begin{pmatrix} 0 \\ 0 \\ 1 \\ 0 \end{pmatrix} \rangle$$

Associate $\vec{\beta}_1$ with $\vec{\delta}_1$, etc. Extending that gives another isomorphism.

$$\begin{pmatrix} a & b \\ c & d \end{pmatrix} = a\vec{\beta}_1 + b\vec{\beta}_2 + c\vec{\beta}_3 + d\vec{\beta}_4 \stackrel{f_3}{\longmapsto} a\vec{\delta}_1 + b\vec{\delta}_2 + c\vec{\delta}_3 + d\vec{\delta}_4 = \begin{pmatrix} a \\ b \\ d \\ c \end{pmatrix}$$

We close with a recap. Recall that the first chapter defines two matrices to be row equivalent if they can be derived from each other by row operations. There we showed that relation is an equivalence and so the collection of matrices is partitioned into classes, where all the matrices that are row equivalent together fall into a single class. Then for insight into which matrices are in each class we gave representatives for the classes, the reduced echelon form matrices.

In this section we have followed that pattern except that the notion here of "the same" is vector space isomorphism. We defined it and established some properties, including that it is an equivalence. Then, as before, we developed a list of class representatives to help us understand the partition—it classifies vector spaces by dimension.

In Chapter Two, with the definition of vector spaces, we seemed to have opened up our studies to many examples of new structures besides the familiar $\mathbb{R}^n$'s. We now know that isn't the case. Any finite-dimensional vector space is actually "the same" as a real space.

Exercises

✓ **2.9** Decide if the spaces are isomorphic.
  (a) $\mathbb{R}^2$, $\mathbb{R}^4$ (b) $\mathcal{P}_5$, $\mathbb{R}^5$ (c) $\mathcal{M}_{2\times 3}$, $\mathbb{R}^6$ (d) $\mathcal{P}_5$, $\mathcal{M}_{2\times 3}$
  (e) $\mathcal{M}_{2\times k}$, $\mathbb{C}^k$

✓ **2.10** Consider the isomorphism $\operatorname{Rep}_B(\cdot)\colon \mathcal{P}_1 \to \mathbb{R}^2$ where $B = \langle 1, 1+x \rangle$. Find the image of each of these elements of the domain.
  (a) $3-2x$; (b) $2+2x$; (c) x

✓ **2.11** Show that if $m \neq n$ then $\mathbb{R}^m \ncong \mathbb{R}^n$.

✓ **2.12** Is $\mathcal{M}_{m\times n} \cong \mathcal{M}_{n\times m}$?

✓ **2.13** Are any two planes through the origin in $\mathbb{R}^3$ isomorphic?

2.14 Find a set of equivalence class representatives other than the set of $\mathbb{R}^n$'s.

2.15 True or false: between any n-dimensional space and $\mathbb{R}^n$ there is exactly one isomorphism.

2.16 Can a vector space be isomorphic to one of its (proper) subspaces?

✓ **2.17** This subsection shows that for any isomorphism, the inverse map is also an isomorphism. This subsection also shows that for a fixed basis B of an n-dimensional vector space V, the map $\operatorname{Rep}_B\colon V \to \mathbb{R}^n$ is an isomorphism. Find the inverse of this map.

✓ **2.18** Prove these facts about matrices.
  (a) The row space of a matrix is isomorphic to the column space of its transpose.

(b) The row space of a matrix is isomorphic to its column space.

2.19 Show that the function from Theorem 2.3 is well-defined.

2.20 Is the proof of Theorem 2.3 valid when $n = 0$?

2.21 For each, decide if it is a set of isomorphism class representatives.

(a) $\{\mathbb{C}^k \mid k \in \mathbb{N}\}$

(b) $\{\mathcal{P}_k \mid k \in \{-1, 0, 1, \ldots\}\}$

(c) $\{\mathcal{M}_{m \times n} \mid m, n \in \mathbb{N}\}$

2.22 Let f be a correspondence between vector spaces V and W (that is, a map that is one-to-one and onto). Show that the spaces V and W are isomorphic via f if and only if there are bases $B \subset V$ and $D \subset W$ such that corresponding vectors have the same coordinates: $\text{Rep}_B(\vec{v}) = \text{Rep}_D(f(\vec{v}))$.

2.23 Consider the isomorphism $\text{Rep}_B \colon \mathcal{P}_3 \to \mathbb{R}^4$.

(a) Vectors in a real space are orthogonal if and only if their dot product is zero. Give a definition of orthogonality for polynomials.

(b) The derivative of a member of $\mathcal{P}_3$ is in $\mathcal{P}_3$. Give a definition of the derivative of a vector in $\mathbb{R}^4$.

✓ 2.24 Does every correspondence between bases, when extended to the spaces, give an isomorphism? That is, suppose that V is a vector space with basis $B = \langle \vec{\beta}_1, \ldots, \vec{\beta}_n \rangle$ and that $f \colon B \to W$ is a correspondence such that $D = \langle f(\vec{\beta}_1), \ldots, f(\vec{\beta}_n) \rangle$ is basis for W. Must $\hat{f} \colon V \to W$ sending $\vec{v} = c_1\vec{\beta}_1 + \cdots + c_n\vec{\beta}_n$ to $\hat{f}(\vec{v}) = c_1\hat{f}(\vec{\beta}_1) + \cdots + c_n\hat{f}(\vec{\beta}_n)$ be an isomorphism?

2.25 *(Requires the subsection on Combining Subspaces, which is optional.)* Suppose that $V = V_1 \oplus V_2$ and that V is isomorphic to the space U under the map f. Show that $U = f(V_1) \oplus f(U_2)$.

2.26 Show that this is not a well-defined function from the rational numbers to the integers: with each fraction, associate the value of its numerator.

II Homomorphisms

The definition of isomorphism has two conditions. In this section we will consider the second one. We will study maps that are required only to preserve structure, maps that are not also required to be correspondences.

Experience shows that these maps are tremendously useful. For one thing we shall see in the second subsection below that while isomorphisms describe how spaces are the same, we can think of these maps as describing how spaces are alike.

II.1 Definition

1.1 Definition A function between vector spaces $h: V \to W$ that preserves addition

$$\text{if } \vec{v}_1, \vec{v}_2 \in V \text{ then } h(\vec{v}_1 + \vec{v}_2) = h(\vec{v}_1) + h(\vec{v}_2)$$

and scalar multiplication

$$\text{if } \vec{v} \in V \text{ and } r \in \mathbb{R} \text{ then } h(r \cdot \vec{v}) = r \cdot h(\vec{v})$$

is a *homomorphism* or *linear map*.

1.2 Example The projection map $\pi: \mathbb{R}^3 \to \mathbb{R}^2$

$$\begin{pmatrix} x \\ y \\ z \end{pmatrix} \stackrel{\pi}{\longmapsto} \begin{pmatrix} x \\ y \end{pmatrix}$$

is a homomorphism. It preserves addition

$$\pi(\begin{pmatrix} x_1 \\ y_1 \\ z_1 \end{pmatrix} + \begin{pmatrix} x_2 \\ y_2 \\ z_2 \end{pmatrix}) = \pi(\begin{pmatrix} x_1 + x_2 \\ y_1 + y_2 \\ z_1 + z_2 \end{pmatrix}) = \begin{pmatrix} x_1 + x_2 \\ y_1 + y_2 \end{pmatrix} = \pi(\begin{pmatrix} x_1 \\ y_1 \\ z_1 \end{pmatrix}) + \pi(\begin{pmatrix} x_2 \\ y_2 \\ z_2 \end{pmatrix})$$

and scalar multiplication.

$$\pi(r \cdot \begin{pmatrix} x_1 \\ y_1 \\ z_1 \end{pmatrix}) = \pi(\begin{pmatrix} rx_1 \\ ry_1 \\ rz_1 \end{pmatrix}) = \begin{pmatrix} rx_1 \\ ry_1 \end{pmatrix} = r \cdot \pi(\begin{pmatrix} x_1 \\ y_1 \\ z_1 \end{pmatrix})$$

This is not an isomorphism since it is not one-to-one. For instance, both $\vec{0}$ and $\vec{e}_3$ in $\mathbb{R}^3$ map to the zero vector in $\mathbb{R}^2$.

1.3 Example The domain and codomain can be other than spaces of column vectors. Both of these are homomorphisms; the verifications are straightforward.

(1) $f_1 \colon \mathcal{P}_2 \to \mathcal{P}_3$ given by

$$a_0 + a_1 x + a_2 x^2 \mapsto a_0 x + (a_1/2) x^2 + (a_2/3) x^3$$

(2) $f_2 \colon \mathcal{M}_{2\times 2} \to \mathbb{R}$ given by

$$\begin{pmatrix} a & b \\ c & d \end{pmatrix} \mapsto a + d$$

1.4 Example Between any two spaces there is a *zero homomorphism*, mapping every vector in the domain to the zero vector in the codomain.

1.5 Example These two suggest why we use the term 'linear map'.

(1) The map $g \colon \mathbb{R}^3 \to \mathbb{R}$ given by

$$\begin{pmatrix} x \\ y \\ z \end{pmatrix} \stackrel{g}{\longmapsto} 3x + 2y - 4.5z$$

is linear, that is, is a homomorphism. The check is easy. In contrast, the map $\hat{g} \colon \mathbb{R}^3 \to \mathbb{R}$ given by

$$\begin{pmatrix} x \\ y \\ z \end{pmatrix} \stackrel{\hat{g}}{\longmapsto} 3x + 2y - 4.5z + 1$$

is not linear. To show this we need only produce a single linear combination that the map does not preserve. Here is one.

$$\hat{g}(\begin{pmatrix} 0 \\ 0 \\ 0 \end{pmatrix} + \begin{pmatrix} 1 \\ 0 \\ 0 \end{pmatrix}) = 4 \qquad \hat{g}(\begin{pmatrix} 0 \\ 0 \\ 0 \end{pmatrix}) + \hat{g}(\begin{pmatrix} 1 \\ 0 \\ 0 \end{pmatrix}) = 5$$

(2) The first of these two maps $t_1, t_2 \colon \mathbb{R}^3 \to \mathbb{R}^2$ is linear while the second is not.

$$\begin{pmatrix} x \\ y \\ z \end{pmatrix} \stackrel{t_1}{\longmapsto} \begin{pmatrix} 5x - 2y \\ x + y \end{pmatrix} \qquad \begin{pmatrix} x \\ y \\ z \end{pmatrix} \stackrel{t_2}{\longmapsto} \begin{pmatrix} 5x - 2y \\ xy \end{pmatrix}$$

Finding a linear combination that the second map does not preserve is easy.

So one way to think of 'homomorphism' is that we are generalizing 'isomorphism' (by dropping the condition that the map is a correspondence), motivated by the observation that many of the properties of isomorphisms have only to do with the map's structure-preservation property. The next two results are examples of this motivation. In the prior section we saw a proof for each that only uses preservation of addition and preservation of scalar multiplication, and therefore applies to homomorphisms.

1.6 Lemma A homomorphism sends the zero vector to the zero vector.

1.7 Lemma The following are equivalent for any map $f\colon V \to W$ between vector spaces.

(1) f is a homomorphism
(2) $f(c_1 \cdot \vec{v}_1 + c_2 \cdot \vec{v}_2) = c_1 \cdot f(\vec{v}_1) + c_2 \cdot f(\vec{v}_2)$ for any $c_1, c_2 \in \mathbb{R}$ and $\vec{v}_1, \vec{v}_2 \in V$
(3) $f(c_1 \cdot \vec{v}_1 + \cdots + c_n \cdot \vec{v}_n) = c_1 \cdot f(\vec{v}_1) + \cdots + c_n \cdot f(\vec{v}_n)$ for any $c_1, \ldots, c_n \in \mathbb{R}$ and $\vec{v}_1, \ldots, \vec{v}_n \in V$

1.8 Example The function $f\colon \mathbb{R}^2 \to \mathbb{R}^4$ given by

$$\begin{pmatrix} x \\ y \end{pmatrix} \stackrel{f}{\longmapsto} \begin{pmatrix} x/2 \\ 0 \\ x+y \\ 3y \end{pmatrix}$$

is linear since it satisfies item (2).

$$\begin{pmatrix} r_1(x_1/2) + r_2(x_2/2) \\ 0 \\ r_1(x_1+y_1) + r_2(x_2+y_2) \\ r_1(3y_1) + r_2(3y_2) \end{pmatrix} = r_1 \begin{pmatrix} x_1/2 \\ 0 \\ x_1+y_1 \\ 3y_1 \end{pmatrix} + r_2 \begin{pmatrix} x_2/2 \\ 0 \\ x_2+y_2 \\ 3y_2 \end{pmatrix}$$

However, some things that hold for isomorphisms fail to hold for homomorphisms. One example is in the proof of Lemma I.2.4, which shows that an isomorphism between spaces gives a correspondence between their bases. Homomorphisms do not give any such correspondence; Example 1.2 shows this and another example is the zero map between two nontrivial spaces. Instead, for homomorphisms we have a weaker but still very useful result.

1.9 Theorem A homomorphism is determined by its action on a basis: if V is a vector space with basis $\langle \vec{\beta}_1, \ldots, \vec{\beta}_n \rangle$, if W is a vector space, and if $\vec{w}_1, \ldots, \vec{w}_n \in W$ (these codomain elements need not be distinct) then there exists a homomorphism from V to W sending each $\vec{\beta}_i$ to $\vec{w}_i$, and that homomorphism is unique.

Proof For any input $\vec{v} \in V$ let its expression with respect to the basis be $\vec{v} = c_1\vec{\beta}_1 + \cdots + c_n\vec{\beta}_n$. Define the associated output by using the same coordinates $h(\vec{v}) = c_1\vec{w}_1 + \cdots + c_n\vec{w}_n$. This is well defined because, with respect to the basis, the representation of each domain vector $\vec{v}$ is unique.

This map is a homomorphism because it preserves linear combinations: where $\vec{v_1} = c_1\vec{\beta}_1 + \cdots + c_n\vec{\beta}_n$ and $\vec{v_2} = d_1\vec{\beta}_1 + \cdots + d_n\vec{\beta}_n$, here is the calculation.

$$\begin{aligned} h(r_1\vec{v}_1 + r_2\vec{v}_2) &= h(\,(r_1c_1 + r_2d_1)\vec{\beta}_1 + \cdots + (r_1c_n + r_2d_n)\vec{\beta}_n\,) \\ &= (r_1c_1 + r_2d_1)\vec{w}_1 + \cdots + (r_1c_n + r_2d_n)\vec{w}_n \\ &= r_1h(\vec{v}_1) + r_2h(\vec{v}_2) \end{aligned}$$

This map is unique because if $\hat{h}\colon V \to W$ is another homomorphism satisfying that $\hat{h}(\vec{\beta}_i) = \vec{w}_i$ for each i then h and $\hat{h}$ have the same effect on all of the vectors in the domain.

$$\begin{aligned} \hat{h}(\vec{v}) = \hat{h}(c_1\vec{\beta}_1 + \cdots + c_n\vec{\beta}_n) &= c_1\hat{h}(\vec{\beta}_1) + \cdots + c_n\hat{h}(\vec{\beta}_n) \\ &= c_1\vec{w}_1 + \cdots + c_n\vec{w}_n = h(\vec{v}) \end{aligned}$$

They have the same action so they are the same function. QED

1.10 Definition Let V and W be vector spaces and let $B = \langle \vec{\beta}_1, \ldots, \vec{\beta}_n \rangle$ be a basis for V. A function defined on that basis $f\colon B \to W$ is *extended linearly* to a function $\hat{f}\colon V \to W$ if for all $\vec{v} \in V$ such that $\vec{v} = c_1\vec{\beta}_1 + \cdots + c_n\vec{\beta}_n$, the action of the map is $\hat{f}(\vec{v}) = c_1 \cdot f(\vec{\beta}_1) + \cdots + c_n \cdot f(\vec{\beta}_n)$.

1.11 Example If we specify a map $h\colon \mathbb{R}^2 \to \mathbb{R}^2$ that acts on the standard basis $\mathcal{E}_2$ in this way

$$h(\begin{pmatrix} 1 \\ 0 \end{pmatrix}) = \begin{pmatrix} -1 \\ 1 \end{pmatrix} \qquad h(\begin{pmatrix} 0 \\ 1 \end{pmatrix}) = \begin{pmatrix} -4 \\ 4 \end{pmatrix}$$

then we have also specified the action of h on any other member of the domain. For instance, the value of h on this argument

$$h(\begin{pmatrix} 3 \\ -2 \end{pmatrix}) = h(3 \cdot \begin{pmatrix} 1 \\ 0 \end{pmatrix} - 2 \cdot \begin{pmatrix} 0 \\ 1 \end{pmatrix}) = 3 \cdot h(\begin{pmatrix} 1 \\ 0 \end{pmatrix}) - 2 \cdot h(\begin{pmatrix} 0 \\ 1 \end{pmatrix}) = \begin{pmatrix} 5 \\ -5 \end{pmatrix}$$

is a direct consequence of the value of h on the basis vectors.

Later in this chapter we shall develop a convenient scheme for computations like this one, using matrices.

1.12 Definition A linear map from a space into itself $t\colon V \to V$ is a *linear transformation*.

1.13 Remark In this book we use 'linear transformation' only in the case where the codomain equals the domain. However, be aware that other sources may instead use it as a synonym for 'homomorphism'.

1.14 Example The map on $\mathbb{R}^2$ that projects all vectors down to the x-axis is a linear transformation.

$$\begin{pmatrix} x \\ y \end{pmatrix} \mapsto \begin{pmatrix} x \\ 0 \end{pmatrix}$$

1.15 Example The derivative map $d/dx\colon \mathcal{P}_n \to \mathcal{P}_n$

$$a_0 + a_1x + \cdots + a_nx^n \overset{d/dx}{\longmapsto} a_1 + 2a_2x + 3a_3x^2 + \cdots + na_nx^{n-1}$$

is a linear transformation as this result from calculus shows: $d(c_1f + c_2g)/dx = c_1\,(df/dx) + c_2\,(dg/dx)$.

1.16 Example The matrix transpose operation

$$\begin{pmatrix} a & b \\ c & d \end{pmatrix} \mapsto \begin{pmatrix} a & c \\ b & d \end{pmatrix}$$

is a linear transformation of $\mathcal{M}_{2\times 2}$. (Transpose is one-to-one and onto and so is in fact an automorphism.)

We finish this subsection about maps by recalling that we can linearly combine maps. For instance, for these maps from $\mathbb{R}^2$ to itself

$$\begin{pmatrix} x \\ y \end{pmatrix} \overset{f}{\longmapsto} \begin{pmatrix} 2x \\ 3x - 2y \end{pmatrix} \quad\text{and}\quad \begin{pmatrix} x \\ y \end{pmatrix} \overset{g}{\longmapsto} \begin{pmatrix} 0 \\ 5x \end{pmatrix}$$

the linear combination $5f - 2g$ is also a transformation of R^2.

$$\begin{pmatrix} x \\ y \end{pmatrix} \overset{5f-2g}{\longmapsto} \begin{pmatrix} 10x \\ 5x - 10y \end{pmatrix}$$

1.17 Lemma For vector spaces V and W, the set of linear functions from V to W is itself a vector space, a subspace of the space of all functions from V to W.

We denote the space of linear maps from V to W by $\mathcal{L}(V, W)$.

PROOF This set is non-empty because it contains the zero homomorphism. So to show that it is a subspace we need only check that it is closed under the operations. Let $f, g\colon V \to W$ be linear. Then the operation of function addition is preserved

$$\begin{aligned}(f+g)(c_1\vec{v}_1+c_2\vec{v}_2) &= f(c_1\vec{v}_1+c_2\vec{v}_2)+g(c_1\vec{v}_1+c_2\vec{v}_2)\\ &= c_1f(\vec{v}_1)+c_2f(\vec{v}_2)+c_1g(\vec{v}_1)+c_2g(\vec{v}_2)\\ &= c_1\bigl(f+g\bigr)(\vec{v}_1)+c_2\bigl(f+g\bigr)(\vec{v}_2)\end{aligned}$$

as is the operation of scalar multiplication of a function.

$$\begin{aligned}(r\cdot f)(c_1\vec{v}_1+c_2\vec{v}_2) &= r(c_1f(\vec{v}_1)+c_2f(\vec{v}_2))\\ &= c_1(r\cdot f)(\vec{v}_1)+c_2(r\cdot f)(\vec{v}_2)\end{aligned}$$

Hence $\mathcal{L}(V, W)$ is a subspace. QED

We started this section by defining 'homomorphism' as a generalization of 'isomorphism', by isolating the structure preservation property. Some of the points about isomorphisms carried over unchanged, while we adapted others.

Note, however, that the idea of 'homomorphism' is in no way somehow secondary to that of 'isomorphism'. In the rest of this chapter we shall work mostly with homomorphisms. This is partly because any statement made about homomorphisms is automatically true about isomorphisms but more because, while the isomorphism concept is more natural, our experience will show that the homomorphism concept is more fruitful and more central to progress.

Exercises

✓ **1.18** Decide if each $h\colon \mathbb{R}^3 \to \mathbb{R}^2$ is linear.

(a) $h(\begin{pmatrix}x\\y\\z\end{pmatrix}) = \begin{pmatrix}x\\x+y+z\end{pmatrix}$ **(b)** $h(\begin{pmatrix}x\\y\\z\end{pmatrix}) = \begin{pmatrix}0\\0\end{pmatrix}$ **(c)** $h(\begin{pmatrix}x\\y\\z\end{pmatrix}) = \begin{pmatrix}1\\1\end{pmatrix}$

(d) $h(\begin{pmatrix}x\\y\\z\end{pmatrix}) = \begin{pmatrix}2x+y\\3y-4z\end{pmatrix}$

✓ **1.19** Decide if each map $h\colon \mathcal{M}_{2\times 2} \to \mathbb{R}$ is linear.

(a) $h(\begin{pmatrix}a&b\\c&d\end{pmatrix}) = a+d$

(b) $h(\begin{pmatrix}a&b\\c&d\end{pmatrix}) = ad-bc$

(c) $h(\begin{pmatrix}a&b\\c&d\end{pmatrix}) = 2a+3b+c-d$

(d) $h(\begin{pmatrix}a&b\\c&d\end{pmatrix}) = a^2+b^2$

✓ **1.20** Show that these are homomorphisms. Are they inverse to each other?

(a) $d/dx\colon \mathcal{P}_3 \to \mathcal{P}_2$ given by $a_0+a_1x+a_2x^2+a_3x^3$ maps to $a_1+2a_2x+3a_3x^2$

(b) $\int\colon \mathcal{P}_2 \to \mathcal{P}_3$ given by $b_0+b_1x+b_2x^2$ maps to $b_0x+(b_1/2)x^2+(b_2/3)x^3$

1.21 Is (perpendicular) projection from $\mathbb{R}^3$ to the xz-plane a homomorphism? Projection to the yz-plane? To the x-axis? The y-axis? The z-axis? Projection to the origin?

1.22 Verify that each map is a homomorphism.

(a) $h\colon \mathcal{P}_3 \to \mathbb{R}^2$ given by

$$ax^2 + bx + c \mapsto \begin{pmatrix} a+b \\ a+c \end{pmatrix}$$

(b) $f\colon \mathbb{R}^2 \to \mathbb{R}^3$ given by

$$\begin{pmatrix} x \\ y \end{pmatrix} \mapsto \begin{pmatrix} 0 \\ x-y \\ 3y \end{pmatrix}$$

1.23 Show that, while the maps from Example 1.3 preserve linear operations, they are not isomorphisms.

1.24 Is an identity map a linear transformation?

✓ 1.25 Stating that a function is 'linear' is different than stating that its graph is a line.

(a) The function $f_1\colon \mathbb{R} \to \mathbb{R}$ given by $f_1(x) = 2x - 1$ has a graph that is a line. Show that it is not a linear function.

(b) The function $f_2\colon \mathbb{R}^2 \to \mathbb{R}$ given by

$$\begin{pmatrix} x \\ y \end{pmatrix} \mapsto x + 2y$$

does not have a graph that is a line. Show that it is a linear function.

✓ 1.26 Part of the definition of a linear function is that it respects addition. Does a linear function respect subtraction?

1.27 Assume that h is a linear transformation of V and that $\langle \vec{\beta}_1, \dots, \vec{\beta}_n \rangle$ is a basis of V. Prove each statement.

(a) If $h(\vec{\beta}_i) = \vec{0}$ for each basis vector then h is the zero map.

(b) If $h(\vec{\beta}_i) = \vec{\beta}_i$ for each basis vector then h is the identity map.

(c) If there is a scalar r such that $h(\vec{\beta}_i) = r \cdot \vec{\beta}_i$ for each basis vector then $h(\vec{v}) = r \cdot \vec{v}$ for all vectors in V.

✓ 1.28 Consider the vector space $\mathbb{R}^+$ where vector addition and scalar multiplication are not the ones inherited from $\mathbb{R}$ but rather are these: $a + b$ is the product of a and b, and $r \cdot a$ is the r-th power of a. (This was shown to be a vector space in an earlier exercise.) Verify that the natural logarithm map $\ln\colon \mathbb{R}^+ \to \mathbb{R}$ is a homomorphism between these two spaces. Is it an isomorphism?

✓ 1.29 Consider this transformation of $\mathbb{R}^2$.

$$\begin{pmatrix} x \\ y \end{pmatrix} \mapsto \begin{pmatrix} x/2 \\ y/3 \end{pmatrix}$$

Find the image under this map of this ellipse.

$$\{ \begin{pmatrix} x \\ y \end{pmatrix} \mid (x^2/4) + (y^2/9) = 1 \}$$

✓ 1.30 Imagine a rope wound around the earth's equator so that it fits snugly (suppose that the earth is a sphere). How much extra rope must we add to raise the circle to a constant six feet off the ground?

✓ **1.31** Verify that this map $h\colon\mathbb{R}^3\to\mathbb{R}$

$$\begin{pmatrix}x\\y\\z\end{pmatrix}\mapsto\begin{pmatrix}x\\y\\z\end{pmatrix}\cdot\begin{pmatrix}3\\-1\\-1\end{pmatrix}=3x-y-z$$

is linear. Generalize.

1.32 Show that every homomorphism from $\mathbb{R}^1$ to $\mathbb{R}^1$ acts via multiplication by a scalar. Conclude that every nontrivial linear transformation of $\mathbb{R}^1$ is an isomorphism. Is that true for transformations of $\mathbb{R}^2$? $\mathbb{R}^n$?

1.33 (a) Show that for any scalars $a_{1,1},\ldots,a_{m,n}$ this map $h\colon\mathbb{R}^n\to\mathbb{R}^m$ is a homomorphism.

$$\begin{pmatrix}x_1\\\vdots\\x_n\end{pmatrix}\mapsto\begin{pmatrix}a_{1,1}x_1+\cdots+a_{1,n}x_n\\\vdots\\a_{m,1}x_1+\cdots+a_{m,n}x_n\end{pmatrix}$$

(b) Show that for each i, the i-th derivative operator d^i/dx^i is a linear transformation of $\mathcal{P}_n$. Conclude that for any scalars $c_k,\ldots,c_0$ this map is a linear transformation of that space.

$$f\mapsto\frac{d^k}{dx^k}f+c_{k-1}\frac{d^{k-1}}{dx^{k-1}}f+\cdots+c_1\frac{d}{dx}f+c_0f$$

1.34 Lemma 1.17 shows that a sum of linear functions is linear and that a scalar multiple of a linear function is linear. Show also that a composition of linear functions is linear.

✓ **1.35** Where $f\colon V\to W$ is linear, suppose that $f(\vec{v}_1)=\vec{w}_1,\ \ldots,\ f(\vec{v}_n)=\vec{w}_n$ for some vectors $\vec{w}_1,\ \ldots,\ \vec{w}_n$ from W.

(a) If the set of $\vec{w}$'s is independent, must the set of $\vec{v}$'s also be independent?
(b) If the set of $\vec{v}$'s is independent, must the set of $\vec{w}$'s also be independent?
(c) If the set of $\vec{w}$'s spans W, must the set of $\vec{v}$'s span V?
(d) If the set of $\vec{v}$'s spans V, must the set of $\vec{w}$'s span W?

1.36 Generalize Example 1.16 by proving that for every appropriate domain and codomain the matrix transpose map is linear. What are the appropriate domains and codomains?

1.37 (a) Where $\vec{u},\vec{v}\in\mathbb{R}^n$, by definition the line segment connecting them is the set $\ell=\{t\cdot\vec{u}+(1-t)\cdot\vec{v}\mid t\in[0..1]\}$. Show that the image, under a homomorphism h, of the segment between $\vec{u}$ and $\vec{v}$ is the segment between $h(\vec{u})$ and $h(\vec{v})$.

(b) A subset of $\mathbb{R}^n$ is *convex* if, for any two points in that set, the line segment joining them lies entirely in that set. (The inside of a sphere is convex while the skin of a sphere is not.) Prove that linear maps from $\mathbb{R}^n$ to $\mathbb{R}^m$ preserve the property of set convexity.

✓ **1.38** Let $h\colon\mathbb{R}^n\to\mathbb{R}^m$ be a homomorphism.

(a) Show that the image under h of a line in $\mathbb{R}^n$ is a (possibly degenerate) line in $\mathbb{R}^m$.

(b) What happens to a k-dimensional linear surface?

1.39 Prove that the restriction of a homomorphism to a subspace of its domain is another homomorphism.

1.40 Assume that $h\colon V \to W$ is linear.

(a) Show that the *range space* of this map $\{h(\vec{v}) \mid \vec{v} \in V\}$ is a subspace of the codomain W.

(b) Show that the *null space* of this map $\{\vec{v} \in V \mid h(\vec{v}) = \vec{0}_W\}$ is a subspace of the domain V.

(c) Show that if U is a subspace of the domain V then its image $\{h(\vec{u}) \mid \vec{u} \in U\}$ is a subspace of the codomain W. This generalizes the first item.

(d) Generalize the second item.

1.41 Consider the set of isomorphisms from a vector space to itself. Is this a subspace of the space $\mathcal{L}(V, V)$ of homomorphisms from the space to itself?

1.42 Does Theorem 1.9 need that $\langle \vec{\beta}_1, \ldots, \vec{\beta}_n \rangle$ is a basis? That is, can we still get a well-defined and unique homomorphism if we drop either the condition that the set of $\vec{\beta}$'s be linearly independent, or the condition that it span the domain?

1.43 Let V be a vector space and assume that the maps $f_1, f_2\colon V \to \mathbb{R}^1$ are linear.

(a) Define a map $F\colon V \to \mathbb{R}^2$ whose component functions are the given linear ones.

$$\vec{v} \mapsto \begin{pmatrix} f_1(\vec{v}) \\ f_2(\vec{v}) \end{pmatrix}$$

Show that F is linear.

(b) Does the converse hold—is any linear map from V to $\mathbb{R}^2$ made up of two linear component maps to $\mathbb{R}^1$?

(c) Generalize.

II.2 Range space and Null space

Isomorphisms and homomorphisms both preserve structure. The difference is that homomorphisms have fewer restrictions, since they needn't be onto and needn't be one-to-one. We will examine what can happen with homomorphisms that cannot happen with isomorphisms.

First consider the fact that homomorphisms need not be onto. Of course, each function is onto some set, namely its range. For example, the injection map $\iota\colon \mathbb{R}^2 \to \mathbb{R}^3$

$$\begin{pmatrix} x \\ y \end{pmatrix} \mapsto \begin{pmatrix} x \\ y \\ 0 \end{pmatrix}$$

is a homomorphism, and is not onto $\mathbb{R}^3$. But it is onto the xy-plane.

2.1 Lemma Under a homomorphism, the image of any subspace of the domain is a subspace of the codomain. In particular, the image of the entire space, the range of the homomorphism, is a subspace of the codomain.

Proof Let $h\colon V \to W$ be linear and let S be a subspace of the domain V. The image $h(S)$ is a subset of the codomain W, which is nonempty because S is nonempty. Thus, to show that $h(S)$ is a subspace of W we need only show that it is closed under linear combinations of two vectors. If $h(\vec{s}_1)$ and $h(\vec{s}_2)$ are members of $h(S)$ then $c_1{\cdot}h(\vec{s}_1)+c_2{\cdot}h(\vec{s}_2) = h(c_1{\cdot}\vec{s}_1)+h(c_2{\cdot}\vec{s}_2) = h(c_1{\cdot}\vec{s}_1+c_2{\cdot}\vec{s}_2)$ is also a member of $h(S)$ because it is the image of $c_1 \cdot \vec{s}_1 + c_2 \cdot \vec{s}_2$ from S. QED

2.2 Definition The *range space* of a homomorphism $h\colon V \to W$ is

$$\mathscr{R}(h) = \{h(\vec{v}) \mid \vec{v} \in V\}$$

sometimes denoted $h(V)$. The dimension of the range space is the map's *rank*.

We shall soon see the connection between the rank of a map and the rank of a matrix.

2.3 Example For the derivative map $d/dx\colon \mathcal{P}_3 \to \mathcal{P}_3$ given by $a_0 + a_1x + a_2x^2 + a_3x^3 \mapsto a_1 + 2a_2x + 3a_3x^2$ the range space $\mathscr{R}(d/dx)$ is the set of quadratic polynomials $\{r + sx + tx^2 \mid r, s, t \in \mathbb{R}\}$. Thus, this map's rank is 3.

2.4 Example With this homomorphism $h\colon \mathcal{M}_{2\times 2} \to \mathcal{P}_3$

$$\begin{pmatrix} a & b \\ c & d \end{pmatrix} \mapsto (a + b + 2d) + cx^2 + cx^3$$

an image vector in the range can have any constant term, must have an x coefficient of zero, and must have the same coefficient of x^2 as of x^3. That is, the range space is $\mathscr{R}(h) = \{r + sx^2 + sx^3 \mid r, s \in \mathbb{R}\}$ and so the rank is 2.

The prior result shows that, in passing from the definition of isomorphism to the more general definition of homomorphism, omitting the onto requirement doesn't make an essential difference. Any homomorphism is onto some space, namely its range.

However, omitting the one-to-one condition does make a difference. A homomorphism may have many elements of the domain that map to one element of the codomain. Below is a bean sketch of a many-to-one map between sets.* It shows three elements of the codomain that are each the image of many members of the domain. (Rather than picture lots of individual $\mapsto$ arrows, each association of many inputs with one output shows only one such arrow.)

* More information on many-to-one maps is in the appendix.

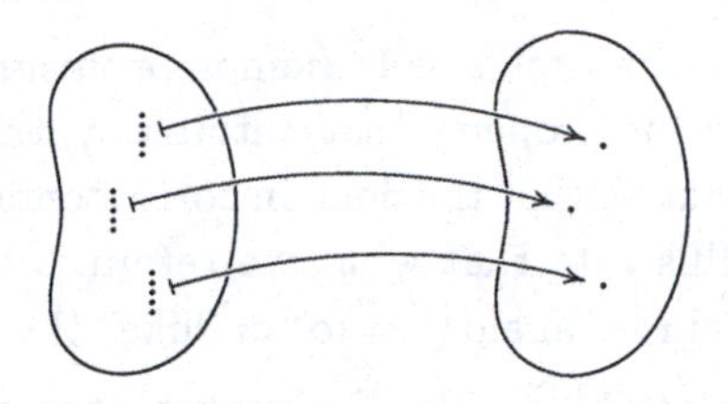

Recall that for any function $h\colon V \to W$, the set of elements of V that map to $\vec{w} \in W$ is the *inverse image* $h^{-1}(\vec{w}) = \{\vec{v} \in V \mid h(\vec{v}) = \vec{w}\}$. Above, the left side shows three inverse image sets.

2.5 Example Consider the projection $\pi\colon \mathbb{R}^3 \to \mathbb{R}^2$

$$\begin{pmatrix} x \\ y \\ z \end{pmatrix} \stackrel{\pi}{\longmapsto} \begin{pmatrix} x \\ y \end{pmatrix}$$

which is a homomorphism that is many-to-one. An inverse image set is a vertical line of vectors in the domain.

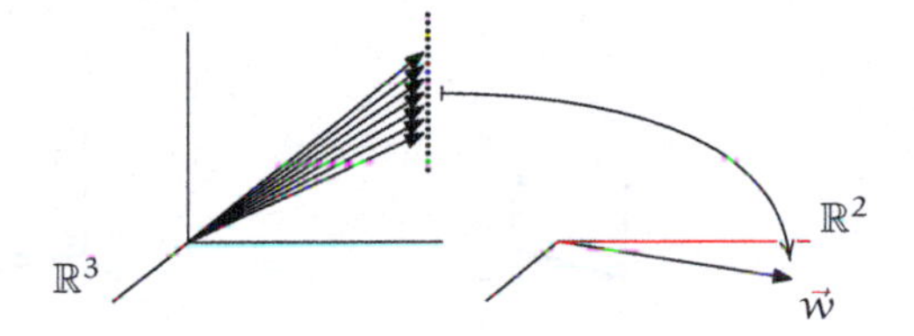

One example is this.

$$\pi^{-1}(\begin{pmatrix} 1 \\ 3 \end{pmatrix}) = \{\begin{pmatrix} 1 \\ 3 \\ z \end{pmatrix} \mid z \in \mathbb{R}\}$$

2.6 Example This homomorphism $h\colon \mathbb{R}^2 \to \mathbb{R}^1$

$$\begin{pmatrix} x \\ y \end{pmatrix} \stackrel{h}{\longmapsto} x + y$$

is also many-to-one. For a fixed $w \in \mathbb{R}^1$ the inverse image $h^{-1}(w)$

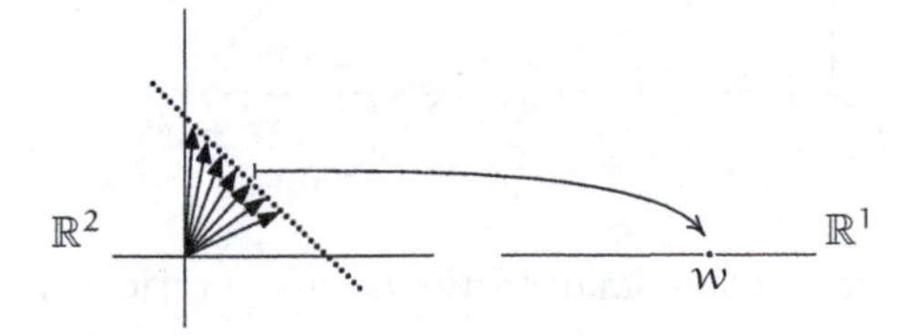

is the set of plane vectors whose components add to w.

In generalizing from isomorphisms to homomorphisms by dropping the one-to-one condition we lose the property that, intuitively, the domain is "the same" as the range. We lose, that is, that the domain corresponds perfectly to the range. The examples below illustrate that what we retain is that a homomorphism describes how the domain is "analogous to" or "like" the range.

2.7 Example We think of $\mathbb{R}^3$ as like $\mathbb{R}^2$ except that vectors have an extra component. That is, we think of the vector with components x, y, and z as like the vector with components x and y. Defining the projection map π makes precise which members of the domain we are thinking of as related to which members of the codomain.

To understanding how the preservation conditions in the definition of homomorphism show that the domain elements are like the codomain elements, start by picturing $\mathbb{R}^2$ as the xy-plane inside of $\mathbb{R}^3$ (the xy plane inside of $\mathbb{R}^3$ is a set of three-tall vectors with a third component of zero and so does not precisely equal the set of two-tall vectors $\mathbb{R}^2$, but this embedding makes the picture much clearer). The preservation of addition property says that vectors in $\mathbb{R}^3$ act like their shadows in the plane.

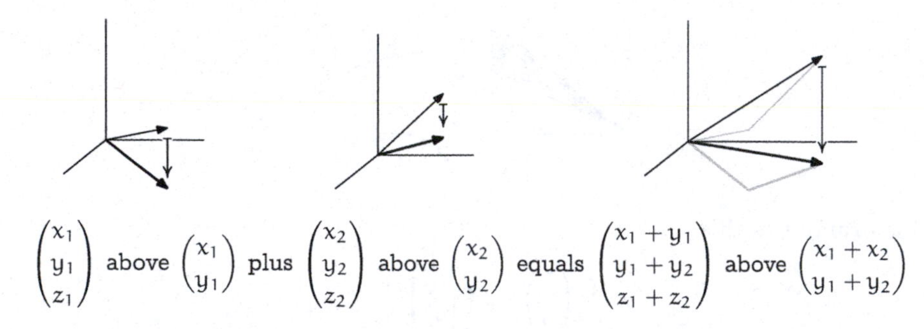

$$\begin{pmatrix} x_1 \\ y_1 \\ z_1 \end{pmatrix} \text{ above } \begin{pmatrix} x_1 \\ y_1 \end{pmatrix} \text{ plus } \begin{pmatrix} x_2 \\ y_2 \\ z_2 \end{pmatrix} \text{ above } \begin{pmatrix} x_2 \\ y_2 \end{pmatrix} \text{ equals } \begin{pmatrix} x_1 + y_1 \\ y_1 + y_2 \\ z_1 + z_2 \end{pmatrix} \text{ above } \begin{pmatrix} x_1 + x_2 \\ y_1 + y_2 \end{pmatrix}$$

Thinking of $\pi(\vec{v})$ as the "shadow" of $\vec{v}$ in the plane gives this restatement: the sum of the shadows $\pi(\vec{v}_1) + \pi(\vec{v}_2)$ equals the shadow of the sum $\pi(\vec{v}_1 + \vec{v}_2)$. Preservation of scalar multiplication is similar.

Drawing the codomain $\mathbb{R}^2$ on the right gives a picture that is uglier but is more faithful to the bean sketch above.

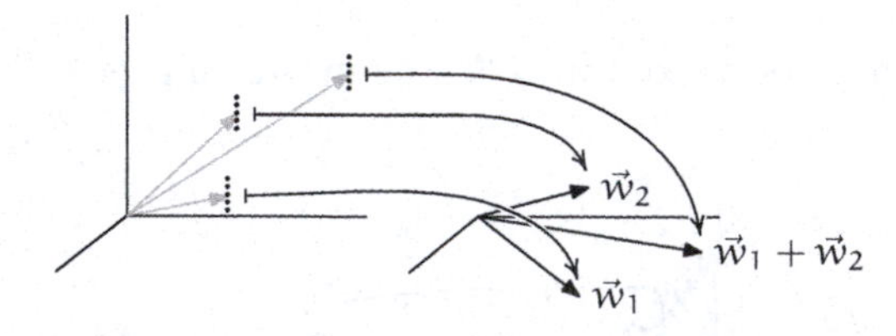

Again, the domain vectors that map to $\vec{w}_1$ lie in a vertical line; one is drawn, in gray. Call any member of this inverse image $\pi^{-1}(\vec{w}_1)$ a "$\vec{w}_1$ vector." Similarly, there is a vertical line of "$\vec{w}_2$ vectors" and a vertical line of "$\vec{w}_1 + \vec{w}_2$ vectors."

Now, saying that π is a homomorphism is recognizing that if $\pi(\vec{v}_1) = \vec{w}_1$ and $\pi(\vec{v}_2) = \vec{w}_2$ then $\pi(\vec{v}_1 + \vec{v}_2) = \pi(\vec{v}_1) + \pi(\vec{v}_2) = \vec{w}_1 + \vec{w}_2$. That is, the classes add: any $\vec{w}_1$ vector plus any $\vec{w}_2$ vector equals a $\vec{w}_1 + \vec{w}_2$ vector. Scalar multiplication is similar.

So although $\mathbb{R}^3$ and $\mathbb{R}^2$ are not isomorphic π describes a way in which they are alike: vectors in $\mathbb{R}^3$ add as do the associated vectors in $\mathbb{R}^2$ — vectors add as their shadows add.

2.8 Example A homomorphism can express an analogy between spaces that is more subtle than the prior one. For the map from Example 2.6

$$\begin{pmatrix} x \\ y \end{pmatrix} \stackrel{h}{\longmapsto} x + y$$

fix two numbers in the range $w_1, w_2 \in \mathbb{R}$. A $\vec{v}_1$ that maps to w_1 has components that add to w_1, so the inverse image $h^{-1}(w_1)$ is the set of vectors with endpoint on the diagonal line $x + y = w_1$. Think of these as "w_1 vectors." Similarly we have "w_2 vectors" and "$w_1 + w_2$ vectors." The addition preservation property says this.

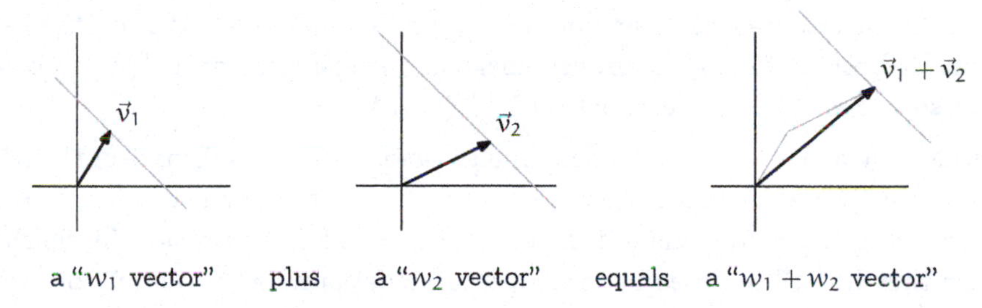

Restated, if we add a w_1 vector to a w_2 vector then h maps the result to a $w_1 + w_2$ vector. Briefly, the sum of the images is the image of the sum. Even more briefly, $h(\vec{v}_1) + h(\vec{v}_2) = h(\vec{v}_1 + \vec{v}_2)$.

2.9 Example The inverse images can be structures other than lines. For the linear map $h\colon \mathbb{R}^3 \to \mathbb{R}^2$

$$\begin{pmatrix} x \\ y \\ z \end{pmatrix} \mapsto \begin{pmatrix} x \\ x \end{pmatrix}$$

the inverse image sets are planes $x = 0$, $x = 1$, etc., perpendicular to the x-axis.

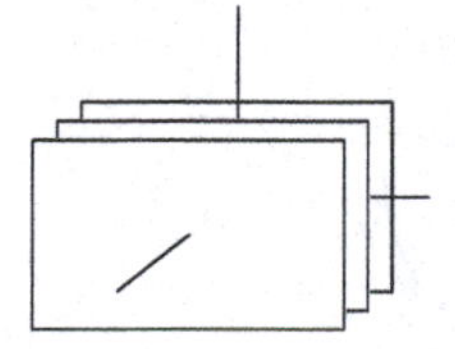

We won't describe how every homomorphism that we will use is an analogy because the formal sense that we make of "alike in that ..." is 'a homomorphism exists such that ...'. Nonetheless, the idea that a homomorphism between two spaces expresses how the domain's vectors fall into classes that act like the range's vectors is a good way to view homomorphisms.

Another reason that we won't treat all of the homomorphisms that we see as above is that many vector spaces are hard to draw, e.g., a space of polynomials. But there is nothing wrong with leveraging spaces that we can draw: from the three examples 2.7, 2.8, and 2.9 we draw two insights.

The first insight is that in all three examples the inverse image of the range's zero vector is a line or plane through the origin. It is therefore a subspace of the domain.

2.10 Lemma For any homomorphism the inverse image of a subspace of the range is a subspace of the domain. In particular, the inverse image of the trivial subspace of the range is a subspace of the domain.

(The examples above consider inverse images of single vectors but this result is about inverse images of sets $h^{-1}(S) = \{\vec{v} \in V \mid h(\vec{v}) \in S\}$. We use the same term for both by taking the inverse image of a single element $h^{-1}(\vec{w})$ to be the inverse image of the one-element set $h^{-1}(\{\vec{w}\})$.)

Proof Let $h\colon V \to W$ be a homomorphism and let S be a subspace of the range space of h. Consider the inverse image of S. It is nonempty because it contains $\vec{0}_V$, since $h(\vec{0}_V) = \vec{0}_W$ and $\vec{0}_W$ is an element of S as S is a subspace. To finish we show that $h^{-1}(S)$ is closed under linear combinations. Let $\vec{v}_1$ and $\vec{v}_2$ be two of its elements, so that $h(\vec{v}_1)$ and $h(\vec{v}_2)$ are elements of S. Then $c_1\vec{v}_1 + c_2\vec{v}_2$ is an element of the inverse image $h^{-1}(S)$ because $h(c_1\vec{v}_1 + c_2\vec{v}_2) = c_1h(\vec{v}_1) + c_2h(\vec{v}_2)$ is a member of S. QED

2.11 Definition The *null space* or *kernel* of a linear map $h\colon V \to W$ is the inverse image of $\vec{0}_W$.

$$\mathcal{N}(h) = h^{-1}(\vec{0}_W) = \{\vec{v} \in V \mid h(\vec{v}) = \vec{0}_W\}$$

The dimension of the null space is the map's *nullity*.

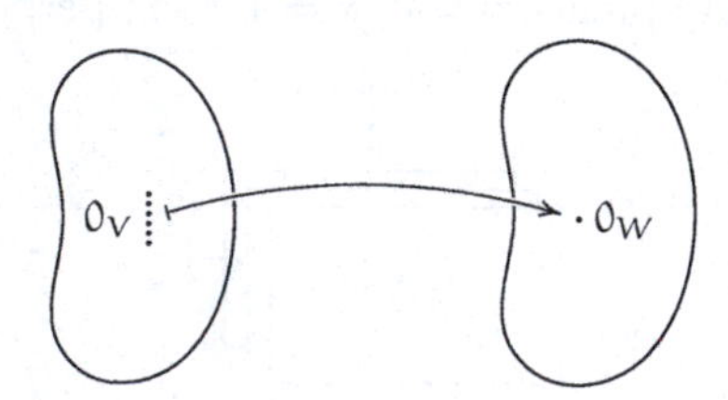

2.12 Example The map from Example 2.3 has this null space $\mathscr{N}(d/dx) = \{a_0 + 0x + 0x^2 + 0x^3 \mid a_0 \in \mathbb{R}\}$ so its nullity is 1.

2.13 Example The map from Example 2.4 has this null space, and nullity 2.

$$\mathscr{N}(h) = \{\begin{pmatrix} a & b \\ 0 & -(a+b)/2 \end{pmatrix} \mid a, b \in \mathbb{R}\}$$

Now for the second insight from the above examples. In Example 2.7 each of the vertical lines squashes down to a single point — in passing from the domain to the range, π takes all of these one-dimensional vertical lines and maps them to a point, leaving the range smaller than the domain by one dimension. Similarly, in Example 2.8 the two-dimensional domain compresses to a one-dimensional range by breaking the domain into the diagonal lines and maps each of those to a single member of the range. Finally, in Example 2.9 the domain breaks into planes which get squashed to a point and so the map starts with a three-dimensional domain but ends two smaller, with a one-dimensional range. (The codomain is two-dimensional but the range is one-dimensional and the dimension of the range is what matters.)

2.14 Theorem A linear map's rank plus its nullity equals the dimension of its domain.

Proof Let $h\colon V \to W$ be linear and let $B_N = \langle \vec{\beta}_1, \ldots, \vec{\beta}_k \rangle$ be a basis for the null space. Expand that to a basis $B_V = \langle \vec{\beta}_1, \ldots, \vec{\beta}_k, \vec{\beta}_{k+1}, \ldots, \vec{\beta}_n \rangle$ for the entire domain, using Corollary Two.III.2.13. We shall show that $B_R = \langle h(\vec{\beta}_{k+1}), \ldots, h(\vec{\beta}_n) \rangle$ is a basis for the range space. Then counting the size of the bases gives the result.

To see that B_R is linearly independent, consider $\vec{0}_W = c_{k+1}h(\vec{\beta}_{k+1}) + \cdots + c_n h(\vec{\beta}_n)$. We have $\vec{0}_W = h(c_{k+1}\vec{\beta}_{k+1} + \cdots + c_n\vec{\beta}_n)$ and so $c_{k+1}\vec{\beta}_{k+1} + \cdots + c_n\vec{\beta}_n$ is in the null space of h. As B_N is a basis for the null space there are scalars $c_1, \ldots, c_k$ satisfying this relationship.

$$c_1\vec{\beta}_1 + \cdots + c_k\vec{\beta}_k = c_{k+1}\vec{\beta}_{k+1} + \cdots + c_n\vec{\beta}_n$$

But this is an equation among members of B_V, which is a basis for V, so each c_i equals 0. Therefore B_R is linearly independent.

To show that B_R spans the range space consider a member of the range space $h(\vec{v})$. Express $\vec{v}$ as a linear combination $\vec{v} = c_1\vec{\beta}_1 + \cdots + c_n\vec{\beta}_n$ of members of B_V. This gives $h(\vec{v}) = h(c_1\vec{\beta}_1 + \cdots + c_n\vec{\beta}_n) = c_1h(\vec{\beta}_1) + \cdots + c_kh(\vec{\beta}_k) + c_{k+1}h(\vec{\beta}_{k+1}) + \cdots + c_nh(\vec{\beta}_n)$ and since $\vec{\beta}_1, \ldots, \vec{\beta}_k$ are in the null space, we have that $h(\vec{v}) = \vec{0} + \cdots + \vec{0} + c_{k+1}h(\vec{\beta}_{k+1}) + \cdots + c_nh(\vec{\beta}_n)$. Thus, $h(\vec{v})$ is a linear combination of members of B_R, and so B_R spans the range space. QED

2.15 Example Where $h\colon \mathbb{R}^3 \to \mathbb{R}^4$ is

$$\begin{pmatrix} x \\ y \\ z \end{pmatrix} \overset{h}{\longmapsto} \begin{pmatrix} x \\ 0 \\ y \\ 0 \end{pmatrix}$$

the range space and null space are

$$\mathscr{R}(h) = \{\begin{pmatrix} a \\ 0 \\ b \\ 0 \end{pmatrix} \mid a, b \in \mathbb{R}\} \quad \text{and} \quad \mathscr{N}(h) = \{\begin{pmatrix} 0 \\ 0 \\ z \end{pmatrix} \mid z \in \mathbb{R}\}$$

and so the rank of h is 2 while the nullity is 1.

2.16 Example If $t\colon \mathbb{R} \to \mathbb{R}$ is the linear transformation $x \mapsto -4x$, then the range is $\mathscr{R}(t) = \mathbb{R}$. The rank is 1 and the nullity is 0.

2.17 Corollary The rank of a linear map is less than or equal to the dimension of the domain. Equality holds if and only if the nullity of the map is 0.

We know that an isomorphism exists between two spaces if and only if the dimension of the range equals the dimension of the domain. We have now seen that for a homomorphism to exist a necessary condition is that the dimension of the range must be less than or equal to the dimension of the domain. For instance, there is no homomorphism from $\mathbb{R}^2$ onto $\mathbb{R}^3$. There are many homomorphisms from $\mathbb{R}^2$ into $\mathbb{R}^3$, but none onto.

The range space of a linear map can be of dimension strictly less than the dimension of the domain and so linearly independent sets in the domain may map to linearly dependent sets in the range. (Example 2.3's derivative transformation on $\mathcal{P}_3$ has a domain of dimension 4 but a range of dimension 3 and the derivative sends $\{1, x, x^2, x^3\}$ to $\{0, 1, 2x, 3x^2\}$). That is, under a homomorphism independence may be lost. In contrast, dependence stays.

2.18 Lemma Under a linear map, the image of a linearly dependent set is linearly dependent.

PROOF Suppose that $c_1\vec{v}_1 + \cdots + c_n\vec{v}_n = \vec{0}_V$ with some c_i nonzero. Apply h to both sides: $h(c_1\vec{v}_1 + \cdots + c_n\vec{v}_n) = c_1h(\vec{v}_1) + \cdots + c_nh(\vec{v}_n)$ and $h(\vec{0}_V) = \vec{0}_W$. Thus we have $c_1h(\vec{v}_1) + \cdots + c_nh(\vec{v}_n) = \vec{0}_W$ with some c_i nonzero. QED

When is independence not lost? The obvious sufficient condition is when the homomorphism is an isomorphism. This condition is also necessary; see

Exercise 36. We will finish this subsection comparing homomorphisms with isomorphisms by observing that a one-to-one homomorphism is an isomorphism from its domain onto its range.

2.19 Example This one-to-one homomorphism $\iota\colon \mathbb{R}^2 \to \mathbb{R}^3$

$$\begin{pmatrix} x \\ y \end{pmatrix} \stackrel{\iota}{\longmapsto} \begin{pmatrix} x \\ y \\ 0 \end{pmatrix}$$

gives a correspondence between $\mathbb{R}^2$ and the xy-plane subset of $\mathbb{R}^3$.

2.20 Theorem Where V is an n-dimensional vector space, these are equivalent statements about a linear map $h\colon V \to W$.

(1) h is one-to-one
(2) h has an inverse from its range to its domain that is a linear map
(3) $\mathscr{N}(h) = \{\vec{0}\,\}$, that is, $\text{nullity}(h) = 0$
(4) $\text{rank}(h) = n$
(5) if $\langle \vec{\beta}_1, \ldots, \vec{\beta}_n \rangle$ is a basis for V then $\langle h(\vec{\beta}_1), \ldots, h(\vec{\beta}_n) \rangle$ is a basis for $\mathscr{R}(h)$

Proof We will first show that (1) $\iff$ (2). We will then show that (1) $\implies$ (3) $\implies$ (4) $\implies$ (5) $\implies$ (2).

For (1) $\implies$ (2), suppose that the linear map h is one-to-one, and therefore has an inverse $h^{-1}\colon \mathscr{R}(h) \to V$. The domain of that inverse is the range of h and thus a linear combination of two members of it has the form $c_1h(\vec{v}_1) + c_2h(\vec{v}_2)$. On that combination, the inverse h^{-1} gives this.

$$\begin{aligned} h^{-1}(c_1h(\vec{v}_1) + c_2h(\vec{v}_2)) &= h^{-1}(h(c_1\vec{v}_1 + c_2\vec{v}_2)) \\ &= h^{-1} \circ h\,(c_1\vec{v}_1 + c_2\vec{v}_2) \\ &= c_1\vec{v}_1 + c_2\vec{v}_2 \\ &= c_1 \cdot h^{-1}(h(\vec{v}_1)) + c_2 \cdot h^{-1}(h(\vec{v}_2)) \end{aligned}$$

Thus if a linear map has an inverse then the inverse must be linear. But this also gives the (2) $\implies$ (1) implication, because the inverse itself must be one-to-one.

Of the remaining implications, (1) $\implies$ (3) holds because any homomorphism maps $\vec{0}_V$ to $\vec{0}_W$, but a one-to-one map sends at most one member of V to $\vec{0}_W$.

Next, (3) $\implies$ (4) is true since rank plus nullity equals the dimension of the domain.

For (4) $\implies$ (5), to show that $\langle h(\vec{\beta}_1), \ldots, h(\vec{\beta}_n) \rangle$ is a basis for the range space we need only show that it is a spanning set, because by assumption the range has dimension n. Consider $h(\vec{v}) \in \mathscr{R}(h)$. Expressing $\vec{v}$ as a linear combination of basis elements produces $h(\vec{v}) = h(c_1\vec{\beta}_1 + c_2\vec{\beta}_2 + \cdots + c_n\vec{\beta}_n)$, which gives that $h(\vec{v}) = c_1h(\vec{\beta}_1) + \cdots + c_nh(\vec{\beta}_n)$, as desired.

Finally, for the (5) $\implies$ (2) implication, assume that $\langle\vec{\beta}_1, \ldots, \vec{\beta}_n\rangle$ is a basis for V so that $\langle h(\vec{\beta}_1), \ldots, h(\vec{\beta}_n)\rangle$ is a basis for $\mathscr{R}(h)$. Then every $\vec{w} \in \mathscr{R}(h)$ has the unique representation $\vec{w} = c_1h(\vec{\beta}_1) + \cdots + c_nh(\vec{\beta}_n)$. Define a map from $\mathscr{R}(h)$ to V by

$$\vec{w} \mapsto c_1\vec{\beta}_1 + c_2\vec{\beta}_2 + \cdots + c_n\vec{\beta}_n$$

(uniqueness of the representation makes this well-defined). Checking that it is linear and that it is the inverse of h are easy. QED

We have seen that a linear map expresses how the structure of the domain is like that of the range. We can think of such a map as organizing the domain space into inverse images of points in the range. In the special case that the map is one-to-one, each inverse image is a single point and the map is an isomorphism between the domain and the range.

Exercises

✓ **2.21** Let $h\colon \mathcal{P}_3 \to \mathcal{P}_4$ be given by $p(x) \mapsto x \cdot p(x)$. Which of these are in the null space? Which are in the range space?

(a) x^3 (b) 0 (c) 7 (d) $12x - 0.5x^3$ (e) $1 + 3x^2 - x^3$

2.22 Find the range space and the rank of each homomorphism.

(a) $h\colon \mathcal{P}_3 \to \mathbb{R}^2$ given by

$$ax^2 + bx + c \mapsto \begin{pmatrix} a+b \\ a+c \end{pmatrix}$$

(b) $f\colon \mathbb{R}^2 \to \mathbb{R}^3$ given by

$$\begin{pmatrix} x \\ y \end{pmatrix} \mapsto \begin{pmatrix} 0 \\ x-y \\ 3y \end{pmatrix}$$

✓ **2.23** Find the range space and rank of each map.

(a) $h\colon \mathbb{R}^2 \to \mathcal{P}_3$ given by

$$\begin{pmatrix} a \\ b \end{pmatrix} \mapsto a + ax + ax^2$$

(b) $h\colon \mathcal{M}_{2\times2} \to \mathbb{R}$ given by

$$\begin{pmatrix} a & b \\ c & d \end{pmatrix} \mapsto a + d$$

(c) $h\colon \mathcal{M}_{2\times2} \to \mathcal{P}_2$ given by

$$\begin{pmatrix} a & b \\ c & d \end{pmatrix} \mapsto a + b + c + dx^2$$

(d) the zero map $Z\colon \mathbb{R}^3 \to \mathbb{R}^4$

✓ **2.24** For each linear map in the prior exercise, find the null space and nullity.

✓ **2.25** Find the nullity of each map below.

(a) $h\colon \mathbb{R}^5 \to \mathbb{R}^8$ of rank five (b) $h\colon \mathcal{P}_3 \to \mathcal{P}_3$ of rank one
(c) $h\colon \mathbb{R}^6 \to \mathbb{R}^3$, an onto map (d) $h\colon \mathcal{M}_{3\times 3} \to \mathcal{M}_{3\times 3}$, onto

✓ **2.26** What is the null space of the differentiation transformation $d/dx\colon \mathcal{P}_n \to \mathcal{P}_n$? What is the null space of the second derivative, as a transformation of $\mathcal{P}_n$? The k-th derivative?

2.27 Example 2.7 restates the first condition in the definition of homomorphism as 'the shadow of a sum is the sum of the shadows'. Restate the second condition in the same style.

2.28 For the homomorphism $h\colon \mathcal{P}_3 \to \mathcal{P}_3$ given by $h(a_0 + a_1x + a_2x^2 + a_3x^3) = a_0 + (a_0 + a_1)x + (a_2 + a_3)x^3$ find these.
(a) $\mathscr{N}(h)$ (b) $h^{-1}(2 - x^3)$ (c) $h^{-1}(1 + x^2)$

✓ **2.29** For the map $f\colon \mathbb{R}^2 \to \mathbb{R}$ given by

$$f(\begin{pmatrix} x \\ y \end{pmatrix}) = 2x + y$$

sketch these inverse image sets: $f^{-1}(-3)$, $f^{-1}(0)$, and $f^{-1}(1)$.

✓ **2.30** Each of these transformations of $\mathcal{P}_3$ is one-to-one. For each, find the inverse.
(a) $a_0 + a_1x + a_2x^2 + a_3x^3 \mapsto a_0 + a_1x + 2a_2x^2 + 3a_3x^3$
(b) $a_0 + a_1x + a_2x^2 + a_3x^3 \mapsto a_0 + a_2x + a_1x^2 + a_3x^3$
(c) $a_0 + a_1x + a_2x^2 + a_3x^3 \mapsto a_1 + a_2x + a_3x^2 + a_0x^3$
(d) $a_0+a_1x+a_2x^2+a_3x^3 \mapsto a_0+(a_0+a_1)x+(a_0+a_1+a_2)x^2+(a_0+a_1+a_2+a_3)x^3$

2.31 Describe the null space and range space of a transformation given by $\vec{v} \mapsto 2\vec{v}$.

2.32 List all pairs $(\text{rank}(h), \text{nullity}(h))$ that are possible for linear maps from $\mathbb{R}^5$ to $\mathbb{R}^3$.

2.33 Does the differentiation map $d/dx\colon \mathcal{P}_n \to \mathcal{P}_n$ have an inverse?

✓ **2.34** Find the nullity of this map $h\colon \mathcal{P}_n \to \mathbb{R}$.

$$a_0 + a_1x + \cdots + a_nx^n \mapsto \int_{x=0}^{x=1} a_0 + a_1x + \cdots + a_nx^n\,dx$$

2.35 (a) Prove that a homomorphism is onto if and only if its rank equals the dimension of its codomain.
(b) Conclude that a homomorphism between vector spaces with the same dimension is one-to-one if and only if it is onto.

2.36 Show that a linear map is one-to-one if and only if it preserves linear independence.

2.37 Corollary 2.17 says that for there to be an onto homomorphism from a vector space V to a vector space W, it is necessary that the dimension of W be less than or equal to the dimension of V. Prove that this condition is also sufficient; use Theorem 1.9 to show that if the dimension of W is less than or equal to the dimension of V, then there is a homomorphism from V to W that is onto.

✓ **2.38** Recall that the null space is a subset of the domain and the range space is a subset of the codomain. Are they necessarily distinct? Is there a homomorphism that has a nontrivial intersection of its null space and its range space?

2.39 Prove that the image of a span equals the span of the images. That is, where $h\colon V\to W$ is linear, prove that if S is a subset of V then $h([S])$ equals $[h(S)]$. This generalizes Lemma 2.1 since it shows that if U is any subspace of V then its image $\{h(\vec{u})\mid\vec{u}\in U\}$ is a subspace of W, because the span of the set U is U.

✓ **2.40** **(a)** Prove that for any linear map $h\colon V\to W$ and any $\vec{w}\in W$, the set $h^{-1}(\vec{w})$ has the form

$$\{\vec{v}+\vec{n}\mid\vec{n}\in\mathscr{N}(h)\}$$

for $\vec{v}\in V$ with $h(\vec{v})=\vec{w}$ (if h is not onto then this set may be empty). Such a set is a *coset* of $\mathscr{N}(h)$ and we denote it as $\vec{v}+\mathscr{N}(h)$.

(b) Consider the map $t\colon\mathbb{R}^2\to\mathbb{R}^2$ given by

$$\begin{pmatrix}x\\y\end{pmatrix}\stackrel{t}{\longmapsto}\begin{pmatrix}ax+by\\cx+dy\end{pmatrix}$$

for some scalars a, b, c, and d. Prove that t is linear.

(c) Conclude from the prior two items that for any linear system of the form

$$\begin{aligned}ax+by&=e\\cx+dy&=f\end{aligned}$$

we can write the solution set (the vectors are members of $\mathbb{R}^2$)

$$\{\vec{p}+\vec{h}\mid\vec{h}\text{ satisfies the associated homogeneous system}\}$$

where $\vec{p}$ is a particular solution of that linear system (if there is no particular solution then the above set is empty).

(d) Show that this map $h\colon\mathbb{R}^n\to\mathbb{R}^m$ is linear

$$\begin{pmatrix}x_1\\\vdots\\x_n\end{pmatrix}\mapsto\begin{pmatrix}a_{1,1}x_1+\cdots+a_{1,n}x_n\\\vdots\\a_{m,1}x_1+\cdots+a_{m,n}x_n\end{pmatrix}$$

for any scalars $a_{1,1}$, ..., $a_{m,n}$. Extend the conclusion made in the prior item.

(e) Show that the k-th derivative map is a linear transformation of $\mathcal{P}_n$ for each k. Prove that this map is a linear transformation of the space

$$f\mapsto\frac{d^k}{dx^k}f+c_{k-1}\frac{d^{k-1}}{dx^{k-1}}f+\cdots+c_1\frac{d}{dx}f+c_0f$$

for any scalars c_k, ..., c_0. Draw a conclusion as above.

2.41 Prove that for any transformation $t\colon V\to V$ that is rank one, the map given by composing the operator with itself $t\circ t\colon V\to V$ satisfies $t\circ t=r\cdot t$ for some real number r.

2.42 Let $h\colon V\to\mathbb{R}$ be a homomorphism, but not the zero homomorphism. Prove that if $\langle\vec{\beta}_1,\ldots,\vec{\beta}_n\rangle$ is a basis for the null space and if $\vec{v}\in V$ is not in the null space then $\langle\vec{v},\vec{\beta}_1,\ldots,\vec{\beta}_n\rangle$ is a basis for the entire domain V.

2.43 Show that for any space V of dimension n, the *dual space*

$$\mathcal{L}(V,\mathbb{R})=\{h\colon V\to\mathbb{R}\mid h\text{ is linear}\}$$

is isomorphic to $\mathbb{R}^n$. It is often denoted V^*. Conclude that $V^*\cong V$.

2.44 Show that any linear map is the sum of maps of rank one.

2.45 Is 'is homomorphic to' an equivalence relation? (*Hint:* the difficulty is to decide on an appropriate meaning for the quoted phrase.)

2.46 Show that the range spaces and null spaces of powers of linear maps $t\colon V \to V$ form descending

$$V \supseteq \mathscr{R}(t) \supseteq \mathscr{R}(t^2) \supseteq \ldots$$

and ascending

$$\{\vec{0}\} \subseteq \mathscr{N}(t) \subseteq \mathscr{N}(t^2) \subseteq \ldots$$

chains. Also show that if k is such that $\mathscr{R}(t^k) = \mathscr{R}(t^{k+1})$ then all following range spaces are equal: $\mathscr{R}(t^k) = \mathscr{R}(t^{k+1}) = \mathscr{R}(t^{k+2}) \ldots$. Similarly, if $\mathscr{N}(t^k) = \mathscr{N}(t^{k+1})$ then $\mathscr{N}(t^k) = \mathscr{N}(t^{k+1}) = \mathscr{N}(t^{k+2}) = \ldots$.

III Computing Linear Maps

The prior section shows that a linear map is determined by its action on a basis. The equation

$$h(\vec{v}) = h(c_1 \cdot \vec{\beta}_1 + \cdots + c_n \cdot \vec{\beta}_n) = c_1 \cdot h(\vec{\beta}_1) + \cdots + c_n \cdot h(\vec{\beta}_n)$$

describes how we get the value of the map on any vector $\vec{v}$ by starting from the value of the map on the vectors $\vec{\beta}_i$ in a basis and extending linearly.

This section gives a convenient scheme based on matrices to use the representations of $h(\vec{\beta}_1), \ldots, h(\vec{\beta}_n)$ to compute, from the representation of a vector in the domain $\mathrm{Rep}_B(\vec{v})$, the representation of that vector's image in the codomain $\mathrm{Rep}_D(h(\vec{v}))$.

III.1 Representing Linear Maps with Matrices

1.1 Example For the spaces $\mathbb{R}^2$ and $\mathbb{R}^3$ fix these bases.

$$B = \langle \begin{pmatrix} 2 \\ 0 \end{pmatrix}, \begin{pmatrix} 1 \\ 4 \end{pmatrix} \rangle \qquad D = \langle \begin{pmatrix} 1 \\ 0 \\ 0 \end{pmatrix}, \begin{pmatrix} 0 \\ -2 \\ 0 \end{pmatrix}, \begin{pmatrix} 1 \\ 0 \\ 1 \end{pmatrix} \rangle$$

Consider the map $h\colon \mathbb{R}^2 \to \mathbb{R}^3$ that is determined by this association.

$$\begin{pmatrix} 2 \\ 0 \end{pmatrix} \stackrel{h}{\longmapsto} \begin{pmatrix} 1 \\ 1 \\ 1 \end{pmatrix} \qquad \begin{pmatrix} 1 \\ 4 \end{pmatrix} \stackrel{h}{\longmapsto} \begin{pmatrix} 1 \\ 2 \\ 0 \end{pmatrix}$$

To compute the action of this map on any vector at all from the domain we first represent the vector $h(\vec{\beta}_1)$

$$\begin{pmatrix} 1 \\ 1 \\ 1 \end{pmatrix} = 0 \begin{pmatrix} 1 \\ 0 \\ 0 \end{pmatrix} - \frac{1}{2} \begin{pmatrix} 0 \\ -2 \\ 0 \end{pmatrix} + 1 \begin{pmatrix} 1 \\ 0 \\ 1 \end{pmatrix} \qquad \mathrm{Rep}_D(h(\vec{\beta}_1)) = \begin{pmatrix} 0 \\ -1/2 \\ 1 \end{pmatrix}_D$$

and the vector $h(\vec{\beta}_2)$.

$$\begin{pmatrix} 1 \\ 2 \\ 0 \end{pmatrix} = 1 \begin{pmatrix} 1 \\ 0 \\ 0 \end{pmatrix} - 1 \begin{pmatrix} 0 \\ -2 \\ 0 \end{pmatrix} + 0 \begin{pmatrix} 1 \\ 0 \\ 1 \end{pmatrix} \qquad \mathrm{Rep}_D(h(\vec{\beta}_2)) = \begin{pmatrix} 1 \\ -1 \\ 0 \end{pmatrix}_D$$

With these, for any member $\vec{v}$ of the domain we can compute $h(\vec{v})$.

$$\begin{aligned}
h(\vec{v}) &= h(c_1 \cdot \begin{pmatrix} 2 \\ 0 \end{pmatrix} + c_2 \cdot \begin{pmatrix} 1 \\ 4 \end{pmatrix}) \\
&= c_1 \cdot h(\begin{pmatrix} 2 \\ 0 \end{pmatrix}) + c_2 \cdot h(\begin{pmatrix} 1 \\ 4 \end{pmatrix}) \\
&= c_1 \cdot (0 \begin{pmatrix} 1 \\ 0 \\ 0 \end{pmatrix} - \frac{1}{2} \begin{pmatrix} 0 \\ -2 \\ 0 \end{pmatrix} + 1 \begin{pmatrix} 1 \\ 0 \\ 1 \end{pmatrix}) + c_2 \cdot (1 \begin{pmatrix} 1 \\ 0 \\ 0 \end{pmatrix} - 1 \begin{pmatrix} 0 \\ -2 \\ 0 \end{pmatrix} + 0 \begin{pmatrix} 1 \\ 0 \\ 1 \end{pmatrix}) \\
&= (0c_1 + 1c_2) \cdot \begin{pmatrix} 1 \\ 0 \\ 0 \end{pmatrix} + (-\frac{1}{2}c_1 - 1c_2) \cdot \begin{pmatrix} 0 \\ -2 \\ 0 \end{pmatrix} + (1c_1 + 0c_2) \cdot \begin{pmatrix} 1 \\ 0 \\ 1 \end{pmatrix}
\end{aligned}$$

Thus,

$$\text{if } \mathrm{Rep}_B(\vec{v}) = \begin{pmatrix} c_1 \\ c_2 \end{pmatrix} \text{ then } \mathrm{Rep}_D(\,h(\vec{v})\,) = \begin{pmatrix} 0c_1 + 1c_2 \\ -(1/2)c_1 - 1c_2 \\ 1c_1 + 0c_2 \end{pmatrix}.$$

For instance,

$$\text{since } \mathrm{Rep}_B(\begin{pmatrix} 4 \\ 8 \end{pmatrix}) = \begin{pmatrix} 1 \\ 2 \end{pmatrix}_B \text{ we have } \mathrm{Rep}_D(\,h(\begin{pmatrix} 4 \\ 8 \end{pmatrix})\,) = \begin{pmatrix} 2 \\ -5/2 \\ 1 \end{pmatrix}.$$

We express computations like the one above with a matrix notation.

$$\begin{pmatrix} 0 & 1 \\ -1/2 & -1 \\ 1 & 0 \end{pmatrix}_{B,D} \begin{pmatrix} c_1 \\ c_2 \end{pmatrix}_B = \begin{pmatrix} 0c_1 + 1c_2 \\ (-1/2)c_1 - 1c_2 \\ 1c_1 + 0c_2 \end{pmatrix}_D$$

In the middle is the argument $\vec{v}$ to the map, represented with respect to the domain's basis B by the column vector with components c_1 and c_2. On the right is the value of the map on that argument $h(\vec{v})$, represented with respect to the codomain's basis D. The matrix on the left is the new thing. We will use it to represent the map and we will think of the above equation as representing an application of the map to the matrix.

That matrix consists of the coefficients from the vector on the right, 0 and 1 from the first row, $-1/2$ and -1 from the second row, and 1 and 0 from the third row. That is, we make it by adjoining the vectors representing the $h(\vec{\beta}_i)$'s.

$$\left(\begin{array}{c|c} \vdots & \vdots \\ \mathrm{Rep}_D(\,h(\vec{\beta}_1)\,) & \mathrm{Rep}_D(\,h(\vec{\beta}_2)\,) \\ \vdots & \vdots \end{array}\right)$$

1.2 Definition Suppose that V and W are vector spaces of dimensions n and m with bases B and D, and that $h\colon V \to W$ is a linear map. If

$$\mathrm{Rep}_D(h(\vec{\beta}_1)) = \begin{pmatrix} h_{1,1} \\ h_{2,1} \\ \vdots \\ h_{m,1} \end{pmatrix}_D \quad \ldots \quad \mathrm{Rep}_D(h(\vec{\beta}_n)) = \begin{pmatrix} h_{1,n} \\ h_{2,n} \\ \vdots \\ h_{m,n} \end{pmatrix}_D$$

then

$$\mathrm{Rep}_{B,D}(h) = \begin{pmatrix} h_{1,1} & h_{1,2} & \ldots & h_{1,n} \\ h_{2,1} & h_{2,2} & \ldots & h_{2,n} \\ & \vdots & & \\ h_{m,1} & h_{m,2} & \ldots & h_{m,n} \end{pmatrix}_{B,D}$$

is the *matrix representation of* h *with respect to* B, D.

In that matrix the number of columns n is the dimension of the map's domain while the number of rows m is the dimension of the codomain.

We use lower case letters for a map, upper case for the matrix, and lower case again for the entries of the matrix. Thus for the map h, the matrix representing it is H, with entries $h_{i,j}$.

1.3 Example If $h\colon \mathbb{R}^3 \to \mathcal{P}_1$ is

$$\begin{pmatrix} a_1 \\ a_2 \\ a_3 \end{pmatrix} \stackrel{h}{\longmapsto} (2a_1 + a_2) + (-a_3)x$$

then where

$$B = \langle \begin{pmatrix} 0 \\ 0 \\ 1 \end{pmatrix}, \begin{pmatrix} 0 \\ 2 \\ 0 \end{pmatrix}, \begin{pmatrix} 2 \\ 0 \\ 0 \end{pmatrix} \rangle \qquad D = \langle 1 + x, -1 + x \rangle$$

the action of h on B is this.

$$\begin{pmatrix} 0 \\ 0 \\ 1 \end{pmatrix} \stackrel{h}{\longmapsto} -x \qquad \begin{pmatrix} 0 \\ 2 \\ 0 \end{pmatrix} \stackrel{h}{\longmapsto} 2 \qquad \begin{pmatrix} 2 \\ 0 \\ 0 \end{pmatrix} \stackrel{h}{\longmapsto} 4$$

A simple calculation

$$\mathrm{Rep}_D(-x) = \begin{pmatrix} -1/2 \\ -1/2 \end{pmatrix}_D \quad \mathrm{Rep}_D(2) = \begin{pmatrix} 1 \\ -1 \end{pmatrix}_D \quad \mathrm{Rep}_D(4) = \begin{pmatrix} 2 \\ -2 \end{pmatrix}_D$$

shows that this is the matrix representing h with respect to the bases.

$$\mathrm{Rep}_{B,D}(h) = \begin{pmatrix} -1/2 & 1 & 2 \\ -1/2 & -1 & -2 \end{pmatrix}_{B,D}$$

1.4 Theorem Assume that V and W are vector spaces of dimensions n and m with bases B and D, and that $h\colon V \to W$ is a linear map. If h is represented by

$$\mathrm{Rep}_{B,D}(h) = \begin{pmatrix} h_{1,1} & h_{1,2} & \ldots & h_{1,n} \\ h_{2,1} & h_{2,2} & \ldots & h_{2,n} \\ & \vdots & & \\ h_{m,1} & h_{m,2} & \ldots & h_{m,n} \end{pmatrix}_{B,D}$$

and $\vec{v} \in V$ is represented by

$$\mathrm{Rep}_B(\vec{v}) = \begin{pmatrix} c_1 \\ c_2 \\ \vdots \\ c_n \end{pmatrix}_B$$

then the representation of the image of $\vec{v}$ is this.

$$\mathrm{Rep}_D(\,h(\vec{v})\,) = \begin{pmatrix} h_{1,1}c_1 + h_{1,2}c_2 + \cdots + h_{1,n}c_n \\ h_{2,1}c_1 + h_{2,2}c_2 + \cdots + h_{2,n}c_n \\ \vdots \\ h_{m,1}c_1 + h_{m,2}c_2 + \cdots + h_{m,n}c_n \end{pmatrix}_D$$

Proof This formalizes Example 1.1. See Exercise 30. QED

1.5 Definition The *matrix-vector product* of a $m \times n$ matrix and a $n \times 1$ vector is this.

$$\begin{pmatrix} a_{1,1} & a_{1,2} & \ldots & a_{1,n} \\ a_{2,1} & a_{2,2} & \ldots & a_{2,n} \\ & \vdots & & \\ a_{m,1} & a_{m,2} & \ldots & a_{m,n} \end{pmatrix} \begin{pmatrix} c_1 \\ \vdots \\ c_n \end{pmatrix} = \begin{pmatrix} a_{1,1}c_1 + \cdots + a_{1,n}c_n \\ a_{2,1}c_1 + \cdots + a_{2,n}c_n \\ \vdots \\ a_{m,1}c_1 + \cdots + a_{m,n}c_n \end{pmatrix}$$

Briefly, application of a linear map is represented by the matrix-vector product of the map's representative and the vector's representative.

1.6 Remark Theorem 1.4 is not surprising, because we chose the matrix representative in Definition 1.2 precisely to make the theorem true — if the theorem were not true then we would adjust the definition to make it so. Nonetheless, we need the verification.

1.7 Example For the matrix from Example 1.3 we can calculate where that map sends this vector.

$$\vec{v} = \begin{pmatrix} 4 \\ 1 \\ 0 \end{pmatrix}$$

With respect to the domain basis B the representation of this vector is

$$\mathrm{Rep}_B(\vec{v}) = \begin{pmatrix} 0 \\ 1/2 \\ 2 \end{pmatrix}_B$$

and so the matrix-vector product gives the representation of the value $h(\vec{v})$ with respect to the codomain basis D.

$$\begin{aligned}
\mathrm{Rep}_D(h(\vec{v})) &= \begin{pmatrix} -1/2 & 1 & 2 \\ -1/2 & -1 & -2 \end{pmatrix}_{B,D} \begin{pmatrix} 0 \\ 1/2 \\ 2 \end{pmatrix}_B \\
&= \begin{pmatrix} (-1/2)\cdot 0 + 1\cdot(1/2) + 2\cdot 2 \\ (-1/2)\cdot 0 - 1\cdot(1/2) - 2\cdot 2 \end{pmatrix}_D = \begin{pmatrix} 9/2 \\ -9/2 \end{pmatrix}_D
\end{aligned}$$

To find $h(\vec{v})$ itself, not its representation, take $(9/2)(1+x) - (9/2)(-1+x) = 9$.

1.8 Example Let $\pi\colon \mathbb{R}^3 \to \mathbb{R}^2$ be projection onto the xy-plane. To give a matrix representing this map, we first fix some bases.

$$B = \langle \begin{pmatrix} 1 \\ 0 \\ 0 \end{pmatrix}, \begin{pmatrix} 1 \\ 1 \\ 0 \end{pmatrix}, \begin{pmatrix} -1 \\ 0 \\ 1 \end{pmatrix} \rangle \qquad D = \langle \begin{pmatrix} 2 \\ 1 \end{pmatrix}, \begin{pmatrix} 1 \\ 1 \end{pmatrix} \rangle$$

For each vector in the domain's basis, find its image under the map.

$$\begin{pmatrix} 1 \\ 0 \\ 0 \end{pmatrix} \stackrel{\pi}{\longmapsto} \begin{pmatrix} 1 \\ 0 \end{pmatrix} \quad \begin{pmatrix} 1 \\ 1 \\ 0 \end{pmatrix} \stackrel{\pi}{\longmapsto} \begin{pmatrix} 1 \\ 1 \end{pmatrix} \quad \begin{pmatrix} -1 \\ 0 \\ 1 \end{pmatrix} \stackrel{\pi}{\longmapsto} \begin{pmatrix} -1 \\ 0 \end{pmatrix}$$

Then find the representation of each image with respect to the codomain's basis.

$$\mathrm{Rep}_D(\begin{pmatrix} 1 \\ 0 \end{pmatrix}) = \begin{pmatrix} 1 \\ -1 \end{pmatrix} \quad \mathrm{Rep}_D(\begin{pmatrix} 1 \\ 1 \end{pmatrix}) = \begin{pmatrix} 0 \\ 1 \end{pmatrix} \quad \mathrm{Rep}_D(\begin{pmatrix} -1 \\ 0 \end{pmatrix}) = \begin{pmatrix} -1 \\ 1 \end{pmatrix}$$

Finally, adjoining these representations gives the matrix representing π with respect to B, D.

$$\mathrm{Rep}_{B,D}(\pi) = \begin{pmatrix} 1 & 0 & -1 \\ -1 & 1 & 1 \end{pmatrix}_{B,D}$$

We can illustrate Theorem 1.4 by computing the matrix-vector product representing this action by the projection map.

$$\pi(\begin{pmatrix}2\\2\\1\end{pmatrix}) = \begin{pmatrix}2\\2\end{pmatrix}$$

Represent the domain vector with respect to the domain's basis

$$\mathrm{Rep}_B(\begin{pmatrix}2\\2\\1\end{pmatrix}) = \begin{pmatrix}1\\2\\1\end{pmatrix}_B$$

to get this matrix-vector product.

$$\mathrm{Rep}_D(\,\pi(\begin{pmatrix}2\\2\\1\end{pmatrix})) = \begin{pmatrix}1&0&-1\\-1&1&1\end{pmatrix}_{B,D}\begin{pmatrix}1\\2\\1\end{pmatrix}_B = \begin{pmatrix}0\\2\end{pmatrix}_D$$

Expanding this into a linear combination of vectors from D

$$0\cdot\begin{pmatrix}2\\1\end{pmatrix} + 2\cdot\begin{pmatrix}1\\1\end{pmatrix} = \begin{pmatrix}2\\2\end{pmatrix}$$

checks that the map's action is indeed reflected in the operation of the matrix. We will sometimes compress these three displayed equations into one.

$$\begin{pmatrix}2\\2\\1\end{pmatrix} = \begin{pmatrix}1\\2\\1\end{pmatrix}_B \overset{h}{\underset{H}{\longmapsto}} \begin{pmatrix}0\\2\end{pmatrix}_D = \begin{pmatrix}2\\2\end{pmatrix}$$

We now have two ways to compute the effect of projection, the straightforward formula that drops each three-tall vector's third component to make a two-tall vector, and the above formula that uses representations and matrix-vector multiplication. The second way may seem complicated compared to the first, but it has advantages. The next example shows that for some maps this new scheme simplifies the formula.

1.9 Example To represent a rotation map $t_\theta\colon \mathbb{R}^2 \to \mathbb{R}^2$ that turns all vectors in the plane counterclockwise through an angle θ

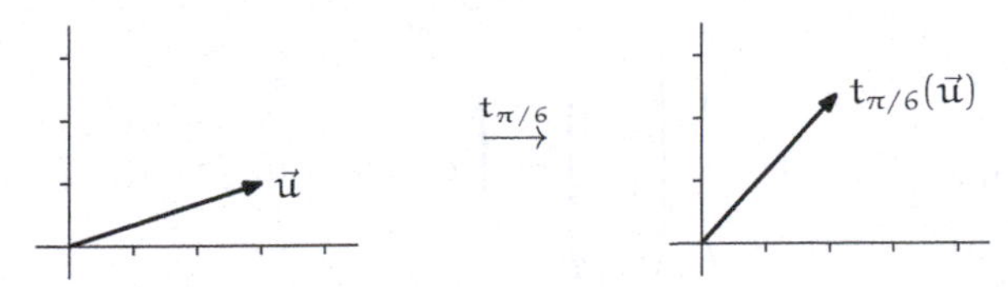

we start by fixing the standard bases $\mathcal{E}_2$ for both the domain and codomain basis, Now find the image under the map of each vector in the domain's basis.

$$\begin{pmatrix}1\\0\end{pmatrix}\overset{t_\theta}{\longmapsto}\begin{pmatrix}\cos\theta\\\sin\theta\end{pmatrix}\qquad\begin{pmatrix}0\\1\end{pmatrix}\overset{t_\theta}{\longmapsto}\begin{pmatrix}-\sin\theta\\\cos\theta\end{pmatrix}\tag{$*$}$$

Represent these images with respect to the codomain's basis. Because this basis is $\mathcal{E}_2$, vectors represent themselves. Adjoin the representations to get the matrix representing the map.

$$\mathrm{Rep}_{\mathcal{E}_2,\mathcal{E}_2}(t_\theta)=\begin{pmatrix}\cos\theta&-\sin\theta\\\sin\theta&\cos\theta\end{pmatrix}$$

The advantage of this scheme is that we get a formula for the image of any vector at all just by knowing in $(*)$ how to represent the image of the two basis vectors. For instance, here we rotate a vector by $\theta=\pi/6$.

$$\begin{pmatrix}3\\-2\end{pmatrix}=\begin{pmatrix}3\\-2\end{pmatrix}_{\mathcal{E}_2}\overset{t_{\pi/6}}{\longmapsto}\begin{pmatrix}\sqrt{3}/2&-1/2\\1/2&\sqrt{3}/2\end{pmatrix}\begin{pmatrix}3\\-2\end{pmatrix}\approx\begin{pmatrix}3.598\\-0.232\end{pmatrix}_{\mathcal{E}_2}=\begin{pmatrix}3.598\\-0.232\end{pmatrix}$$

More generally, we have a formula for rotation by $\theta=\pi/6$.

$$\begin{pmatrix}x\\y\end{pmatrix}\overset{t_{\pi/6}}{\longmapsto}\begin{pmatrix}\sqrt{3}/2&-1/2\\1/2&\sqrt{3}/2\end{pmatrix}\begin{pmatrix}x\\y\end{pmatrix}=\begin{pmatrix}(\sqrt{3}/2)x-(1/2)y\\(1/2)x+(\sqrt{3}/2)y\end{pmatrix}$$

1.10 Example In the definition of matrix-vector product the width of the matrix equals the height of the vector. Hence, this product is not defined.

$$\begin{pmatrix}1&0&0\\4&3&1\end{pmatrix}\begin{pmatrix}1\\0\end{pmatrix}$$

It is undefined for a reason: the three-wide matrix represents a map with a three-dimensional domain while the two-tall vector represents a member of a two-dimensional space. So the vector cannot be in the domain of the map.

Nothing in Definition 1.5 forces us to view matrix-vector product in terms of representations. We can get some insights by focusing on how the entries combine.

A good way to view matrix-vector product is that it is formed from the dot products of the rows of the matrix with the column vector.

$$\begin{pmatrix}&\vdots&&\\a_{i,1}&a_{i,2}&\ldots&a_{i,n}\\&\vdots&&\end{pmatrix}\begin{pmatrix}c_1\\c_2\\\vdots\\c_n\end{pmatrix}=\begin{pmatrix}\vdots\\a_{i,1}c_1+a_{i,2}c_2+\cdots+a_{i,n}c_n\\\vdots\end{pmatrix}$$

Looked at in this row-by-row way, this new operation generalizes dot product.

We can also view the operation column-by-column.

$$\begin{pmatrix} h_{1,1} & h_{1,2} & \ldots & h_{1,n} \\ h_{2,1} & h_{2,2} & \ldots & h_{2,n} \\ & \vdots & & \\ h_{m,1} & h_{m,2} & \ldots & h_{m,n} \end{pmatrix} \begin{pmatrix} c_1 \\ c_2 \\ \vdots \\ c_n \end{pmatrix} = \begin{pmatrix} h_{1,1}c_1 + h_{1,2}c_2 + \cdots + h_{1,n}c_n \\ h_{2,1}c_1 + h_{2,2}c_2 + \cdots + h_{2,n}c_n \\ \vdots \\ h_{m,1}c_1 + h_{m,2}c_2 + \cdots + h_{m,n}c_n \end{pmatrix}$$

$$= c_1 \begin{pmatrix} h_{1,1} \\ h_{2,1} \\ \vdots \\ h_{m,1} \end{pmatrix} + \cdots + c_n \begin{pmatrix} h_{1,n} \\ h_{2,n} \\ \vdots \\ h_{m,n} \end{pmatrix}$$

The result is the columns of the matrix weighted by the entries of the vector.

1.11 Example

$$\begin{pmatrix} 1 & 0 & -1 \\ 2 & 0 & 3 \end{pmatrix} \begin{pmatrix} 2 \\ -1 \\ 1 \end{pmatrix} = 2 \begin{pmatrix} 1 \\ 2 \end{pmatrix} - 1 \begin{pmatrix} 0 \\ 0 \end{pmatrix} + 1 \begin{pmatrix} -1 \\ 3 \end{pmatrix} = \begin{pmatrix} 1 \\ 7 \end{pmatrix}$$

This way of looking at matrix-vector product brings us back to the objective stated at the start of this section, to compute $h(c_1\vec{\beta}_1 + \cdots + c_n\vec{\beta}_n)$ as $c_1h(\vec{\beta}_1) + \cdots + c_nh(\vec{\beta}_n)$.

We began this section by noting that the equality of these two enables us to compute the action of h on any argument knowing only $h(\vec{\beta}_1)$, ..., $h(\vec{\beta}_n)$. We have developed this into a scheme to compute the action of the map by taking the matrix-vector product of the matrix representing the map with the vector representing the argument. In this way, with respect to any bases, for any linear map there is a matrix representation. The next subsection will show the converse, that if we fix bases then for any matrix there is an associated linear map.

Exercises

✓ **1.12** Multiply the matrix

$$\begin{pmatrix} 1 & 3 & 1 \\ 0 & -1 & 2 \\ 1 & 1 & 0 \end{pmatrix}$$

by each vector, or state "not defined."

(a) $\begin{pmatrix} 2 \\ 1 \\ 0 \end{pmatrix}$ (b) $\begin{pmatrix} -2 \\ -2 \end{pmatrix}$ (c) $\begin{pmatrix} 0 \\ 0 \\ 0 \end{pmatrix}$

1.13 Perform, if possible, each matrix-vector multiplication.

(a) $\begin{pmatrix}2&1\\3&-1/2\end{pmatrix}\begin{pmatrix}4\\2\end{pmatrix}$ (b) $\begin{pmatrix}1&1&0\\-2&1&0\end{pmatrix}\begin{pmatrix}1\\3\\1\end{pmatrix}$ (c) $\begin{pmatrix}1&1\\-2&1\end{pmatrix}\begin{pmatrix}1\\3\\1\end{pmatrix}$

✓ **1.14** Solve this matrix equation.

$$\begin{pmatrix}2&1&1\\0&1&3\\1&-1&2\end{pmatrix}\begin{pmatrix}x\\y\\z\end{pmatrix}=\begin{pmatrix}8\\4\\4\end{pmatrix}$$

✓ **1.15** For a homomorphism from $\mathcal{P}_2$ to $\mathcal{P}_3$ that sends

$$1\mapsto 1+x,\quad x\mapsto 1+2x,\quad\text{and}\quad x^2\mapsto x-x^3$$

where does $1-3x+2x^2$ go?

✓ **1.16** Assume that $h\colon\mathbb{R}^2\to\mathbb{R}^3$ is determined by this action.

$$\begin{pmatrix}1\\0\end{pmatrix}\mapsto\begin{pmatrix}2\\2\\0\end{pmatrix}\qquad\begin{pmatrix}0\\1\end{pmatrix}\mapsto\begin{pmatrix}0\\1\\-1\end{pmatrix}$$

Using the standard bases, find

(a) the matrix representing this map;

(b) a general formula for $h(\vec{v})$.

1.17 Represent the homomorphism $h\colon\mathbb{R}^3\to\mathbb{R}^2$ given by this formula and with respect to these bases.

$$\begin{pmatrix}x\\y\\z\end{pmatrix}\mapsto\begin{pmatrix}x+y\\x+z\end{pmatrix}\qquad B=\langle\begin{pmatrix}1\\1\\1\end{pmatrix},\begin{pmatrix}1\\1\\0\end{pmatrix},\begin{pmatrix}1\\0\\0\end{pmatrix}\rangle\quad D=\langle\begin{pmatrix}1\\0\end{pmatrix},\begin{pmatrix}0\\2\end{pmatrix}\rangle$$

✓ **1.18** Let $d/dx\colon\mathcal{P}_3\to\mathcal{P}_3$ be the derivative transformation.

(a) Represent d/dx with respect to B,B where $B=\langle 1,x,x^2,x^3\rangle$.

(b) Represent d/dx with respect to B,D where $D=\langle 1,2x,3x^2,4x^3\rangle$.

✓ **1.19** Represent each linear map with respect to each pair of bases.

(a) $d/dx\colon\mathcal{P}_n\to\mathcal{P}_n$ with respect to B,B where $B=\langle 1,x,\ldots,x^n\rangle$, given by

$$a_0+a_1x+a_2x^2+\cdots+a_nx^n\mapsto a_1+2a_2x+\cdots+na_nx^{n-1}$$

(b) $\int\colon\mathcal{P}_n\to\mathcal{P}_{n+1}$ with respect to B_n,B_{n+1} where $B_i=\langle 1,x,\ldots,x^i\rangle$, given by

$$a_0+a_1x+a_2x^2+\cdots+a_nx^n\mapsto a_0x+\frac{a_1}{2}x^2+\cdots+\frac{a_n}{n+1}x^{n+1}$$

(c) $\int_0^1\colon\mathcal{P}_n\to\mathbb{R}$ with respect to $B,\mathcal{E}_1$ where $B=\langle 1,x,\ldots,x^n\rangle$ and $\mathcal{E}_1=\langle 1\rangle$, given by

$$a_0+a_1x+a_2x^2+\cdots+a_nx^n\mapsto a_0+\frac{a_1}{2}+\cdots+\frac{a_n}{n+1}$$

(d) $\text{eval}_3\colon\mathcal{P}_n\to\mathbb{R}$ with respect to $B,\mathcal{E}_1$ where $B=\langle 1,x,\ldots,x^n\rangle$ and $\mathcal{E}_1=\langle 1\rangle$, given by

$$a_0+a_1x+a_2x^2+\cdots+a_nx^n\mapsto a_0+a_1\cdot 3+a_2\cdot 3^2+\cdots+a_n\cdot 3^n$$

(e) $\text{slide}_{-1}\colon\mathcal{P}_n\to\mathcal{P}_n$ with respect to B,B where $B=\langle 1,x,\ldots,x^n\rangle$, given by

$$a_0+a_1x+a_2x^2+\cdots+a_nx^n\mapsto a_0+a_1\cdot(x+1)+\cdots+a_n\cdot(x+1)^n$$

1.20 Represent the identity map on any nontrivial space with respect to B,B, where B is any basis.

1.21 Represent, with respect to the natural basis, the transpose transformation on the space $\mathcal{M}_{2\times 2}$ of 2×2 matrices.

1.22 Assume that $B = \langle \vec{\beta}_1, \vec{\beta}_2, \vec{\beta}_3, \vec{\beta}_4 \rangle$ is a basis for a vector space. Represent with respect to B, B the transformation that is determined by each.

(a) $\vec{\beta}_1 \mapsto \vec{\beta}_2$, $\vec{\beta}_2 \mapsto \vec{\beta}_3$, $\vec{\beta}_3 \mapsto \vec{\beta}_4$, $\vec{\beta}_4 \mapsto \vec{0}$

(b) $\vec{\beta}_1 \mapsto \vec{\beta}_2$, $\vec{\beta}_2 \mapsto \vec{0}$, $\vec{\beta}_3 \mapsto \vec{\beta}_4$, $\vec{\beta}_4 \mapsto \vec{0}$

(c) $\vec{\beta}_1 \mapsto \vec{\beta}_2$, $\vec{\beta}_2 \mapsto \vec{\beta}_3$, $\vec{\beta}_3 \mapsto \vec{0}$, $\vec{\beta}_4 \mapsto \vec{0}$

1.23 Example 1.9 shows how to represent the rotation transformation of the plane with respect to the standard basis. Express these other transformations also with respect to the standard basis.

(a) the *dilation* map d_s, which multiplies all vectors by the same scalar s

(b) the *reflection* map f_ℓ, which reflects all all vectors across a line ℓ through the origin

✓ **1.24** Consider a linear transformation of $\mathbb{R}^2$ determined by these two.

$$\begin{pmatrix}1\\1\end{pmatrix} \mapsto \begin{pmatrix}2\\0\end{pmatrix} \qquad \begin{pmatrix}1\\0\end{pmatrix} \mapsto \begin{pmatrix}-1\\0\end{pmatrix}$$

(a) Represent this transformation with respect to the standard bases.

(b) Where does the transformation send this vector?

$$\begin{pmatrix}0\\5\end{pmatrix}$$

(c) Represent this transformation with respect to these bases.

$$B = \langle \begin{pmatrix}1\\-1\end{pmatrix}, \begin{pmatrix}1\\1\end{pmatrix} \rangle \qquad D = \langle \begin{pmatrix}2\\2\end{pmatrix}, \begin{pmatrix}-1\\1\end{pmatrix} \rangle$$

(d) Using B from the prior item, represent the transformation with respect to B, B.

1.25 Suppose that $h\colon V \to W$ is one-to-one so that by Theorem 2.20, for any basis $B = \langle \vec{\beta}_1, \ldots, \vec{\beta}_n \rangle \subset V$ the image $h(B) = \langle h(\vec{\beta}_1), \ldots, h(\vec{\beta}_n) \rangle$ is a basis for W.

(a) Represent the map h with respect to $B, h(B)$.

(b) For a member $\vec{v}$ of the domain, where the representation of $\vec{v}$ has components $c_1, \ldots, c_n$, represent the image vector $h(\vec{v})$ with respect to the image basis $h(B)$.

1.26 Give a formula for the product of a matrix and $\vec{e}_i$, the column vector that is all zeroes except for a single one in the i-th position.

✓ **1.27** For each vector space of functions of one real variable, represent the derivative transformation with respect to B, B.

(a) $\{a\cos x + b\sin x \mid a, b \in \mathbb{R}\}$, $B = \langle \cos x, \sin x \rangle$

(b) $\{ae^x + be^{2x} \mid a, b \in \mathbb{R}\}$, $B = \langle e^x, e^{2x} \rangle$

(c) $\{a + bx + ce^x + dxe^x \mid a, b, c, d \in \mathbb{R}\}$, $B = \langle 1, x, e^x, xe^x \rangle$

1.28 Find the range of the linear transformation of $\mathbb{R}^2$ represented with respect to the standard bases by each matrix.

(a) $\begin{pmatrix}1 & 0\\0 & 0\end{pmatrix}$ (b) $\begin{pmatrix}0 & 0\\3 & 2\end{pmatrix}$ (c) a matrix of the form $\begin{pmatrix}a & b\\2a & 2b\end{pmatrix}$

✓ **1.29** Can one matrix represent two different linear maps? That is, can $\text{Rep}_{B,D}(h) = \text{Rep}_{\hat{B},\hat{D}}(\hat{h})$?

1.30 Prove Theorem 1.4.

✓ 1.31 Example 1.9 shows how to represent rotation of all vectors in the plane through an angle θ about the origin, with respect to the standard bases.

(a) Rotation of all vectors in three-space through an angle θ about the x-axis is a transformation of $\mathbb{R}^3$. Represent it with respect to the standard bases. Arrange the rotation so that to someone whose feet are at the origin and whose head is at $(1,0,0)$, the movement appears clockwise.

(b) Repeat the prior item, only rotate about the y-axis instead. (Put the person's head at $\vec{e}_2$.)

(c) Repeat, about the z-axis.

(d) Extend the prior item to $\mathbb{R}^4$. (*Hint:* we can restate 'rotate about the z-axis' as 'rotate parallel to the xy-plane'.)

1.32 (Schur's Triangularization Lemma)

(a) Let U be a subspace of V and fix bases $B_U \subseteq B_V$. What is the relationship between the representation of a vector from U with respect to B_U and the representation of that vector (viewed as a member of V) with respect to B_V?

(b) What about maps?

(c) Fix a basis $B = \langle \vec{\beta}_1, \dots, \vec{\beta}_n \rangle$ for V and observe that the spans

$$[\varnothing] = \{\vec{0}\} \subset [\{\vec{\beta}_1\}] \subset [\{\vec{\beta}_1, \vec{\beta}_2\}] \subset \quad \cdots \quad \subset [B] = V$$

form a strictly increasing chain of subspaces. Show that for any linear map $h\colon V \to W$ there is a chain $W_0 = \{\vec{0}\} \subseteq W_1 \subseteq \cdots \subseteq W_m = W$ of subspaces of W such that

$$h([\{\vec{\beta}_1, \dots, \vec{\beta}_i\}]) \subset W_i$$

for each i.

(d) Conclude that for every linear map $h\colon V \to W$ there are bases B, D so the matrix representing h with respect to B, D is upper-triangular (that is, each entry $h_{i,j}$ with $i > j$ is zero).

(e) Is an upper-triangular representation unique?

III.2 Any Matrix Represents a Linear Map

The prior subsection shows that the action of a linear map h is described by a matrix H, with respect to appropriate bases, in this way.

$$\vec{v} = \begin{pmatrix} v_1 \\ \vdots \\ v_n \end{pmatrix}_B \quad \overset{h}{\underset{H}{\longmapsto}} \quad h(\vec{v}) = \begin{pmatrix} h_{1,1}v_1 + \cdots + h_{1,n}v_n \\ \vdots \\ h_{m,1}v_1 + \cdots + h_{m,n}v_n \end{pmatrix}_D \qquad (*)$$

Here we will show the converse, that each matrix represents a linear map.

So we start with a matrix

$$H = \begin{pmatrix} h_{1,1} & h_{1,2} & \ldots & h_{1,n} \\ h_{2,1} & h_{2,2} & \ldots & h_{2,n} \\ & \vdots & & \\ h_{m,1} & h_{m,2} & \ldots & h_{m,n} \end{pmatrix}$$

and we will describe how it defines a map h. We require that the map be represented by the matrix so first note that in $(*)$ the dimension of the map's domain is the number of columns n of the matrix and the dimension of the codomain is the number of rows m. Thus, for h's domain fix an n-dimensional vector space V and for the codomain fix an m-dimensional space W. Also fix bases $B = \langle \vec{\beta}_1, \ldots, \vec{\beta}_n \rangle$ and $D = \langle \vec{\delta}_1, \ldots, \vec{\delta}_m \rangle$ for those spaces.

Now let $h \colon V \to W$ be: where $\vec{v}$ in the domain has the representation

$$\mathrm{Rep}_B(\vec{v}) = \begin{pmatrix} v_1 \\ \vdots \\ v_n \end{pmatrix}_B$$

then its image $h(\vec{v})$ is the member the codomain with this representation.

$$\mathrm{Rep}_D(\, h(\vec{v})\,) = \begin{pmatrix} h_{1,1}v_1 + \cdots + h_{1,n}v_n \\ \vdots \\ h_{m,1}v_1 + \cdots + h_{m,n}v_n \end{pmatrix}_D$$

That is, to compute the action of h on any $\vec{v} \in V$, first express $\vec{v}$ with respect to the basis $\vec{v} = v_1\vec{\beta}_1 + \cdots + v_n\vec{\beta}_n$ and then $h(\vec{v}) = (h_{1,1}v_1 + \cdots + h_{1,n}v_n) \cdot \vec{\delta}_1 + \cdots + (h_{m,1}v_1 + \cdots + h_{m,n}v_n) \cdot \vec{\delta}_m$.

Above we have made some choices; for instance V can be any n-dimensional space and B could be any basis for V, so H does not define a unique function. However, note once we have fixed V, B, W, and D then h is well-defined since $\vec{v}$ has a unique representation with respect to the basis B and the calculation of $\vec{w}$ from its representation is also uniquely determined.

2.1 Example Consider this matrix.

$$H = \begin{pmatrix} 1 & 2 \\ 3 & 4 \\ 5 & 6 \end{pmatrix}$$

It is 3×2 so any map that it defines must carry a dimension 2 domain to a dimension 3 codomain. We can choose the domain and codomain to be $\mathbb{R}^2$ and $\mathcal{P}_2$, with these bases.

$$B = \langle \begin{pmatrix} 1 \\ 1 \end{pmatrix}, \begin{pmatrix} 1 \\ -1 \end{pmatrix} \rangle \qquad D = \langle x^2, x^2 + x, x^2 + x + 1 \rangle$$

Then let $h\colon \mathbb{R}^2 \to \mathcal{P}_2$ be the function defined by H. We will compute the image under h of this member of the domain.

$$\vec{v} = \begin{pmatrix} -3 \\ 2 \end{pmatrix}$$

The computation is straightforward.

$$\mathrm{Rep}_D(h(\vec{v})) = H \cdot \mathrm{Rep}_B(\vec{v}) = \begin{pmatrix} 1 & 2 \\ 3 & 4 \\ 5 & 6 \end{pmatrix} \begin{pmatrix} -1/2 \\ -5/2 \end{pmatrix} = \begin{pmatrix} -11/2 \\ -23/2 \\ -35/2 \end{pmatrix}$$

From its representation, computation of $\vec{w}$ is routine $(-11/2)(x^2) - (23/2)(x^2 + x) - (35/2)(x^2 + x + 1) = (-69/2)x^2 - (58/2)x - (35/2)$.

2.2 Theorem Any matrix represents a homomorphism between vector spaces of appropriate dimensions, with respect to any pair of bases.

Proof We must check that for any matrix H and any domain and codomain bases B, D, the defined map h is linear. If $\vec{v}, \vec{u} \in V$ are such that

$$\mathrm{Rep}_B(\vec{v}) = \begin{pmatrix} v_1 \\ \vdots \\ v_n \end{pmatrix} \qquad \mathrm{Rep}_B(\vec{u}) = \begin{pmatrix} u_1 \\ \vdots \\ u_n \end{pmatrix}$$

and $c, d \in \mathbb{R}$ then the calculation

$$\begin{aligned} h(c\vec{v} + d\vec{u}) = {} & \bigl(h_{1,1}(cv_1 + du_1) + \cdots + h_{1,n}(cv_n + du_n)\bigr) \cdot \vec{\delta}_1 + \\ & \cdots + \bigl(h_{m,1}(cv_1 + du_1) + \cdots + h_{m,n}(cv_n + du_n)\bigr) \cdot \vec{\delta}_m \\ = {} & c \cdot h(\vec{v}) + d \cdot h(\vec{u}) \end{aligned}$$

supplies that check. QED

2.3 Example Even if the domain and codomain are the same, the map that the matrix represents depends on the bases that we choose. If

$$H = \begin{pmatrix} 1 & 0 \\ 0 & 0 \end{pmatrix}, \quad B_1 = D_1 = \langle \begin{pmatrix} 1 \\ 0 \end{pmatrix}, \begin{pmatrix} 0 \\ 1 \end{pmatrix} \rangle, \quad \text{and} \quad B_2 = D_2 = \langle \begin{pmatrix} 0 \\ 1 \end{pmatrix}, \begin{pmatrix} 1 \\ 0 \end{pmatrix} \rangle,$$

then $h_1\colon \mathbb{R}^2 \to \mathbb{R}^2$ represented by H with respect to B_1, D_1 maps

$$\begin{pmatrix} c_1 \\ c_2 \end{pmatrix} = \begin{pmatrix} c_1 \\ c_2 \end{pmatrix}_{B_1} \mapsto \begin{pmatrix} c_1 \\ 0 \end{pmatrix}_{D_1} = \begin{pmatrix} c_1 \\ 0 \end{pmatrix}$$

while $h_2\colon \mathbb{R}^2 \to \mathbb{R}^2$ represented by H with respect to B_2, D_2 is this map.

$$\begin{pmatrix} c_1 \\ c_2 \end{pmatrix} = \begin{pmatrix} c_2 \\ c_1 \end{pmatrix}_{B_2} \mapsto \begin{pmatrix} c_2 \\ 0 \end{pmatrix}_{D_2} = \begin{pmatrix} 0 \\ c_2 \end{pmatrix}$$

These are different functions. The first is projection onto the x-axis while the second is projection onto the y-axis.

This result means that when convenient we can work solely with matrices, just doing the computations without having to worry whether a matrix of interest represents a linear map on some pair of spaces.

When we are working with a matrix but we do not have particular spaces or bases in mind then we can take the domain and codomain to be $\mathbb{R}^n$ and $\mathbb{R}^m$, with the standard bases. This is convenient because with the standard bases vector representation is transparent — the representation of $\vec{v}$ is $\vec{v}$. (In this case the column space of the matrix equals the range of the map and consequently the column space of H is often denoted by $\mathscr{R}(H)$.)

Given a matrix, to come up with an associated map we can choose among many domain and codomain spaces, and many bases for those. So a matrix can represent many maps. We finish this section by illustrating how the matrix can give us information about the associated maps.

2.4 Theorem The rank of a matrix equals the rank of any map that it represents.

Proof Suppose that the matrix H is $m \times n$. Fix domain and codomain spaces V and W of dimension n and m with bases $B = \langle \vec{\beta}_1, \dots, \vec{\beta}_n \rangle$ and D. Then H represents some linear map h between those spaces with respect to these bases whose range space

$$\begin{aligned} \{h(\vec{v}) \mid \vec{v} \in V\} &= \{h(c_1\vec{\beta}_1 + \cdots + c_n\vec{\beta}_n) \mid c_1, \dots, c_n \in \mathbb{R}\} \\ &= \{c_1 h(\vec{\beta}_1) + \cdots + c_n h(\vec{\beta}_n) \mid c_1, \dots, c_n \in \mathbb{R}\} \end{aligned}$$

is the span $[\{h(\vec{\beta}_1), \dots, h(\vec{\beta}_n)\}]$. The rank of the map h is the dimension of this range space.

The rank of the matrix is the dimension of its column space, the span of the set of its columns $[\{\text{Rep}_D(h(\vec{\beta}_1)), \dots, \text{Rep}_D(h(\vec{\beta}_n))\}]$.

To see that the two spans have the same dimension, recall from the proof of Lemma I.2.5 that if we fix a basis then representation with respect to that basis gives an isomorphism $\text{Rep}_D\colon W \to \mathbb{R}^m$. Under this isomorphism there is a linear relationship among members of the range space if and only if the same relationship holds in the column space, e.g, $\vec{0} = c_1 \cdot h(\vec{\beta}_1) + \cdots + c_n \cdot h(\vec{\beta}_n)$ if and only if $\vec{0} = c_1 \cdot \text{Rep}_D(h(\vec{\beta}_1)) + \cdots + c_n \cdot \text{Rep}_D(h(\vec{\beta}_n))$. Hence, a subset of the range space is linearly independent if and only if the corresponding subset of the column space is linearly independent. Therefore the size of the largest linearly independent subset of the range space equals the size of the largest linearly independent subset of the column space, and so the two spaces have the same dimension. QED

That settles the apparent ambiguity in our use of the same word 'rank' to apply both to matrices and to maps.

2.5 Example Any map represented by

$$\begin{pmatrix} 1 & 2 & 2 \\ 1 & 2 & 1 \\ 0 & 0 & 3 \\ 0 & 0 & 2 \end{pmatrix}$$

must have three-dimensional domain and a four-dimensional codomain. In addition, because the rank of this matrix is two (we can spot this by eye or get it with Gauss's Method), any map represented by this matrix has a two-dimensional range space.

2.6 Corollary Let h be a linear map represented by a matrix H. Then h is onto if and only if the rank of H equals the number of its rows, and h is one-to-one if and only if the rank of H equals the number of its columns.

Proof For the onto half, the dimension of the range space of h is the rank of h, which equals the rank of H by the theorem. Since the dimension of the codomain of h equals the number of rows in H, if the rank of H equals the number of rows then the dimension of the range space equals the dimension of the codomain. But a subspace with the same dimension as its superspace must equal that superspace (because any basis for the range space is a linearly independent subset of the codomain whose size is equal to the dimension of the codomain, and thus so this basis for the range space must also be a basis for the codomain).

For the other half, a linear map is one-to-one if and only if it is an isomorphism between its domain and its range, that is, if and only if its domain has the same dimension as its range. The number of columns in H is the dimension of h's domain and by the theorem the rank of H equals the dimension of h's range. QED

2.7 Definition A linear map that is one-to-one and onto is *nonsingular*, otherwise it is *singular*. That is, a linear map is nonsingular if and only if it is an isomorphism.

2.8 Remark Some authors use 'nonsingular' as a synonym for one-to-one while others use it the way that we have here. The difference is slight because any map is onto its range space, so a one-to-one map is an isomorphism with its range.

In the first chapter we defined a matrix to be nonsingular if it is square and is the matrix of coefficients of a linear system with a unique solution. The next result justifies our dual use of the term.

2.9 Lemma A nonsingular linear map is represented by a square matrix. A square matrix represents nonsingular maps if and only if it is a nonsingular matrix. Thus, a matrix represents isomorphisms if and only if it is square and nonsingular.

PROOF Assume that the map $h\colon V \to W$ is nonsingular. Corollary 2.6 says that for any matrix H representing that map, because h is onto the number of rows of H equals the rank of H, and because h is one-to-one the number of columns of H is also equal to the rank of H. Hence H is square.

Next assume that H is square, $n \times n$. The matrix H is nonsingular if and only if its row rank is n, which is true if and only if H's rank is n by Theorem Two.III.3.11, which is true if and only if h's rank is n by Theorem 2.4, which is true if and only if h is an isomorphism by Theorem I.2.3. (This last holds because the domain of h is n-dimensional as it is the number of columns in H.) QED

2.10 Example Any map from $\mathbb{R}^2$ to $\mathcal{P}_1$ represented with respect to any pair of bases by

$$\begin{pmatrix} 1 & 2 \\ 0 & 3 \end{pmatrix}$$

is nonsingular because this matrix has rank two.

2.11 Example Any map $g\colon V \to W$ represented by

$$\begin{pmatrix} 1 & 2 \\ 3 & 6 \end{pmatrix}$$

is singular because this matrix is singular.

We've now seen that the relationship between maps and matrices goes both ways: for a particular pair of bases, any linear map is represented by a matrix and any matrix describes a linear map. That is, by fixing spaces and bases we get a correspondence between maps and matrices. In the rest of this chapter we will explore this correspondence. For instance, we've defined for linear maps the operations of addition and scalar multiplication and we shall see what the corresponding matrix operations are. We shall also see the matrix operation that represent the map operation of composition. And, we shall see how to find the matrix that represents a map's inverse.

Exercises

✓ **2.12** Let h be the linear map defined by this matrix on the domain $\mathcal{P}_1$ and codomain $\mathbb{R}^2$ with respect to the given bases.

$$H = \begin{pmatrix} 2 & 1 \\ 4 & 2 \end{pmatrix} \quad B = \langle 1+x, x \rangle,\ D = \langle \begin{pmatrix} 1 \\ 1 \end{pmatrix}, \begin{pmatrix} 1 \\ 0 \end{pmatrix} \rangle$$

What is the image under h of the vector $\vec{v} = 2x - 1$?

✓ **2.13** Decide if each vector lies in the range of the map from $\mathbb{R}^3$ to $\mathbb{R}^2$ represented with respect to the standard bases by the matrix.

(a) $\begin{pmatrix} 1 & 1 & 3 \\ 0 & 1 & 4 \end{pmatrix}, \begin{pmatrix} 1 \\ 3 \end{pmatrix}$ (b) $\begin{pmatrix} 2 & 0 & 3 \\ 4 & 0 & 6 \end{pmatrix}, \begin{pmatrix} 1 \\ 1 \end{pmatrix}$

✓ **2.14** Consider this matrix, representing a transformation of $\mathbb{R}^2$, and these bases for that space.

$$\frac{1}{2} \cdot \begin{pmatrix} 1 & 1 \\ -1 & 1 \end{pmatrix} \quad B = \langle \begin{pmatrix} 0 \\ 1 \end{pmatrix}, \begin{pmatrix} 1 \\ 0 \end{pmatrix} \rangle \quad D = \langle \begin{pmatrix} 1 \\ 1 \end{pmatrix}, \begin{pmatrix} 1 \\ -1 \end{pmatrix} \rangle$$

(a) To what vector in the codomain is the first member of B mapped?
(b) The second member?
(c) Where is a general vector from the domain (a vector with components x and y) mapped? That is, what transformation of $\mathbb{R}^2$ is represented with respect to B, D by this matrix?

2.15 What transformation of $F = \{a\cos\theta + b\sin\theta \mid a, b \in \mathbb{R}\}$ is represented with respect to $B = \langle \cos\theta - \sin\theta, \sin\theta \rangle$ and $D = \langle \cos\theta + \sin\theta, \cos\theta \rangle$ by this matrix?

$$\begin{pmatrix} 0 & 0 \\ 1 & 0 \end{pmatrix}$$

✓ **2.16** Decide whether $1 + 2x$ is in the range of the map from $\mathbb{R}^3$ to $\mathcal{P}_2$ represented with respect to $\mathcal{E}_3$ and $\langle 1, 1+x^2, x \rangle$ by this matrix.

$$\begin{pmatrix} 1 & 3 & 0 \\ 0 & 1 & 0 \\ 1 & 0 & 1 \end{pmatrix}$$

2.17 Example 2.11 gives a matrix that is nonsingular and is therefore associated with maps that are nonsingular.

(a) Find the set of column vectors representing the members of the null space of any map represented by this matrix.
(b) Find the nullity of any such map.
(c) Find the set of column vectors representing the members of the range space of any map represented by this matrix.
(d) Find the rank of any such map.
(e) Check that rank plus nullity equals the dimension of the domain.

✓ **2.18** Take each matrix to represent $h\colon \mathbb{R}^m \to \mathbb{R}^n$ with respect to the standard bases. For each (i) state m and n. Then set up an augmented matrix with the given matrix on the left and a vector representing a range space element on the right (e.g., if the codomain is $\mathbb{R}^3$ then in the right-hand column put the three entries a, b, and c). Perform Gauss-Jordan reduction. Use that to (ii) find $\mathscr{R}(h)$ and $\text{rank}(h)$ (and state whether the underlying map is onto), and (iii) find $\mathscr{N}(h)$ and $\text{nullity}(h)$ (and state whether the underlying map is one-to-one).

(a) $\begin{pmatrix} 2 & 1 \\ -1 & 3 \end{pmatrix}$

(b) $\begin{pmatrix} 0 & 1 & 3 \\ 2 & 3 & 4 \\ -2 & -1 & 2 \end{pmatrix}$

(c) $\begin{pmatrix} 1 & 1 \\ 2 & 1 \\ 3 & 1 \end{pmatrix}$

2.19 Use the method from the prior exercise on each.

(a) $\begin{pmatrix} 1 & 0 & -1 \\ 2 & 1 & 0 \\ 2 & 2 & 2 \end{pmatrix}$

(b) Verify that the map represented by this matrix is an isomorphism.

$$\begin{pmatrix} 2 & 1 & 0 \\ 3 & 1 & 1 \\ 7 & 2 & 1 \end{pmatrix}$$

2.20 This is an alternative proof of Lemma 2.9. Given an $n\times n$ matrix H, fix a domain V and codomain W of appropriate dimension n, and bases B, D for those spaces, and consider the map h represented by the matrix.

(a) Show that h is onto if and only if there is at least one $\text{Rep}_B(\vec{v})$ associated by H with each $\text{Rep}_D(\vec{w})$.

(b) Show that h is one-to-one if and only if there is at most one $\text{Rep}_B(\vec{v})$ associated by H with each $\text{Rep}_D(\vec{w})$.

(c) Consider the linear system $H\cdot\text{Rep}_B(\vec{v}) = \text{Rep}_D(\vec{w})$. Show that H is nonsingular if and only if there is exactly one solution $\text{Rep}_B(\vec{v})$ for each $\text{Rep}_D(\vec{w})$.

✓ **2.21** Because the rank of a matrix equals the rank of any map it represents, if one matrix represents two different maps $H = \text{Rep}_{B,D}(h) = \text{Rep}_{\hat{B},\hat{D}}(\hat{h})$ (where $h, \hat{h}\colon V \to W$) then the dimension of the range space of h equals the dimension of the range space of $\hat{h}$. Must these equal-dimensioned range spaces actually be the same?

✓ **2.22** Let V be an n-dimensional space with bases B and D. Consider a map that sends, for $\vec{v} \in V$, the column vector representing $\vec{v}$ with respect to B to the column vector representing $\vec{v}$ with respect to D. Show that map is a linear transformation of $\mathbb{R}^n$.

2.23 Example 2.3 shows that changing the pair of bases can change the map that a matrix represents, even though the domain and codomain remain the same. Could the map ever not change? Is there a matrix H, vector spaces V and W, and associated pairs of bases B_1, D_1 and B_2, D_2 (with $B_1 \neq B_2$ or $D_1 \neq D_2$ or both) such that the map represented by H with respect to B_1, D_1 equals the map represented by H with respect to B_2, D_2?

✓ **2.24** A square matrix is a *diagonal* matrix if it is all zeroes except possibly for the entries on its upper-left to lower-right diagonal—its 1,1 entry, its 2,2 entry, etc. Show that a linear map is an isomorphism if there are bases such that, with respect to those bases, the map is represented by a diagonal matrix with no zeroes on the diagonal.

2.25 Describe geometrically the action on $\mathbb{R}^2$ of the map represented with respect to the standard bases $\mathcal{E}_2, \mathcal{E}_2$ by this matrix.

$$\begin{pmatrix} 3 & 0 \\ 0 & 2 \end{pmatrix}$$

Do the same for these.

$$\begin{pmatrix} 1 & 0 \\ 0 & 0 \end{pmatrix} \quad \begin{pmatrix} 0 & 1 \\ 1 & 0 \end{pmatrix} \quad \begin{pmatrix} 1 & 3 \\ 0 & 1 \end{pmatrix}$$

2.26 The fact that for any linear map the rank plus the nullity equals the dimension of the domain shows that a necessary condition for the existence of a homomorphism between two spaces, onto the second space, is that there be no gain in dimension. That is, where $h\colon V \to W$ is onto, the dimension of W must be less than or equal to the dimension of V.

(a) Show that this (strong) converse holds: no gain in dimension implies that there is a homomorphism and, further, any matrix with the correct size and correct rank represents such a map.

(b) Are there bases for $\mathbb{R}^3$ such that this matrix

$$H = \begin{pmatrix} 1 & 0 & 0 \\ 2 & 0 & 0 \\ 0 & 1 & 0 \end{pmatrix}$$

represents a map from $\mathbb{R}^3$ to $\mathbb{R}^3$ whose range is the xy plane subspace of $\mathbb{R}^3$?

2.27 Let V be an n-dimensional space and suppose that $\vec{x} \in \mathbb{R}^n$. Fix a basis B for V and consider the map $h_{\vec{x}}\colon V \to \mathbb{R}$ given $\vec{v} \mapsto \vec{x} \cdot \operatorname{Rep}_B(\vec{v})$ by the dot product.

(a) Show that this map is linear.

(b) Show that for any linear map $g\colon V \to \mathbb{R}$ there is an $\vec{x} \in \mathbb{R}^n$ such that $g = h_{\vec{x}}$.

(c) In the prior item we fixed the basis and varied the $\vec{x}$ to get all possible linear maps. Can we get all possible linear maps by fixing an $\vec{x}$ and varying the basis?

2.28 Let V, W, X be vector spaces with bases B, C, D.

(a) Suppose that $h\colon V \to W$ is represented with respect to B, C by the matrix H. Give the matrix representing the scalar multiple rh (where $r \in \mathbb{R}$) with respect to B, C by expressing it in terms of H.

(b) Suppose that $h, g\colon V \to W$ are represented with respect to B, C by H and G. Give the matrix representing $h + g$ with respect to B, C by expressing it in terms of H and G.

(c) Suppose that $h\colon V \to W$ is represented with respect to B, C by H and $g\colon W \to X$ is represented with respect to C, D by G. Give the matrix representing $g \circ h$ with respect to B, D by expressing it in terms of H and G.

IV Matrix Operations

The prior section shows how matrices represent linear maps. We now explore how this representation interacts with things that we already know. First we will see how the representation of a scalar product $r \cdot f$ of a linear map relates to the representation of f, and also how the representation of a sum $f + g$ relates to the representations of the two summands. Later we will do the same comparison for the map operations of composition and inverse.

IV.1 Sums and Scalar Products

1.1 Example Let $f\colon V \to W$ be a linear function represented with respect to some bases by this matrix.

$$\text{Rep}_{B,D}(f) = \begin{pmatrix} 1 & 0 \\ 1 & 1 \end{pmatrix}$$

Consider the map that is the scalar multiple $5f\colon V \to W$. We will relate the representation $\text{Rep}_{B,D}(5f)$ with $\text{Rep}_{B,D}(f)$.

Let f associate $\vec{v} \mapsto \vec{w}$ with these representations.

$$\text{Rep}_B(\vec{v}) = \begin{pmatrix} v_1 \\ v_2 \end{pmatrix} \qquad \text{Rep}_D(\vec{w}) = \begin{pmatrix} w_1 \\ w_2 \end{pmatrix}$$

Where the codomain's basis is $D = \langle \vec{\delta}_1, \vec{\delta}_2 \rangle$, that representation gives that the output vector is $\vec{w} = w_1\vec{\delta}_1 + w_2\vec{\delta}_2$.

The action of the map $5f$ is $\vec{v} \mapsto 5\vec{w}$ and $5\vec{w} = 5 \cdot (w_1\vec{\delta}_1 + w_2\vec{\delta}_2) = (5w_1)\vec{\delta}_1 + (5w_2)\vec{\delta}_2$. So $5f$ associates the input vector $\vec{v}$ with the output vector having this representation.

$$\text{Rep}_D(5\vec{w}) = \begin{pmatrix} 5w_1 \\ 5w_2 \end{pmatrix}$$

Changing from the map f to the map $5f$ has the effect on the representation of the output vector of multiplying each entry by 5.

Because of that, $\text{Rep}_{B,D}(5f)$ is this matrix.

$$\text{Rep}_{B,D}(5f) \cdot \begin{pmatrix} v_1 \\ v_2 \end{pmatrix} = \begin{pmatrix} 5v_1 \\ 5v_1 + 5v_2 \end{pmatrix} \qquad \text{Rep}_{B,D}(5f) = \begin{pmatrix} 5 & 0 \\ 5 & 5 \end{pmatrix}$$

Therefore, going from the matrix representing f to the one representing $5f$ means multiplying all the matrix entries by 5.

1.2 Example We can do a similar exploration for the sum of two maps. Suppose that two linear maps with the same domain and codomain $f, g\colon \mathbb{R}^2 \to \mathbb{R}^2$ are represented with respect to bases B and D by these matrices.

$$\operatorname{Rep}_{B,D}(f) = \begin{pmatrix} 1 & 3 \\ 2 & 0 \end{pmatrix} \qquad \operatorname{Rep}_{B,D}(g) = \begin{pmatrix} -2 & -1 \\ 2 & 4 \end{pmatrix}$$

Recall the definition of sum: if f does $\vec{v} \mapsto \vec{u}$ and g does $\vec{v} \mapsto \vec{w}$ then $f+g$ is the function whose action is $\vec{v} \mapsto \vec{u} + \vec{w}$. Let these be the representations of the input and output vectors.

$$\operatorname{Rep}_B(\vec{v}) = \begin{pmatrix} v_1 \\ v_2 \end{pmatrix} \qquad \operatorname{Rep}_D(\vec{u}) = \begin{pmatrix} u_1 \\ u_2 \end{pmatrix} \quad \operatorname{Rep}_D(\vec{w}) = \begin{pmatrix} w_1 \\ w_2 \end{pmatrix}$$

Where $D = \langle \vec{\delta}_1, \vec{\delta}_2 \rangle$ we have $\vec{u} + \vec{w} = (u_1\vec{\delta}_1 + u_2\vec{\delta}_2) + (w_1\vec{\delta}_1 + w_2\vec{\delta}_2) = (u_1 + w_1)\vec{\delta}_1 + (u_2 + w_2)\vec{\delta}_2$ and so this is the representation of the vector sum.

$$\operatorname{Rep}_D(\vec{u} + \vec{w}) = \begin{pmatrix} u_1 + w_1 \\ u_2 + w_2 \end{pmatrix}$$

Thus, since these represent the actions of of the maps f and g on the input $\vec{v}$

$$\begin{pmatrix} 1 & 3 \\ 2 & 0 \end{pmatrix} \begin{pmatrix} v_1 \\ v_2 \end{pmatrix} = \begin{pmatrix} v_1 + 3v_2 \\ 2v_1 \end{pmatrix} \qquad \begin{pmatrix} -2 & -1 \\ 2 & 4 \end{pmatrix} \begin{pmatrix} v_1 \\ v_2 \end{pmatrix} = \begin{pmatrix} -2v_1 - v_2 \\ 2v_1 + 4v_2 \end{pmatrix}$$

adding the entries represents the action of the map $f+g$.

$$\operatorname{Rep}_{B,D}(f+g) \cdot \begin{pmatrix} v_1 \\ v_2 \end{pmatrix} = \begin{pmatrix} -v_1 + 2v_2 \\ 4v_1 + 4v_2 \end{pmatrix}$$

Therefore, we compute the matrix representing the function sum by adding the entries of the matrices representing the functions.

$$\operatorname{Rep}_{B,D}(f+g) = \begin{pmatrix} -1 & 2 \\ 4 & 4 \end{pmatrix}$$

1.3 Definition The *scalar multiple* of a matrix is the result of entry-by-entry scalar multiplication. The *sum* of two same-sized matrices is their entry-by-entry sum.

These operations extend the first chapter's operations of addition and scalar multiplication of vectors.

We need a result that proves these matrix operations do what the examples suggest that they do.

1.4 Theorem Let $h, g\colon V \to W$ be linear maps represented with respect to bases B, D by the matrices H and G and let r be a scalar. Then with respect to B, D the map $r \cdot h\colon V \to W$ is represented by rH and the map $h + g\colon V \to W$ is represented by $H + G$.

PROOF Generalize the examples. This is Exercise 9. QED

1.5 Remark These two operations on matrices are simple. But we did not define them in this way because they are simple. We defined them in this way because they represent function addition and function scalar multiplication. That is, our program is to define matrix operations by referencing function operations. Simplicity is a pleasant bonus.

We will see this again in the next subsection, where we will define the operation of multiplying matrices. Since we've just defined matrix scalar multiplication and matrix sum to be entry-by-entry operations, a naive thought is to define matrix multiplication to be the entry-by-entry product. In theory we could do whatever we please but we will instead be practical and combine the entries in the way that represents the function operation of composition.

A special case of scalar multiplication is multiplication by zero. For any map $0 \cdot h$ is the zero homomorphism and for any matrix $0 \cdot H$ is the matrix with all entries zero.

1.6 Definition A *zero matrix* has all entries 0. We write $Z_{n\times m}$ or simply Z (another common notation is $0_{n\times m}$ or just 0).

1.7 Example The zero map from any three-dimensional space to any two-dimensional space is represented by the 2×3 zero matrix

$$Z = \begin{pmatrix} 0 & 0 & 0 \\ 0 & 0 & 0 \end{pmatrix}$$

no matter what domain and codomain bases we use.

Exercises

✓ 1.8 Perform the indicated operations, if defined.

(a) $\begin{pmatrix} 5 & -1 & 2 \\ 6 & 1 & 1 \end{pmatrix} + \begin{pmatrix} 2 & 1 & 4 \\ 3 & 0 & 5 \end{pmatrix}$

(b) $6 \cdot \begin{pmatrix} 2 & -1 & -1 \\ 1 & 2 & 3 \end{pmatrix}$

(c) $\begin{pmatrix} 2 & 1 \\ 0 & 3 \end{pmatrix} + \begin{pmatrix} 2 & 1 \\ 0 & 3 \end{pmatrix}$

(d) $4\begin{pmatrix} 1 & 2 \\ 3 & -1 \end{pmatrix} + 5\begin{pmatrix} -1 & 4 \\ -2 & 1 \end{pmatrix}$

(e) $3\begin{pmatrix}2&1\\3&0\end{pmatrix}+2\begin{pmatrix}1&1&4\\3&0&5\end{pmatrix}$

1.9 Prove Theorem 1.4.

(a) Prove that matrix addition represents addition of linear maps.

(b) Prove that matrix scalar multiplication represents scalar multiplication of linear maps.

✓ 1.10 Prove each, assuming that the operations are defined, where G, H, and J are matrices, where Z is the zero matrix, and where r and s are scalars.

(a) Matrix addition is commutative $G + H = H + G$.

(b) Matrix addition is associative $G + (H + J) = (G + H) + J$.

(c) The zero matrix is an additive identity $G + Z = G$.

(d) $0 \cdot G = Z$

(e) $(r + s)G = rG + sG$

(f) Matrices have an additive inverse $G + (-1) \cdot G = Z$.

(g) $r(G + H) = rG + rH$

(h) $(rs)G = r(sG)$

1.11 Fix domain and codomain spaces. In general, one matrix can represent many different maps with respect to different bases. However, prove that a zero matrix represents only a zero map. Are there other such matrices?

✓ 1.12 Let V and W be vector spaces of dimensions n and m. Show that the space $\mathcal{L}(V, W)$ of linear maps from V to W is isomorphic to $\mathcal{M}_{m\times n}$.

✓ 1.13 Show that it follows from the prior questions that for any six transformations $t_1, \ldots, t_6 \colon \mathbb{R}^2 \to \mathbb{R}^2$ there are scalars $c_1, \ldots, c_6 \in \mathbb{R}$ such that $c_1t_1 + \cdots + c_6t_6$ is the zero map. (*Hint:* the six is slightly misleading.)

1.14 The *trace* of a square matrix is the sum of the entries on the main diagonal (the 1, 1 entry plus the 2, 2 entry, etc.; we will see the significance of the trace in Chapter Five). Show that $\text{trace}(H + G) = \text{trace}(H) + \text{trace}(G)$. Is there a similar result for scalar multiplication?

1.15 Recall that the *transpose* of a matrix M is another matrix, whose i, j entry is the j, i entry of M. Verify these identities.

(a) $(G + H)^T = G^T + H^T$

(b) $(r \cdot H)^T = r \cdot H^T$

✓ 1.16 A square matrix is *symmetric* if each i, j entry equals the j, i entry, that is, if the matrix equals its transpose.

(a) Prove that for any square H, the matrix $H + H^T$ is symmetric. Does every symmetric matrix have this form?

(b) Prove that the set of $n\times n$ symmetric matrices is a subspace of $\mathcal{M}_{n\times n}$.

✓ 1.17 (a) How does matrix rank interact with scalar multiplication — can a scalar product of a rank n matrix have rank less than n? Greater?

(b) How does matrix rank interact with matrix addition — can a sum of rank n matrices have rank less than n? Greater?

IV.2 Matrix Multiplication

After representing addition and scalar multiplication of linear maps in the prior subsection, the natural next operation to consider is function composition.

2.1 Lemma The composition of linear maps is linear.

PROOF *(Note: this argument has already appeared, as part of the proof of Theorem I.2.2.)* Let $h\colon V \to W$ and $g\colon W \to U$ be linear. The calculation

$$\begin{aligned} g \circ h\,(c_1 \cdot \vec{v}_1 + c_2 \cdot \vec{v}_2) &= g\big(\,h(c_1 \cdot \vec{v}_1 + c_2 \cdot \vec{v}_2)\,\big) = g\big(\,c_1 \cdot h(\vec{v}_1) + c_2 \cdot h(\vec{v}_2)\,\big) \\ &= c_1 \cdot g\big(h(\vec{v}_1)\big) + c_2 \cdot g\big(h(\vec{v}_2)\big) = c_1 \cdot (g \circ h)(\vec{v}_1) + c_2 \cdot (g \circ h)(\vec{v}_2) \end{aligned}$$

shows that $g \circ h\colon V \to U$ preserves linear combinations, and so is linear. QED

As we did with the operation of matrix addition and scalar multiplication, we will see how the representation of the composite relates to the representations of the compositors by first considering an example.

2.2 Example Let $h\colon \mathbb{R}^4 \to \mathbb{R}^2$ and $g\colon \mathbb{R}^2 \to \mathbb{R}^3$, fix bases $B \subset \mathbb{R}^4$, $C \subset \mathbb{R}^2$, $D \subset \mathbb{R}^3$, and let these be the representations.

$$H = \text{Rep}_{B,C}(h) = \begin{pmatrix} 4 & 6 & 8 & 2 \\ 5 & 7 & 9 & 3 \end{pmatrix}_{B,C} \qquad G = \text{Rep}_{C,D}(g) = \begin{pmatrix} 1 & 1 \\ 0 & 1 \\ 1 & 0 \end{pmatrix}_{C,D}$$

To represent the composition $g \circ h\colon \mathbb{R}^4 \to \mathbb{R}^3$ we start with a $\vec{v}$, represent h of $\vec{v}$, and then represent g of that. The representation of $h(\vec{v})$ is the product of h's matrix and $\vec{v}$'s vector.

$$\text{Rep}_C(\,h(\vec{v})\,) = \begin{pmatrix} 4 & 6 & 8 & 2 \\ 5 & 7 & 9 & 3 \end{pmatrix}_{B,C} \begin{pmatrix} v_1 \\ v_2 \\ v_3 \\ v_4 \end{pmatrix}_B = \begin{pmatrix} 4v_1 + 6v_2 + 8v_3 + 2v_4 \\ 5v_1 + 7v_2 + 9v_3 + 3v_4 \end{pmatrix}_C$$

The representation of $g(\,h(\vec{v})\,)$ is the product of g's matrix and $h(\vec{v})$'s vector.

$$\begin{aligned} \text{Rep}_D(\,g(h(\vec{v}))\,) &= \begin{pmatrix} 1 & 1 \\ 0 & 1 \\ 1 & 0 \end{pmatrix}_{C,D} \begin{pmatrix} 4v_1 + 6v_2 + 8v_3 + 2v_4 \\ 5v_1 + 7v_2 + 9v_3 + 3v_4 \end{pmatrix}_C \\ &= \begin{pmatrix} 1 \cdot (4v_1 + 6v_2 + 8v_3 + 2v_4) + 1 \cdot (5v_1 + 7v_2 + 9v_3 + 3v_4) \\ 0 \cdot (4v_1 + 6v_2 + 8v_3 + 2v_4) + 1 \cdot (5v_1 + 7v_2 + 9v_3 + 3v_4) \\ 1 \cdot (4v_1 + 6v_2 + 8v_3 + 2v_4) + 0 \cdot (5v_1 + 7v_2 + 9v_3 + 3v_4) \end{pmatrix}_D \end{aligned}$$

Distributing and regrouping on the $\vec{v}$'s gives

$$= \begin{pmatrix} (1\cdot 4+1\cdot 5)v_1+(1\cdot 6+1\cdot 7)v_2+(1\cdot 8+1\cdot 9)v_3+(1\cdot 2+1\cdot 3)v_4 \\ (0\cdot 4+1\cdot 5)v_1+(0\cdot 6+1\cdot 7)v_2+(0\cdot 8+1\cdot 9)v_3+(0\cdot 2+1\cdot 3)v_4 \\ (1\cdot 4+0\cdot 5)v_1+(1\cdot 6+0\cdot 7)v_2+(1\cdot 8+0\cdot 9)v_3+(1\cdot 2+0\cdot 3)v_4 \end{pmatrix}_D$$

which is this matrix-vector product.

$$= \begin{pmatrix} 1\cdot 4+1\cdot 5 & 1\cdot 6+1\cdot 7 & 1\cdot 8+1\cdot 9 & 1\cdot 2+1\cdot 3 \\ 0\cdot 4+1\cdot 5 & 0\cdot 6+1\cdot 7 & 0\cdot 8+1\cdot 9 & 0\cdot 2+1\cdot 3 \\ 1\cdot 4+0\cdot 5 & 1\cdot 6+0\cdot 7 & 1\cdot 8+0\cdot 9 & 1\cdot 2+0\cdot 3 \end{pmatrix}_{B,D} \begin{pmatrix} v_1 \\ v_2 \\ v_3 \\ v_4 \end{pmatrix}_D$$

The matrix representing $g \circ h$ has the rows of G combined with the columns of H.

2.3 Definition The *matrix-multiplicative product* of the $m \times r$ matrix G and the $r \times n$ matrix H is the $m \times n$ matrix P, where

$$p_{i,j} = g_{i,1}h_{1,j} + g_{i,2}h_{2,j} + \cdots + g_{i,r}h_{r,j}$$

so that the i,j-th entry of the product is the dot product of the i-th row of the first matrix with the j-th column of the second.

$$GH = \begin{pmatrix} & \vdots & & \\ g_{i,1} & g_{i,2} & \cdots & g_{i,r} \\ & \vdots & & \end{pmatrix} \begin{pmatrix} & h_{1,j} & \\ \cdots & h_{2,j} & \cdots \\ & \vdots & \\ & h_{r,j} & \end{pmatrix} = \begin{pmatrix} & \vdots & \\ \cdots & p_{i,j} & \cdots \\ & \vdots & \end{pmatrix}$$

2.4 Example

$$\begin{pmatrix} 2 & 0 \\ 4 & 6 \\ 8 & 2 \end{pmatrix} \begin{pmatrix} 1 & 3 \\ 5 & 7 \end{pmatrix} = \begin{pmatrix} 2\cdot 1+0\cdot 5 & 2\cdot 3+0\cdot 7 \\ 4\cdot 1+6\cdot 5 & 4\cdot 3+6\cdot 7 \\ 8\cdot 1+2\cdot 5 & 8\cdot 3+2\cdot 7 \end{pmatrix} = \begin{pmatrix} 2 & 6 \\ 34 & 54 \\ 18 & 38 \end{pmatrix}$$

2.5 Example Some products are not defined, such as the product of a 2×3 matrix with a 2×2, because the number of columns in the first matrix must equal the number of rows in the second. But the product of two $n\times n$ matrices is always defined. Here are two 2×2's.

$$\begin{pmatrix} 1 & 2 \\ 3 & 4 \end{pmatrix} \begin{pmatrix} -1 & 0 \\ 2 & -2 \end{pmatrix} = \begin{pmatrix} 1\cdot(-1)+2\cdot 2 & 1\cdot 0+2\cdot(-2) \\ 3\cdot(-1)+4\cdot 2 & 3\cdot 0+4\cdot(-2) \end{pmatrix} = \begin{pmatrix} 3 & -4 \\ 5 & -8 \end{pmatrix}$$

2.6 Example The matrices from Example 2.2 combine in this way.

$$\begin{pmatrix}1&1\\0&1\\1&0\end{pmatrix}\begin{pmatrix}4&6&8&2\\5&7&9&3\end{pmatrix} = \begin{pmatrix}1\cdot4+1\cdot5&1\cdot6+1\cdot7&1\cdot8+1\cdot9&1\cdot2+1\cdot3\\0\cdot4+1\cdot5&0\cdot6+1\cdot7&0\cdot8+1\cdot9&0\cdot2+1\cdot3\\1\cdot4+0\cdot5&1\cdot6+0\cdot7&1\cdot8+0\cdot9&1\cdot2+0\cdot3\end{pmatrix} = \begin{pmatrix}9&13&17&5\\5&7&9&3\\4&6&8&2\end{pmatrix}$$

2.7 Theorem A composition of linear maps is represented by the matrix product of the representatives.

PROOF This argument generalizes Example 2.2. Let $h\colon V\to W$ and $g\colon W\to X$ be represented by H and G with respect to bases $B\subset V$, $C\subset W$, and $D\subset X$, of sizes n, r, and m. For any $\vec{v}\in V$ the k-th component of $\operatorname{Rep}_C(\,h(\vec{v})\,)$ is

$$h_{k,1}v_1+\cdots+h_{k,n}v_n$$

and so the i-th component of $\operatorname{Rep}_D(\,g\circ h\,(\vec{v})\,)$ is this.

$$g_{i,1}\cdot(h_{1,1}v_1+\cdots+h_{1,n}v_n)+g_{i,2}\cdot(h_{2,1}v_1+\cdots+h_{2,n}v_n)\\ +\cdots+g_{i,r}\cdot(h_{r,1}v_1+\cdots+h_{r,n}v_n)$$

Distribute and regroup on the v's.

$$=(g_{i,1}h_{1,1}+g_{i,2}h_{2,1}+\cdots+g_{i,r}h_{r,1})\cdot v_1\\ +\cdots+(g_{i,1}h_{1,n}+g_{i,2}h_{2,n}+\cdots+g_{i,r}h_{r,n})\cdot v_n$$

Finish by recognizing that the coefficient of each v_j

$$g_{i,1}h_{1,j}+g_{i,2}h_{2,j}+\cdots+g_{i,r}h_{r,j}$$

matches the definition of the i,j entry of the product GH. QED

This *arrow diagram* pictures the relationship between maps and matrices ('wrt' abbreviates 'with respect to').

$$\begin{array}{ccccc} & & W_{wrt\ C} & & \\ & {}^{h}\nearrow_{H} & & {}^{g}\searrow_{G} & \\ V_{wrt\ B} & & \xrightarrow[GH]{g\circ h} & & X_{wrt\ D}\end{array}$$

Above the arrows, the maps show that the two ways of going from V to X, straight over via the composition or else in two steps by way of W, have the same effect

$$\vec{v} \overset{g\circ h}{\longmapsto} g(h(\vec{v})) \qquad \vec{v} \overset{h}{\longmapsto} h(\vec{v}) \overset{g}{\longmapsto} g(h(\vec{v}))$$

(this is just the definition of composition). Below the arrows, the matrices indicate that multiplying GH into the column vector $\text{Rep}_B(\vec{v})$ has the same effect as multiplying the column vector first by H and then multiplying the result by G.

$$\text{Rep}_{B,D}(g \circ h) = GH \qquad \text{Rep}_{C,D}(g)\ \text{Rep}_{B,C}(h) = GH$$

As mentioned in Example 2.5, because the number of columns on the left does not equal the number of rows on the right, the product as here of a 2×3 matrix with a 2×2 matrix is not defined.

$$\begin{pmatrix} -1 & 2 & 0 \\ 0 & 10 & 1.1 \end{pmatrix}\begin{pmatrix} 0 & 0 \\ 0 & 2 \end{pmatrix}$$

The definition requires that the sizes match because we want that the underlying function composition is possible.

$$\text{dimension } n \text{ space} \overset{h}{\longrightarrow} \text{dimension } r \text{ space} \overset{g}{\longrightarrow} \text{dimension } m \text{ space} \qquad (*)$$

Thus, matrix product combines an $m\times r$ matrix G with an $r\times n$ matrix F to yield the $m\times n$ result GF. Briefly: $m\times r$ times $r\times n$ equals $m\times n$.

2.8 Remark The order in which we write things can be confusing. In '$m\times r$ times $r\times n$ equals $m\times n$' the number written first m is the dimension of g's codomain and is thus the number that appears last in the map dimension description $(*)$. The explanation is that while h is done first and is followed by g, we write the composition as $g\circ h$, with g on the left, from the notation $g(h(\vec{v}))$. That carries over to matrices, so that $g\circ h$ is represented by GH.

We can get insight into matrix-matrix product operation by studying how the entries combine. For instance, an alternative way to understand why we require above that the sizes match is that the row of the left-hand matrix must have the same number of entries as the column of the right-hand matrix, or else some entry will be left without a matching entry from the other matrix.

Another aspect of the combinatorics of matrix multiplication, in the sum defining the i,j entry, is brought out here by the boxing the equal subscripts.

$$p_{i,j} = g_{i,\boxed{1}}h_{\boxed{1},j} + g_{i,\boxed{2}}h_{\boxed{2},j} + \cdots + g_{i,\boxed{r}}h_{\boxed{r},j}$$

The highlighted subscripts on the g's are column indices while those on the h's are for rows. That is, the summation takes place over the columns of G but over the rows of H—the definition treats left differently than right. So we may reasonably suspect that GH can be unequal to HG.

2.9 Example Matrix multiplication is not commutative.

$$\begin{pmatrix}1&2\\3&4\end{pmatrix}\begin{pmatrix}5&6\\7&8\end{pmatrix}=\begin{pmatrix}19&22\\43&50\end{pmatrix}\qquad\begin{pmatrix}5&6\\7&8\end{pmatrix}\begin{pmatrix}1&2\\3&4\end{pmatrix}=\begin{pmatrix}23&34\\31&46\end{pmatrix}$$

2.10 Example Commutativity can fail more dramatically:

$$\begin{pmatrix}5&6\\7&8\end{pmatrix}\begin{pmatrix}1&2&0\\3&4&0\end{pmatrix}=\begin{pmatrix}23&34&0\\31&46&0\end{pmatrix}$$

while

$$\begin{pmatrix}1&2&0\\3&4&0\end{pmatrix}\begin{pmatrix}5&6\\7&8\end{pmatrix}$$

isn't even defined.

2.11 Remark The fact that matrix multiplication is not commutative can seem odd at first, perhaps because most mathematical operations in prior courses are commutative. But matrix multiplication represents function composition and function composition is not commutative: if $f(x) = 2x$ and $g(x) = x + 1$ then $g \circ f(x) = 2x + 1$ while $f \circ g(x) = 2(x + 1) = 2x + 2$.

Except for the lack of commutativity, matrix multiplication is algebraically well-behaved. The next result gives some nice properties and more are in Exercise 25 and Exercise 26.

2.12 Theorem If F, G, and H are matrices, and the matrix products are defined, then the product is associative $(FG)H = F(GH)$ and distributes over matrix addition $F(G + H) = FG + FH$ and $(G + H)F = GF + HF$.

Proof Associativity holds because matrix multiplication represents function composition, which is associative: the maps $(f \circ g) \circ h$ and $f \circ (g \circ h)$ are equal as both send $\vec{v}$ to $f(g(h(\vec{v})))$.

Distributivity is similar. For instance, the first one goes $f \circ (g + h)\,(\vec{v}) = f\big((g + h)(\vec{v})\big) = f\big(g(\vec{v}) + h(\vec{v})\big) = f(g(\vec{v})) + f(h(\vec{v})) = f \circ g(\vec{v}) + f \circ h(\vec{v})$ (the third equality uses the linearity of f). Right-distributivity goes the same way. QED

2.13 Remark We could instead prove that result by slogging through indices. For example, for associativity the i,j entry of $(FG)H$ is

$$\begin{aligned}&(f_{i,1}g_{1,1}+f_{i,2}g_{2,1}+\cdots+f_{i,r}g_{r,1})h_{1,j}\\&\quad+(f_{i,1}g_{1,2}+f_{i,2}g_{2,2}+\cdots+f_{i,r}g_{r,2})h_{2,j}\\&\quad\vdots\\&\quad+(f_{i,1}g_{1,s}+f_{i,2}g_{2,s}+\cdots+f_{i,r}g_{r,s})h_{s,j}\end{aligned}$$

where F, G, and H are $m\times r$, $r\times s$, and $s\times n$ matrices. Distribute

$$\begin{aligned}&f_{i,1}g_{1,1}h_{1,j}+f_{i,2}g_{2,1}h_{1,j}+\cdots+f_{i,r}g_{r,1}h_{1,j}\\&\quad+f_{i,1}g_{1,2}h_{2,j}+f_{i,2}g_{2,2}h_{2,j}+\cdots+f_{i,r}g_{r,2}h_{2,j}\\&\quad\vdots\\&\quad+f_{i,1}g_{1,s}h_{s,j}+f_{i,2}g_{2,s}h_{s,j}+\cdots+f_{i,r}g_{r,s}h_{s,j}\end{aligned}$$

and regroup around the f's

$$\begin{aligned}&f_{i,1}(g_{1,1}h_{1,j}+g_{1,2}h_{2,j}+\cdots+g_{1,s}h_{s,j})\\&\quad+f_{i,2}(g_{2,1}h_{1,j}+g_{2,2}h_{2,j}+\cdots+g_{2,s}h_{s,j})\\&\quad\vdots\\&\quad+f_{i,r}(g_{r,1}h_{1,j}+g_{r,2}h_{2,j}+\cdots+g_{r,s}h_{s,j})\end{aligned}$$

to get the i,j entry of $F(GH)$.

Contrast the two proofs. The index-heavy argument is hard to understand in that while the calculations are easy to check, the arithmetic seems unconnected to any idea. The argument in the proof is shorter and also says why this property "really" holds. This illustrates the comments made at the start of the chapter on vector spaces — at least sometimes an argument from higher-level constructs is clearer.

We have now seen how to represent the composition of linear maps. The next subsection will continue to explore this operation.

Exercises

✓ **2.14** Compute, or state "not defined".

(a) $\begin{pmatrix}3&1\\-4&2\end{pmatrix}\begin{pmatrix}0&5\\0&0.5\end{pmatrix}$ (b) $\begin{pmatrix}1&1&-1\\4&0&3\end{pmatrix}\begin{pmatrix}2&-1&-1\\3&1&1\\3&1&1\end{pmatrix}$

(c) $\begin{pmatrix}2&-7\\7&4\end{pmatrix}\begin{pmatrix}1&0&5\\-1&1&1\\3&8&4\end{pmatrix}$ (d) $\begin{pmatrix}5&2\\3&1\end{pmatrix}\begin{pmatrix}-1&2\\3&-5\end{pmatrix}$

✓ **2.15** Where

$$A=\begin{pmatrix}1&-1\\2&0\end{pmatrix}\quad B=\begin{pmatrix}5&2\\4&4\end{pmatrix}\quad C=\begin{pmatrix}-2&3\\-4&1\end{pmatrix}$$

compute or state 'not defined'.

(a) AB (b) $(AB)C$ (c) BC (d) $A(BC)$

2.16 Which products are defined?

(a) 3×2 times 2×3 (b) 2×3 times 3×2 (c) 2×2 times 3×3
(d) 3×3 times 2×2

✓ **2.17** Give the size of the product or state "not defined".

(a) a 2×3 matrix times a 3×1 matrix
(b) a 1×12 matrix times a 12×1 matrix
(c) a 2×3 matrix times a 2×1 matrix
(d) a 2×2 matrix times a 2×2 matrix

✓ **2.18** Find the system of equations resulting from starting with

$$\begin{aligned}h_{1,1}x_1+h_{1,2}x_2+h_{1,3}x_3&=d_1\\h_{2,1}x_1+h_{2,2}x_2+h_{2,3}x_3&=d_2\end{aligned}$$

and making this change of variable (i.e., substitution).

$$\begin{aligned}x_1&=g_{1,1}y_1+g_{1,2}y_2\\x_2&=g_{2,1}y_1+g_{2,2}y_2\\x_3&=g_{3,1}y_1+g_{3,2}y_2\end{aligned}$$

✓ **2.19** Consider the two linear functions $h\colon\mathbb{R}^3\to\mathcal{P}_2$ and $g\colon\mathcal{P}_2\to\mathcal{M}_{2\times2}$ given as here.

$$\begin{pmatrix}a\\b\\c\end{pmatrix}\mapsto(a+b)x^2+(2a+2b)x+c\qquad px^2+qx+r\mapsto\begin{pmatrix}p&p-2q\\q&0\end{pmatrix}$$

Use these bases for the spaces.

$$B=\langle\begin{pmatrix}1\\1\\1\end{pmatrix},\begin{pmatrix}0\\1\\1\end{pmatrix},\begin{pmatrix}0\\0\\1\end{pmatrix}\rangle\qquad C=\langle 1+x,1-x,x^2\rangle$$

$$D=\langle\begin{pmatrix}1&0\\0&0\end{pmatrix},\begin{pmatrix}0&2\\0&0\end{pmatrix},\begin{pmatrix}0&0\\3&0\end{pmatrix},\begin{pmatrix}0&0\\0&4\end{pmatrix}\rangle$$

(a) Give the formula for the composition map $g\circ h\colon\mathbb{R}^3\to\mathcal{M}_{2\times2}$ derived directly from the above definition.
(b) Represent h and g with respect to the appropriate bases.
(c) Represent the map $g\circ h$ computed in the first part with respect to the appropriate bases.
(d) Check that the product of the two matrices from the second part is the matrix from the third part.

2.20 As Definition 2.3 points out, the matrix product operation generalizes the dot product. Is the dot product of a $1\times n$ row vector and a $n\times 1$ column vector the same as their matrix-multiplicative product?

✓ **2.21** Represent the derivative map on $\mathcal{P}_n$ with respect to B,B where B is the natural basis $\langle 1,x,\ldots,x^n\rangle$. Show that the product of this matrix with itself is defined; what map does it represent?

2.22 [Cleary] Match each type of matrix with all these descriptions that could fit: (i) can be multiplied by its transpose to make a 1×1 matrix, (ii) is similar to the 3×3 matrix of all zeros, (iii) can represent a linear map from $\mathbb{R}^3$ to $\mathbb{R}^2$ that is not onto, (iv) can represent an isomorphism from $\mathbb{R}^3$ to $\mathcal{P}^2$.

(a) a 2×3 matrix whose rank is 1
(b) a 3×3 matrix that is nonsingular
(c) a 2×2 matrix that is singular
(d) an $n\times1$ column vector

2.23 Show that composition of linear transformations on $\mathbb{R}^1$ is commutative. Is this true for any one-dimensional space?

2.24 Why is matrix multiplication not defined as entry-wise multiplication? That would be easier, and commutative too.

2.25 (a) Prove that $H^pH^q = H^{p+q}$ and $(H^p)^q = H^{pq}$ for positive integers p, q.
(b) Prove that $(rH)^p = r^p \cdot H^p$ for any positive integer p and scalar $r \in \mathbb{R}$.

✓ **2.26** (a) How does matrix multiplication interact with scalar multiplication: is $r(GH) = (rG)H$? Is $G(rH) = r(GH)$?
(b) How does matrix multiplication interact with linear combinations: is $F(rG + sH) = r(FG) + s(FH)$? Is $(rF + sG)H = rFH + sGH$?

2.27 We can ask how the matrix product operation interacts with the transpose operation.
(a) Show that $(GH)^{\mathsf{T}} = H^{\mathsf{T}}G^{\mathsf{T}}$.
(b) A square matrix is *symmetric* if each i, j entry equals the j, i entry, that is, if the matrix equals its own transpose. Show that the matrices HH^{T} and $H^{\mathsf{T}}H$ are symmetric.

✓ **2.28** Rotation of vectors in $\mathbb{R}^3$ about an axis is a linear map. Show that linear maps do not commute by showing geometrically that rotations do not commute.

2.29 In the proof of Theorem 2.12 we used some maps. What are the domains and codomains?

2.30 How does matrix rank interact with matrix multiplication?
(a) Can the product of rank n matrices have rank less than n? Greater?
(b) Show that the rank of the product of two matrices is less than or equal to the minimum of the rank of each factor.

2.31 Is 'commutes with' an equivalence relation among $n\times n$ matrices?

✓ **2.32** *(We will use this exercise in the Matrix Inverses exercises.)* Here is another property of matrix multiplication that might be puzzling at first sight.
(a) Prove that the composition of the projections $\pi_x, \pi_y \colon \mathbb{R}^3 \to \mathbb{R}^3$ onto the x and y axes is the zero map despite that neither one is itself the zero map.
(b) Prove that the composition of the derivatives d^2/dx^2, $d^3/dx^3 \colon \mathcal{P}_4 \to \mathcal{P}_4$ is the zero map despite that neither is the zero map.
(c) Give a matrix equation representing the first fact.
(d) Give a matrix equation representing the second.

When two things multiply to give zero despite that neither is zero we say that each is a *zero divisor*.

2.33 Show that, for square matrices, $(S + T)(S - T)$ need not equal $S^2 - T^2$.

✓ **2.34** Represent the identity transformation $\mathrm{id}\colon V \to V$ with respect to B, B for any basis B. This is the *identity matrix* I. Show that this matrix plays the role in matrix multiplication that the number 1 plays in real number multiplication: $HI = IH = H$ (for all matrices H for which the product is defined).

2.35 In real number algebra, quadratic equations have at most two solutions. That is not so with matrix algebra. Show that the 2×2 matrix equation $T^2 = I$ has more than two solutions, where I is the identity matrix (this matrix has ones in its $1,1$ and $2,2$ entries and zeroes elsewhere; see Exercise 34).

2.36 (a) Prove that for any 2×2 matrix T there are scalars $c_0, \ldots, c_4$ that are not all 0 such that the combination $c_4T^4 + c_3T^3 + c_2T^2 + c_1T + c_0I$ is the zero matrix (where I is the 2×2 identity matrix, with 1's in its $1,1$ and $2,2$ entries and zeroes elsewhere; see Exercise 34).

(b) Let $p(x)$ be a polynomial $p(x) = c_nx^n + \cdots + c_1x + c_0$. If T is a square matrix we define $p(T)$ to be the matrix $c_nT^n + \cdots + c_1T + c_0I$ (where I is the appropriately-sized identity matrix). Prove that for any square matrix there is a polynomial such that $p(T)$ is the zero matrix.

(c) The *minimal polynomial* $m(x)$ of a square matrix is the polynomial of least degree, and with leading coefficient 1, such that $m(T)$ is the zero matrix. Find the minimal polynomial of this matrix.

$$\begin{pmatrix} \sqrt{3}/2 & -1/2 \\ 1/2 & \sqrt{3}/2 \end{pmatrix}$$

(This is the representation with respect to $\mathcal{E}_2, \mathcal{E}_2$, the standard basis, of a rotation through $\pi/6$ radians counterclockwise.)

2.37 The infinite-dimensional space $\mathcal{P}$ of all finite-degree polynomials gives a memorable example of the non-commutativity of linear maps. Let $d/dx\colon \mathcal{P} \to \mathcal{P}$ be the usual derivative and let $s\colon \mathcal{P} \to \mathcal{P}$ be the *shift* map.

$$a_0 + a_1x + \cdots + a_nx^n \stackrel{s}{\longmapsto} 0 + a_0x + a_1x^2 + \cdots + a_nx^{n+1}$$

Show that the two maps don't commute $d/dx \circ s \neq s \circ d/dx$; in fact, not only is $(d/dx \circ s) - (s \circ d/dx)$ not the zero map, it is the identity map.

2.38 Recall the notation for the sum of the sequence of numbers $a_1, a_2, \ldots, a_n$.

$$\sum_{i=1}^{n} a_i = a_1 + a_2 + \cdots + a_n$$

In this notation, the i, j entry of the product of G and H is this.

$$p_{i,j} = \sum_{k=1}^{r} g_{i,k}h_{k,j}$$

Using this notation,

(a) reprove that matrix multiplication is associative;

(b) reprove Theorem 2.7.

IV.3 Mechanics of Matrix Multiplication

We can consider matrix multiplication as a mechanical process, putting aside for the moment any implications about the underlying maps.

The striking thing about this operation is the way that rows and columns combine. The i,j entry of the matrix product is the dot product of row i of the left matrix with column j of the right one. For instance, here a second row and a third column combine to make a $2,3$ entry.

$$\begin{pmatrix}1&1\\ \boxed{0\quad 1}\\ 1&0\end{pmatrix}\begin{pmatrix}4&6&\boxed{8}&2\\5&7&\boxed{9}&3\end{pmatrix}=\begin{pmatrix}9&13&17&5\\5&7&\boxed{9}&3\\4&6&8&2\end{pmatrix}$$

We can view this as the left matrix acting by multiplying its rows into the columns of the right matrix. Or, it is the right matrix using its columns to act on the rows of the left matrix. Below, we will examine actions from the left and from the right for some simple matrices.

Simplest is the zero matrix.

3.1 Example Multiplying by a zero matrix from the left or from the right results in a zero matrix.

$$\begin{pmatrix}0&0\\0&0\end{pmatrix}\begin{pmatrix}1&3&2\\-1&1&-1\end{pmatrix}=\begin{pmatrix}0&0&0\\0&0&0\end{pmatrix}\qquad\begin{pmatrix}2&3\\1&4\end{pmatrix}\begin{pmatrix}0&0\\0&0\end{pmatrix}=\begin{pmatrix}0&0\\0&0\end{pmatrix}$$

The next easiest matrices are the ones with a single nonzero entry.

3.2 Definition A matrix with all 0's except for a 1 in the i,j entry is an i,j *unit* matrix (or *matrix unit*).

3.3 Example This is the $1,2$ unit matrix with three rows and two columns, multiplying from the left.

$$\begin{pmatrix}0&1\\0&0\\0&0\end{pmatrix}\begin{pmatrix}5&6\\7&8\end{pmatrix}=\begin{pmatrix}7&8\\0&0\\0&0\end{pmatrix}$$

Acting from the left, an i,j unit matrix copies row j of the multiplicand into row i of the result. From the right an i,j unit matrix picks out column i of the multiplicand and copies it into column j of the result.

$$\begin{pmatrix}1&2&3\\4&5&6\\7&8&9\end{pmatrix}\begin{pmatrix}0&1\\0&0\\0&0\end{pmatrix}=\begin{pmatrix}0&1\\0&4\\0&7\end{pmatrix}$$

3.4 Example Rescaling unit matrices simply rescales the result. This is the action from the left of the matrix that is twice the one in the prior example.

$$\begin{pmatrix} 0 & 2 \\ 0 & 0 \\ 0 & 0 \end{pmatrix} \begin{pmatrix} 5 & 6 \\ 7 & 8 \end{pmatrix} = \begin{pmatrix} 14 & 16 \\ 0 & 0 \\ 0 & 0 \end{pmatrix}$$

Next in complication are matrices with two nonzero entries.

3.5 Example There are two cases. If a left-multiplier has entries in different rows then their actions don't interact.

$$\begin{aligned} \begin{pmatrix} 1 & 0 & 0 \\ 0 & 0 & 2 \\ 0 & 0 & 0 \end{pmatrix} \begin{pmatrix} 1 & 2 & 3 \\ 4 & 5 & 6 \\ 7 & 8 & 9 \end{pmatrix} &= \left(\begin{pmatrix} 1 & 0 & 0 \\ 0 & 0 & 0 \\ 0 & 0 & 0 \end{pmatrix} + \begin{pmatrix} 0 & 0 & 0 \\ 0 & 0 & 2 \\ 0 & 0 & 0 \end{pmatrix} \right) \begin{pmatrix} 1 & 2 & 3 \\ 4 & 5 & 6 \\ 7 & 8 & 9 \end{pmatrix} \\ &= \begin{pmatrix} 1 & 2 & 3 \\ 0 & 0 & 0 \\ 0 & 0 & 0 \end{pmatrix} + \begin{pmatrix} 0 & 0 & 0 \\ 14 & 16 & 18 \\ 0 & 0 & 0 \end{pmatrix} \\ &= \begin{pmatrix} 1 & 2 & 3 \\ 14 & 16 & 18 \\ 0 & 0 & 0 \end{pmatrix} \end{aligned}$$

But if the left-multiplier's nonzero entries are in the same row then that row of the result is a combination.

$$\begin{aligned} \begin{pmatrix} 1 & 0 & 2 \\ 0 & 0 & 0 \\ 0 & 0 & 0 \end{pmatrix} \begin{pmatrix} 1 & 2 & 3 \\ 4 & 5 & 6 \\ 7 & 8 & 9 \end{pmatrix} &= \left(\begin{pmatrix} 1 & 0 & 0 \\ 0 & 0 & 0 \\ 0 & 0 & 0 \end{pmatrix} + \begin{pmatrix} 0 & 0 & 2 \\ 0 & 0 & 0 \\ 0 & 0 & 0 \end{pmatrix} \right) \begin{pmatrix} 1 & 2 & 3 \\ 4 & 5 & 6 \\ 7 & 8 & 9 \end{pmatrix} \\ &= \begin{pmatrix} 1 & 2 & 3 \\ 0 & 0 & 0 \\ 0 & 0 & 0 \end{pmatrix} + \begin{pmatrix} 14 & 16 & 18 \\ 0 & 0 & 0 \\ 0 & 0 & 0 \end{pmatrix} \\ &= \begin{pmatrix} 15 & 18 & 21 \\ 0 & 0 & 0 \\ 0 & 0 & 0 \end{pmatrix} \end{aligned}$$

Right-multiplication acts in the same way, but with columns.

3.6 Example Consider the columns of the product of two 2×2 matrices.

$$\begin{pmatrix} g_{1,1} & g_{1,2} \\ g_{2,1} & g_{2,2} \end{pmatrix} \begin{pmatrix} h_{1,1} & h_{1,2} \\ h_{2,1} & h_{2,2} \end{pmatrix} = \begin{pmatrix} g_{1,1}h_{1,1} + g_{1,2}h_{2,1} & g_{1,1}h_{1,2} + g_{1,2}h_{2,2} \\ g_{2,1}h_{1,1} + g_{2,2}h_{2,1} & g_{2,1}h_{1,2} + g_{2,2}h_{2,2} \end{pmatrix}$$

Each column is the result of multiplying G by the corresponding column of H.

$$G \begin{pmatrix} h_{1,1} \\ h_{2,1} \end{pmatrix} = \begin{pmatrix} g_{1,1}h_{1,1} + g_{1,2}h_{2,1} \\ g_{2,1}h_{1,1} + g_{2,2}h_{2,1} \end{pmatrix} \qquad G \begin{pmatrix} h_{1,2} \\ h_{2,2} \end{pmatrix} = \begin{pmatrix} g_{1,1}h_{1,2} + g_{1,2}h_{2,2} \\ g_{2,1}h_{1,2} + g_{2,2}h_{2,2} \end{pmatrix}$$

3.7 Lemma In a product of two matrices G and H, the columns of GH are formed by taking G times the columns of H

$$G \cdot \begin{pmatrix} \vdots & & \vdots \\ \vec{h}_1 & \cdots & \vec{h}_n \\ \vdots & & \vdots \end{pmatrix} = \begin{pmatrix} \vdots & & \vdots \\ G \cdot \vec{h}_1 & \cdots & G \cdot \vec{h}_n \\ \vdots & & \vdots \end{pmatrix}$$

and the rows of GH are formed by taking the rows of G times H

$$\begin{pmatrix} \cdots & \vec{g}_1 & \cdots \\ & \vdots & \\ \cdots & \vec{g}_r & \cdots \end{pmatrix} \cdot H = \begin{pmatrix} \cdots & \vec{g}_1 \cdot H & \cdots \\ & \vdots & \\ \cdots & \vec{g}_r \cdot H & \cdots \end{pmatrix}$$

(ignoring the extra parentheses).

Proof We will check that in a product of $2{\times}2$ matrices, the rows of the product equal the product of the rows of G with the entire matrix H.

$$\begin{aligned} \begin{pmatrix} g_{1,1} & g_{1,2} \\ g_{2,1} & g_{2,2} \end{pmatrix} \begin{pmatrix} h_{1,1} & h_{1,2} \\ h_{2,1} & h_{2,2} \end{pmatrix} &= \begin{pmatrix} (g_{1,1} \;\; g_{1,2})H \\ (g_{2,1} \;\; g_{2,2})H \end{pmatrix} \\ &= \begin{pmatrix} (g_{1,1}h_{1,1} + g_{1,2}h_{2,1} \;\; g_{1,1}h_{1,2} + g_{1,2}h_{2,2}) \\ (g_{2,1}h_{1,1} + g_{2,2}h_{2,1} \;\; g_{2,1}h_{1,2} + g_{2,2}h_{2,2}) \end{pmatrix} \end{aligned}$$

We leave the more general check as an exercise. QED

An application of those observations is that there is a matrix that just copies out the rows and columns.

3.8 Definition The *main diagonal* (or *principle diagonal* or *diagonal*) of a square matrix goes from the upper left to the lower right.

3.9 Definition An *identity matrix* is square and every entry is 0 except for 1's in the main diagonal.

$$I_{n{\times}n} = \begin{pmatrix} 1 & 0 & \ldots & 0 \\ 0 & 1 & \ldots & 0 \\ & & \vdots & \\ 0 & 0 & \ldots & 1 \end{pmatrix}$$

3.10 Example Here is the $2{\times}2$ identity matrix leaving its multiplicand unchanged when it acts from the right.

$$\begin{pmatrix} 1 & -2 \\ 0 & -2 \\ 1 & -1 \\ 4 & 3 \end{pmatrix} \begin{pmatrix} 1 & 0 \\ 0 & 1 \end{pmatrix} = \begin{pmatrix} 1 & -2 \\ 0 & -2 \\ 1 & -1 \\ 4 & 3 \end{pmatrix}$$

3.11 Example Here the 3×3 identity leaves its multiplicand unchanged both from the left

$$\begin{pmatrix} 1 & 0 & 0 \\ 0 & 1 & 0 \\ 0 & 0 & 1 \end{pmatrix} \begin{pmatrix} 2 & 3 & 6 \\ 1 & 3 & 8 \\ -7 & 1 & 0 \end{pmatrix} = \begin{pmatrix} 2 & 3 & 6 \\ 1 & 3 & 8 \\ -7 & 1 & 0 \end{pmatrix}$$

and from the right.

$$\begin{pmatrix} 2 & 3 & 6 \\ 1 & 3 & 8 \\ -7 & 1 & 0 \end{pmatrix} \begin{pmatrix} 1 & 0 & 0 \\ 0 & 1 & 0 \\ 0 & 0 & 1 \end{pmatrix} = \begin{pmatrix} 2 & 3 & 6 \\ 1 & 3 & 8 \\ -7 & 1 & 0 \end{pmatrix}$$

In short, an identity matrix is the identity element of the set of $n\times n$ matrices with respect to the operation of matrix multiplication.

We can generalize the identity matrix by relaxing the ones to arbitrary reals. The resulting matrix rescales whole rows or columns.

3.12 Definition A *diagonal matrix* is square and has 0's off the main diagonal.

$$\begin{pmatrix} a_{1,1} & 0 & \ldots & 0 \\ 0 & a_{2,2} & \ldots & 0 \\ & \vdots & & \\ 0 & 0 & \ldots & a_{n,n} \end{pmatrix}$$

3.13 Example From the left, the action of multiplication by a diagonal matrix is to rescales the rows.

$$\begin{pmatrix} 2 & 0 \\ 0 & -1 \end{pmatrix} \begin{pmatrix} 2 & 1 & 4 & -1 \\ -1 & 3 & 4 & 4 \end{pmatrix} = \begin{pmatrix} 4 & 2 & 8 & -2 \\ 1 & -3 & -4 & -4 \end{pmatrix}$$

From the right such a matrix rescales the columns.

$$\begin{pmatrix} 1 & 2 & 1 \\ 2 & 2 & 2 \end{pmatrix} \begin{pmatrix} 3 & 0 & 0 \\ 0 & 2 & 0 \\ 0 & 0 & -2 \end{pmatrix} = \begin{pmatrix} 3 & 4 & -2 \\ 6 & 4 & -4 \end{pmatrix}$$

We can also generalize identity matrices by putting a single one in each row and column in ways other than putting them down the diagonal.

3.14 Definition A *permutation matrix* is square and is all 0's except for a single 1 in each row and column.

3.15 Example From the left these matrices permute rows.

$$\begin{pmatrix} 0 & 0 & 1 \\ 1 & 0 & 0 \\ 0 & 1 & 0 \end{pmatrix} \begin{pmatrix} 1 & 2 & 3 \\ 4 & 5 & 6 \\ 7 & 8 & 9 \end{pmatrix} = \begin{pmatrix} 7 & 8 & 9 \\ 1 & 2 & 3 \\ 4 & 5 & 6 \end{pmatrix}$$

From the right they permute columns.

$$\begin{pmatrix} 1 & 2 & 3 \\ 4 & 5 & 6 \\ 7 & 8 & 9 \end{pmatrix} \begin{pmatrix} 0 & 0 & 1 \\ 1 & 0 & 0 \\ 0 & 1 & 0 \end{pmatrix} = \begin{pmatrix} 2 & 3 & 1 \\ 5 & 6 & 4 \\ 8 & 9 & 7 \end{pmatrix}$$

We finish this subsection by applying these observations to get matrices that perform Gauss's Method and Gauss-Jordan reduction. We have already seen how to produce a matrix that rescales rows, and a row swapper.

3.16 Example Multiplying by this matrix rescales the second row by three.

$$\begin{pmatrix} 1 & 0 & 0 \\ 0 & 3 & 0 \\ 0 & 0 & 1 \end{pmatrix} \begin{pmatrix} 0 & 2 & 1 & 1 \\ 0 & 1/3 & 1 & -1 \\ 1 & 0 & 2 & 0 \end{pmatrix} = \begin{pmatrix} 0 & 2 & 1 & 1 \\ 0 & 1 & 3 & -3 \\ 1 & 0 & 2 & 0 \end{pmatrix}$$

3.17 Example This multiplication swaps the first and third rows.

$$\begin{pmatrix} 0 & 0 & 1 \\ 0 & 1 & 0 \\ 1 & 0 & 0 \end{pmatrix} \begin{pmatrix} 0 & 2 & 1 & 1 \\ 0 & 1 & 3 & -3 \\ 1 & 0 & 2 & 0 \end{pmatrix} = \begin{pmatrix} 1 & 0 & 2 & 0 \\ 0 & 1 & 3 & -3 \\ 0 & 2 & 1 & 1 \end{pmatrix}$$

To see how to perform a row combination, we observe something about those two examples. The matrix that rescales the second row by a factor of three arises in this way from the identity.

$$\begin{pmatrix} 1 & 0 & 0 \\ 0 & 1 & 0 \\ 0 & 0 & 1 \end{pmatrix} \xrightarrow{3\rho_2} \begin{pmatrix} 1 & 0 & 0 \\ 0 & 3 & 0 \\ 0 & 0 & 1 \end{pmatrix}$$

Similarly, the matrix that swaps first and third rows arises in this way.

$$\begin{pmatrix} 1 & 0 & 0 \\ 0 & 1 & 0 \\ 0 & 0 & 1 \end{pmatrix} \xrightarrow{\rho_1 \leftrightarrow \rho_3} \begin{pmatrix} 0 & 0 & 1 \\ 0 & 1 & 0 \\ 1 & 0 & 0 \end{pmatrix}$$

3.18 Example The 3×3 matrix that arises as

$$\begin{pmatrix}1&0&0\\0&1&0\\0&0&1\end{pmatrix}\xrightarrow{-2\rho_2+\rho_3}\begin{pmatrix}1&0&0\\0&1&0\\0&-2&1\end{pmatrix}$$

will, when it acts from the left, perform the combination operation $-2\rho_2+\rho_3$.

$$\begin{pmatrix}1&0&0\\0&1&0\\0&-2&1\end{pmatrix}\begin{pmatrix}1&0&2&0\\0&1&3&-3\\0&2&1&1\end{pmatrix}=\begin{pmatrix}1&0&2&0\\0&1&3&-3\\0&0&-5&7\end{pmatrix}$$

3.19 Definition The *elementary reduction matrices* (or just *elementary matrices*) result from applying a one Gaussian operation to an identity matrix.

(1) $I \xrightarrow{k\rho_i} M_i(k)$ for $k\neq 0$

(2) $I \xrightarrow{\rho_i\leftrightarrow\rho_j} P_{i,j}$ for $i\neq j$

(3) $I \xrightarrow{k\rho_i+\rho_j} C_{i,j}(k)$ for $i\neq j$

3.20 Lemma Matrix multiplication can do Gaussian reduction.

(1) If $H \xrightarrow{k\rho_i} G$ then $M_i(k)H=G$.

(2) If $H \xrightarrow{\rho_i\leftrightarrow\rho_j} G$ then $P_{i,j}H=G$.

(3) If $H \xrightarrow{k\rho_i+\rho_j} G$ then $C_{i,j}(k)H=G$.

Proof Clear. QED

3.21 Example This is the first system, from the first chapter, on which we performed Gauss's Method.

$$\begin{aligned}3x_3&=9\\x_1+5x_2-2x_3&=2\\(1/3)x_1+2x_2\qquad&=3\end{aligned}$$

We can reduce it with matrix multiplication. Swap the first and third rows,

$$\begin{pmatrix}0&0&1\\0&1&0\\1&0&0\end{pmatrix}\left(\begin{array}{ccc|c}0&0&3&9\\1&5&-2&2\\1/3&2&0&3\end{array}\right)=\left(\begin{array}{ccc|c}1/3&2&0&3\\1&5&-2&2\\0&0&3&9\end{array}\right)$$

triple the first row,

$$\begin{pmatrix} 3 & 0 & 0 \\ 0 & 1 & 0 \\ 0 & 0 & 1 \end{pmatrix} \left(\begin{array}{ccc|c} 1/3 & 2 & 0 & 3 \\ 1 & 5 & -2 & 2 \\ 0 & 0 & 3 & 9 \end{array}\right) = \left(\begin{array}{ccc|c} 1 & 6 & 0 & 9 \\ 1 & 5 & -2 & 2 \\ 0 & 0 & 3 & 9 \end{array}\right)$$

and then add -1 times the first row to the second.

$$\begin{pmatrix} 1 & 0 & 0 \\ -1 & 1 & 0 \\ 0 & 0 & 1 \end{pmatrix} \left(\begin{array}{ccc|c} 1 & 6 & 0 & 9 \\ 1 & 5 & -2 & 2 \\ 0 & 0 & 3 & 9 \end{array}\right) = \left(\begin{array}{ccc|c} 1 & 6 & 0 & 9 \\ 0 & -1 & -2 & -7 \\ 0 & 0 & 3 & 9 \end{array}\right)$$

Now back substitution will give the solution.

3.22 Example Gauss-Jordan reduction works the same way. For the matrix ending the prior example, first turn the leading entries to ones,

$$\begin{pmatrix} 1 & 0 & 0 \\ 0 & -1 & 0 \\ 0 & 0 & 1/3 \end{pmatrix} \left(\begin{array}{ccc|c} 1 & 6 & 0 & 9 \\ 0 & -1 & -2 & -7 \\ 0 & 0 & 3 & 9 \end{array}\right) = \left(\begin{array}{ccc|c} 1 & 6 & 0 & 9 \\ 0 & 1 & 2 & 7 \\ 0 & 0 & 1 & 3 \end{array}\right)$$

then clear the third column, and then the second column.

$$\begin{pmatrix} 1 & -6 & 0 \\ 0 & 1 & 0 \\ 0 & 0 & 1 \end{pmatrix} \begin{pmatrix} 1 & 0 & 0 \\ 0 & 1 & -2 \\ 0 & 0 & 1 \end{pmatrix} \left(\begin{array}{ccc|c} 1 & 6 & 0 & 9 \\ 0 & 1 & 2 & 7 \\ 0 & 0 & 1 & 3 \end{array}\right) = \left(\begin{array}{ccc|c} 1 & 0 & 0 & 3 \\ 0 & 1 & 0 & 1 \\ 0 & 0 & 1 & 3 \end{array}\right)$$

3.23 Corollary For any matrix H there are elementary reduction matrices $R_1, \ldots, R_r$ such that $R_r \cdot R_{r-1} \cdots R_1 \cdot H$ is in reduced echelon form.

Until now we have taken the point of view that our primary objects of study are vector spaces and the maps between them, and we seemed to have adopted matrices only for computational convenience. This subsection show that this isn't the entire story.

Understanding matrices operations by understanding the mechanics of how the entries combine is also useful. In the rest of this book we shall continue to focus on maps as the primary objects but we will be pragmatic — if the matrix point of view gives some clearer idea then we will go with it.

Exercises

✓ **3.24** Predict the result of each multiplication by an elementary reduction matrix, and then check by multiplying it out.

(a) $\begin{pmatrix}3&0\\0&1\end{pmatrix}\begin{pmatrix}1&2\\3&4\end{pmatrix}$ (b) $\begin{pmatrix}1&0\\0&2\end{pmatrix}\begin{pmatrix}1&2\\3&4\end{pmatrix}$ (c) $\begin{pmatrix}1&0\\-2&1\end{pmatrix}\begin{pmatrix}1&2\\3&4\end{pmatrix}$
(d) $\begin{pmatrix}1&2\\3&4\end{pmatrix}\begin{pmatrix}1&-1\\0&1\end{pmatrix}$ (e) $\begin{pmatrix}1&2\\3&4\end{pmatrix}\begin{pmatrix}0&1\\1&0\end{pmatrix}$

3.25 Predict the result of each multiplication by a diagonal matrix, and then check by multiplying it out.

(a) $\begin{pmatrix}3&0\\0&1\end{pmatrix}\begin{pmatrix}1&2\\3&4\end{pmatrix}$ (b) $\begin{pmatrix}4&0\\0&2\end{pmatrix}\begin{pmatrix}1&2\\3&4\end{pmatrix}$

3.26 Produce each.

(a) a 3×3 matrix that, acting from the left, swaps rows one and two
(b) a 2×2 matrix that, acting from the right, swaps column one and two

✓ **3.27** Show how to use matrix multiplication to bring this matrix to echelon form.

$$\begin{pmatrix}1&2&1&0\\2&3&1&-1\\7&11&4&-3\end{pmatrix}$$

3.28 Find the product of this matrix with its transpose.

$$\begin{pmatrix}\cos\theta&-\sin\theta\\\sin\theta&\cos\theta\end{pmatrix}$$

✓ **3.29** The need to take linear combinations of rows and columns in tables of numbers arises often in practice. For instance, this is a map of part of Vermont and New York.

In part because of Lake Champlain, there are no roads directly connecting some pairs of towns. For instance, there is no way to go from Winooski to Grand Isle without going through Colchester. (To simplify the graph many other roads and towns have been omitted. From top to bottom of this map is about forty miles.)

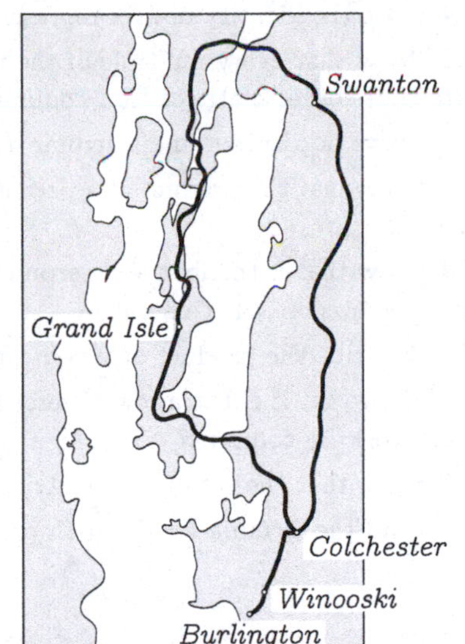

(a) The *adjacency matrix* of a map is the square matrix whose i,j entry is the number of roads from city i to city j (all (i,i) entries are 0). Produce the adjacency matrix of this map, with the cities in alphabetical order.
(b) A matrix is *symmetric* if it equals its transpose. Show that an adjacency matrix is symmetric. (These are all two-way streets. Vermont doesn't have many one-way streets.)
(c) What is the significance of the square of the incidence matrix? The cube?

✓ **3.30** This table gives the number of hours of each type done by each worker, and the associated pay rates. Use matrices to compute the wages due.

	regular	*overtime*
Alan	40	12
Betty	35	6
Catherine	40	18
Donald	28	0

	wage
regular	$25.00
overtime	$45.00

Remark. This illustrates that in practice we often want to compute linear combinations of rows and columns in a context where we really aren't interested in any associated linear maps.

3.31 Express this nonsingular matrix as a product of elementary reduction matrices.

$$T = \begin{pmatrix} 1 & 2 & 0 \\ 2 & -1 & 0 \\ 3 & 1 & 2 \end{pmatrix}$$

3.32 Express

$$\begin{pmatrix} 1 & 0 \\ -3 & 3 \end{pmatrix}$$

as the product of two elementary reduction matrices.

✓ **3.33** Prove that the diagonal matrices form a subspace of $\mathcal{M}_{n\times n}$. What is its dimension?

3.34 Does the identity matrix represent the identity map if the bases are unequal?

3.35 Show that every multiple of the identity commutes with every square matrix. Are there other matrices that commute with all square matrices?

3.36 Prove or disprove: nonsingular matrices commute.

✓ **3.37** Show that the product of a permutation matrix and its transpose is an identity matrix.

3.38 Show that if the first and second rows of G are equal then so are the first and second rows of GH. Generalize.

3.39 Describe the product of two diagonal matrices.

✓ **3.40** Show that if G has a row of zeros then GH (if defined) has a row of zeros. Does that work for columns?

3.41 Show that the set of unit matrices forms a basis for $\mathcal{M}_{n\times m}$.

3.42 Find the formula for the n-th power of this matrix.

$$\begin{pmatrix} 1 & 1 \\ 1 & 0 \end{pmatrix}$$

✓ **3.43** The *trace* of a square matrix is the sum of the entries on its diagonal (its significance appears in Chapter Five). Show that $\operatorname{Tr}(GH) = \operatorname{Tr}(HG)$.

✓ **3.44** A square matrix is *upper triangular* if its only nonzero entries lie above, or on, the diagonal. Show that the product of two upper triangular matrices is upper triangular. Does this hold for lower triangular also?

3.45 A square matrix is a *Markov matrix* if each entry is between zero and one and the sum along each row is one. Prove that a product of Markov matrices is Markov.

3.46 Give an example of two matrices of the same rank and size with squares of differing rank.

3.47 On a computer multiplications have traditionally been more costly than additions, so people have tried to in reduce the number of multiplications used to compute a matrix product.

(a) How many real number multiplications do we need in the formula we gave for the product of a $m \times r$ matrix and a $r \times n$ matrix?

(b) Matrix multiplication is associative, so all associations yield the same result. The cost in number of multiplications, however, varies. Find the association requiring the fewest real number multiplications to compute the matrix product of a 5×10 matrix, a 10×20 matrix, a 20×5 matrix, and a 5×1 matrix.

(c) *(Very hard.)* Find a way to multiply two 2×2 matrices using only seven multiplications instead of the eight suggested by the naive approach.

? **3.48** [Putnam, 1990, A-5] If A and B are square matrices of the same size such that $ABAB = 0$, does it follow that $BABA = 0$?

3.49 [Am. Math. Mon., Dec. 1966] Demonstrate these four assertions to get an alternate proof that column rank equals row rank.

(a) $\vec{y} \cdot \vec{y} = 0$ iff $\vec{y} = \vec{0}$.

(b) $A\vec{x} = \vec{0}$ iff $A^T A\vec{x} = \vec{0}$.

(c) $\dim(\mathscr{R}(A)) = \dim(\mathscr{R}(A^T A))$.

(d) $\text{col rank}(A) = \text{col rank}(A^T) = \text{row rank}(A)$.

3.50 [Ackerson] Prove (where A is an $n \times n$ matrix and so defines a transformation of any n-dimensional space V with respect to B, B where B is a basis) that $\dim(\mathscr{R}(A) \cap \mathscr{N}(A)) = \dim(\mathscr{R}(A)) - \dim(\mathscr{R}(A^2))$. Conclude

(a) $\mathscr{N}(A) \subset \mathscr{R}(A)$ iff $\dim(\mathscr{N}(A)) = \dim(\mathscr{R}(A)) - \dim(\mathscr{R}(A^2))$;

(b) $\mathscr{R}(A) \subseteq \mathscr{N}(A)$ iff $A^2 = 0$;

(c) $\mathscr{R}(A) = \mathscr{N}(A)$ iff $A^2 = 0$ and $\dim(\mathscr{N}(A)) = \dim(\mathscr{R}(A))$;

(d) $\dim(\mathscr{R}(A) \cap \mathscr{N}(A)) = 0$ iff $\dim(\mathscr{R}(A)) = \dim(\mathscr{R}(A^2))$;

(e) *(Requires the Direct Sum subsection, which is optional.)* $V = \mathscr{R}(A) \oplus \mathscr{N}(A)$ iff $\dim(\mathscr{R}(A)) = \dim(\mathscr{R}(A^2))$.

IV.4 Inverses

We finish this section by considering how to represent the inverse of a linear map. We first recall some things about inverses. Where $\pi\colon \mathbb{R}^3 \to \mathbb{R}^2$ is the projection map and $\iota\colon \mathbb{R}^2 \to \mathbb{R}^3$ is the embedding

$$\begin{pmatrix} x \\ y \\ z \end{pmatrix} \stackrel{\pi}{\longmapsto} \begin{pmatrix} x \\ y \end{pmatrix} \qquad \begin{pmatrix} x \\ y \end{pmatrix} \stackrel{\iota}{\longmapsto} \begin{pmatrix} x \\ y \\ 0 \end{pmatrix}$$

then the composition $\pi \circ \iota$ is the identity map $\pi \circ \iota = \text{id}$ on $\mathbb{R}^2$.

$$\begin{pmatrix} x \\ y \end{pmatrix} \overset{\iota}{\longmapsto} \begin{pmatrix} x \\ y \\ 0 \end{pmatrix} \overset{\pi}{\longmapsto} \begin{pmatrix} x \\ y \end{pmatrix}$$

We say that ι is a *right inverse* of π or, what is the same thing, that π is a *left inverse* of ι. However, composition in the other order $\iota \circ \pi$ doesn't give the identity map—here is a vector that is not sent to itself under $\iota \circ \pi$.

$$\begin{pmatrix} 0 \\ 0 \\ 1 \end{pmatrix} \overset{\pi}{\longmapsto} \begin{pmatrix} 0 \\ 0 \end{pmatrix} \overset{\iota}{\longmapsto} \begin{pmatrix} 0 \\ 0 \\ 0 \end{pmatrix}$$

In fact, π has no left inverse at all. For, if f were to be a left inverse of π then we would have

$$\begin{pmatrix} x \\ y \\ z \end{pmatrix} \overset{\pi}{\longmapsto} \begin{pmatrix} x \\ y \end{pmatrix} \overset{f}{\longmapsto} \begin{pmatrix} x \\ y \\ z \end{pmatrix}$$

for all of the infinitely many z's. But a function f cannot send a single argument $\binom{x}{y}$ to more than one value.

So a function can have a right inverse but no left inverse, or a left inverse but no right inverse. A function can also fail to have an inverse on either side; one example is the zero transformation on $\mathbb{R}^2$.

Some functions have a *two-sided inverse*, another function that is the inverse both from the left and from the right. For instance, the transformation given by $\vec{v} \mapsto 2 \cdot \vec{v}$ has the two-sided inverse $\vec{v} \mapsto (1/2) \cdot \vec{v}$. The appendix shows that a function has a two-sided inverse if and only if it is both one-to-one and onto. The appendix also shows that if a function f has a two-sided inverse then it is unique, so we call it 'the' inverse and write f^{-1}.

In addition, recall that we have shown in Theorem II.2.20 that if a linear map has a two-sided inverse then that inverse is also linear.

Thus, our goal in this subsection is, where a linear h has an inverse, to find the relationship between $\text{Rep}_{B,D}(h)$ and $\text{Rep}_{D,B}(h^{-1})$.

4.1 Definition A matrix G is a *left inverse matrix* of the matrix H if GH is the identity matrix. It is a *right inverse* if HG is the identity. A matrix H with a two-sided inverse is an *invertible matrix*. That two-sided inverse is denoted H^{-1}.

Because of the correspondence between linear maps and matrices, statements about map inverses translate into statements about matrix inverses.

4.2 Lemma If a matrix has both a left inverse and a right inverse then the two are equal.

4.3 Theorem A matrix is invertible if and only if it is nonsingular.

Proof *(For both results.)* Given a matrix H, fix spaces of appropriate dimension for the domain and codomain and fix bases for these spaces. With respect to these bases, H represents a map h. The statements are true about the map and therefore they are true about the matrix. QED

4.4 Lemma A product of invertible matrices is invertible: if G and H are invertible and GH is defined then GH is invertible and $(GH)^{-1} = H^{-1}G^{-1}$.

Proof Because the two matrices are invertible they are square, and because their product is defined they must both be $n \times n$. Fix spaces and bases — say, $\mathbb{R}^n$ with the standard bases — to get maps $g, h\colon \mathbb{R}^n \to \mathbb{R}^n$ that are associated with the matrices, $G = \text{Rep}_{\mathcal{E}_n,\mathcal{E}_n}(g)$ and $H = \text{Rep}_{\mathcal{E}_n,\mathcal{E}_n}(h)$.

Consider $h^{-1}g^{-1}$. By the prior paragraph this composition is defined. This map is a two-sided inverse of gh since $(h^{-1}g^{-1})(gh) = h^{-1}(\text{id})h = h^{-1}h = \text{id}$ and $(gh)(h^{-1}g^{-1}) = g(\text{id})g^{-1} = gg^{-1} = \text{id}$. The matrices representing the maps reflect this equality. QED

This is the arrow diagram giving the relationship between map inverses and matrix inverses. It is a special case of the diagram relating function composition to matrix multiplication.

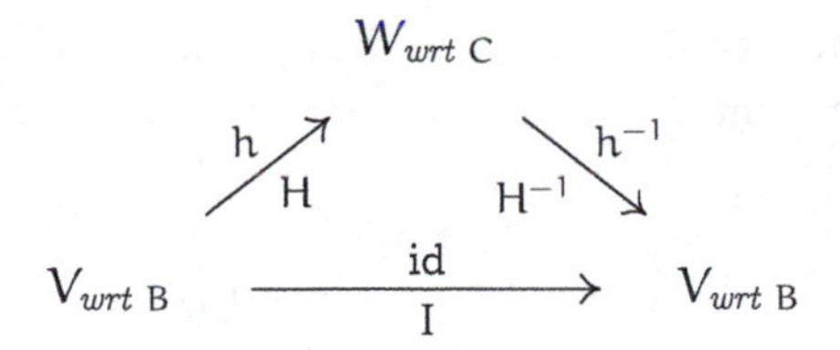

Beyond its place in our program of seeing how to represent map operations, another reason for our interest in inverses comes from linear systems. A linear system is equivalent to a matrix equation, as here.

$$\begin{aligned} x_1 + x_2 &= 3 \\ 2x_1 - x_2 &= 2 \end{aligned} \iff \begin{pmatrix} 1 & 1 \\ 2 & -1 \end{pmatrix} \begin{pmatrix} x_1 \\ x_2 \end{pmatrix} = \begin{pmatrix} 3 \\ 2 \end{pmatrix}$$

By fixing spaces and bases (for instance, $\mathbb{R}^2, \mathbb{R}^2$ with the standard bases), we take the matrix H to represent a map h. The matrix equation then becomes this linear map equation.

$$h(\vec{x}) = \vec{d}$$

If we had a left inverse map g then we could apply it to both sides $g \circ h(\vec{x}) = g(\vec{d})$ to get $\vec{x} = g(\vec{d})$. Restating in terms of the matrices, we want to multiply by the inverse matrix $\mathrm{Rep}_{C,B}(g) \cdot \mathrm{Rep}_C(\vec{d})$ to get $\mathrm{Rep}_B(\vec{x})$.

4.5 Example We can find a left inverse for the matrix just given

$$\begin{pmatrix} m & n \\ p & q \end{pmatrix} \begin{pmatrix} 1 & 1 \\ 2 & -1 \end{pmatrix} = \begin{pmatrix} 1 & 0 \\ 0 & 1 \end{pmatrix}$$

by using Gauss's Method to solve the resulting linear system.

$$\begin{array}{rl} m + 2n \qquad\qquad & = 1 \\ m - \ n \qquad\qquad & = 0 \\ p + 2q & = 0 \\ p - \ q & = 1 \end{array}$$

Answer: $m = 1/3$, $n = 1/3$, $p = 2/3$, and $q = -1/3$. (This matrix is actually the two-sided inverse of H; the check is easy.) With it, we can solve the system from the prior example.

$$\begin{pmatrix} x \\ y \end{pmatrix} = \begin{pmatrix} 1/3 & 1/3 \\ 2/3 & -1/3 \end{pmatrix} \begin{pmatrix} 3 \\ 2 \end{pmatrix} = \begin{pmatrix} 5/3 \\ 4/3 \end{pmatrix}$$

4.6 Remark Why do inverse matrices when we have Gauss's Method? Beyond the conceptual appeal of representing the map inverse operation, solving linear systems this way has two advantages.

First, once we have done the work of finding an inverse then solving a system with the same coefficients but different constants is fast: if we change the constants on the right of the system above then we get a related problem

$$\begin{pmatrix} 1 & 1 \\ 2 & -1 \end{pmatrix} \begin{pmatrix} x \\ y \end{pmatrix} = \begin{pmatrix} 5 \\ 1 \end{pmatrix}$$

that our inverse method solves quickly.

$$\begin{pmatrix} x \\ y \end{pmatrix} = \begin{pmatrix} 1/3 & 1/3 \\ 2/3 & -1/3 \end{pmatrix} \begin{pmatrix} 5 \\ 1 \end{pmatrix} = \begin{pmatrix} 2 \\ 3 \end{pmatrix}$$

Another advantage of inverses is that we can explore a system's sensitivity to changes in the constants. For example, tweaking the 3 on the right of the prior example's system to

$$\begin{pmatrix} 1 & 1 \\ 2 & -1 \end{pmatrix} \begin{pmatrix} x_1 \\ x_2 \end{pmatrix} = \begin{pmatrix} 3.01 \\ 2 \end{pmatrix}$$

and solving with the inverse

$$\begin{pmatrix} 1/3 & 1/3 \\ 2/3 & -1/3 \end{pmatrix} \begin{pmatrix} 3.01 \\ 2 \end{pmatrix} = \begin{pmatrix} (1/3)(3.01)+(1/3)(2) \\ (2/3)(3.01)-(1/3)(2) \end{pmatrix}$$

shows that the first component of the solution changes by 1/3 of the tweak, while the second component moves by 2/3 of the tweak. This is *sensitivity analysis*. We could use it to decide how accurately we must specify the data in a linear model to ensure that the solution has a desired accuracy.

4.7 Lemma A matrix H is invertible if and only if it can be written as the product of elementary reduction matrices. We can compute the inverse by applying to the identity matrix the same row steps, in the same order, that Gauss-Jordan reduce H.

Proof The matrix H is invertible if and only if it is nonsingular and thus Gauss-Jordan reduces to the identity. By Corollary 3.23 we can do this reduction with elementary matrices.

$$R_r \cdot R_{r-1} \ldots R_1 \cdot H = I \qquad (*)$$

For the first sentence of the result, note that elementary matrices are invertible because elementary row operations are reversible, and that their inverses are also elementary. Apply R_r^{-1} from the left to both sides of $(*)$. Then apply R_{r-1}^{-1}, etc. The result gives H as the product of elementary matrices $H = R_1^{-1} \cdots R_r^{-1} \cdot I$. (The I there covers the case $r = 0$.)

For the second sentence, group $(*)$ as $(R_r \cdot R_{r-1} \ldots R_1) \cdot H = I$ and recognize what's in the parentheses as the inverse $H^{-1} = R_r \cdot R_{r-1} \ldots R_1 \cdot I$. Restated: applying R_1 to the identity, followed by R_2, etc., yields the inverse of H. QED

4.8 Example To find the inverse of

$$\begin{pmatrix} 1 & 1 \\ 2 & -1 \end{pmatrix}$$

do Gauss-Jordan reduction, meanwhile performing the same operations on the identity. For clerical convenience we write the matrix and the identity side-by-side and do the reduction steps together.

$$\left(\begin{array}{cc|cc} 1 & 1 & 1 & 0 \\ 2 & -1 & 0 & 1 \end{array}\right) \xrightarrow{-2\rho_1+\rho_2} \left(\begin{array}{cc|cc} 1 & 1 & 1 & 0 \\ 0 & -3 & -2 & 1 \end{array}\right)$$
$$\xrightarrow{-1/3\rho_2} \left(\begin{array}{cc|cc} 1 & 1 & 1 & 0 \\ 0 & 1 & 2/3 & -1/3 \end{array}\right)$$
$$\xrightarrow{-\rho_2+\rho_1} \left(\begin{array}{cc|cc} 1 & 0 & 1/3 & 1/3 \\ 0 & 1 & 2/3 & -1/3 \end{array}\right)$$

This calculation has found the inverse.

$$\begin{pmatrix} 1 & 1 \\ 2 & -1 \end{pmatrix}^{-1} = \begin{pmatrix} 1/3 & 1/3 \\ 2/3 & -1/3 \end{pmatrix}$$

4.9 Example This one happens to start with a row swap.

$$\left(\begin{array}{ccc|ccc} 0 & 3 & -1 & 1 & 0 & 0 \\ 1 & 0 & 1 & 0 & 1 & 0 \\ 1 & -1 & 0 & 0 & 0 & 1 \end{array}\right) \xrightarrow{\rho_1 \leftrightarrow \rho_2} \left(\begin{array}{ccc|ccc} 1 & 0 & 1 & 0 & 1 & 0 \\ 0 & 3 & -1 & 1 & 0 & 0 \\ 1 & -1 & 0 & 0 & 0 & 1 \end{array}\right)$$

$$\xrightarrow{-\rho_1 + \rho_3} \left(\begin{array}{ccc|ccc} 1 & 0 & 1 & 0 & 1 & 0 \\ 0 & 3 & -1 & 1 & 0 & 0 \\ 0 & -1 & -1 & 0 & -1 & 1 \end{array}\right)$$

$$\vdots$$

$$\longrightarrow \left(\begin{array}{ccc|ccc} 1 & 0 & 0 & 1/4 & 1/4 & 3/4 \\ 0 & 1 & 0 & 1/4 & 1/4 & -1/4 \\ 0 & 0 & 1 & -1/4 & 3/4 & -3/4 \end{array}\right)$$

4.10 Example This algorithm detects a non-invertible matrix when the left half won't reduce to the identity.

$$\left(\begin{array}{cc|cc} 1 & 1 & 1 & 0 \\ 2 & 2 & 0 & 1 \end{array}\right) \xrightarrow{-2\rho_1 + \rho_2} \left(\begin{array}{cc|cc} 1 & 1 & 1 & 0 \\ 0 & 0 & -2 & 1 \end{array}\right)$$

With this procedure we can give a formula for the inverse of a general 2×2 matrix, which is worth memorizing.

4.11 Corollary The inverse for a 2×2 matrix exists and equals

$$\begin{pmatrix} a & b \\ c & d \end{pmatrix}^{-1} = \frac{1}{ad - bc} \begin{pmatrix} d & -b \\ -c & a \end{pmatrix}$$

if and only if $ad - bc \neq 0$.

Proof This computation is Exercise 21. QED

We have seen in this subsection, as in the subsection on Mechanics of Matrix Multiplication, how to exploit the correspondence between linear maps and matrices. We can fruitfully study both maps and matrices, translating back and forth to use whichever is handiest.

Over the course of this entire section we have developed an algebra system for matrices. We can compare it with the familiar algebra of real numbers.

Matrix addition and subtraction work in much the same way as the real number operations except that they only combine same-sized matrices. Scalar multiplication is in some ways an extension of real number multiplication. We also have a matrix multiplication operation and its inverse that are somewhat like the familiar real number operations (associativity, and distributivity over addition, for example), but there are differences (failure of commutativity). This section provides an example that algebra systems other than the usual real number one can be interesting and useful.

Exercises

4.12 Supply the intermediate steps in Example 4.9.

✓ **4.13** Use Corollary 4.11 to decide if each matrix has an inverse.

(a) $\begin{pmatrix} 2 & 1 \\ -1 & 1 \end{pmatrix}$ (b) $\begin{pmatrix} 0 & 4 \\ 1 & -3 \end{pmatrix}$ (c) $\begin{pmatrix} 2 & -3 \\ -4 & 6 \end{pmatrix}$

✓ **4.14** For each invertible matrix in the prior problem, use Corollary 4.11 to find its inverse.

✓ **4.15** Find the inverse, if it exists, by using the Gauss-Jordan Method. Check the answers for the 2×2 matrices with Corollary 4.11.

(a) $\begin{pmatrix} 3 & 1 \\ 0 & 2 \end{pmatrix}$ (b) $\begin{pmatrix} 2 & 1/2 \\ 3 & 1 \end{pmatrix}$ (c) $\begin{pmatrix} 2 & -4 \\ -1 & 2 \end{pmatrix}$ (d) $\begin{pmatrix} 1 & 1 & 3 \\ 0 & 2 & 4 \\ -1 & 1 & 0 \end{pmatrix}$

(e) $\begin{pmatrix} 0 & 1 & 5 \\ 0 & -2 & 4 \\ 2 & 3 & -2 \end{pmatrix}$ (f) $\begin{pmatrix} 2 & 2 & 3 \\ 1 & -2 & -3 \\ 4 & -2 & -3 \end{pmatrix}$

✓ **4.16** What matrix has this one for its inverse?

$$\begin{pmatrix} 1 & 3 \\ 2 & 5 \end{pmatrix}$$

4.17 How does the inverse operation interact with scalar multiplication and addition of matrices?

(a) What is the inverse of rH?

(b) Is $(H+G)^{-1} = H^{-1} + G^{-1}$?

✓ **4.18** Is $(T^k)^{-1} = (T^{-1})^k$?

4.19 Is H^{-1} invertible?

4.20 For each real number θ let $t_\theta\colon \mathbb{R}^2 \to \mathbb{R}^2$ be represented with respect to the standard bases by this matrix.

$$\begin{pmatrix} \cos\theta & -\sin\theta \\ \sin\theta & \cos\theta \end{pmatrix}$$

Show that $t_{\theta_1+\theta_2} = t_{\theta_1} \cdot t_{\theta_2}$. Show also that ${t_\theta}^{-1} = t_{-\theta}$.

4.21 Do the calculations for the proof of Corollary 4.11.

4.22 Show that this matrix

$$H = \begin{pmatrix} 1 & 0 & 1 \\ 0 & 1 & 0 \end{pmatrix}$$

has infinitely many right inverses. Show also that it has no left inverse.

4.23 In the review of inverses example, starting this subsection, how many left inverses has ι?

4.24 If a matrix has infinitely many right-inverses, can it have infinitely many left-inverses? Must it have?

4.25 Assume that $g\colon V\to W$ is linear. One of these is true, the other is false. Which is which?

(a) If $f\colon W\to V$ is a left inverse of g then f must be linear.

(b) If $f\colon W\to V$ is a right inverse of g then f must be linear.

✓ **4.26** Assume that H is invertible and that HG is the zero matrix. Show that G is a zero matrix.

4.27 Prove that if H is invertible then the inverse commutes with a matrix $GH^{-1}=H^{-1}G$ if and only if H itself commutes with that matrix $GH=HG$.

✓ **4.28** Show that if T is square and if T^4 is the zero matrix then $(I-T)^{-1}=I+T+T^2+T^3$. Generalize.

✓ **4.29** Let D be diagonal. Describe D^2, D^3, ..., etc. Describe D^{-1}, D^{-2}, ..., etc. Define D^0 appropriately.

4.30 Prove that any matrix row-equivalent to an invertible matrix is also invertible.

4.31 *The first question below appeared as Exercise 30.*

(a) Show that the rank of the product of two matrices is less than or equal to the minimum of the rank of each.

(b) Show that if T and S are square then $TS=I$ if and only if $ST=I$.

4.32 Show that the inverse of a permutation matrix is its transpose.

4.33 *The first two parts of this question appeared as Exercise 27.*

(a) Show that $(GH)^{\mathsf{T}}=H^{\mathsf{T}}G^{\mathsf{T}}$.

(b) A square matrix is *symmetric* if each i,j entry equals the j,i entry (that is, if the matrix equals its transpose). Show that the matrices HH^{T} and $H^{\mathsf{T}}H$ are symmetric.

(c) Show that the inverse of the transpose is the transpose of the inverse.

(d) Show that the inverse of a symmetric matrix is symmetric.

✓ **4.34** *The items starting this question appeared as Exercise 32.*

(a) Prove that the composition of the projections $\pi_x,\pi_y\colon\mathbb{R}^3\to\mathbb{R}^3$ is the zero map despite that neither is the zero map.

(b) Prove that the composition of the derivatives $d^2/dx^2,\ d^3/dx^3\colon\mathcal{P}_4\to\mathcal{P}_4$ is the zero map despite that neither map is the zero map.

(c) Give matrix equations representing each of the prior two items.

When two things multiply to give zero despite that neither is zero, each is said to be a *zero divisor*. Prove that no zero divisor is invertible.

4.35 In real number algebra, there are exactly two numbers, 1 and -1, that are their own multiplicative inverse. Does $H^2=I$ have exactly two solutions for 2×2 matrices?

4.36 Is the relation 'is a two-sided inverse of' transitive? Reflexive? Symmetric?

4.37 [Am. Math. Mon., Nov. 1951] Prove: if the sum of the elements of each row of a square matrix is k, then the sum of the elements in each row of the inverse matrix is $1/k$.

V Change of Basis

Representations vary with the bases. For instance, with respect to the bases $\mathcal{E}_2$ and

$$B = \langle \begin{pmatrix} 1 \\ 1 \end{pmatrix}, \begin{pmatrix} 1 \\ -1 \end{pmatrix} \rangle$$

$\vec{e}_1 \in \mathbb{R}^2$ has these different representations.

$$\text{Rep}_{\mathcal{E}_2}(\vec{e}_1) = \begin{pmatrix} 1 \\ 0 \end{pmatrix} \qquad \text{Rep}_B(\vec{e}_1) = \begin{pmatrix} 1/2 \\ 1/2 \end{pmatrix}$$

The same holds for maps: with respect to the basis pairs $\mathcal{E}_2, \mathcal{E}_2$ and $\mathcal{E}_2, B$, the identity map has these representations.

$$\text{Rep}_{\mathcal{E}_2,\mathcal{E}_2}(\text{id}) = \begin{pmatrix} 1 & 0 \\ 0 & 1 \end{pmatrix} \qquad \text{Rep}_{\mathcal{E}_2,B}(\text{id}) = \begin{pmatrix} 1/2 & 1/2 \\ 1/2 & -1/2 \end{pmatrix}$$

This section shows how to translate among the representations. That is, we will compute how the representations vary as the bases vary.

V.1 Changing Representations of Vectors

In converting $\text{Rep}_B(\vec{v})$ to $\text{Rep}_D(\vec{v})$ the underlying vector $\vec{v}$ doesn't change. Thus, the translation between these two ways of expressing the vector is accomplished by the identity map on the space, described so that the domain space vectors are represented with respect to B and the codomain space vectors are represented with respect to D.

$$\begin{array}{c} V_{wrt\ B} \\ \text{id}\big\downarrow \\ V_{wrt\ D} \end{array}$$

(This diagram is vertical to fit with the ones in the next subsection.)

1.1 Definition The *change of basis matrix* for bases $B, D \subset V$ is the representation of the identity map $\text{id}\colon V \to V$ with respect to those bases.

$$\text{Rep}_{B,D}(\text{id}) = \begin{pmatrix} \vdots & & \vdots \\ \text{Rep}_D(\vec{\beta}_1) & \cdots & \text{Rep}_D(\vec{\beta}_n) \\ \vdots & & \vdots \end{pmatrix}$$

1.2 Remark A better name would be 'change of representation matrix' but the above name is standard.

The next result supports the definition.

1.3 Lemma Left-multiplication by the change of basis matrix for B, D converts a representation with respect to B to one with respect to D. Conversely, if left-multiplication by a matrix changes bases $M \cdot \mathrm{Rep}_B(\vec{v}) = \mathrm{Rep}_D(\vec{v})$ then M is a change of basis matrix.

Proof The first sentence holds because matrix-vector multiplication represents a map application and so $\mathrm{Rep}_{B,D}(\mathrm{id}) \cdot \mathrm{Rep}_B(\vec{v}) = \mathrm{Rep}_D(\,\mathrm{id}(\vec{v})\,) = \mathrm{Rep}_D(\vec{v})$ for each $\vec{v}$. For the second sentence, with respect to B, D the matrix M represents a linear map whose action is to map each vector to itself, and is therefore the identity map. QED

1.4 Example With these bases for $\mathbb{R}^2$,

$$B = \langle \begin{pmatrix} 2 \\ 1 \end{pmatrix}, \begin{pmatrix} 1 \\ 0 \end{pmatrix} \rangle \qquad D = \langle \begin{pmatrix} -1 \\ 1 \end{pmatrix}, \begin{pmatrix} 1 \\ 1 \end{pmatrix} \rangle$$

because

$$\mathrm{Rep}_D(\,\mathrm{id}(\begin{pmatrix} 2 \\ 1 \end{pmatrix})) = \begin{pmatrix} -1/2 \\ 3/2 \end{pmatrix}_D \qquad \mathrm{Rep}_D(\,\mathrm{id}(\begin{pmatrix} 1 \\ 0 \end{pmatrix})) = \begin{pmatrix} -1/2 \\ 1/2 \end{pmatrix}_D$$

the change of basis matrix is this.

$$\mathrm{Rep}_{B,D}(\mathrm{id}) = \begin{pmatrix} -1/2 & -1/2 \\ 3/2 & 1/2 \end{pmatrix}$$

For instance, this is the representation of $\vec{e}_2$

$$\mathrm{Rep}_B(\begin{pmatrix} 0 \\ 1 \end{pmatrix}) = \begin{pmatrix} 1 \\ -2 \end{pmatrix}$$

and the matrix does the conversion.

$$\begin{pmatrix} -1/2 & -1/2 \\ 3/2 & 1/2 \end{pmatrix} \begin{pmatrix} 1 \\ -2 \end{pmatrix} = \begin{pmatrix} 1/2 \\ 1/2 \end{pmatrix}$$

Checking that vector on the right is $\mathrm{Rep}_D(\vec{e}_2)$ is easy.

We finish this subsection by recognizing the change of basis matrices as a familiar set.

1.5 Lemma A matrix changes bases if and only if it is nonsingular.

Proof For the 'only if' direction, if left-multiplication by a matrix changes bases then the matrix represents an invertible function, simply because we can invert the function by changing the bases back. Because it represents a function that is invertible, the matrix itself is invertible, and so is nonsingular.

For 'if' we will show that any nonsingular matrix M performs a change of basis operation from any given starting basis B (having n vectors, where the matrix is $n \times n$) to some ending basis.

If the matrix is the identity I then the statement is obvious. Otherwise because the matrix is nonsingular Corollary IV.3.23 says there are elementary reduction matrices such that $R_r \cdots R_1 \cdot M = I$ with $r \geqslant 1$. Elementary matrices are invertible and their inverses are also elementary so multiplying both sides of that equation from the left by R_r^{-1}, then by R_{r-1}^{-1}, etc., gives M as a product of elementary matrices $M = R_1^{-1} \cdots R_r^{-1}$.

We will be done if we show that elementary matrices change a given basis to another basis, since then R_r^{-1} changes B to some other basis B_r and R_{r-1}^{-1} changes B_r to some B_{r-1}, etc. We will cover the three types of elementary matrices separately; recall the notation for the three.

$$M_i(k) \begin{pmatrix} c_1 \\ \vdots \\ c_i \\ \vdots \\ c_n \end{pmatrix} = \begin{pmatrix} c_1 \\ \vdots \\ kc_i \\ \vdots \\ c_n \end{pmatrix} \quad P_{i,j} \begin{pmatrix} c_1 \\ \vdots \\ c_i \\ \vdots \\ c_j \\ \vdots \\ c_n \end{pmatrix} = \begin{pmatrix} c_1 \\ \vdots \\ c_j \\ \vdots \\ c_i \\ \vdots \\ c_n \end{pmatrix} \quad C_{i,j}(k) \begin{pmatrix} c_1 \\ \vdots \\ c_i \\ \vdots \\ c_j \\ \vdots \\ c_n \end{pmatrix} = \begin{pmatrix} c_1 \\ \vdots \\ c_i \\ \vdots \\ kc_i + c_j \\ \vdots \\ c_n \end{pmatrix}$$

Applying a row-multiplication matrix $M_i(k)$ changes a representation with respect to $\langle \vec{\beta}_1, \ldots, \vec{\beta}_i, \ldots, \vec{\beta}_n \rangle$ to one with respect to $\langle \vec{\beta}_1, \ldots, (1/k)\vec{\beta}_i, \ldots, \vec{\beta}_n \rangle$.

$$\begin{aligned} \vec{v} = c_1 \cdot \vec{\beta}_1 + \cdots + c_i \cdot \vec{\beta}_i + \cdots + c_n \cdot \vec{\beta}_n \\ \mapsto c_1 \cdot \vec{\beta}_1 + \cdots + kc_i \cdot (1/k)\vec{\beta}_i + \cdots + c_n \cdot \vec{\beta}_n = \vec{v} \end{aligned}$$

The second one is a basis because the first is a basis and because of the $k \neq 0$ restriction in the definition of a row-multiplication matrix. Similarly, left-multiplication by a row-swap matrix $P_{i,j}$ changes a representation with respect to the basis $\langle \vec{\beta}_1, \ldots, \vec{\beta}_i, \ldots, \vec{\beta}_j, \ldots, \vec{\beta}_n \rangle$ into one with respect to this basis

$\langle\vec{\beta}_1,\ldots,\vec{\beta}_j,\ldots,\vec{\beta}_i,\ldots,\vec{\beta}_n\rangle$.

$$\begin{aligned}\vec{v} &= c_1\cdot\vec{\beta}_1+\cdots+c_i\cdot\vec{\beta}_i+\cdots+c_j\vec{\beta}_j+\cdots+c_n\cdot\vec{\beta}_n\\ &\mapsto c_1\cdot\vec{\beta}_1+\cdots+c_j\cdot\vec{\beta}_j+\cdots+c_i\cdot\vec{\beta}_i+\cdots+c_n\cdot\vec{\beta}_n=\vec{v}\end{aligned}$$

And, a representation with respect to $\langle\vec{\beta}_1,\ldots,\vec{\beta}_i,\ldots,\vec{\beta}_j,\ldots,\vec{\beta}_n\rangle$ changes via left-multiplication by a row-combination matrix $C_{i,j}(k)$ into a representation with respect to $\langle\vec{\beta}_1,\ldots,\vec{\beta}_i-k\vec{\beta}_j,\ldots,\vec{\beta}_j,\ldots,\vec{\beta}_n\rangle$

$$\begin{aligned}\vec{v} &= c_1\cdot\vec{\beta}_1+\cdots+c_i\cdot\vec{\beta}_i+c_j\vec{\beta}_j+\cdots+c_n\cdot\vec{\beta}_n\\ &\mapsto c_1\cdot\vec{\beta}_1+\cdots+c_i\cdot(\vec{\beta}_i-k\vec{\beta}_j)+\cdots+(kc_i+c_j)\cdot\vec{\beta}_j+\cdots+c_n\cdot\vec{\beta}_n=\vec{v}\end{aligned}$$

(the definition of $C_{i,j}(k)$ specifies that $i\neq j$ and $k\neq 0$). QED

1.6 Corollary A matrix is nonsingular if and only if it represents the identity map with respect to some pair of bases.

Exercises

✓ **1.7** In $\mathbb{R}^2$, where

$$D=\langle\begin{pmatrix}2\\1\end{pmatrix},\begin{pmatrix}-2\\4\end{pmatrix}\rangle$$

find the change of basis matrices from D to $\mathcal{E}_2$ and from $\mathcal{E}_2$ to D. Multiply the two.

✓ **1.8** Find the change of basis matrix for $B,D\subseteq\mathbb{R}^2$.

(a) $B=\mathcal{E}_2$, $D=\langle\vec{e}_2,\vec{e}_1\rangle$ (b) $B=\mathcal{E}_2$, $D=\langle\begin{pmatrix}1\\2\end{pmatrix},\begin{pmatrix}1\\4\end{pmatrix}\rangle$

(c) $B=\langle\begin{pmatrix}1\\2\end{pmatrix},\begin{pmatrix}1\\4\end{pmatrix}\rangle$, $D=\mathcal{E}_2$ (d) $B=\langle\begin{pmatrix}-1\\1\end{pmatrix},\begin{pmatrix}2\\2\end{pmatrix}\rangle$, $D=\langle\begin{pmatrix}0\\4\end{pmatrix},\begin{pmatrix}1\\3\end{pmatrix}\rangle$

✓ **1.9** Find the change of basis matrix for each $B,D\subseteq\mathcal{P}_2$.

(a) $B=\langle 1,x,x^2\rangle, D=\langle x^2,1,x\rangle$ (b) $B=\langle 1,x,x^2\rangle, D=\langle 1,1+x,1+x+x^2\rangle$
(c) $B=\langle 2,2x,x^2\rangle, D=\langle 1+x^2,1-x^2,x+x^2\rangle$

1.10 For the bases in Exercise 8, find the change of basis matrix in the other direction, from D to B.

✓ **1.11** Decide if each changes bases on $\mathbb{R}^2$. To what basis is $\mathcal{E}_2$ changed?

(a) $\begin{pmatrix}5&0\\0&4\end{pmatrix}$ (b) $\begin{pmatrix}2&1\\3&1\end{pmatrix}$ (c) $\begin{pmatrix}-1&4\\2&-8\end{pmatrix}$ (d) $\begin{pmatrix}1&-1\\1&1\end{pmatrix}$

1.12 For each space find the matrix changing a vector representation with respect to B to one with respect to D.

(a) $V=\mathbb{R}^3$, $B=\mathcal{E}_3$, $D=\langle\begin{pmatrix}1\\2\\3\end{pmatrix},\begin{pmatrix}1\\1\\1\end{pmatrix},\begin{pmatrix}0\\1\\-1\end{pmatrix}\rangle$

(b) $V=\mathbb{R}^3$, $B=\langle\begin{pmatrix}1\\2\\3\end{pmatrix},\begin{pmatrix}1\\1\\1\end{pmatrix},\begin{pmatrix}0\\1\\-1\end{pmatrix}\rangle$, $D=\mathcal{E}_3$

(c) $V = \mathcal{P}_2$, $B = \langle x^2, x^2 + x, x^2 + x + 1\rangle$, $D = \langle 2, -x, x^2\rangle$

1.13 Find bases such that this matrix represents the identity map with respect to those bases.

$$\begin{pmatrix} 3 & 1 & 4 \\ 2 & -1 & 1 \\ 0 & 0 & 4 \end{pmatrix}$$

1.14 Consider the vector space of real-valued functions with basis $\langle \sin(x), \cos(x)\rangle$. Show that $\langle 2\sin(x) + \cos(x), 3\cos(x)\rangle$ is also a basis for this space. Find the change of basis matrix in each direction.

1.15 Where does this matrix

$$\begin{pmatrix} \cos(2\theta) & \sin(2\theta) \\ \sin(2\theta) & -\cos(2\theta) \end{pmatrix}$$

send the standard basis for $\mathbb{R}^2$? Any other bases? *Hint.* Consider the inverse.

✓ 1.16 What is the change of basis matrix with respect to B, B?

1.17 Prove that a matrix changes bases if and only if it is invertible.

1.18 Finish the proof of Lemma 1.5.

✓ 1.19 Let H be an $n \times n$ nonsingular matrix. What basis of $\mathbb{R}^n$ does H change to the standard basis?

✓ 1.20 (a) In $\mathcal{P}_3$ with basis $B = \langle 1+x, 1-x, x^2+x^3, x^2-x^3\rangle$ we have this representation.

$$\text{Rep}_B(1 - x + 3x^2 - x^3) = \begin{pmatrix} 0 \\ 1 \\ 1 \\ 2 \end{pmatrix}_B$$

Find a basis D giving this different representation for the same polynomial.

$$\text{Rep}_D(1 - x + 3x^2 - x^3) = \begin{pmatrix} 1 \\ 0 \\ 2 \\ 0 \end{pmatrix}_D$$

(b) State and prove that we can change any nonzero vector representation to any other.

Hint. The proof of Lemma 1.5 is constructive—it not only says the bases change, it shows how they change.

1.21 Let V, W be vector spaces, and let $B, \hat{B}$ be bases for V and $D, \hat{D}$ be bases for W. Where $h\colon V \to W$ is linear, find a formula relating $\text{Rep}_{B,D}(h)$ to $\text{Rep}_{\hat{B},\hat{D}}(h)$.

✓ 1.22 Show that the columns of an $n \times n$ change of basis matrix form a basis for $\mathbb{R}^n$. Do all bases appear in that way: can the vectors from any $\mathbb{R}^n$ basis make the columns of a change of basis matrix?

✓ 1.23 Find a matrix having this effect.

$$\begin{pmatrix} 1 \\ 3 \end{pmatrix} \mapsto \begin{pmatrix} 4 \\ -1 \end{pmatrix}$$

That is, find a M that left-multiplies the starting vector to yield the ending vector. Is there a matrix having these two effects?

(a) $\begin{pmatrix}1\\3\end{pmatrix}\mapsto\begin{pmatrix}1\\1\end{pmatrix}$ $\begin{pmatrix}2\\-1\end{pmatrix}\mapsto\begin{pmatrix}-1\\-1\end{pmatrix}$ (b) $\begin{pmatrix}1\\3\end{pmatrix}\mapsto\begin{pmatrix}1\\1\end{pmatrix}$ $\begin{pmatrix}2\\6\end{pmatrix}\mapsto\begin{pmatrix}-1\\-1\end{pmatrix}$

Give a necessary and sufficient condition for there to be a matrix such that $\vec{v}_1\mapsto\vec{w}_1$ and $\vec{v}_2\mapsto\vec{w}_2$.

V.2 Changing Map Representations

The first subsection shows how to convert the representation of a vector with respect to one basis to the representation of that same vector with respect to another basis. We next convert the representation of a map with respect to one pair of bases to the representation with respect to a different pair — we convert from $\text{Rep}_{B,D}(h)$ to $\text{Rep}_{\hat{B},\hat{D}}(h)$. Here is the arrow diagram.

$$\begin{array}{ccc} V_{wrt\ B} & \xrightarrow[H]{h} & W_{wrt\ D} \\ {\scriptstyle id}\big\downarrow & & {\scriptstyle id}\big\downarrow \\ V_{wrt\ \hat{B}} & \xrightarrow[\hat{H}]{h} & W_{wrt\ \hat{D}} \end{array}$$

To move from the lower-left to the lower-right we can either go straight over, or else up to V_B then over to W_D and then down. So we can calculate $\hat{H}=\text{Rep}_{\hat{B},\hat{D}}(h)$ either by directly using $\hat{B}$ and $\hat{D}$, or else by first changing bases with $\text{Rep}_{\hat{B},B}(\text{id})$ then multiplying by $H=\text{Rep}_{B,D}(h)$ and then changing bases with $\text{Rep}_{D,\hat{D}}(\text{id})$.

$$\hat{H}=\text{Rep}_{D,\hat{D}}(\text{id})\cdot H\cdot\text{Rep}_{\hat{B},B}(\text{id}) \qquad (*)$$

2.1 Example The matrix

$$T=\begin{pmatrix}\cos(\pi/6)&-\sin(\pi/6)\\ \sin(\pi/6)&\cos(\pi/6)\end{pmatrix}=\begin{pmatrix}\sqrt{3}/2&-1/2\\ 1/2&\sqrt{3}/2\end{pmatrix}$$

represents, with respect to $\mathcal{E}_2,\mathcal{E}_2$, the transformation $t\colon\mathbb{R}^2\to\mathbb{R}^2$ that rotates vectors through the counterclockwise angle of $\pi/6$ radians.

$$\begin{pmatrix}1\\3\end{pmatrix}\ \stackrel{t_{\pi/6}}{\longrightarrow}\ \begin{pmatrix}(-3+\sqrt{3})/2\\(1+3\sqrt{3})/2\end{pmatrix}$$

We can translate T to a representation with respect to these

$$\hat{B}=\langle\begin{pmatrix}1\\1\end{pmatrix}\begin{pmatrix}0\\2\end{pmatrix}\rangle\qquad\hat{D}=\langle\begin{pmatrix}-1\\0\end{pmatrix}\begin{pmatrix}2\\3\end{pmatrix}\rangle$$

by using the arrow diagram above.

$$\begin{array}{ccc} \mathbb{R}^2_{wrt\ \mathcal{E}_2} & \xrightarrow[\mathrm{T}]{\mathrm{t}} & \mathbb{R}^2_{wrt\ \mathcal{E}_2} \\ \mathrm{id}\big\downarrow & & \mathrm{id}\big\downarrow \\ \mathbb{R}^2_{wrt\ \hat{\mathrm{B}}} & \xrightarrow[\hat{\mathrm{T}}]{\mathrm{t}} & \mathbb{R}^2_{wrt\ \hat{\mathrm{D}}} \end{array}$$

The picture illustrates that we can compute $\hat{\mathrm{T}}$ either directly by going along the square's bottom, or as in formula (∗) by going up on the left, then across the top, and then down on the right, with $\hat{\mathrm{T}} = \mathrm{Rep}_{\mathcal{E}_2,\hat{\mathrm{D}}}(\mathrm{id}) \cdot \mathrm{T} \cdot \mathrm{Rep}_{\hat{\mathrm{B}},\mathcal{E}_2}(\mathrm{id})$. (Note again that the matrix multiplication reads right to left, as the three functions are composed and function composition reads right to left.)

Find the matrix for the left-hand side, the matrix $\mathrm{Rep}_{\hat{\mathrm{B}},\mathcal{E}_2}(\mathrm{id})$, in the usual way: find the effect of the identity matrix on the starting basis $\hat{\mathrm{B}}$ — which of course is no effect at all — and then represent those basis elements with respect to the ending basis $\mathcal{E}_3$.

$$\mathrm{Rep}_{\hat{\mathrm{B}},\mathcal{E}_2}(\mathrm{id}) = \begin{pmatrix} 1 & 0 \\ 1 & 2 \end{pmatrix}$$

This calculation is easy when the ending basis is the standard one.

There are two ways to compute the matrix for going down the square's right side, $\mathrm{Rep}_{\mathcal{E}_2,\hat{\mathrm{D}}}(\mathrm{id})$. We could calculate it directly as we did for the other change of basis matrix. Or, we could instead calculate it as the inverse of the matrix for going up $\mathrm{Rep}_{\hat{\mathrm{D}},\mathcal{E}_2}(\mathrm{id})$. Find that matrix is easy, and we have a formula for the 2×2 inverse so that's what is in the equation below.

$$\begin{aligned} \mathrm{Rep}_{\hat{\mathrm{B}},\hat{\mathrm{D}}}(\mathrm{t}) &= \begin{pmatrix} -1 & 2 \\ 0 & 3 \end{pmatrix}^{-1} \begin{pmatrix} \sqrt{3}/2 & -1/2 \\ 1/2 & \sqrt{3}/2 \end{pmatrix} \begin{pmatrix} 1 & 0 \\ 1 & 2 \end{pmatrix} \\ &= \begin{pmatrix} (5-\sqrt{3})/6 & (3+2\sqrt{3})/3 \\ (1+\sqrt{3})/6 & \sqrt{3}/3 \end{pmatrix} \end{aligned}$$

The matrix is messier but the map that it represents is the same. For instance, to replicate the effect of t in the picture, start with $\hat{\mathrm{B}}$,

$$\mathrm{Rep}_{\hat{\mathrm{B}}}(\begin{pmatrix} 1 \\ 3 \end{pmatrix}) = \begin{pmatrix} 1 \\ 1 \end{pmatrix}_{\hat{\mathrm{B}}}$$

apply $\hat{\mathrm{T}}$,

$$\begin{pmatrix} (5-\sqrt{3})/6 & (3+2\sqrt{3})/3 \\ (1+\sqrt{3})/6 & \sqrt{3}/3 \end{pmatrix}_{\hat{\mathrm{B}},\hat{\mathrm{D}}} \begin{pmatrix} 1 \\ 1 \end{pmatrix}_{\hat{\mathrm{B}}} = \begin{pmatrix} (11+3\sqrt{3})/6 \\ (1+3\sqrt{3})/6 \end{pmatrix}_{\hat{\mathrm{D}}}$$

and check it against $\hat{D}$.

$$\frac{11+3\sqrt{3}}{6}\cdot\begin{pmatrix}-1\\0\end{pmatrix}+\frac{1+3\sqrt{3}}{6}\cdot\begin{pmatrix}2\\3\end{pmatrix}=\begin{pmatrix}(-3+\sqrt{3})/2\\(1+3\sqrt{3})/2\end{pmatrix}$$

2.2 Example Changing bases can make the matrix simpler. On $\mathbb{R}^3$ the map

$$\begin{pmatrix}x\\y\\z\end{pmatrix}\stackrel{t}{\longmapsto}\begin{pmatrix}y+z\\x+z\\x+y\end{pmatrix}$$

is represented with respect to the standard basis in this way.

$$\text{Rep}_{\mathcal{E}_3,\mathcal{E}_3}(t)=\begin{pmatrix}0&1&1\\1&0&1\\1&1&0\end{pmatrix}$$

Representing it with respect to

$$B=\langle\begin{pmatrix}1\\-1\\0\end{pmatrix},\begin{pmatrix}1\\1\\-2\end{pmatrix},\begin{pmatrix}1\\1\\1\end{pmatrix}\rangle$$

gives a matrix that is diagonal.

$$\text{Rep}_{B,B}(t)=\begin{pmatrix}-1&0&0\\0&-1&0\\0&0&2\end{pmatrix}$$

Naturally we usually prefer representations that are easier to understand. We say that a map or matrix has been *diagonalized* when we find a basis B such that the representation is diagonal with respect to B, B, that is, with respect to the same starting basis as ending basis. Chapter Five finds which maps and matrices are diagonalizable.

The rest of this subsection develops the easier case of finding two bases B, D such that a representation is simple. Recall that the prior subsection shows that a matrix is a change of basis matrix if and only if it is nonsingular.

2.3 Definition Same-sized matrices H and $\hat{H}$ are *matrix equivalent* if there are nonsingular matrices P and Q such that $\hat{H}=PHQ$.

2.4 Corollary Matrix equivalent matrices represent the same map, with respect to appropriate pairs of bases.

PROOF This is immediate from equation $(*)$ above. QED

Exercise 21 checks that matrix equivalence is an equivalence relation. Thus it partitions the set of matrices into matrix equivalence classes.

All matrices:

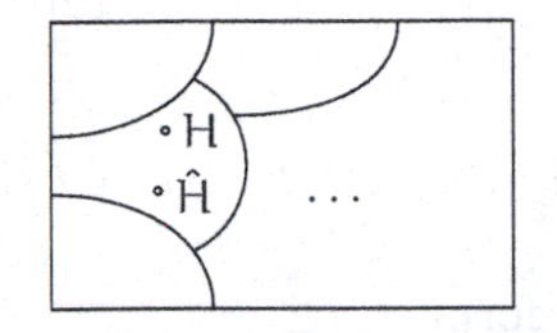

H matrix equivalent to Ĥ

We can get insight into the classes by comparing matrix equivalence with row equivalence (remember that matrices are row equivalent when they can be reduced to each other by row operations). In $\hat{H} = PHQ$, the matrices P and Q are nonsingular and thus each is a product of elementary reduction matrices by Lemma IV.4.7. Left-multiplication by the reduction matrices making up P performs row operations. Right-multiplication by the reduction matrices making up Q performs column operations. Hence, matrix equivalence is a generalization of row equivalence — two matrices are row equivalent if one can be converted to the other by a sequence of row reduction steps, while two matrices are matrix equivalent if one can be converted to the other by a sequence of row reduction steps followed by a sequence of column reduction steps.

Consequently, if matrices are row equivalent then they are also matrix equivalent since we can take Q to be the identity matrix. The converse, however, does not hold: two matrices can be matrix equivalent but not row equivalent.

2.5 Example These two are matrix equivalent

$$\begin{pmatrix} 1 & 0 \\ 0 & 0 \end{pmatrix} \quad \begin{pmatrix} 1 & 1 \\ 0 & 0 \end{pmatrix}$$

because the second reduces to the first by the column operation of taking -1 times the first column and adding to the second. They are not row equivalent because they have different reduced echelon forms (both are already in reduced form).

We close this section by giving a set of representatives for the matrix equivalence classes.

2.6 Theorem Any $m{\times}n$ matrix of rank k is matrix equivalent to the $m{\times}n$ matrix that is all zeros except that the first k diagonal entries are ones.

$$\begin{pmatrix} 1 & 0 & \ldots & 0 & 0 & \ldots & 0 \\ 0 & 1 & \ldots & 0 & 0 & \ldots & 0 \\ & \vdots \\ 0 & 0 & \ldots & 1 & 0 & \ldots & 0 \\ 0 & 0 & \ldots & 0 & 0 & \ldots & 0 \\ & \vdots \\ 0 & 0 & \ldots & 0 & 0 & \ldots & 0 \end{pmatrix}$$

This is a *block partial-identity* form.

$$\left(\begin{array}{c|c} I & Z \\ \hline Z & Z \end{array}\right)$$

Proof Gauss-Jordan reduce the given matrix and combine all the row reduction matrices to make P. Then use the leading entries to do column reduction and finish by swapping the columns to put the leading ones on the diagonal. Combine the column reduction matrices into Q. QED

2.7 Example We illustrate the proof by finding P and Q for this matrix.

$$\begin{pmatrix} 1 & 2 & 1 & -1 \\ 0 & 0 & 1 & -1 \\ 2 & 4 & 2 & -2 \end{pmatrix}$$

First Gauss-Jordan row-reduce.

$$\begin{pmatrix} 1 & -1 & 0 \\ 0 & 1 & 0 \\ 0 & 0 & 1 \end{pmatrix} \begin{pmatrix} 1 & 0 & 0 \\ 0 & 1 & 0 \\ -2 & 0 & 1 \end{pmatrix} \begin{pmatrix} 1 & 2 & 1 & -1 \\ 0 & 0 & 1 & -1 \\ 2 & 4 & 2 & -2 \end{pmatrix} = \begin{pmatrix} 1 & 2 & 0 & 0 \\ 0 & 0 & 1 & -1 \\ 0 & 0 & 0 & 0 \end{pmatrix}$$

Then column-reduce, which involves right-multiplication.

$$\begin{pmatrix} 1 & 2 & 0 & 0 \\ 0 & 0 & 1 & -1 \\ 0 & 0 & 0 & 0 \end{pmatrix} \begin{pmatrix} 1 & -2 & 0 & 0 \\ 0 & 1 & 0 & 0 \\ 0 & 0 & 1 & 0 \\ 0 & 0 & 0 & 1 \end{pmatrix} \begin{pmatrix} 1 & 0 & 0 & 0 \\ 0 & 1 & 0 & 0 \\ 0 & 0 & 1 & 1 \\ 0 & 0 & 0 & 1 \end{pmatrix} = \begin{pmatrix} 1 & 0 & 0 & 0 \\ 0 & 0 & 1 & 0 \\ 0 & 0 & 0 & 0 \end{pmatrix}$$

Finish by swapping columns.

$$\begin{pmatrix} 1 & 0 & 0 & 0 \\ 0 & 0 & 1 & 0 \\ 0 & 0 & 0 & 0 \end{pmatrix} \begin{pmatrix} 1 & 0 & 0 & 0 \\ 0 & 0 & 1 & 0 \\ 0 & 1 & 0 & 0 \\ 0 & 0 & 0 & 1 \end{pmatrix} = \begin{pmatrix} 1 & 0 & 0 & 0 \\ 0 & 1 & 0 & 0 \\ 0 & 0 & 0 & 0 \end{pmatrix}$$

Finally, combine the left-multipliers together as P and the right-multipliers together as Q to get PHQ.

$$\begin{pmatrix} 1 & -1 & 0 \\ 0 & 1 & 0 \\ -2 & 0 & 1 \end{pmatrix} \begin{pmatrix} 1 & 2 & 1 & -1 \\ 0 & 0 & 1 & -1 \\ 2 & 4 & 2 & -2 \end{pmatrix} \begin{pmatrix} 1 & 0 & -2 & 0 \\ 0 & 0 & 1 & 0 \\ 0 & 1 & 0 & 1 \\ 0 & 0 & 0 & 1 \end{pmatrix} = \begin{pmatrix} 1 & 0 & 0 & 0 \\ 0 & 1 & 0 & 0 \\ 0 & 0 & 0 & 0 \end{pmatrix}$$

2.8 Corollary Matrix equivalence classes are characterized by rank: two same-sized matrices are matrix equivalent if and only if they have the same rank.

Proof Two same-sized matrices with the same rank are equivalent to the same block partial-identity matrix. QED

2.9 Example The 2×2 matrices have only three possible ranks: zero, one, or two. Thus there are three matrix equivalence classes.

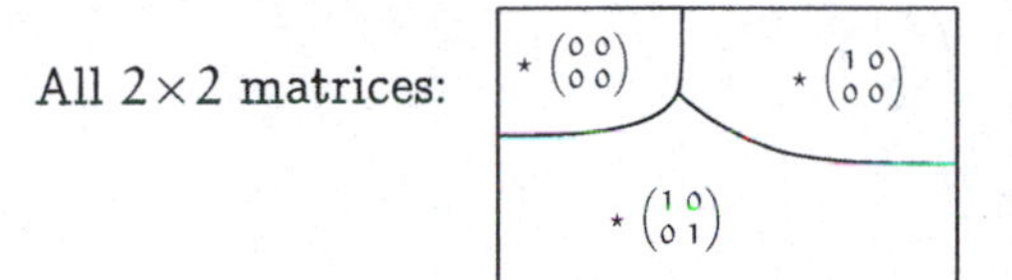

Each class consists of all of the 2×2 matrices with the same rank. There is only one rank zero matrix. The other two classes have infinitely many members; we've shown only the canonical representative.

One nice thing about the representative in Theorem 2.6 is that we can completely understand the linear map when it is expressed in this way: where the bases are $B = \langle\vec{\beta}_1, \ldots, \vec{\beta}_n\rangle$ and $D = \langle\vec{\delta}_1, \ldots, \vec{\delta}_m\rangle$ then the map's action is

$$c_1\vec{\beta}_1 + \cdots + c_k\vec{\beta}_k + c_{k+1}\vec{\beta}_{k+1} + \cdots + c_n\vec{\beta}_n \mapsto c_1\vec{\delta}_1 + \cdots + c_k\vec{\delta}_k + \vec{0} + \cdots + \vec{0}$$

where k is the rank. Thus we can view any linear map as a projection.

$$\begin{pmatrix} c_1 \\ \vdots \\ c_k \\ c_{k+1} \\ \vdots \\ c_n \end{pmatrix}_B \longmapsto \begin{pmatrix} c_1 \\ \vdots \\ c_k \\ 0 \\ \vdots \\ 0 \end{pmatrix}_D$$

Exercises

✓ **2.10** Decide if these matrices are matrix equivalent.

(a) $\begin{pmatrix} 1 & 3 & 0 \\ 2 & 3 & 0 \end{pmatrix}, \begin{pmatrix} 2 & 2 & 1 \\ 0 & 5 & -1 \end{pmatrix}$

(b) $\begin{pmatrix}0&3\\1&1\end{pmatrix}$, $\begin{pmatrix}4&0\\0&5\end{pmatrix}$

(c) $\begin{pmatrix}1&3\\2&6\end{pmatrix}$, $\begin{pmatrix}1&3\\2&-6\end{pmatrix}$

✓ **2.11** Find the canonical representative of the matrix equivalence class of each matrix.

(a) $\begin{pmatrix}2&1&0\\4&2&0\end{pmatrix}$ (b) $\begin{pmatrix}0&1&0&2\\1&1&0&4\\3&3&3&-1\end{pmatrix}$

2.12 Suppose that, with respect to

$$B = \mathcal{E}_2 \qquad D = \langle \begin{pmatrix}1\\1\end{pmatrix}, \begin{pmatrix}1\\-1\end{pmatrix} \rangle$$

the transformation $t\colon \mathbb{R}^2 \to \mathbb{R}^2$ is represented by this matrix.

$$\begin{pmatrix}1&2\\3&4\end{pmatrix}$$

Use change of basis matrices to represent t with respect to each pair.

(a) $\hat{B} = \langle \begin{pmatrix}0\\1\end{pmatrix}, \begin{pmatrix}1\\1\end{pmatrix} \rangle$, $\hat{D} = \langle \begin{pmatrix}-1\\0\end{pmatrix}, \begin{pmatrix}2\\1\end{pmatrix} \rangle$

(b) $\hat{B} = \langle \begin{pmatrix}1\\2\end{pmatrix}, \begin{pmatrix}1\\0\end{pmatrix} \rangle$, $\hat{D} = \langle \begin{pmatrix}1\\2\end{pmatrix}, \begin{pmatrix}2\\1\end{pmatrix} \rangle$

2.13 What sizes are P and Q in the equation $\hat{H} = PHQ$?

✓ **2.14** Consider the spaces $V = \mathcal{P}_2$ and $W = \mathcal{M}_{2\times 2}$, with these bases.

$$B = \langle 1, 1+x, 1+x^2 \rangle \qquad D = \langle \begin{pmatrix}0&0\\0&1\end{pmatrix}, \begin{pmatrix}0&0\\1&1\end{pmatrix}, \begin{pmatrix}0&1\\1&1\end{pmatrix}, \begin{pmatrix}1&1\\1&1\end{pmatrix} \rangle$$

$$\hat{B} = \langle 1, x, x^2 \rangle \qquad \hat{D} = \langle \begin{pmatrix}-1&0\\0&0\end{pmatrix}, \begin{pmatrix}0&-1\\0&0\end{pmatrix}, \begin{pmatrix}0&0\\1&0\end{pmatrix}, \begin{pmatrix}0&0\\0&1\end{pmatrix} \rangle$$

We will find P and Q to convert the representation of a map with respect to B, D to one with respect to $\hat{B}, \hat{D}$

(a) Draw the appropriate arrow diagram.

(b) Compute P and Q.

✓ **2.15** Find the P and Q to express H via PHQ as a block partial identity matrix.

$$H = \begin{pmatrix}2&1&1\\3&-1&0\\1&3&2\end{pmatrix}$$

✓ **2.16** Use Theorem 2.6 to show that a square matrix is nonsingular if and only if it is equivalent to an identity matrix.

✓ **2.17** Show that, where A is a nonsingular square matrix, if P and Q are nonsingular square matrices such that $PAQ = I$ then $QP = A^{-1}$.

✓ **2.18** Why does Theorem 2.6 not show that every matrix is diagonalizable (see Example 2.2)?

2.19 Must matrix equivalent matrices have matrix equivalent transposes?

2.20 What happens in Theorem 2.6 if $k = 0$?

2.21 Show that matrix equivalence is an equivalence relation.

✓ 2.22 Show that a zero matrix is alone in its matrix equivalence class. Are there other matrices like that?

2.23 What are the matrix equivalence classes of matrices of transformations on $\mathbb{R}^1$? $\mathbb{R}^3$?

2.24 How many matrix equivalence classes are there?

2.25 Are matrix equivalence classes closed under scalar multiplication? Addition?

2.26 Let $t\colon \mathbb{R}^n \to \mathbb{R}^n$ represented by T with respect to $\mathcal{E}_n, \mathcal{E}_n$.

(a) Find $\operatorname{Rep}_{B,B}(t)$ in this specific case.

$$T = \begin{pmatrix} 1 & 1 \\ 3 & -1 \end{pmatrix} \qquad B = \langle \begin{pmatrix} 1 \\ 2 \end{pmatrix}, \begin{pmatrix} -1 \\ -1 \end{pmatrix} \rangle$$

(b) Describe $\operatorname{Rep}_{B,B}(t)$ in the general case where $B = \langle \vec{\beta}_1, \ldots, \vec{\beta}_n \rangle$.

2.27 (a) Let V have bases B_1 and B_2 and suppose that W has the basis D. Where $h\colon V \to W$, find the formula that computes $\operatorname{Rep}_{B_2,D}(h)$ from $\operatorname{Rep}_{B_1,D}(h)$.

(b) Repeat the prior question with one basis for V and two bases for W.

2.28 (a) If two matrices are matrix equivalent and invertible, must their inverses be matrix equivalent?

(b) If two matrices have matrix equivalent inverses, must the two be matrix equivalent?

(c) If two matrices are square and matrix equivalent, must their squares be matrix equivalent?

(d) If two matrices are square and have matrix equivalent squares, must they be matrix equivalent?

✓ 2.29 Square matrices are *similar* if they represent the same transformation, but each with respect to the same ending as starting basis. That is, $\operatorname{Rep}_{B_1,B_1}(t)$ is similar to $\operatorname{Rep}_{B_2,B_2}(t)$.

(a) Give a definition of matrix similarity like that of Definition 2.3.

(b) Prove that similar matrices are matrix equivalent.

(c) Show that similarity is an equivalence relation.

(d) Show that if T is similar to $\hat{T}$ then T^2 is similar to $\hat{T}^2$, the cubes are similar, etc. *Contrast with the prior exercise.*

(e) Prove that there are matrix equivalent matrices that are not similar.

VI Projection

This section is optional. It is a prerequisite only for the final two sections of Chapter Five, and some Topics.

We have described projection from $\mathbb{R}^3$ into its xy-plane subspace as a shadow map. This shows why but it also shows that some shadows fall upward.

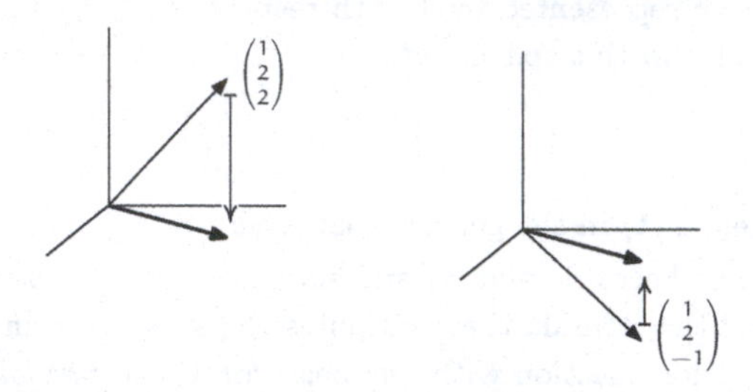

So perhaps a better description is: the projection of $\vec{v}$ is the vector $\vec{p}$ in the plane with the property that someone standing on $\vec{p}$ and looking straight up or down — that is, looking orthogonally to the plane — sees the tip of $\vec{v}$. In this section we will generalize this to other projections, orthogonal and non-orthogonal.

VI.1 Orthogonal Projection Into a Line

We first consider orthogonal projection of a vector $\vec{v}$ into a line ℓ. This shows a figure walking out on the line to a point $\vec{p}$ such that the tip of $\vec{v}$ is directly above them, where "above" does not mean parallel to the y-axis but instead means orthogonal to the line.

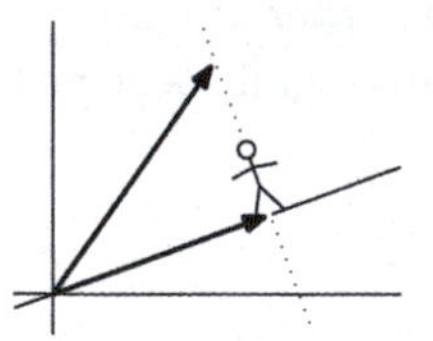

Since the line is the span of some vector $\ell = \{c \cdot \vec{s} \mid c \in \mathbb{R}\}$, we have a coefficient $c_{\vec{p}}$ with the property that $\vec{v} - c_{\vec{p}}\vec{s}$ is orthogonal to $c_{\vec{p}}\vec{s}$.

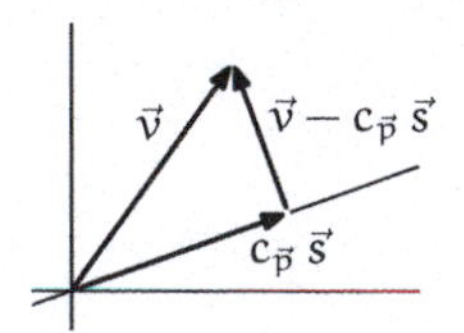

To solve for this coefficient, observe that because $\vec{v} - c_{\vec{p}}\vec{s}$ is orthogonal to a scalar multiple of $\vec{s}$, it must be orthogonal to $\vec{s}$ itself. Then $(\vec{v} - c_{\vec{p}}\vec{s}) \cdot \vec{s} = 0$ gives that $c_{\vec{p}} = \vec{v} \cdot \vec{s} / \vec{s} \cdot \vec{s}$.

1.1 Definition The *orthogonal projection of $\vec{v}$ into the line spanned by a nonzero $\vec{s}$* is this vector.

$$\text{proj}_{[\vec{s}]}(\vec{v}) = \frac{\vec{v} \cdot \vec{s}}{\vec{s} \cdot \vec{s}} \cdot \vec{s}$$

(That says 'spanned by $\vec{s}$' instead the more formal 'span of the set $\{\vec{s}\,\}$'. This more casual phrase is common.)

1.2 Example To orthogonally project the vector $\binom{2}{3}$ into the line $y = 2x$, first pick a direction vector for the line.

$$\vec{s} = \begin{pmatrix} 1 \\ 2 \end{pmatrix}$$

The calculation is easy.

$$\frac{\begin{pmatrix} 2 \\ 3 \end{pmatrix} \cdot \begin{pmatrix} 1 \\ 2 \end{pmatrix}}{\begin{pmatrix} 1 \\ 2 \end{pmatrix} \cdot \begin{pmatrix} 1 \\ 2 \end{pmatrix}} \cdot \begin{pmatrix} 1 \\ 2 \end{pmatrix} = \frac{8}{5} \cdot \begin{pmatrix} 1 \\ 2 \end{pmatrix} = \begin{pmatrix} 8/5 \\ 16/5 \end{pmatrix}$$

1.3 Example In $\mathbb{R}^3$, the orthogonal projection of a general vector

$$\begin{pmatrix} x \\ y \\ z \end{pmatrix}$$

into the y-axis is

$$\frac{\begin{pmatrix} x \\ y \\ z \end{pmatrix} \cdot \begin{pmatrix} 0 \\ 1 \\ 0 \end{pmatrix}}{\begin{pmatrix} 0 \\ 1 \\ 0 \end{pmatrix} \cdot \begin{pmatrix} 0 \\ 1 \\ 0 \end{pmatrix}} \cdot \begin{pmatrix} 0 \\ 1 \\ 0 \end{pmatrix} = \begin{pmatrix} 0 \\ y \\ 0 \end{pmatrix}$$

which matches our intuitive expectation.

The picture above showing the figure walking out on the line until $\vec{v}$'s tip is overhead is one way to think of the orthogonal projection of a vector into a line. We finish this subsection with two other ways.

1.4 Example A railroad car left on an east-west track without its brake is pushed by a wind blowing toward the northeast at fifteen miles per hour; what speed will the car reach?

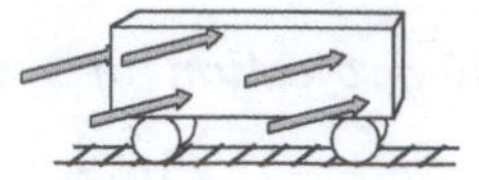

For the wind we use a vector of length 15 that points toward the northeast.

$$\vec{v} = \begin{pmatrix} 15\sqrt{1/2} \\ 15\sqrt{1/2} \end{pmatrix}$$

The car is only affected by the part of the wind blowing in the east-west direction — the part of $\vec{v}$ in the direction of the x-axis is this (the picture has the same perspective as the railroad car picture above).

north

east

$$\vec{p} = \begin{pmatrix} 15\sqrt{1/2} \\ 0 \end{pmatrix}$$

So the car will reach a velocity of $15\sqrt{1/2}$ miles per hour toward the east.

Thus, another way to think of the picture that precedes the definition is that it shows $\vec{v}$ as decomposed into two parts, the part $\vec{p}$ with the line, and the part that is orthogonal to the line (shown above on the north-south axis). These two are non-interacting in the sense that the east-west car is not at all affected by the north-south part of the wind (see Exercise 10). So we can think of the orthogonal projection of $\vec{v}$ into the line spanned by $\vec{s}$ as the part of $\vec{v}$ that lies in the direction of $\vec{s}$.

Still another useful way to think of orthogonal projection into a line is to have the person stand on the vector, not the line. This person holds a rope looped over the line. As they pull, the loop slides on the line.

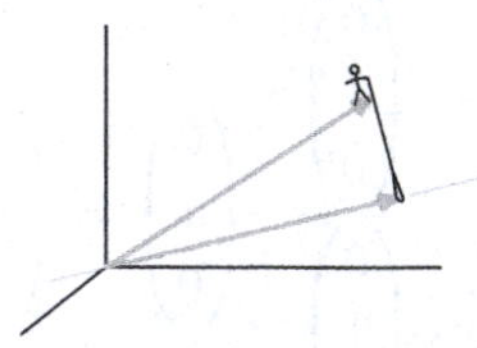

When it is tight, the rope is orthogonal to the line. That is, we can think of the projection $\vec{p}$ as being the vector in the line that is closest to $\vec{v}$ (see Exercise 16).

1.5 Example A submarine is tracking a ship moving along the line $y = 3x + 2$. Torpedo range is one-half mile. If the sub stays where it is, at the origin on the chart below, will the ship pass within range?

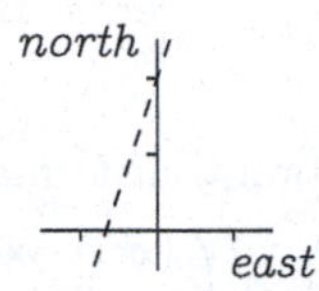

The formula for projection into a line does not immediately apply because the line doesn't pass through the origin, and so isn't the span of any $\vec{s}$. To adjust for this, we start by shifting the entire map down two units. Now the line is $y = 3x$, a subspace. We project to get the point $\vec{p}$ on the line closest to

$$\vec{v} = \begin{pmatrix} 0 \\ -2 \end{pmatrix}$$

the sub's shifted position.

$$\vec{p} = \frac{\begin{pmatrix} 0 \\ -2 \end{pmatrix} \cdot \begin{pmatrix} 1 \\ 3 \end{pmatrix}}{\begin{pmatrix} 1 \\ 3 \end{pmatrix} \cdot \begin{pmatrix} 1 \\ 3 \end{pmatrix}} \cdot \begin{pmatrix} 1 \\ 3 \end{pmatrix} = \begin{pmatrix} -3/5 \\ -9/5 \end{pmatrix}$$

The distance between $\vec{v}$ and $\vec{p}$ is about 0.63 miles. The ship will never be in range.

Exercises

✓ **1.6** Project the first vector orthogonally into the line spanned by the second vector.

(a) $\begin{pmatrix} 2 \\ 1 \end{pmatrix}, \begin{pmatrix} 3 \\ -2 \end{pmatrix}$ (b) $\begin{pmatrix} 2 \\ 1 \end{pmatrix}, \begin{pmatrix} 3 \\ 0 \end{pmatrix}$ (c) $\begin{pmatrix} 1 \\ 1 \\ 4 \end{pmatrix}, \begin{pmatrix} 1 \\ 2 \\ -1 \end{pmatrix}$ (d) $\begin{pmatrix} 1 \\ 1 \\ 4 \end{pmatrix}, \begin{pmatrix} 3 \\ 3 \\ 12 \end{pmatrix}$

✓ **1.7** Project the vector orthogonally into the line.

(a) $\begin{pmatrix} 2 \\ -1 \\ 4 \end{pmatrix}, \{c \begin{pmatrix} -3 \\ 1 \\ -3 \end{pmatrix} \mid c \in \mathbb{R}\}$ (b) $\begin{pmatrix} -1 \\ -1 \end{pmatrix}$, the line $y = 3x$

1.8 Although pictures guided our development of Definition 1.1, we are not restricted to spaces that we can draw. In $\mathbb{R}^4$ project this vector into this line.

$$\vec{v} = \begin{pmatrix} 1 \\ 2 \\ 1 \\ 3 \end{pmatrix} \qquad \ell = \{c \cdot \begin{pmatrix} -1 \\ 1 \\ -1 \\ 1 \end{pmatrix} \mid c \in \mathbb{R}\}$$

✓ **1.9** Definition 1.1 uses two vectors $\vec{s}$ and $\vec{v}$. Consider the transformation of $\mathbb{R}^2$ resulting from fixing

$$\vec{s} = \begin{pmatrix} 3 \\ 1 \end{pmatrix}$$

and projecting $\vec{v}$ into the line that is the span of $\vec{s}$. Apply it to these vectors.

(a) $\begin{pmatrix}1\\2\end{pmatrix}$ (b) $\begin{pmatrix}0\\4\end{pmatrix}$

Show that in general the projection transformation is this.

$$\begin{pmatrix}x_1\\x_2\end{pmatrix} \mapsto \begin{pmatrix}(x_1+3x_2)/10\\(3x_1+9x_2)/10\end{pmatrix}$$

Express the action of this transformation with a matrix.

1.10 Example 1.4 suggests that projection breaks $\vec{v}$ into two parts, $\text{proj}_{[\vec{s}\,]}(\vec{v}\,)$ and $\vec{v}-\text{proj}_{[\vec{s}\,]}(\vec{v}\,)$, that are non-interacting. Recall that the two are orthogonal. Show that any two nonzero orthogonal vectors make up a linearly independent set.

1.11 (a) What is the orthogonal projection of $\vec{v}$ into a line if $\vec{v}$ is a member of that line?

(b) Show that if $\vec{v}$ is not a member of the line then the set $\{\vec{v},\vec{v}-\text{proj}_{[\vec{s}\,]}(\vec{v}\,)\}$ is linearly independent.

1.12 Definition 1.1 requires that $\vec{s}$ be nonzero. Why? What is the right definition of the orthogonal projection of a vector into the (degenerate) line spanned by the zero vector?

1.13 Are all vectors the projection of some other vector into some line?

✓ 1.14 Show that the projection of $\vec{v}$ into the line spanned by $\vec{s}$ has length equal to the absolute value of the number $\vec{v}\cdot\vec{s}$ divided by the length of the vector $\vec{s}$.

1.15 Find the formula for the distance from a point to a line.

1.16 Find the scalar c such that the point (cs_1,cs_2) is a minimum distance from the point (v_1,v_2) by using calculus (i.e., consider the distance function, set the first derivative equal to zero, and solve). Generalize to $\mathbb{R}^n$.

✓ 1.17 Prove that the orthogonal projection of a vector into a line is shorter than the vector.

✓ 1.18 Show that the definition of orthogonal projection into a line does not depend on the spanning vector: if $\vec{s}$ is a nonzero multiple of $\vec{q}$ then $(\vec{v}\cdot\vec{s}/\vec{s}\cdot\vec{s}\,)\cdot\vec{s}$ equals $(\vec{v}\cdot\vec{q}/\vec{q}\cdot\vec{q}\,)\cdot\vec{q}$.

✓ 1.19 Consider the function mapping the plane to itself that takes a vector to its projection into the line $y=x$. These two each show that the map is linear, the first one in a way that is coordinate-bound (that is, it fixes a basis and then computes) and the second in a way that is more conceptual.

(a) Produce a matrix that describes the function's action.

(b) Show that we can obtain this map by first rotating everything in the plane $\pi/4$ radians clockwise, then projecting into the x-axis, and then rotating $\pi/4$ radians counterclockwise.

1.20 For $\vec{a},\vec{b}\in\mathbb{R}^n$ let $\vec{v}_1$ be the projection of $\vec{a}$ into the line spanned by $\vec{b}$, let $\vec{v}_2$ be the projection of $\vec{v}_1$ into the line spanned by $\vec{a}$, let $\vec{v}_3$ be the projection of $\vec{v}_2$ into the line spanned by $\vec{b}$, etc., back and forth between the spans of $\vec{a}$ and $\vec{b}$. That is, $\vec{v}_{i+1}$ is the projection of $\vec{v}_i$ into the span of $\vec{a}$ if $i+1$ is even, and into the span of $\vec{b}$ if $i+1$ is odd. Must that sequence of vectors eventually settle down — must there be a sufficiently large i such that $\vec{v}_{i+2}$ equals $\vec{v}_i$ and $\vec{v}_{i+3}$ equals $\vec{v}_{i+1}$? If so, what is the earliest such i?

VI.2 Gram-Schmidt Orthogonalization

The prior subsection suggests that projecting $\vec{v}$ into the line spanned by $\vec{s}$ decomposes that vector into two parts

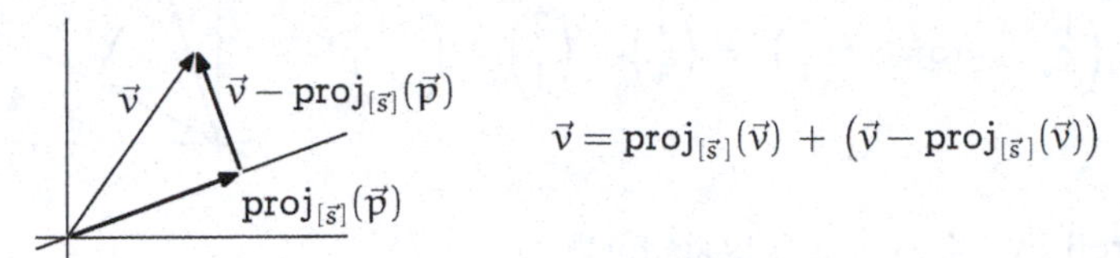

that are orthogonal and so are "non-interacting." We now develop that suggestion.

2.1 Definition Vectors $\vec{v}_1,\dots,\vec{v}_k\in\mathbb{R}^n$ are *mutually orthogonal* when any two are orthogonal: if $i\neq j$ then the dot product $\vec{v}_i\cdot\vec{v}_j$ is zero.

2.2 Theorem If the vectors in a set $\{\vec{v}_1,\dots,\vec{v}_k\}\subset\mathbb{R}^n$ are mutually orthogonal and nonzero then that set is linearly independent.

PROOF Consider $\vec{0}=c_1\vec{v}_1+c_2\vec{v}_2+\cdots+c_k\vec{v}_k$. For $i\in\{1,..,k\}$, taking the dot product of $\vec{v}_i$ with both sides of the equation $\vec{v}_i\cdot(c_1\vec{v}_1+c_2\vec{v}_2+\cdots+c_k\vec{v}_k)=\vec{v}_i\cdot\vec{0}$, which gives $c_i\cdot(\vec{v}_i\cdot\vec{v}_i)=0$, shows that $c_i=0$ since $\vec{v}_i\neq\vec{0}$. QED

2.3 Corollary In a k dimensional vector space, if the vectors in a size k set are mutually orthogonal and nonzero then that set is a basis for the space.

PROOF Any linearly independent size k subset of a k dimensional space is a basis. QED

Of course, the converse of Corollary 2.3 does not hold — not every basis of every subspace of $\mathbb{R}^n$ has mutually orthogonal vectors. However, we can get the partial converse that for every subspace of $\mathbb{R}^n$ there is at least one basis consisting of mutually orthogonal vectors.

2.4 Example The members $\vec{\beta}_1$ and $\vec{\beta}_2$ of this basis for $\mathbb{R}^2$ are not orthogonal.

$$B=\langle\begin{pmatrix}4\\2\end{pmatrix},\begin{pmatrix}1\\3\end{pmatrix}\rangle$$

We will derive from B a new basis for the space $\langle\vec{\kappa}_1,\vec{\kappa}_2\rangle$ consisting of mutually orthogonal vectors. The first member of the new basis is just $\vec{\beta}_1$.

$$\vec{\kappa}_1=\begin{pmatrix}4\\2\end{pmatrix}$$

For the second member of the new basis, we subtract from $\vec{\beta}_2$ the part in the direction of $\vec{\kappa}_1$. This leaves the part of $\vec{\beta}_2$ that is orthogonal to $\vec{\kappa}_1$.

$$\vec{\kappa}_2 = \begin{pmatrix}1\\3\end{pmatrix} - \mathrm{proj}_{[\vec{\kappa}_1]}(\begin{pmatrix}1\\3\end{pmatrix}) = \begin{pmatrix}1\\3\end{pmatrix} - \begin{pmatrix}2\\1\end{pmatrix} = \begin{pmatrix}-1\\2\end{pmatrix}$$

By the corollary $\langle\vec{\kappa}_1, \vec{\kappa}_2\rangle$ is a basis for $\mathbb{R}^2$.

2.5 Definition An *orthogonal basis* for a vector space is a basis of mutually orthogonal vectors.

2.6 Example To produce from this basis for $\mathbb{R}^3$

$$B = \langle \begin{pmatrix}1\\1\\1\end{pmatrix}, \begin{pmatrix}0\\2\\0\end{pmatrix}, \begin{pmatrix}1\\0\\3\end{pmatrix} \rangle$$

an orthogonal basis, start by taking the first vector unchanged.

$$\vec{\kappa}_1 = \begin{pmatrix}1\\1\\1\end{pmatrix}$$

Get $\vec{\kappa}_2$ by subtracting from $\vec{\beta}_2$ its part in the direction of $\vec{\kappa}_1$.

$$\vec{\kappa}_2 = \begin{pmatrix}0\\2\\0\end{pmatrix} - \mathrm{proj}_{[\vec{\kappa}_1]}(\begin{pmatrix}0\\2\\0\end{pmatrix}) = \begin{pmatrix}0\\2\\0\end{pmatrix} - \begin{pmatrix}2/3\\2/3\\2/3\end{pmatrix} = \begin{pmatrix}-2/3\\4/3\\-2/3\end{pmatrix}$$

Find $\vec{\kappa}_3$ by subtracting from $\vec{\beta}_3$ the part in the direction of $\vec{\kappa}_1$ and also the part in the direction of $\vec{\kappa}_2$.

$$\vec{\kappa}_3 = \begin{pmatrix}1\\0\\3\end{pmatrix} - \mathrm{proj}_{[\vec{\kappa}_1]}(\begin{pmatrix}1\\0\\3\end{pmatrix}) - \mathrm{proj}_{[\vec{\kappa}_2]}(\begin{pmatrix}1\\0\\3\end{pmatrix}) = \begin{pmatrix}-1\\0\\1\end{pmatrix}$$

As above, the corollary gives that the result is a basis for $\mathbb{R}^3$.

$$\langle \begin{pmatrix}1\\1\\1\end{pmatrix}, \begin{pmatrix}-2/3\\4/3\\-2/3\end{pmatrix}, \begin{pmatrix}-1\\0\\1\end{pmatrix} \rangle$$

2.7 Theorem (Gram-Schmidt orthogonalization) If $\langle\vec{\beta}_1,\ldots\vec{\beta}_k\rangle$ is a basis for a subspace of $\mathbb{R}^n$ then the vectors

$$\begin{aligned}\vec{\kappa}_1&=\vec{\beta}_1\\ \vec{\kappa}_2&=\vec{\beta}_2-\mbox{proj}_{[\vec{\kappa}_1]}(\vec{\beta}_2)\\ \vec{\kappa}_3&=\vec{\beta}_3-\mbox{proj}_{[\vec{\kappa}_1]}(\vec{\beta}_3)-\mbox{proj}_{[\vec{\kappa}_2]}(\vec{\beta}_3)\\ &\vdots\\ \vec{\kappa}_k&=\vec{\beta}_k-\mbox{proj}_{[\vec{\kappa}_1]}(\vec{\beta}_k)-\cdots-\mbox{proj}_{[\vec{\kappa}_{k-1}]}(\vec{\beta}_k)\end{aligned}$$

form an orthogonal basis for the same subspace.

2.8 Remark This is restricted to $\mathbb{R}^n$ only because we have not given a definition of orthogonality for other spaces.

Proof We will use induction to check that each $\vec{\kappa}_i$ is nonzero, is in the span of $\langle\vec{\beta}_1,\ldots\vec{\beta}_i\rangle$, and is orthogonal to all preceding vectors $\vec{\kappa}_1\cdot\vec{\kappa}_i=\cdots=\vec{\kappa}_{i-1}\cdot\vec{\kappa}_i=0$. Then Corollary 2.3 gives that $\langle\vec{\kappa}_1,\ldots\vec{\kappa}_k\rangle$ is a basis for the same space as is the starting basis.

We shall only cover the cases up to $i=3$, to give the sense of the argument. The full argument is Exercise 25.

The $i=1$ case is trivial; taking $\vec{\kappa}_1$ to be $\vec{\beta}_1$ makes it a nonzero vector since $\vec{\beta}_1$ is a member of a basis, it is obviously in the span of $\langle\vec{\beta}_1\rangle$, and the 'orthogonal to all preceding vectors' condition is satisfied vacuously.

In the $i=2$ case the expansion

$$\vec{\kappa}_2=\vec{\beta}_2-\mbox{proj}_{[\vec{\kappa}_1]}(\vec{\beta}_2)=\vec{\beta}_2-\frac{\vec{\beta}_2\cdot\vec{\kappa}_1}{\vec{\kappa}_1\cdot\vec{\kappa}_1}\cdot\vec{\kappa}_1=\vec{\beta}_2-\frac{\vec{\beta}_2\cdot\vec{\kappa}_1}{\vec{\kappa}_1\cdot\vec{\kappa}_1}\cdot\vec{\beta}_1$$

shows that $\vec{\kappa}_2\neq\vec{0}$ or else this would be a non-trivial linear dependence among the $\vec{\beta}$'s (it is nontrivial because the coefficient of $\vec{\beta}_2$ is 1). It also shows that $\vec{\kappa}_2$ is in the span of $\langle\vec{\beta}_1,\vec{\beta}_2\rangle$. And, $\vec{\kappa}_2$ is orthogonal to the only preceding vector

$$\vec{\kappa}_1\cdot\vec{\kappa}_2=\vec{\kappa}_1\cdot(\vec{\beta}_2-\mbox{proj}_{[\vec{\kappa}_1]}(\vec{\beta}_2))=0$$

because this projection is orthogonal.

The $i=3$ case is the same as the $i=2$ case except for one detail. As in the $i=2$ case, expand the definition.

$$\begin{aligned}\vec{\kappa}_3&=\vec{\beta}_3-\frac{\vec{\beta}_3\cdot\vec{\kappa}_1}{\vec{\kappa}_1\cdot\vec{\kappa}_1}\cdot\vec{\kappa}_1-\frac{\vec{\beta}_3\cdot\vec{\kappa}_2}{\vec{\kappa}_2\cdot\vec{\kappa}_2}\cdot\vec{\kappa}_2\\ &=\vec{\beta}_3-\frac{\vec{\beta}_3\cdot\vec{\kappa}_1}{\vec{\kappa}_1\cdot\vec{\kappa}_1}\cdot\vec{\beta}_1-\frac{\vec{\beta}_3\cdot\vec{\kappa}_2}{\vec{\kappa}_2\cdot\vec{\kappa}_2}\cdot\bigl(\vec{\beta}_2-\frac{\vec{\beta}_2\cdot\vec{\kappa}_1}{\vec{\kappa}_1\cdot\vec{\kappa}_1}\cdot\vec{\beta}_1\bigr)\end{aligned}$$

By the first line $\vec{\kappa}_3 \neq \vec{0}$, since $\vec{\beta}_3$ isn't in the span $[\vec{\beta}_1, \vec{\beta}_2]$ and therefore by the inductive hypothesis it isn't in the span $[\vec{\kappa}_1, \vec{\kappa}_2]$. By the second line $\vec{\kappa}_3$ is in the span of the first three $\vec{\beta}$'s. Finally, the calculation below shows that $\vec{\kappa}_3$ is orthogonal to $\vec{\kappa}_1$.

$$\begin{aligned}\vec{\kappa}_1 \cdot \vec{\kappa}_3 &= \vec{\kappa}_1 \cdot \bigl(\,\vec{\beta}_3 - \mathrm{proj}_{[\vec{\kappa}_1]}(\vec{\beta}_3) - \mathrm{proj}_{[\vec{\kappa}_2]}(\vec{\beta}_3)\,\bigr) \\ &= \vec{\kappa}_1 \cdot \bigl(\vec{\beta}_3 - \mathrm{proj}_{[\vec{\kappa}_1]}(\vec{\beta}_3)\bigr) - \vec{\kappa}_1 \cdot \mathrm{proj}_{[\vec{\kappa}_2]}(\vec{\beta}_3) \\ &= 0\end{aligned}$$

(Here is the difference with the $i = 2$ case: as happened for $i = 2$ the first term is 0 because this projection is orthogonal, but here the second term in the second line is 0 because $\vec{\kappa}_1$ is orthogonal to $\vec{\kappa}_2$ and so is orthogonal to any vector in the line spanned by $\vec{\kappa}_2$.) A similar check shows that $\vec{\kappa}_3$ is also orthogonal to $\vec{\kappa}_2$. QED

In addition to having the vectors in the basis be orthogonal, we can also *normalize* each vector by dividing by its length, to end with an *orthonormal basis.*.

2.9 Example From the orthogonal basis of Example 2.6, normalizing produces this orthonormal basis.

$$\langle \begin{pmatrix} 1/\sqrt{3} \\ 1/\sqrt{3} \\ 1/\sqrt{3} \end{pmatrix}, \begin{pmatrix} -1/\sqrt{6} \\ 2/\sqrt{6} \\ -1/\sqrt{6} \end{pmatrix}, \begin{pmatrix} -1/\sqrt{2} \\ 0 \\ 1/\sqrt{2} \end{pmatrix} \rangle$$

Besides its intuitive appeal, and its analogy with the standard basis $\mathcal{E}_n$ for $\mathbb{R}^n$, an orthonormal basis also simplifies some computations. Exercise 19 is an example.

Exercises

2.10 Perform the Gram-Schmidt process on each of these bases for $\mathbb{R}^2$.

(a) $\langle \begin{pmatrix} 1 \\ 1 \end{pmatrix}, \begin{pmatrix} 2 \\ 1 \end{pmatrix} \rangle$ (b) $\langle \begin{pmatrix} 0 \\ 1 \end{pmatrix}, \begin{pmatrix} -1 \\ 3 \end{pmatrix} \rangle$ (c) $\langle \begin{pmatrix} 0 \\ 1 \end{pmatrix}, \begin{pmatrix} -1 \\ 0 \end{pmatrix} \rangle$

Then turn those orthogonal bases into orthonormal bases.

✓ 2.11 Perform the Gram-Schmidt process on each of these bases for $\mathbb{R}^3$.

(a) $\langle \begin{pmatrix} 2 \\ 2 \\ 2 \end{pmatrix}, \begin{pmatrix} 1 \\ 0 \\ -1 \end{pmatrix}, \begin{pmatrix} 0 \\ 3 \\ 1 \end{pmatrix} \rangle$ (b) $\langle \begin{pmatrix} 1 \\ -1 \\ 0 \end{pmatrix}, \begin{pmatrix} 0 \\ 1 \\ 0 \end{pmatrix}, \begin{pmatrix} 2 \\ 3 \\ 1 \end{pmatrix} \rangle$

Then turn those orthogonal bases into orthonormal bases.

✓ 2.12 Find an orthonormal basis for this subspace of $\mathbb{R}^3$: the plane $x - y + z = 0$.

2.13 Find an orthonormal basis for this subspace of $\mathbb{R}^4$.

$$\{ \begin{pmatrix} x \\ y \\ z \\ w \end{pmatrix} \mid x - y - z + w = 0 \text{ and } x + z = 0 \}$$

2.14 Show that any linearly independent subset of $\mathbb{R}^n$ can be orthogonalized without changing its span.

2.15 What happens if we try to apply the Gram-Schmidt process to a finite set that is not a basis?

✓ **2.16** What happens if we apply the Gram-Schmidt process to a basis that is already orthogonal?

2.17 Let $\langle \vec{\kappa}_1, \ldots, \vec{\kappa}_k \rangle$ be a set of mutually orthogonal vectors in $\mathbb{R}^n$.

(a) Prove that for any $\vec{v}$ in the space, the vector $\vec{v} - (\text{proj}_{[\vec{\kappa}_1]}(\vec{v}) + \cdots + \text{proj}_{[\vec{v}_k]}(\vec{v}))$ is orthogonal to each of $\vec{\kappa}_1, \ldots, \vec{\kappa}_k$.

(b) Illustrate the prior item in $\mathbb{R}^3$ by using $\vec{e}_1$ as $\vec{\kappa}_1$, using $\vec{e}_2$ as $\vec{\kappa}_2$, and taking $\vec{v}$ to have components 1, 2, and 3.

(c) Show that $\text{proj}_{[\vec{\kappa}_1]}(\vec{v}) + \cdots + \text{proj}_{[\vec{v}_k]}(\vec{v})$ is the vector in the span of the set of $\vec{\kappa}$'s that is closest to $\vec{v}$. *Hint.* To the illustration done for the prior part, add a vector $d_1\vec{\kappa}_1 + d_2\vec{\kappa}_2$ and apply the Pythagorean Theorem to the resulting triangle.

2.18 Find a nonzero vector in $\mathbb{R}^3$ that is orthogonal to both of these.

$$\begin{pmatrix} 1 \\ 5 \\ -1 \end{pmatrix} \quad \begin{pmatrix} 2 \\ 2 \\ 0 \end{pmatrix}$$

✓ **2.19** One advantage of orthogonal bases is that they simplify finding the representation of a vector with respect to that basis.

(a) For this vector and this non-orthogonal basis for $\mathbb{R}^2$

$$\vec{v} = \begin{pmatrix} 2 \\ 3 \end{pmatrix} \qquad B = \langle \begin{pmatrix} 1 \\ 1 \end{pmatrix}, \begin{pmatrix} 1 \\ 0 \end{pmatrix} \rangle$$

first represent the vector with respect to the basis. Then project the vector into the span of each basis vector $[\vec{\beta}_1]$ and $[\vec{\beta}_2]$.

(b) With this orthogonal basis for $\mathbb{R}^2$

$$K = \langle \begin{pmatrix} 1 \\ 1 \end{pmatrix}, \begin{pmatrix} 1 \\ -1 \end{pmatrix} \rangle$$

represent the same vector $\vec{v}$ with respect to the basis. Then project the vector into the span of each basis vector. Note that the coefficients in the representation and the projection are the same.

(c) Let $K = \langle \vec{\kappa}_1, \ldots, \vec{\kappa}_k \rangle$ be an orthogonal basis for some subspace of $\mathbb{R}^n$. Prove that for any $\vec{v}$ in the subspace, the i-th component of the representation $\text{Rep}_K(\vec{v})$ is the scalar coefficient $(\vec{v} \cdot \vec{\kappa}_i)/(\vec{\kappa}_i \cdot \vec{\kappa}_i)$ from $\text{proj}_{[\vec{\kappa}_i]}(\vec{v})$.

(d) Prove that $\vec{v} = \text{proj}_{[\vec{\kappa}_1]}(\vec{v}) + \cdots + \text{proj}_{[\vec{\kappa}_k]}(\vec{v})$.

2.20 *Bessel's Inequality.* Consider these orthonormal sets

$$B_1 = \{\vec{e}_1\} \quad B_2 = \{\vec{e}_1, \vec{e}_2\} \quad B_3 = \{\vec{e}_1, \vec{e}_2, \vec{e}_3\} \quad B_4 = \{\vec{e}_1, \vec{e}_2, \vec{e}_3, \vec{e}_4\}$$

along with the vector $\vec{v} \in \mathbb{R}^4$ whose components are 4, 3, 2, and 1.

(a) Find the coefficient c_1 for the projection of $\vec{v}$ into the span of the vector in B_1. Check that $\|\vec{v}\|^2 \geqslant |c_1|^2$.

(b) Find the coefficients c_1 and c_2 for the projection of $\vec{v}$ into the spans of the two vectors in B_2. Check that $\|\vec{v}\|^2 \geqslant |c_1|^2 + |c_2|^2$.

(c) Find c_1, c_2, and c_3 associated with the vectors in B_3, and c_1, c_2, c_3, and c_4 for the vectors in B_4. Check that $\|\vec{v}\|^2 \geqslant |c_1|^2 + \cdots + |c_3|^2$ and that $\|\vec{v}\|^2 \geqslant |c_1|^2 + \cdots + |c_4|^2$.

Show that this holds in general: where $\{\vec{\kappa}_1, \ldots, \vec{\kappa}_k\}$ is an orthonormal set and c_i is coefficient of the projection of a vector $\vec{v}$ from the space then $\|\vec{v}\|^2 \geqslant |c_1|^2 + \cdots + |c_k|^2$. *Hint.* One way is to look at the inequality $0 \leqslant \|\vec{v} - (c_1\vec{\kappa}_1 + \cdots + c_k\vec{\kappa}_k)\|^2$ and expand the c's.

2.21 Prove or disprove: every vector in $\mathbb{R}^n$ is in some orthogonal basis.

2.22 Show that the columns of an $n \times n$ matrix form an orthonormal set if and only if the inverse of the matrix is its transpose. Produce such a matrix.

2.23 Does the proof of Theorem 2.2 fail to consider the possibility that the set of vectors is empty (i.e., that $k = 0$)?

2.24 Theorem 2.7 describes a change of basis from any basis $B = \langle \vec{\beta}_1, \ldots, \vec{\beta}_k \rangle$ to one that is orthogonal $K = \langle \vec{\kappa}_1, \ldots, \vec{\kappa}_k \rangle$. Consider the change of basis matrix $\text{Rep}_{B,K}(\text{id})$.

(a) Prove that the matrix $\text{Rep}_{K,B}(\text{id})$ changing bases in the direction opposite to that of the theorem has an upper triangular shape — all of its entries below the main diagonal are zeros.

(b) Prove that the inverse of an upper triangular matrix is also upper triangular (if the matrix is invertible, that is). This shows that the matrix $\text{Rep}_{B,K}(\text{id})$ changing bases in the direction described in the theorem is upper triangular.

2.25 Complete the induction argument in the proof of Theorem 2.7.

VI.3 Projection Into a Subspace

This subsection uses material from the optional earlier subsection on Combining Subspaces.

The prior subsections project a vector into a line by decomposing it into two parts: the part in the line $\text{proj}_{[\vec{s}]}(\vec{v})$ and the rest $\vec{v} - \text{proj}_{[\vec{s}]}(\vec{v})$. To generalize projection to arbitrary subspaces we will follow this decomposition idea.

3.1 Definition Let a vector space be a direct sum $V = M \oplus N$. Then for any $\vec{v} \in V$ with $\vec{v} = \vec{m} + \vec{n}$ where $\vec{m} \in M$, $\vec{n} \in N$, the *projection of* $\vec{v}$ *into* M *along* N is $\text{proj}_{M,N}(\vec{v}) = \vec{m}$.

This definition applies in spaces where we don't have a ready definition of orthogonal. (Definitions of orthogonality for spaces other than the $\mathbb{R}^n$ are perfectly possible but we haven't seen any in this book.)

3.2 Example The space $\mathcal{M}_{2\times 2}$ of 2×2 matrices is the direct sum of these two.

$$M=\{\begin{pmatrix}a&b\\0&0\end{pmatrix}\mid a,b\in\mathbb{R}\}\qquad N=\{\begin{pmatrix}0&0\\c&d\end{pmatrix}\mid c,d\in\mathbb{R}\}$$

To project

$$A=\begin{pmatrix}3&1\\0&4\end{pmatrix}$$

into M along N, we first fix bases for the two subspaces.

$$B_M=\langle\begin{pmatrix}1&0\\0&0\end{pmatrix},\begin{pmatrix}0&1\\0&0\end{pmatrix}\rangle\qquad B_N=\langle\begin{pmatrix}0&0\\1&0\end{pmatrix},\begin{pmatrix}0&0\\0&1\end{pmatrix}\rangle$$

Their concatenation

$$B=B_M\frown B_N=\langle\begin{pmatrix}1&0\\0&0\end{pmatrix},\begin{pmatrix}0&1\\0&0\end{pmatrix},\begin{pmatrix}0&0\\1&0\end{pmatrix},\begin{pmatrix}0&0\\0&1\end{pmatrix}\rangle$$

is a basis for the entire space because $\mathcal{M}_{2\times 2}$ is the direct sum. So we can use it to represent A.

$$\begin{pmatrix}3&1\\0&4\end{pmatrix}=3\cdot\begin{pmatrix}1&0\\0&0\end{pmatrix}+1\cdot\begin{pmatrix}0&1\\0&0\end{pmatrix}+0\cdot\begin{pmatrix}0&0\\1&0\end{pmatrix}+4\cdot\begin{pmatrix}0&0\\0&1\end{pmatrix}$$

The projection of A into M along N keeps the M part and drops the N part.

$$\text{proj}_{M,N}(\begin{pmatrix}3&1\\0&4\end{pmatrix})=3\cdot\begin{pmatrix}1&0\\0&0\end{pmatrix}+1\cdot\begin{pmatrix}0&1\\0&0\end{pmatrix}=\begin{pmatrix}3&1\\0&0\end{pmatrix}$$

3.3 Example Both subscripts on $\text{proj}_{M,N}(\vec{v})$ are significant. The first subscript M matters because the result of the projection is a member of M. For an example showing that the second one matters, fix this plane subspace of $\mathbb{R}^3$ and its basis.

$$M=\{\begin{pmatrix}x\\y\\z\end{pmatrix}\mid y-2z=0\}\qquad B_M=\langle\begin{pmatrix}1\\0\\0\end{pmatrix},\begin{pmatrix}0\\2\\1\end{pmatrix}\rangle$$

We will compare the projections of this element of $\mathbb{R}^3$

$$\vec{v}=\begin{pmatrix}2\\2\\5\end{pmatrix}$$

into M along these two subspaces (verification that $\mathbb{R}^3 = M \oplus N$ and $\mathbb{R}^3 = M \oplus \hat{N}$ is routine).

$$N = \{k \begin{pmatrix} 0 \\ 0 \\ 1 \end{pmatrix} \mid k \in \mathbb{R}\} \qquad \hat{N} = \{k \begin{pmatrix} 0 \\ 1 \\ -2 \end{pmatrix} \mid k \in \mathbb{R}\}$$

Here are natural bases for N and $\hat{N}$.

$$B_N = \langle \begin{pmatrix} 0 \\ 0 \\ 1 \end{pmatrix} \rangle \qquad B_{\hat{N}} = \langle \begin{pmatrix} 0 \\ 1 \\ -2 \end{pmatrix} \rangle$$

To project into M along N, represent $\vec{v}$ with respect to the concatenation $B_M \frown B_N$

$$\begin{pmatrix} 2 \\ 2 \\ 5 \end{pmatrix} = 2 \cdot \begin{pmatrix} 1 \\ 0 \\ 0 \end{pmatrix} + 1 \cdot \begin{pmatrix} 0 \\ 2 \\ 1 \end{pmatrix} + 4 \cdot \begin{pmatrix} 0 \\ 0 \\ 1 \end{pmatrix}$$

and drop the N term.

$$\text{proj}_{M,N}(\vec{v}) = 2 \cdot \begin{pmatrix} 1 \\ 0 \\ 0 \end{pmatrix} + 1 \cdot \begin{pmatrix} 0 \\ 2 \\ 1 \end{pmatrix} = \begin{pmatrix} 2 \\ 2 \\ 1 \end{pmatrix}$$

To project into M along $\hat{N}$ represent $\vec{v}$ with respect to $B_M \frown B_{\hat{N}}$

$$\begin{pmatrix} 2 \\ 2 \\ 5 \end{pmatrix} = 2 \cdot \begin{pmatrix} 1 \\ 0 \\ 0 \end{pmatrix} + (9/5) \cdot \begin{pmatrix} 0 \\ 2 \\ 1 \end{pmatrix} - (8/5) \cdot \begin{pmatrix} 0 \\ 1 \\ -2 \end{pmatrix}$$

and omit the $\hat{N}$ part.

$$\text{proj}_{M,\hat{N}}(\vec{v}) = 2 \cdot \begin{pmatrix} 1 \\ 0 \\ 0 \end{pmatrix} + (9/5) \cdot \begin{pmatrix} 0 \\ 2 \\ 1 \end{pmatrix} = \begin{pmatrix} 2 \\ 18/5 \\ 9/5 \end{pmatrix}$$

So projecting along different subspaces can give different results.

These pictures compare the two maps. Both show that the projection is indeed 'into' the plane and 'along' the line.

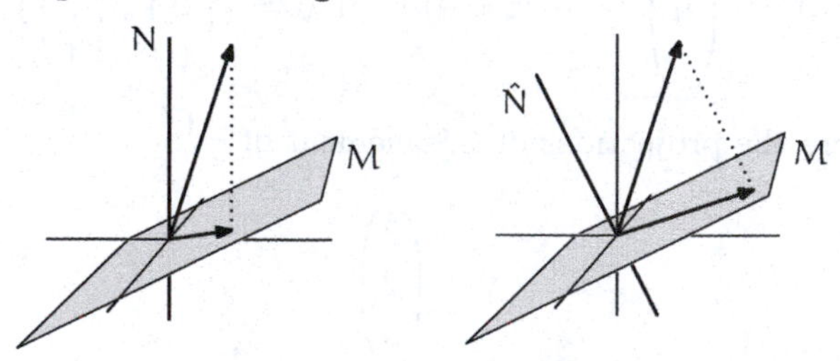

Notice that the projection along N is not orthogonal since there are members of the plane M that are not orthogonal to the dotted line. But the projection along $\hat{N}$ is orthogonal.

We have seen two projection operations, orthogonal projection into a line as well as this subsections's projection into an M and along an N, and we naturally ask whether they are related. The right-hand picture above suggests the answer — orthogonal projection into a line is a special case of this subsection's projection; it is projection along a subspace perpendicular to the line.

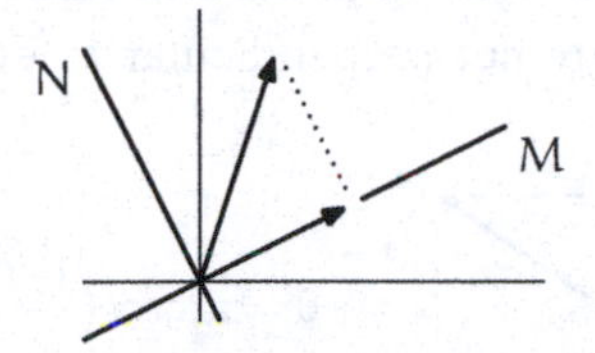

3.4 Definition The *orthogonal complement* of a subspace M of $\mathbb{R}^n$ is

$$M^{\perp} = \{\vec{v} \in \mathbb{R}^n \mid \vec{v} \text{ is perpendicular to all vectors in } M\}$$

(read "M perp"). The *orthogonal projection* $\text{proj}_M(\vec{v})$ of a vector is its projection into M along $M^{\perp}$.

3.5 Example In $\mathbb{R}^3$, to find the orthogonal complement of the plane

$$P = \{\begin{pmatrix} x \\ y \\ z \end{pmatrix} \mid 3x + 2y - z = 0\}$$

we start with a basis for P.

$$B = \langle \begin{pmatrix} 1 \\ 0 \\ 3 \end{pmatrix}, \begin{pmatrix} 0 \\ 1 \\ 2 \end{pmatrix} \rangle$$

Any $\vec{v}$ perpendicular to every vector in B is perpendicular to every vector in the span of B (the proof of this is Exercise 19). Therefore, the subspace $P^{\perp}$ consists of the vectors that satisfy these two conditions.

$$\begin{pmatrix} 1 \\ 0 \\ 3 \end{pmatrix} \cdot \begin{pmatrix} v_1 \\ v_2 \\ v_3 \end{pmatrix} = 0 \qquad \begin{pmatrix} 0 \\ 1 \\ 2 \end{pmatrix} \cdot \begin{pmatrix} v_1 \\ v_2 \\ v_3 \end{pmatrix} = 0$$

Those conditions give a linear system.

$$P^{\perp} = \{\begin{pmatrix} v_1 \\ v_2 \\ v_3 \end{pmatrix} \mid \begin{pmatrix} 1 & 0 & 3 \\ 0 & 1 & 2 \end{pmatrix} \begin{pmatrix} v_1 \\ v_2 \\ v_3 \end{pmatrix} = \begin{pmatrix} 0 \\ 0 \end{pmatrix}\}$$

We are thus left with finding the null space of the map represented by the matrix, that is, with calculating the solution set of the homogeneous linear system.

$$\begin{aligned} v_1 \quad\; + 3v_3 &= 0 \\ v_2 + 2v_3 &= 0 \end{aligned} \quad\Longrightarrow\quad P^{\perp} = \{k\begin{pmatrix} -3 \\ -2 \\ 1 \end{pmatrix} \mid k \in \mathbb{R}\}$$

3.6 Example Where M is the xy-plane subspace of $\mathbb{R}^3$, what is $M^{\perp}$? A common first reaction is that $M^{\perp}$ is the yz-plane but that's not right because some vectors from the yz-plane are not perpendicular to every vector in the xy-plane.

$$\begin{pmatrix} 1 \\ 1 \\ 0 \end{pmatrix} \not\perp \begin{pmatrix} 0 \\ 3 \\ 2 \end{pmatrix} \qquad \theta = \arccos(\frac{1\cdot 0 + 1\cdot 3 + 0\cdot 2}{\sqrt{2}\cdot\sqrt{13}}) \approx 0.94 \text{ rad}$$

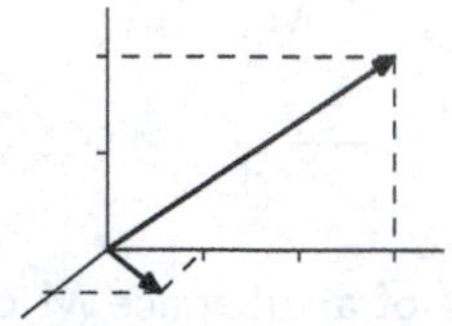

Instead $M^{\perp}$ is the z-axis, since proceeding as in the prior example and taking the natural basis for the xy-plane gives this.

$$M^{\perp} = \{\begin{pmatrix} x \\ y \\ z \end{pmatrix} \mid \begin{pmatrix} 1 & 0 & 0 \\ 0 & 1 & 0 \end{pmatrix}\begin{pmatrix} x \\ y \\ z \end{pmatrix} = \begin{pmatrix} 0 \\ 0 \end{pmatrix}\} = \{\begin{pmatrix} x \\ y \\ z \end{pmatrix} \mid x = 0 \text{ and } y = 0\}$$

3.7 Lemma If M is a subspace of $\mathbb{R}^n$ then its orthogonal complement $M^{\perp}$ is also a subspace. The space is the direct sum of the two $\mathbb{R}^n = M \oplus M^{\perp}$. And, for any $\vec{v} \in \mathbb{R}^n$ the vector $\vec{v} - \text{proj}_M(\vec{v})$ is perpendicular to every vector in M.

PROOF First, the orthogonal complement $M^{\perp}$ is a subspace of $\mathbb{R}^n$ because, as noted in the prior two examples, it is a null space.

Next, to show that the space is the direct sum of the two, start with any basis $B_M = \langle \vec{\mu}_1, \dots, \vec{\mu}_k \rangle$ for M and expand it to a basis for the entire space. Apply the Gram-Schmidt process to get an orthogonal basis $K = \langle \vec{\kappa}_1, \dots, \vec{\kappa}_n \rangle$ for $\mathbb{R}^n$. This K is the concatenation of two bases: $\langle \vec{\kappa}_1, \dots, \vec{\kappa}_k \rangle$ with the same number of members k as B_M, and $\langle \vec{\kappa}_{k+1}, \dots, \vec{\kappa}_n \rangle$. The first is a basis for M so if we show that the second is a basis for $M^{\perp}$ then we will have that the entire space is the direct sum.

Exercise 19 from the prior subsection proves this about any orthogonal basis: each vector $\vec{v}$ in the space is the sum of its orthogonal projections into the lines spanned by the basis vectors.

$$\vec{v} = \text{proj}_{[\vec{\kappa}_1]}(\vec{v}) + \cdots + \text{proj}_{[\vec{\kappa}_n]}(\vec{v}) \qquad (*)$$

To check this, represent the vector as $\vec{v} = r_1\vec{\kappa}_1 + \cdots + r_n\vec{\kappa}_n$, apply $\vec{\kappa}_i$ to both sides $\vec{v} \cdot \vec{\kappa}_i = (r_1\vec{\kappa}_1 + \cdots + r_n\vec{\kappa}_n) \cdot \vec{\kappa}_i = r_1 \cdot 0 + \cdots + r_i \cdot (\vec{\kappa}_i \cdot \vec{\kappa}_i) + \cdots + r_n \cdot 0$, and solve to get $r_i = (\vec{v} \cdot \vec{\kappa}_i)/(\vec{\kappa}_i \cdot \vec{\kappa}_i)$, as desired.

Since obviously any member of the span of $\langle \vec{\kappa}_{k+1}, \ldots, \vec{\kappa}_n \rangle$ is orthogonal to any vector in M, to show that this is a basis for $M^\perp$ we need only show the other containment — that any $\vec{w} \in M^\perp$ is in the span of this basis. The prior paragraph does this. Any $\vec{w} \in M^\perp$ gives this on projections into basis vectors from M: $\text{proj}_{[\vec{\kappa}_1]}(\vec{w}) = \vec{0}, \ldots, \text{proj}_{[\vec{\kappa}_k]}(\vec{w}) = \vec{0}$. Therefore equation $(*)$ gives that $\vec{w}$ is a linear combination of $\vec{\kappa}_{k+1}, \ldots, \vec{\kappa}_n$. Thus this is a basis for $M^\perp$ and $\mathbb{R}^n$ is the direct sum of the two.

The final sentence of the lemma is proved in much the same way. Write $\vec{v} = \text{proj}_{[\vec{\kappa}_1]}(\vec{v}) + \cdots + \text{proj}_{[\vec{\kappa}_n]}(\vec{v})$. Then $\text{proj}_M(\vec{v})$ keeps only the M part and drops the $M^\perp$ part $\text{proj}_M(\vec{v}) = \text{proj}_{[\vec{\kappa}_{k+1}]}(\vec{v}) + \cdots + \text{proj}_{[\vec{\kappa}_k]}(\vec{v})$. Therefore $\vec{v} - \text{proj}_M(\vec{v})$ consists of a linear combination of elements of $M^\perp$ and so is perpendicular to every vector in M. QED

Given a subspace, we could compute the orthogonal projection into that subspace by following the steps of that proof: finding a basis, expanding it to a basis for the entire space, applying Gram-Schmidt to get an orthogonal basis, and projecting into each linear subspace. However we will instead use a convenient formula.

3.8 Theorem Let M be a subspace of $\mathbb{R}^n$ with basis $\langle \vec{\beta}_1, \ldots, \vec{\beta}_k \rangle$ and let A be the matrix whose columns are the $\vec{\beta}$'s. Then for any $\vec{v} \in \mathbb{R}^n$ the orthogonal projection is $\text{proj}_M(\vec{v}) = c_1\vec{\beta}_1 + \cdots + c_k\vec{\beta}_k$, where the coefficients c_i are the entries of the vector $(A^T A)^{-1} A^T \cdot \vec{v}$. That is, $\text{proj}_M(\vec{v}) = A(A^T A)^{-1} A^T \cdot \vec{v}$.

Proof The vector $\text{proj}_M(\vec{v})$ is a member of M and so is a linear combination of basis vectors $c_1 \cdot \vec{\beta}_1 + \cdots + c_k \cdot \vec{\beta}_k$. Since A's columns are the $\vec{\beta}$'s, there is a $\vec{c} \in \mathbb{R}^k$ such that $\text{proj}_M(\vec{v}) = A\vec{c}$. To find $\vec{c}$ note that the vector $\vec{v} - \text{proj}_M(\vec{v})$ is perpendicular to each member of the basis so

$$\vec{0} = A^T(\vec{v} - A\vec{c}) = A^T\vec{v} - A^T A\vec{c}$$

and solving gives this (showing that $A^T A$ is invertible is an exercise).

$$\vec{c} = (A^T A)^{-1} A^T \cdot \vec{v}$$

Therefore $\text{proj}_M(\vec{v}) = A \cdot \vec{c} = A(A^T A)^{-1} A^T \cdot \vec{v}$, as required. QED

3.9 Example To orthogonally project this vector into this subspace

$$\vec{v} = \begin{pmatrix} 1 \\ -1 \\ 1 \end{pmatrix} \qquad P = \{ \begin{pmatrix} x \\ y \\ z \end{pmatrix} \mid x + z = 0 \}$$

first make a matrix whose columns are a basis for the subspace

$$A=\begin{pmatrix}0&1\\1&0\\0&-1\end{pmatrix}$$

and then compute.

$$\begin{aligned}A(A^{\mathsf T}A)^{-1}A^{\mathsf T}&=\begin{pmatrix}0&1\\1&0\\0&-1\end{pmatrix}\begin{pmatrix}1&0\\0&1/2\end{pmatrix}\begin{pmatrix}0&1&0\\1&0&-1\end{pmatrix}\\&=\begin{pmatrix}1/2&0&-1/2\\0&1&0\\-1/2&0&1/2\end{pmatrix}\end{aligned}$$

With the matrix, calculating the orthogonal projection of any vector into P is easy.

$$\mathrm{proj}_P(\vec v)=\begin{pmatrix}1/2&0&-1/2\\0&1&0\\-1/2&0&1/2\end{pmatrix}\begin{pmatrix}1\\-1\\1\end{pmatrix}=\begin{pmatrix}0\\-1\\0\end{pmatrix}$$

Note, as a check, that this result is indeed in P.

Exercises

✓ **3.10** Project the vectors into M along N.

(a) $\begin{pmatrix}3\\-2\end{pmatrix}$, $M=\{\begin{pmatrix}x\\y\end{pmatrix}\mid x+y=0\}$, $N=\{\begin{pmatrix}x\\y\end{pmatrix}\mid -x-2y=0\}$

(b) $\begin{pmatrix}1\\2\end{pmatrix}$, $M=\{\begin{pmatrix}x\\y\end{pmatrix}\mid x-y=0\}$, $N=\{\begin{pmatrix}x\\y\end{pmatrix}\mid 2x+y=0\}$

(c) $\begin{pmatrix}3\\0\\1\end{pmatrix}$, $M=\{\begin{pmatrix}x\\y\\z\end{pmatrix}\mid x+y=0\}$, $N=\{c\cdot\begin{pmatrix}1\\0\\1\end{pmatrix}\mid c\in\mathbb{R}\}$

✓ **3.11** Find $M^{\perp}$.

(a) $M=\{\begin{pmatrix}x\\y\end{pmatrix}\mid x+y=0\}$ (b) $M=\{\begin{pmatrix}x\\y\end{pmatrix}\mid -2x+3y=0\}$

(c) $M=\{\begin{pmatrix}x\\y\end{pmatrix}\mid x-y=0\}$ (d) $M=\{\vec 0\}$ (e) $M=\{\begin{pmatrix}x\\y\end{pmatrix}\mid x=0\}$

(f) $M=\{\begin{pmatrix}x\\y\\z\end{pmatrix}\mid -x+3y+z=0\}$ (g) $M=\{\begin{pmatrix}x\\y\\z\end{pmatrix}\mid x=0 \text{ and } y+z=0\}$

3.12 This subsection shows how to project orthogonally in two ways, the method of Example 3.2 and 3.3, and the method of Theorem 3.8. To compare them, consider the plane P specified by $3x+2y-z=0$ in $\mathbb{R}^3$.

(a) Find a basis for P.

(b) Find $P^{\perp}$ and a basis for $P^{\perp}$.

(c) Represent this vector with respect to the concatenation of the two bases from the prior item.

$$\vec{v} = \begin{pmatrix} 1 \\ 1 \\ 2 \end{pmatrix}$$

(d) Find the orthogonal projection of $\vec{v}$ into P by keeping only the P part from the prior item.

(e) Check that against the result from applying Theorem 3.8.

✓ **3.13** We have three ways to find the orthogonal projection of a vector into a line, the Definition 1.1 way from the first subsection of this section, the Example 3.2 and 3.3 way of representing the vector with respect to a basis for the space and then keeping the M part, and the way of Theorem 3.8. For these cases, do all three ways.

(a) $\vec{v} = \begin{pmatrix} 1 \\ -3 \end{pmatrix}$, $M = \{\begin{pmatrix} x \\ y \end{pmatrix} \mid x + y = 0\}$

(b) $\vec{v} = \begin{pmatrix} 0 \\ 1 \\ 2 \end{pmatrix}$, $M = \{\begin{pmatrix} x \\ y \\ z \end{pmatrix} \mid x + z = 0 \text{ and } y = 0\}$

3.14 Check that the operation of Definition 3.1 is well-defined. That is, in Example 3.2 and 3.3, doesn't the answer depend on the choice of bases?

3.15 What is the orthogonal projection into the trivial subspace?

3.16 What is the projection of $\vec{v}$ into M along N if $\vec{v} \in M$?

3.17 Show that if $M \subseteq \mathbb{R}^n$ is a subspace with orthonormal basis $\langle \vec{\kappa}_1, \ldots, \vec{\kappa}_n \rangle$ then the orthogonal projection of $\vec{v}$ into M is this.

$$(\vec{v} \bullet \vec{\kappa}_1) \cdot \vec{\kappa}_1 + \cdots + (\vec{v} \bullet \vec{\kappa}_n) \cdot \vec{\kappa}_n$$

✓ **3.18** Prove that the map $p\colon V \to V$ is the projection into M along N if and only if the map $\mathrm{id} - p$ is the projection into N along M. (Recall the definition of the difference of two maps: $(\mathrm{id} - p)(\vec{v}) = \mathrm{id}(\vec{v}) - p(\vec{v}) = \vec{v} - p(\vec{v})$.)

✓ **3.19** Show that if a vector is perpendicular to every vector in a set then it is perpendicular to every vector in the span of that set.

3.20 True or false: the intersection of a subspace and its orthogonal complement is trivial.

3.21 Show that the dimensions of orthogonal complements add to the dimension of the entire space.

✓ **3.22** Suppose that $\vec{v}_1, \vec{v}_2 \in \mathbb{R}^n$ are such that for all complements $M, N \subseteq \mathbb{R}^n$, the projections of $\vec{v}_1$ and $\vec{v}_2$ into M along N are equal. Must $\vec{v}_1$ equal $\vec{v}_2$? (If so, what if we relax the condition to: all orthogonal projections of the two are equal?)

✓ **3.23** Let M, N be subspaces of $\mathbb{R}^n$. The perp operator acts on subspaces; we can ask how it interacts with other such operations.

(a) Show that two perps cancel: $(M^{\perp})^{\perp} = M$.

(b) Prove that $M \subseteq N$ implies that $N^{\perp} \subseteq M^{\perp}$.

(c) Show that $(M + N)^{\perp} = M^{\perp} \cap N^{\perp}$.

✓ **3.24** The material in this subsection allows us to express a geometric relationship that we have not yet seen between the range space and the null space of a linear map.

(a) Represent $f\colon \mathbb{R}^3 \to \mathbb{R}$ given by

$$\begin{pmatrix} v_1 \\ v_2 \\ v_3 \end{pmatrix} \mapsto 1v_1 + 2v_2 + 3v_3$$

with respect to the standard bases and show that

$$\begin{pmatrix} 1 \\ 2 \\ 3 \end{pmatrix}$$

is a member of the perp of the null space. Prove that $\mathscr{N}(f)^{\perp}$ is equal to the span of this vector.

(b) Generalize that to apply to any $f\colon \mathbb{R}^n \to \mathbb{R}$.

(c) Represent $f\colon \mathbb{R}^3 \to \mathbb{R}^2$

$$\begin{pmatrix} v_1 \\ v_2 \\ v_3 \end{pmatrix} \mapsto \begin{pmatrix} 1v_1 + 2v_2 + 3v_3 \\ 4v_1 + 5v_2 + 6v_3 \end{pmatrix}$$

with respect to the standard bases and show that

$$\begin{pmatrix} 1 \\ 2 \\ 3 \end{pmatrix}, \begin{pmatrix} 4 \\ 5 \\ 6 \end{pmatrix}$$

are both members of the perp of the null space. Prove that $\mathscr{N}(f)^{\perp}$ is the span of these two. (*Hint.* See the third item of Exercise 23.)

(d) Generalize that to apply to any $f\colon \mathbb{R}^n \to \mathbb{R}^m$.

In [Strang 93] this is called the *Fundamental Theorem of Linear Algebra*

3.25 Define a *projection* to be a linear transformation $t\colon V \to V$ with the property that repeating the projection does nothing more than does the projection alone: $(t \circ t)(\vec{v}) = t(\vec{v})$ for all $\vec{v} \in V$.

(a) Show that orthogonal projection into a line has that property.

(b) Show that projection along a subspace has that property.

(c) Show that for any such t there is a basis $B = \langle \vec{\beta}_1, \dots, \vec{\beta}_n \rangle$ for V such that

$$t(\vec{\beta}_i) = \begin{cases} \vec{\beta}_i & i = 1, 2, \dots, r \\ \vec{0} & i = r+1, r+2, \dots, n \end{cases}$$

where r is the rank of t.

(d) Conclude that every projection is a projection along a subspace.

(e) Also conclude that every projection has a representation

$$\text{Rep}_{B,B}(t) = \left(\begin{array}{c|c} I & Z \\ \hline Z & Z \end{array}\right)$$

in block partial-identity form.

3.26 A square matrix is *symmetric* if each i, j entry equals the j, i entry (i.e., if the matrix equals its transpose). Show that the projection matrix $A(A^{\mathsf{T}}A)^{-1}A^{\mathsf{T}}$ is symmetric. [Strang 80] *Hint.* Find properties of transposes by looking in the index under 'transpose'.

Topic

Line of Best Fit

This Topic requires the formulas from the subsections on Orthogonal Projection Into a Line and Projection Into a Subspace.

Scientists are often presented with a system that has no solution and they must find an answer anyway. More precisely, they must find a best answer. For instance, this is the result of flipping a penny, including some intermediate numbers.

number of flips	30	60	90
number of heads	16	34	51

Because of the randomness in this experiment we expect that the ratio of heads to flips will fluctuate around a penny's long-term ratio of 50-50. So the system for such an experiment likely has no solution, and that's what happened here.

$$\begin{aligned} 30m &= 16 \\ 60m &= 34 \\ 90m &= 51 \end{aligned}$$

That is, the vector of data that we collected is not in the subspace where theory has it.

$$\begin{pmatrix} 16 \\ 34 \\ 51 \end{pmatrix} \notin \{m \begin{pmatrix} 30 \\ 60 \\ 90 \end{pmatrix} \mid m \in \mathbb{R}\}$$

However, we have to do something so we look for the m that most nearly works. An orthogonal projection of the data vector into the line subspace gives this best guess.

$$\frac{\begin{pmatrix} 16 \\ 34 \\ 51 \end{pmatrix} \cdot \begin{pmatrix} 30 \\ 60 \\ 90 \end{pmatrix}}{\begin{pmatrix} 30 \\ 60 \\ 90 \end{pmatrix} \cdot \begin{pmatrix} 30 \\ 60 \\ 90 \end{pmatrix}} \cdot \begin{pmatrix} 30 \\ 60 \\ 90 \end{pmatrix} = \frac{7110}{12600} \cdot \begin{pmatrix} 30 \\ 60 \\ 90 \end{pmatrix}$$

The estimate ($m = 7110/12600 \approx 0.56$) is a bit more than one half, but not much more than half, so probably the penny is fair enough.

The line with the slope $m \approx 0.56$ is the *line of best fit* for this data.

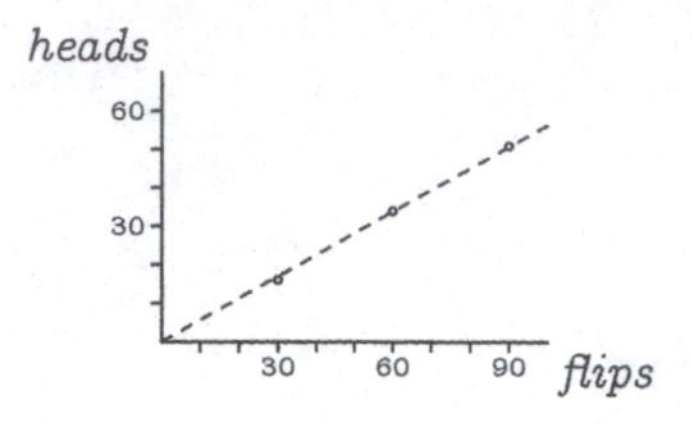

Minimizing the distance between the given vector and the vector used as the right-hand side minimizes the total of these vertical lengths, and consequently we say that the line comes from *fitting by least-squares*.

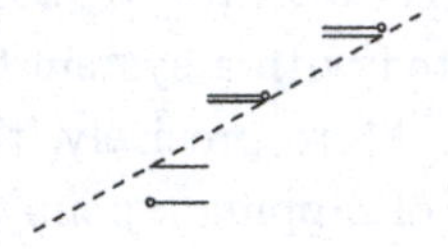

This diagram exaggerates the vertical scale by a factor of ten to make the lengths more visible.

In the above equation the line must pass through $(0, 0)$, because we take it to be the line whose slope is this coin's true proportion of heads to flips. We can also handle cases where the line need not pass through the origin.

Here is the progression of world record times for the men's mile race [Oakley & Baker]. In the early 1900's many people wondered when, or if, this record would fall below the four minute mark. Here are the times that were in force on January first of each decade through the first half of that century.

year	1870	1880	1890	1900	1910	1920	1930	1940	1950
secs	268.8	264.5	258.4	255.6	255.6	252.6	250.4	246.4	241.4

We can use this to give a circa 1950 prediction of the date for 240 seconds, and then compare that to the actual date. As with the penny data, these numbers do not lie in a perfect line. That is, this system does not have an exact solution for the slope and intercept.

$$\begin{aligned} b + 1870m &= 268.8 \\ b + 1880m &= 264.5 \\ &\vdots \\ b + 1950m &= 241.4 \end{aligned}$$

We find a best approximation by using orthogonal projection.

(*Comments on the data.* Restricting to the times at the start of each decade reduces the data entry burden, smooths the data to some extent, and gives much the same result as entering all of the dates and records. There are different sequences of times from competing standards bodies but the ones here are from [Wikipedia, Mens Mile]. We've started the plot at 1870 because at one point there were two classes of records, called 'professional' and 'amateur', and after a while the first class stopped being active so we've followed the second class.)

Write the linear system's matrix of coefficients and also its vector of constants, the world record times.

$$A = \begin{pmatrix} 1 & 1870 \\ 1 & 1880 \\ \vdots & \vdots \\ 1 & 1950 \end{pmatrix} \qquad \vec{v} = \begin{pmatrix} 268.8 \\ 264.5 \\ \vdots \\ 241.4 \end{pmatrix}$$

The ending result in the subsection on Projection into a Subspace gives the formula for the the coefficients b and m that make the linear combination of A's columns as close as possible to $\vec{v}$. Those coefficients are the entries of the vector $(A^{\mathsf{T}}A)^{-1}A^{\mathsf{T}} \cdot \vec{v}$.

Sage can do the computation for us.

```
sage: year = [1870, 1880, 1890, 1900, 1910, 1920, 1930, 1940, 1950]
sage: secs = [268.8, 264.5, 258.4, 255.6, 255.6, 252.6, 250.4, 246.4, 241.4]
sage: var('a, b, t')
(a, b, t)
sage: model(t) = a*t+b
sage: data = zip(year, secs)
sage: fit = find_fit(data, model, solution_dict=True)
sage: model.subs(fit)
t |--> -0.3048333333333295*t + 837.0872222222147
sage: g=points(data)+plot(model.subs(fit),(t,1860,1960),color='red',
....:                    figsize=3,fontsize=7,typeset='latex')
sage: g.save("four_minute_mile.pdf")
sage: g
```

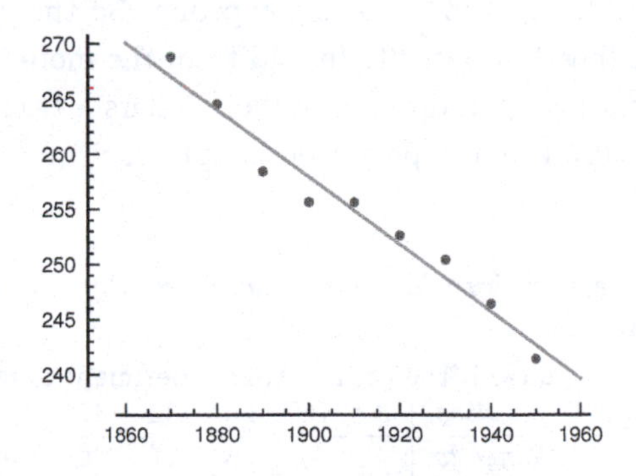

The progression makes a surprisingly good line. From the slope and intercept we predict 1958.73; the actual date of Roger Bannister's record was 1954-May-06.

The final example compares team salaries from US major league baseball against the number of wins the team had, for the year 2002. In this year the

Oakland Athletics used mathematical techniques to optimize the players that they fielded for the money that they could spend, as told in the film *Moneyball*. (Salaries are in millions of dollars and the number of wins is out of 162 games).

To do the computations we again use *Sage*.

```
sage: sal = [40, 40, 39, 42, 45, 42, 62, 34, 41, 57, 58, 63, 47, 75, 57, 78, 80, 50, 60, 93,
....:        77, 55, 95, 103, 79, 76, 108, 126, 95, 106]
sage: wins = [103, 94, 83, 79, 78, 72, 99, 55, 66, 81, 80, 84, 62, 97, 73, 95, 93, 56, 67,
....:        101, 78, 55, 92, 98, 74, 67, 93, 103, 75, 72]
sage: var('a, b, t')
(a, b, t)
sage: model(t) = a*t+b
sage: data = zip(sal,wins)
sage: fit = find_fit(data, model, solution_dict=True)
sage: model.subs(fit)
t |--> 0.2634981251436269*t + 63.06477642781477
sage: p = points(data,size=25)+plot(model.subs(fit),(t,30,130),color='red',typeset='latex')
sage: p.save('moneyball.pdf')
```

The graph is below. The team in the upper left, who paid little for many wins, is the Oakland A's.

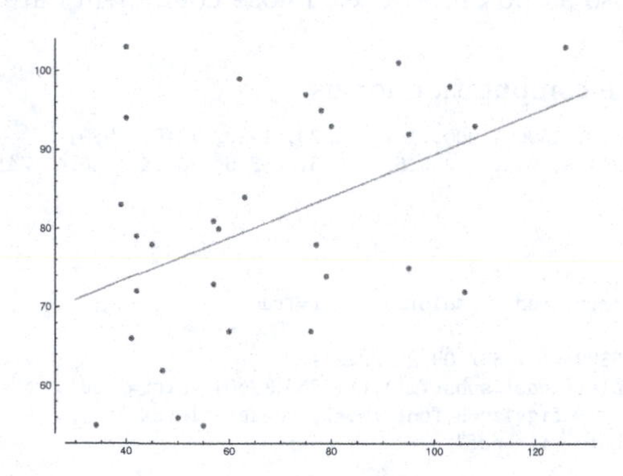

Judging this line by eye would be error-prone. So the equations give us a certainty about the 'best' in best fit. In addition, the model's equation tells us roughly that by spending an additional million dollars a team owner can expect to buy 1/4 of a win (and that expectation is not very sure, thank goodness).

Exercises

The calculations here are best done on a computer. Some of the problems require data from the Internet.

1 Use least-squares to judge if the coin in this experiment is fair.

flips	8	16	24	32	40
heads	4	9	13	17	20

2 For the men's mile record, rather than give each of the many records and its exact date, we've "smoothed" the data somewhat by taking a periodic sample. Do the longer calculation and compare the conclusions.

3 Find the line of best fit for the men's 1500 meter run. How does the slope compare with that for the men's mile? (The distances are close; a mile is about 1609 meters.)

4 Find the line of best fit for the records for women's mile.

5 Do the lines of best fit for the men's and women's miles cross?

6 *(This illustrates that there are data sets for which a linear model is not right, and that the line of best fit doesn't in that case have any predictive value.)* In a highway restaurant a trucker told me that his boss often sends him by a roundabout route, using more gas but paying lower bridge tolls. He said that New York State calibrates the toll for each bridge across the Hudson, playing off the extra gas to get there from New York City against a lower crossing cost, to encourage people to go upstate. This table, from [Cost Of Tolls] and [Google Maps], lists for each toll crossing of the Hudson River, the distance to drive from Times Square in miles and the cost in US dollars for a passenger car (if a crossings has a one-way toll then it shows half that number).

Crossing	*Distance*	*Toll*
Lincoln Tunnel	2	6.00
Holland Tunnel	7	6.00
George Washington Bridge	8	6.00
Verrazano-Narrows Bridge	16	6.50
Tappan Zee Bridge	27	2.50
Bear Mountain Bridge	47	1.00
Newburgh-Beacon Bridge	67	1.00
Mid-Hudson Bridge	82	1.00
Kingston-Rhinecliff Bridge	102	1.00
Rip Van Winkle Bridge	120	1.00

Find the line of best fit and graph the data to show that the driver was practicing on my credulity.

7 When the space shuttle Challenger exploded in 1986, one of the criticisms made of NASA's decision to launch was in the way they did the analysis of number of O-ring failures versus temperature (O-ring failure caused the explosion). Four O-ring failures would be fatal. NASA had data from 24 previous flights.

temp °*F*	53	75	57	58	63	70	70	66	67	67	67
failures	3	2	1	1	1	1	1	0	0	0	0

68	69	70	70	72	73	75	76	76	78	79	80	81
0	0	0	0	0	0	0	0	0	0	0	0	0

The temperature that day was forecast to be 31°F.

(a) NASA based the decision to launch partially on a chart showing only the flights that had at least one O-ring failure. Find the line that best fits these seven flights. On the basis of this data, predict the number of O-ring failures when the temperature is 31, and when the number of failures will exceed four.

(b) Find the line that best fits all 24 flights. On the basis of this extra data, predict the number of O-ring failures when the temperature is 31, and when the number of failures will exceed four.

Which do you think is the more accurate method of predicting? (An excellent discussion is in [Dalal, et. al.].)

8 This table lists the average distance from the sun to each of the first seven planets, using Earth's average as a unit.

Mercury	Venus	Earth	Mars	Jupiter	Saturn	Uranus
0.39	0.72	1.00	1.52	5.20	9.54	19.2

(a) Plot the number of the planet (Mercury is 1, etc.) versus the distance. Note that it does not look like a line, and so finding the line of best fit is not fruitful.

(b) It does, however look like an exponential curve. Therefore, plot the number of the planet versus the logarithm of the distance. Does this look like a line?

(c) The asteroid belt between Mars and Jupiter is what is left of a planet that broke apart. Renumber so that Jupiter is 6, Saturn is 7, and Uranus is 8, and plot against the log again. Does this look better?

(d) Use least squares on that data to predict the location of Neptune.

(e) Repeat to predict where Pluto is.

(f) Is the formula accurate for Neptune and Pluto?

This method was used to help discover Neptune (although the second item is misleading about the history; actually, the discovery of Neptune in position 9 prompted people to look for the "missing planet" in position 5). See [Gardner, 1970]

Topic

Geometry of Linear Maps

These pairs of pictures contrast the geometric action of the nonlinear maps $f_1(x) = e^x$ and $f_2(x) = x^2$

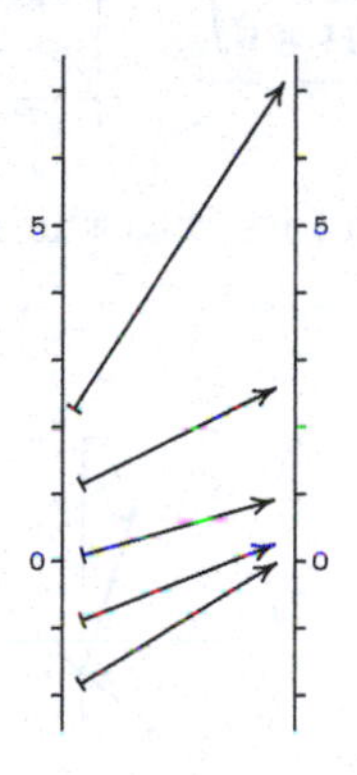

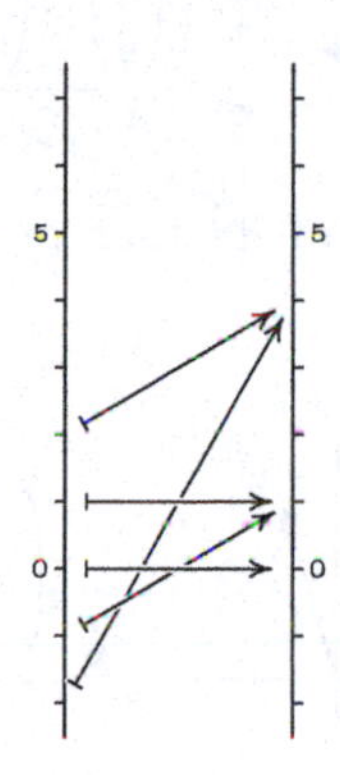

with the linear maps $h_1(x) = 2x$ and $h_2(x) = -x$.

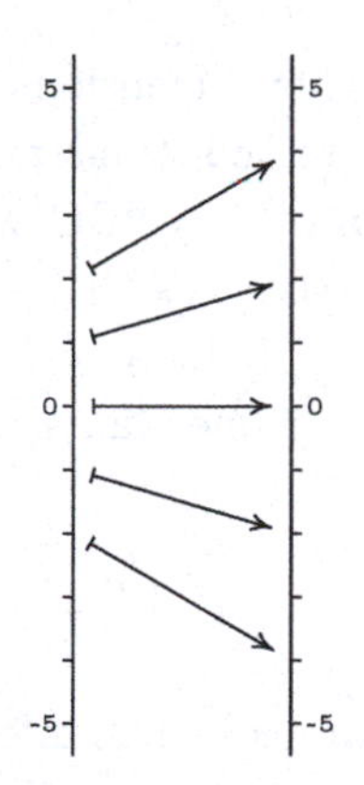

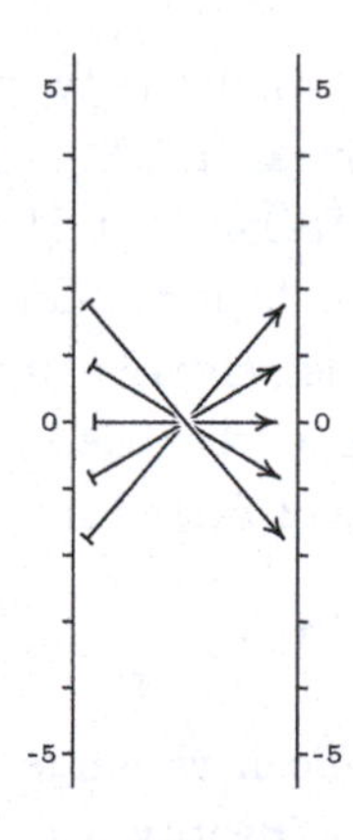

Each of the four pictures shows the domain $\mathbb{R}$ on the left mapped to the codomain $\mathbb{R}$ on the right. Arrows trace where each map sends $x = 0$, $x = 1$, $x = 2$, $x = -1$, and $x = -2$.

The nonlinear maps distort the domain in transforming it into the range. For instance, $f_1(1)$ is further from $f_1(2)$ than it is from $f_1(0)$ — this map spreads the domain out unevenly so that a domain interval near $x = 2$ is spread apart more than is a domain interval near $x = 0$. The linear maps are nicer, more regular, in that for each map all of the domain spreads by the same factor. The map h_1 on the left spreads all intervals apart to be twice as wide while on the right h_2 keeps intervals the same length but reverses their orientation, as with the rising interval from 1 to 2 being transformed to the falling interval from -1 to -2.

The only linear maps from $\mathbb{R}$ to $\mathbb{R}$ are multiplications by a scalar but in higher dimensions more can happen. For instance, this linear transformation of $\mathbb{R}^2$ rotates vectors counterclockwise.

$$\begin{pmatrix} x \\ y \end{pmatrix} \mapsto \begin{pmatrix} x\cos\theta - y\sin\theta \\ x\sin\theta + y\cos\theta \end{pmatrix}$$

The transformation of $\mathbb{R}^3$ that projects vectors into the xz-plane is also not simply a rescaling.

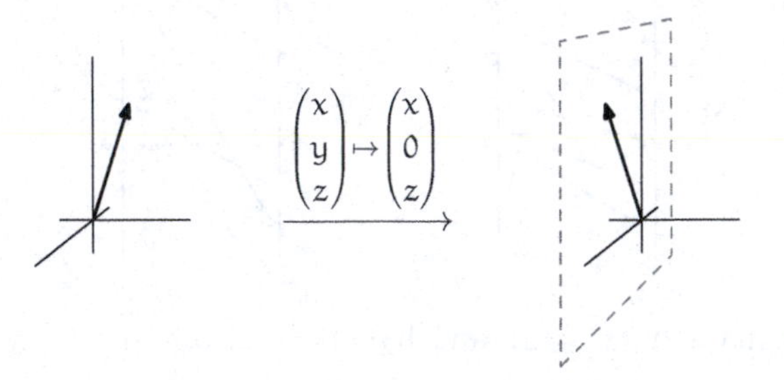

Despite this additional variety, even in higher dimensions linear maps behave nicely. Consider a linear $h\colon \mathbb{R}^n \to \mathbb{R}^m$ and use the standard bases to represent it by a matrix H. Recall that H factors into $H = PBQ$ where P and Q are nonsingular and B is a partial-identity matrix. Recall also that nonsingular matrices factor into elementary matrices $PBQ = T_nT_{n-1}\cdots T_sBT_{s-1}\cdots T_1$, which are matrices that come from the identity I after one Gaussian row operation, so each T matrix is one of these three kinds

$$I \xrightarrow{k\rho_i} M_i(k) \qquad I \xrightarrow{\rho_i\leftrightarrow\rho_j} P_{i,j} \qquad I \xrightarrow{k\rho_i+\rho_j} C_{i,j}(k)$$

with $i \neq j$, $k \neq 0$. So if we understand the geometric effect of a linear map described by a partial-identity matrix and the effect of the linear maps described by the elementary matrices then we will in some sense completely understand the effect of any linear map. (The pictures below stick to transformations of $\mathbb{R}^2$ for ease of drawing but the principles extend for maps from any $\mathbb{R}^n$ to any $\mathbb{R}^m$.)

The geometric effect of the linear transformation represented by a partial-identity matrix is projection.

$$\begin{pmatrix} x \\ y \\ z \end{pmatrix} \xrightarrow{\begin{pmatrix} 1 & 0 & 0 \\ 0 & 1 & 0 \\ 0 & 0 & 0 \end{pmatrix}} \begin{pmatrix} x \\ y \\ 0 \end{pmatrix}$$

The geometric effect of the $M_i(k)$ matrices is to stretch vectors by a factor of k along the i-th axis. This map stretches by a factor of 3 along the x-axis.

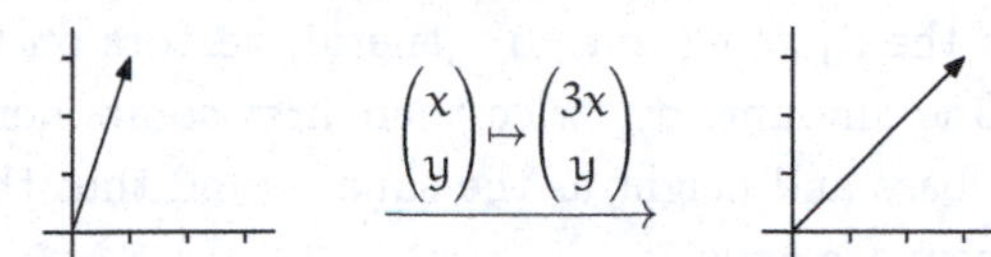

If $0 \leqslant k < 1$ or if $k < 0$ then the i-th component goes the other way, here to the left.

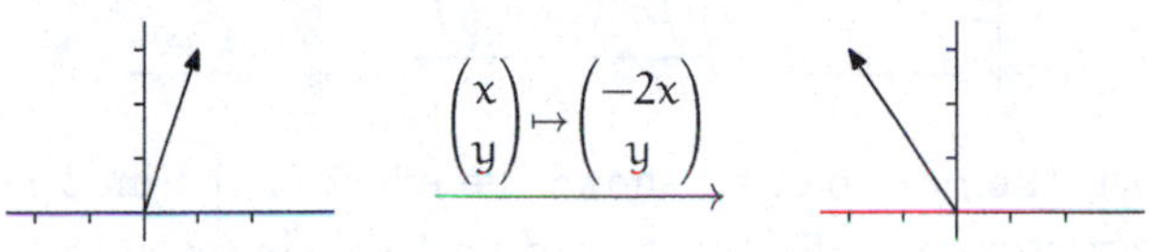

Either of these stretches is a *dilation.*

A transformation represented by a $P_{i,j}$ matrix interchanges the i-th and j-th axes. This is *reflection* about the line $x_i = x_j$.

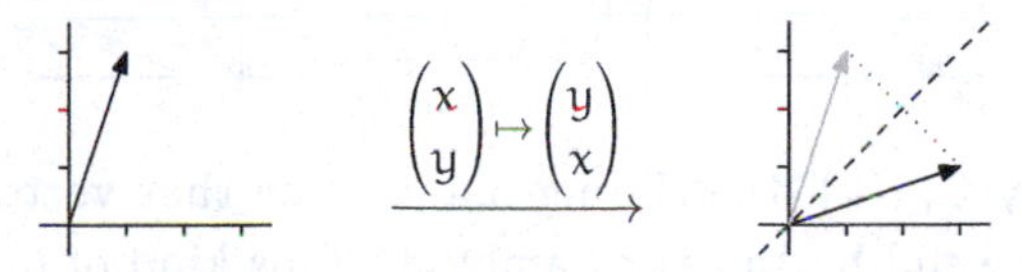

Permutations involving more than two axes decompose into a combination of swaps of pairs of axes; see Exercise 5.

The remaining matrices have the form $C_{i,j}(k)$. For instance $C_{1,2}(2)$ performs $2\rho_1 + \rho_2$.

$$\begin{pmatrix} x \\ y \end{pmatrix} \xrightarrow{\begin{pmatrix} 1 & 0 \\ 2 & 1 \end{pmatrix}} \begin{pmatrix} x \\ 2x + y \end{pmatrix}$$

In the picture below, the vector $\vec{u}$ with the first component of 1 is affected less than the vector $\vec{v}$ with the first component of 2. The vector $\vec{u}$ is mapped to a $h(\vec{u})$ that is only 2 higher than $\vec{u}$ while $h(\vec{v})$ is 4 higher than $\vec{v}$.

$$\begin{pmatrix} x \\ y \end{pmatrix} \mapsto \begin{pmatrix} x \\ 2x + y \end{pmatrix}$$

Any vector with a first component of 1 would be affected in the same way as $\vec{u}$: it would slide up by 2. And any vector with a first component of 2 would slide up 4, as was $\vec{v}$. That is, the transformation represented by $C_{i,j}(k)$ affects vectors depending on their i-th component.

Another way to see this point is to consider the action of this map on the unit square. In the next picture, vectors with a first component of 0, such as the origin, are not pushed vertically at all but vectors with a positive first component slide up. Here, all vectors with a first component of 1, the entire right side of the square, slide to the same extent. In general, vectors on the same vertical line slide by the same amount, by twice their first component. The resulting shape has the same base and height as the square (and thus the same area) but the right angle corners are gone.

$$\begin{pmatrix} x \\ y \end{pmatrix} \mapsto \begin{pmatrix} x \\ 2x+y \end{pmatrix}$$

For contrast, the next picture shows the effect of the map represented by $C_{2,1}(2)$. Here vectors are affected according to their second component: $\binom{x}{y}$ slides horizontally by twice y.

$$\begin{pmatrix} x \\ y \end{pmatrix} \mapsto \begin{pmatrix} x+2y \\ y \end{pmatrix}$$

In general, for any $C_{i,j}(k)$, the sliding happens so that vectors with the same i-th component are slid by the same amount. This kind of map is a *shear*.

With that we understand the geometric effect of the four types of matrices on the right-hand side of $H = T_n T_{n-1} \cdots T_j B T_{j-1} \cdots T_1$ and so in some sense we understand the action of any matrix H. Thus, even in higher dimensions the geometry of linear maps is easy: it is built by putting together a number of components, each of which acts in a simple way.

We will apply this understanding in two ways. The first way is to prove something general about the geometry of linear maps. Recall that under a linear map, the image of a subspace is a subspace and thus the linear transformation h represented by H maps lines through the origin to lines through the origin. (The dimension of the image space cannot be greater than the dimension of the domain space, so a line can't map onto, say, a plane.) We will show that h maps any line — not just one through the origin — to a line. The proof is simple: the partial-identity projection B and the elementary T_i's each turn a line input into a line output; verifying the four cases is Exercise 6. Therefore their composition also preserves lines.

The second way that we will apply the geometric understanding of linear maps is to elucidate a point from Calculus. Below is a picture of the action of the one-variable real function $y(x) = x^2 + x$. As with the nonlinear functions pictured earlier, the geometric effect of this map is irregular in that at different domain points it has different effects; for example as the input x goes from 2 to -2, the associated output $f(x)$ at first decreases, then pauses for an instant, and then increases.

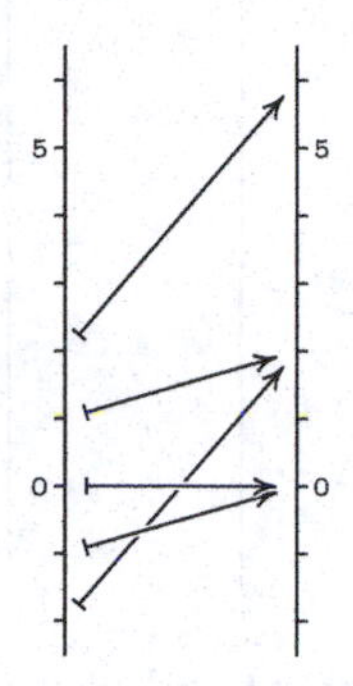

But in Calculus we focus less on the map overall and more on the local effect of the map. Below we look closely at what this map does near $x = 1$. The derivative is $dy/dx = 2x + 1$ so that near $x = 1$ we have $\Delta y \approx 3 \cdot \Delta x$. That is, in a neighborhood of $x = 1$, in carrying the domain over this map causes it to grow by a factor of 3—it is, locally, approximately, a dilation. The picture below shows this as a small interval in the domain $(1 - \Delta x \,..\, 1 + \Delta x)$ carried over to an interval in the codomain $(2 - \Delta y \,..\, 2 + \Delta y)$ that is three times as wide.

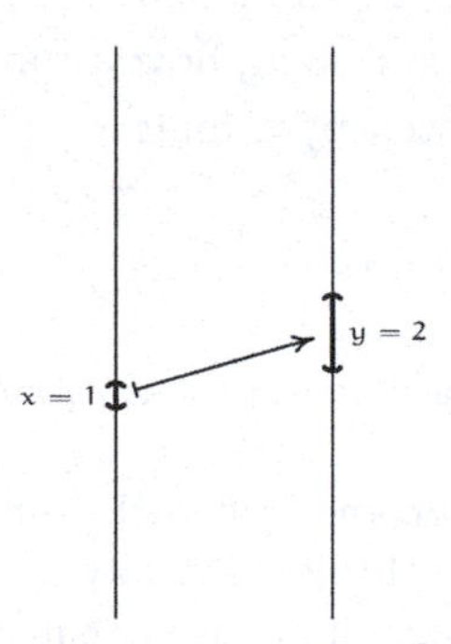

In higher dimensions the core idea is the same but more can happen. For a function $y\colon \mathbb{R}^n \to \mathbb{R}^m$ and a point $\vec{x} \in \mathbb{R}^n$, the derivative is defined to be the linear map $h\colon \mathbb{R}^n \to \mathbb{R}^m$ that best approximates how y changes near $y(\vec{x})$. So the geometry described above directly applies to the derivative.

We close by remarking how this point of view makes clear an often misunderstood result about derivatives, the Chain Rule. Recall that, under suitable conditions on the two functions, the derivative of the composition is this.

$$\frac{d\,(g\circ f)}{dx}(x)=\frac{dg}{dx}(f(x))\cdot\frac{df}{dx}(x)$$

For instance the derivative of $\sin(x^2+3x)$ is $\cos(x^2+3x)\cdot(2x+3)$.

Where does this come from? Consider $f,g\colon\mathbb{R}\to\mathbb{R}$.

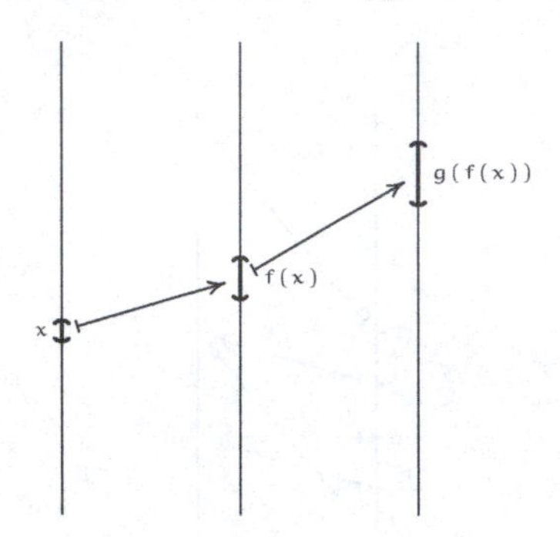

The first map f dilates the neighborhood of x by a factor of

$$\frac{df}{dx}(x)$$

and the second map g follows that by dilating a neighborhood of $f(x)$ by a factor of

$$\frac{dg}{dx}(\,f(x)\,)$$

and when combined, the composition dilates by the product of the two. In higher dimensions the map expressing how a function changes near a point is a linear map, and is represented by a matrix. The Chain Rule multiplies the matrices.

Exercises

1 Let $h\colon\mathbb{R}^2\to\mathbb{R}^2$ be the transformation that rotates vectors clockwise by $\pi/4$ radians.

(a) Find the matrix H representing h with respect to the standard bases. Use Gauss's Method to reduce H to the identity.

(b) Translate the row reduction to a matrix equation $T_jT_{j-1}\cdots T_1H=I$ (the prior item shows both that H is similar to I, and that we need no column operations to derive I from H).

(c) Solve this matrix equation for H.

(d) Sketch how H is a combination of dilations, flips, skews, and projections (the identity is a trivial projection).

2 What combination of dilations, flips, skews, and projections produces a rotation counterclockwise by $2\pi/3$ radians?

3 What combination of dilations, flips, skews, and projections produces the map $h\colon \mathbb{R}^3 \to \mathbb{R}^3$ represented with respect to the standard bases by this matrix?

$$\begin{pmatrix} 1 & 2 & 1 \\ 3 & 6 & 0 \\ 1 & 2 & 2 \end{pmatrix}$$

4 Show that any linear transformation of $\mathbb{R}^1$ is the map that multiplies by a scalar $x \mapsto kx$.

5 Show that for any permutation (that is, reordering) p of the numbers $1, \ldots, n$, the map

$$\begin{pmatrix} x_1 \\ x_2 \\ \vdots \\ x_n \end{pmatrix} \mapsto \begin{pmatrix} x_{p(1)} \\ x_{p(2)} \\ \vdots \\ x_{p(n)} \end{pmatrix}$$

can be done with a composition of maps, each of which only swaps a single pair of coordinates. *Hint:* you can use induction on n. (*Remark:* in the fourth chapter we will show this and we will also show that the parity of the number of swaps used is determined by p. That is, although a particular permutation could be expressed in two different ways with two different numbers of swaps, either both ways use an even number of swaps, or both use an odd number.)

6 Show that linear maps preserve the linear structures of a space.

(a) Show that for any linear map from $\mathbb{R}^n$ to $\mathbb{R}^m$, the image of any line is a line. The image may be a degenerate line, that is, a single point.

(b) Show that the image of any linear surface is a linear surface. This generalizes the result that under a linear map the image of a subspace is a subspace.

(c) Linear maps preserve other linear ideas. Show that linear maps preserve "betweeness": if the point B is between A and C then the image of B is between the image of A and the image of C.

7 Use a picture like the one that appears in the discussion of the Chain Rule to answer: if a function $f\colon \mathbb{R} \to \mathbb{R}$ has an inverse, what's the relationship between how the function — locally, approximately — dilates space, and how its inverse dilates space (assuming, of course, that it has an inverse)?

Topic

Magic Squares

A Chinese legend tells the story of a flood by the Lo river. People offered sacrifices to appease the river. Each time a turtle emerged, walked around the sacrifice, and returned to the water. Fuh-Hi, the founder of Chinese civilization, interpreted this to mean that the river was still cranky. Fortunately, a child noticed that on its shell the turtle had the pattern on the left below, which is today called Lo Shu ("river scroll").

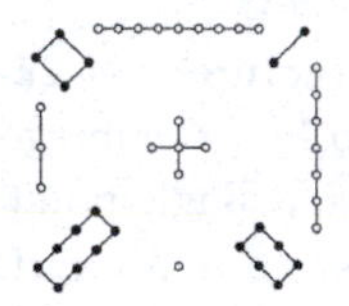

4	9	2
3	5	7
8	1	6

The dots make the matrix on the right where the rows, columns, and diagonals add to 15. Now that the people knew how much to sacrifice, the river's anger cooled.

A square matrix is *magic* if each row, column, and diagonal add to the same number, the matrix's *magic number*.

Another magic square appears in the engraving *Melencolia I* by Dürer.

One interpretation is that it depicts melancholy, a depressed state. The figure, genius, has a wealth of fascinating things to explore including the compass, the geometrical solid, the scale, and the hourglass. But the figure is unmoved; all of the things lie unused. One of the potential delights, in the upper right, is a 4×4 matrix whose rows, columns, and diagonals add to 34.

16	3	2	13
5	10	11	8
9	6	7	12
4	15	14	1

The middle entries on the bottom row give 1514, the date of the engraving.

The above two squares are arrangements of $1 \ldots n^2$. They are *normal*. The 1×1 square whose sole entry is 1 is normal, Exercise 2 shows that there is no normal 2×2 magic square, and there are normal magic squares of every other size; see [Wikipedia, Magic Square]. Finding how many normal magic squares there are of each size is an unsolved problem; see [Online Encyclopedia of Integer Sequences].

If we don't require that the squares be normal then we can say much more. Every 1×1 square is magic, trivially. If the rows, columns, and diagonals of a 2×2 matrix

$$\begin{pmatrix} a & b \\ c & d \end{pmatrix}$$

add to s then $a+b=s$, $c+d=s$, $a+c=s$, $b+d=s$, $a+d=s$, and $b+c=s$. Exercise 2 shows that this system has the unique solution $a=b=c=d=s/2$. So the set of 2×2 magic squares is a one-dimensional subspace of $\mathcal{M}_{2\times 2}$.

A sum of two same-sized magic squares is magic and a scalar multiple of a magic square is magic so the set of $n\times n$ magic squares $\mathscr{M}_n$ is a vector space, a subspace of $\mathcal{M}_{n\times n}$. This Topic shows that for $n \geqslant 3$ the dimension of $\mathscr{M}_n$ is n^2-n. The set $\mathscr{M}_{n,0}$ of $n\times n$ magic squares with magic number 0 is another subspace and we will verify the formula for its dimension also: n^2-2n-1 when $n \geqslant 3$.

We will first prove that $\dim \mathscr{M}_n = \dim \mathscr{M}_{n,0} + 1$. Define the *trace* of a matrix to be the sum down its upper-left to lower-right diagonal $\mathrm{Tr}(M) = m_{1,1}+\cdots+m_{n,n}$. Consider the restriction of the trace to the magic squares $\mathrm{Tr}\colon \mathscr{M}_n \to \mathbb{R}$. The null space $\mathscr{N}(\mathrm{Tr})$ is the set of magic squares with magic number zero $\mathscr{M}_{n,0}$. Observe that the trace is onto because for any r in the codomain $\mathbb{R}$ the $n\times n$ matrix whose entries are all r/n is a magic square with magic number r. Theorem Two.II.2.14 says that for any linear map the dimension of the domain equals the dimension of the range space plus the dimension of the null space, the map's rank plus its nullity. Here the domain is $\mathscr{M}_n$, the range space is $\mathbb{R}$ and the null space is $\mathscr{M}_{n,0}$, so we have that $\dim \mathscr{M}_n = 1 + \dim \mathscr{M}_{n,0}$.

We will finish by finding the dimension of the vector space $\mathscr{M}_{n,0}$. For $n = 1$ the dimension is clearly 0. Exercise 3 shows that $\dim \mathscr{M}_{n,0}$ is also 0 for $n = 2$.

That leaves showing that $\dim \mathscr{M}_{n,0} = n^2 - 2n - 1$ for $n \geqslant 3$. The fact that the squares in this vector space are magic gives us a linear system of restrictions, and the fact that they have magic number zero makes this system homogeneous: for instance consider the 3×3 case. The restriction that the rows, columns, and diagonals of

$$\begin{pmatrix} a & b & c \\ d & e & f \\ g & h & i \end{pmatrix}$$

add to zero gives this $(2n+2)\times n^2$ linear system.

$$\begin{array}{rcl}
a+b+c & & =0 \\
d+e+f & & =0 \\
g+h+i & & =0 \\
a+d+g & & =0 \\
b+e+h & & =0 \\
c+f+i & & =0 \\
a+e+i & & =0 \\
c+e+g & & =0
\end{array}$$

We will find the dimension of the space by finding the number of free variables in the linear system.

The matrix of coefficients for the particular cases of $n = 3$ and $n = 4$ are below, with the rows and columns numbered to help in reading the proof. With respect to the standard basis, each represents a linear map $h\colon \mathbb{R}^{n^2} \to \mathbb{R}^{2n+2}$. The domain has dimension n^2 so if we show that the rank of the matrix is $2n+1$ then we will have what we want, that the dimension of the null space $\mathscr{M}_{n,0}$ is $n^2 - (2n+1)$.

	1	2	3	4	5	6	7	8	9
$\vec{\rho}_1$	1	1	1	0	0	0	0	0	0
$\vec{\rho}_2$	0	0	0	1	1	1	0	0	0
$\vec{\rho}_3$	0	0	0	0	0	0	1	1	1
$\vec{\rho}_4$	1	0	0	1	0	0	1	0	0
$\vec{\rho}_5$	0	1	0	0	1	0	0	1	0
$\vec{\rho}_6$	0	0	1	0	0	1	0	0	1
$\vec{\rho}_7$	1	0	0	0	1	0	0	0	1
$\vec{\rho}_8$	0	0	1	0	1	0	1	0	0

	1	2	3	4	5	6	7	8	9	10	11	12	13	14	15	16
$\vec{\rho}_1$	1	1	1	1	0	0	0	0	0	0	0	0	0	0	0	0
$\vec{\rho}_2$	0	0	0	0	1	1	1	1	0	0	0	0	0	0	0	0
$\vec{\rho}_3$	0	0	0	0	0	0	0	0	1	1	1	1	0	0	0	0
$\vec{\rho}_4$	0	0	0	0	0	0	0	0	0	0	0	0	1	1	1	1
$\vec{\rho}_5$	1	0	0	0	1	0	0	0	1	0	0	0	1	0	0	0
$\vec{\rho}_6$	0	1	0	0	0	1	0	0	0	1	0	0	0	1	0	0
$\vec{\rho}_7$	0	0	1	0	0	0	1	0	0	0	1	0	0	0	1	0
$\vec{\rho}_8$	0	0	0	1	0	0	0	1	0	0	0	1	0	0	0	1
$\vec{\rho}_9$	1	0	0	0	0	1	0	0	0	0	1	0	0	0	0	1
$\vec{\rho}_{10}$	0	0	0	1	0	0	1	0	0	1	0	0	1	0	0	0

We want to show that the rank of the matrix of coefficients, the number of rows in a maximal linearly independent set, is $2n+1$. The first n rows of the matrix of coefficients add to the same vector as the second n rows, the vector of all ones. So a maximal linearly independent must omit at least one row. We will show that the set of all rows but the first $\{\vec{\rho}_2 \ldots \vec{\rho}_{2n+2}\}$ is linearly independent. So consider this linear relationship.

$$c_2\vec{\rho}_2 + \cdots + c_{2n}\vec{\rho}_{2n} + c_{2n+1}\vec{\rho}_{2n+1} + c_{2n+2}\vec{\rho}_{2n+2} = \vec{0} \qquad (*)$$

Now it gets messy. Focus on the lower left of the tables. Observe that in the final two rows, in the first n columns, is a subrow that is all zeros except that it starts with a one in column 1 and a subrow that is all zeros except that it ends with a one in column n.

First, with $\vec{\rho}_1$ omitted, both column 1 and column n contain only two ones. Since the only rows in $(*)$ with nonzero column 1 entries are rows $\vec{\rho}_{n+1}$ and $\vec{\rho}_{2n+1}$, which have ones, we must have $c_{2n+1} = -c_{n+1}$. Likewise considering the n-th entries of the vectors in $(*)$ gives that $c_{2n+2} = -c_{2n}$.

Next consider the columns between those two—in the $n = 3$ table this includes only column 2 while in the $n = 4$ table it includes both columns 2 and 3. Each such column has a single one. That is, for each column index $j \in \{2 \ldots n-2\}$ the column consists of only zeros except for a one in row $n+j$, and hence $c_{n+j} = 0$.

On to the next block of columns, from $n+1$ through $2n$. Column $n+1$ has only two ones (because $n \geqslant 3$ the ones in the last two rows do not fall in the first column of this block). Thus $c_2 = -c_{n+1}$ and therefore $c_2 = c_{2n+1}$. Likewise, from column $2n$ we conclude that $c_2 = -c_{2n}$ and so $c_2 = c_{2n+2}$.

Because $n \geqslant 3$ there is at least one column between column $n+1$ and column $2n-1$. In at least one of those columns a one appears in $\vec{\rho}_{2n+1}$. If a one also appears in that column in $\vec{\rho}_{2n+2}$ then we have $c_2 = -(c_{2n+1} + c_{2n+2})$ since

$c_{n+j} = 0$ for $j \in \{2 \ldots n-2\}$. If a one does not appear in that column in $\vec{\rho}_{2n+2}$ then we have $c_2 = -c_{2n+1}$. In either case $c_2 = 0$, and thus $c_{2n+1} = c_{2n+2} = 0$ and $c_{n+1} = c_{2n} = 0$.

If the next block of n-many columns is not the last then similarly conclude from its first column that $c_3 = c_{n+1} = 0$.

Keep this up until we reach the last block of columns, those numbered $(n-1)n+1$ through n^2. Because $c_{n+1} = \cdots = c_{2n} = 0$ column n^2 gives that $c_n = -c_{2n+1} = 0$.

Therefore the rank of the matrix is $2n+1$, as required.

The classic source on normal magic squares is [Ball & Coxeter]. More on the Lo Shu square is at [Wikipedia, Lo Shu Square]. The proof given here began with [Ward].

Exercises

1 Let M be a 3×3 magic square with magic number s.
  (a) Prove that the sum of M's entries is $3s$.
  (b) Prove that $s = 3 \cdot m_{2,2}$.
  (c) Prove that $m_{2,2}$ is the average of the entries in its row, its column, and in each diagonal.
  (d) Prove that $m_{2,2}$ is the median of M's entries.

2 Solve the system $a+b=s$, $c+d=s$, $a+c=s$, $b+d=s$, $a+d=s$, and $b+c=s$.

3 Show that $\dim \mathscr{M}_{2,0} = 0$.

4 Let the *trace* function be $\text{Tr}(M) = m_{1,1} + \cdots + m_{n,n}$. Define also the sum down the other diagonal $\text{Tr}^*(M) = m_{1,n} + \cdots + m_{n,1}$.
  (a) Show that the two functions $\text{Tr}, \text{Tr}^* \colon \mathcal{M}_{n\times n} \to \mathbb{R}$ are linear.
  (b) Show that the function $\theta \colon \mathcal{M}_{n\times n} \to \mathbb{R}^2$ given by $\theta(M) = (\text{Tr}(M), \text{Tr}^*(m))$ is linear.
  (c) Generalize the prior item.

5 A square matrix is *semimagic* if the rows and columns add to the same value, that is, if we drop the condition on the diagonals.
  (a) Show that the set of semimagic squares $\mathscr{H}_n$ is a subspace of $\mathcal{M}_{n\times n}$.
  (b) Show that the set $\mathscr{H}_{n,0}$ of $n\times n$ semimagic squares with magic number 0 is also a subspace of $\mathcal{M}_{n\times n}$.

Topic

Markov Chains

Here is a simple game: a player bets on coin tosses, a dollar each time, and the game ends either when the player has no money or is up to five dollars. If the player starts with three dollars, what is the chance that the game takes at least five flips? Twenty-five flips?

At any point, this player has either \$0, or \$1, ..., or \$5. We say that the player is in the *state* s_0, s_1, ..., or s_5. In the game the player moves from state to state. For instance, a player now in state s_3 has on the next flip a 0.5 chance of moving to state s_2 and a 0.5 chance of moving to s_4. The boundary states are different; a player never leaves state s_0 or state s_5.

Let $p_i(n)$ be the probability that the player is in state s_i after n flips. Then for instance the probability of being in state s_0 after flip $n+1$ is $p_0(n+1) = p_0(n) + 0.5 \cdot p_1(n)$. This equation summarizes.

$$\begin{pmatrix} 1.0 & 0.5 & 0.0 & 0.0 & 0.0 & 0.0 \\ 0.0 & 0.0 & 0.5 & 0.0 & 0.0 & 0.0 \\ 0.0 & 0.5 & 0.0 & 0.5 & 0.0 & 0.0 \\ 0.0 & 0.0 & 0.5 & 0.0 & 0.5 & 0.0 \\ 0.0 & 0.0 & 0.0 & 0.5 & 0.0 & 0.0 \\ 0.0 & 0.0 & 0.0 & 0.0 & 0.5 & 1.0 \end{pmatrix} \begin{pmatrix} p_0(n) \\ p_1(n) \\ p_2(n) \\ p_3(n) \\ p_4(n) \\ p_5(n) \end{pmatrix} = \begin{pmatrix} p_0(n+1) \\ p_1(n+1) \\ p_2(n+1) \\ p_3(n+1) \\ p_4(n+1) \\ p_5(n+1) \end{pmatrix}$$

Sage will compute the evolution of this game.

```
sage: M = matrix(RDF, [[1.0, 0.5, 0.0, 0.0, 0.0, 0.0],
....:                  [0.5, 0.0, 0.5, 0.0, 0.0, 0.0],
....:                  [0.0, 0.5, 0.0, 0.5, 0.0, 0.0],
....:                  [0.0, 0.0, 0.5, 0.0, 0.5, 0.0],
....:                  [0.0, 0.0, 0.0, 0.5, 0.0, 0.5],
....:                  [0.0, 0.0, 0.0, 0.0, 0.5, 1.0]])
sage: M = M.transpose()
sage: v0 = vector(RDF, [0.0, 0.0, 0.0, 1.0, 0.0, 0.0])
sage: v1 = v0*M
sage: v1
(0.0, 0.0, 0.5, 0.0, 0.5, 0.0)
sage: v2 = v1*M
sage: v2
(0.0, 0.25, 0.0, 0.5, 0.0, 0.25)
```

(Two notes: (1) *Sage* can use various number systems to make the matrix entries and here we have used Real Double Float, and (2) *Sage* likes to do matrix multiplication from the right, as $\vec{v}M$ instead of our usual $M\vec{v}$, so we needed to take the matrix's transpose.)

These are components of the resulting vectors.

$n=0$	$n=1$	$n=2$	$n=3$	$n=4$	$\cdots$	$n=24$
0	0	0	0.125	0.125		0.39600
0	0	0.25	0	0.1875		0.00276
0	0.5	0	0.375	0		0
1	0	0.5	0	0.3125		0.00447
0	0.5	0	0.25	0		0
0	0	0.25	0.25	0.375		0.59676

This game is not likely to go on for long since the player quickly moves to an ending state. For instance, after the fourth flip there is already a 0.50 probability that the game is over.

This is a *Markov chain*. Each vector is a *probability vector*, whose entries are nonnegative real numbers that sum to 1. The matrix is a *transition matrix* or *stochastic matrix*, whose entries are nonnegative reals and whose columns sum to 1.

A characteristic feature of a Markov chain model is that it is *historyless* in that the next state depends only on the current state, not on any prior ones. Thus, a player who arrives at s_2 by starting in state s_3 and then going to state s_2 has exactly the same chance of moving next to s_3 as does a player whose history was to start in s_3 then go to s_4 then to s_3 and then to s_2.

Here is a Markov chain from sociology. A study ([Macdonald & Ridge], p. 202) divided occupations in the United Kingdom into three levels: executives and professionals, supervisors and skilled manual workers, and unskilled workers. They asked about two thousand men, "At what level are you, and at what level was your father when you were fourteen years old?" Here the Markov model assumption about history may seem reasonable — we may guess that while a parent's occupation has a direct influence on the occupation of the child, the grandparent's occupation likely has no such direct influence. This summarizes the study's conclusions.

$$\begin{pmatrix} .60 & .29 & .16 \\ .26 & .37 & .27 \\ .14 & .34 & .57 \end{pmatrix} \begin{pmatrix} p_U(n) \\ p_M(n) \\ p_L(n) \end{pmatrix} = \begin{pmatrix} p_U(n+1) \\ p_M(n+1) \\ p_L(n+1) \end{pmatrix}$$

For instance, looking at the middle class for the next generation, a child of an upper class worker has a 0.26 probability of becoming middle class, a child of a middle class worker has a 0.37 chance of being middle class, and a child of a lower class worker has a 0.27 probability of becoming middle class.

Sage will compute the successive stages of this system (the current class distribution is $\vec{v}_0$).

```
sage: M = matrix(RDF, [[0.60, 0.29, 0.16],
....:                  [0.26, 0.37, 0.27],
....:                  [0.14, 0.34, 0.57]])
sage: M = M.transpose()
sage: v0 = vector(RDF, [0.12, 0.32, 0.56])
sage: v0*M
(0.2544, 0.3008, 0.4448)
sage: v0*M^2
(0.31104, 0.297536, 0.391424)
sage: v0*M^3
(0.33553728, 0.2966432, 0.36781952)
```

Here are the next five generations. They show upward mobility, especially in the first generation. In particular, lower class shrinks a good bit.

$n=0$	$n=1$	$n=2$	$n=3$	$n=4$	$n=5$
.12	.25	.31	.34	.35	.35
.32	.30	.30	.30	.30	.30
.56	.44	.39	.37	.36	.35

One more example. In professional American baseball there are two leagues, the American League and the National League. At the end of the annual season the team winning the American League and the team winning the National League play the World Series. The winner is the first team to take four games. That means that a series is in one of twenty-four states: 0-0 (no games won yet by either team), 1-0 (one game won for the American League team and no games for the National League team), etc.

Consider a series with a probability p that the American League team wins each game. We have this.

$$\begin{pmatrix} 0 & 0 & 0 & 0 & \ldots \\ p & 0 & 0 & 0 & \ldots \\ 1-p & 0 & 0 & 0 & \ldots \\ 0 & p & 0 & 0 & \ldots \\ 0 & 1-p & p & 0 & \ldots \\ 0 & 0 & 1-p & 0 & \ldots \\ \vdots & \vdots & \vdots & \vdots \end{pmatrix} \begin{pmatrix} p_{0\text{-}0}(n) \\ p_{1\text{-}0}(n) \\ p_{0\text{-}1}(n) \\ p_{2\text{-}0}(n) \\ p_{1\text{-}1}(n) \\ p_{0\text{-}2}(n) \\ \vdots \end{pmatrix} = \begin{pmatrix} p_{0\text{-}0}(n+1) \\ p_{1\text{-}0}(n+1) \\ p_{0\text{-}1}(n+1) \\ p_{2\text{-}0}(n+1) \\ p_{1\text{-}1}(n+1) \\ p_{0\text{-}2}(n+1) \\ \vdots \end{pmatrix}$$

An especially interesting special case is when the teams are evenly matched, $p = 0.50$. This table below lists the resulting components of the $n = 0$ through $n = 7$ vectors.

Note that evenly-matched teams are likely to have a long series — there is a probability of 0.625 that the series goes at least six games.

	$n=0$	$n=1$	$n=2$	$n=3$	$n=4$	$n=5$	$n=6$	$n=7$
$0-0$	1	0	0	0	0	0	0	0
$1-0$	0	0.5	0	0	0	0	0	0
$0-1$	0	0.5	0	0	0	0	0	0
$2-0$	0	0	0.25	0	0	0	0	0
$1-1$	0	0	0.5	0	0	0	0	0
$0-2$	0	0	0.25	0	0	0	0	0
$3-0$	0	0	0	0.125	0	0	0	0
$2-1$	0	0	0	0.375	0	0	0	0
$1-2$	0	0	0	0.375	0	0	0	0
$0-3$	0	0	0	0.125	0	0	0	0
$4-0$	0	0	0	0	0.0625	0.0625	0.0625	0.0625
$3-1$	0	0	0	0	0.25	0	0	0
$2-2$	0	0	0	0	0.375	0	0	0
$1-3$	0	0	0	0	0.25	0	0	0
$0-4$	0	0	0	0	0.0625	0.0625	0.0625	0.0625
$4-1$	0	0	0	0	0	0.125	0.125	0.125
$3-2$	0	0	0	0	0	0.3125	0	0
$2-3$	0	0	0	0	0	0.3125	0	0
$1-4$	0	0	0	0	0	0.125	0.125	0.125
$4-2$	0	0	0	0	0	0	0.15625	0.15625
$3-3$	0	0	0	0	0	0	0.3125	0
$2-4$	0	0	0	0	0	0	0.15625	0.15625
$4-3$	0	0	0	0	0	0	0	0.15625
$3-4$	0	0	0	0	0	0	0	0.15625

Markov chains are a widely used application of matrix operations. They also give us an example of the use of matrices where we do not consider the significance of the maps represented by the matrices. For more on Markov chains, there are many sources such as [Kemeny & Snell] and [Iosifescu].

Exercises

1 These questions refer to the coin-flipping game.

(a) Check the computations in the table at the end of the first paragraph.

(b) Consider the second row of the vector table. Note that this row has alternating 0's. Must $p_1(j)$ be 0 when j is odd? Prove that it must be, or produce a counterexample.

(c) Perform a computational experiment to estimate the chance that the player ends at five dollars, starting with one dollar, two dollars, and four dollars.

2 [Feller] We consider throws of a die, and say the system is in state s_i if the largest number yet appearing on the die was i.

(a) Give the transition matrix.

(b) Start the system in state s_1, and run it for five throws. What is the vector at the end?

3 [Kelton] There has been much interest in whether industries in the United States are moving from the Northeast and North Central regions to the South and West, motivated by the warmer climate, by lower wages, and by less unionization. Here is the transition matrix for large firms in Electric and Electronic Equipment.

	NE	*NC*	*S*	*W*	*Z*
NE	0.787	0	0	0.111	0.102
NC	0	0.966	0.034	0	0
S	0	0.063	0.937	0	0
W	0	0	0.074	0.612	0.314
Z	0.021	0.009	0.005	0.010	0.954

For example, a firm in the Northeast region will be in the West region next year with probability 0.111. (The Z entry is a "birth-death" state. For instance, with probability 0.102 a large Electric and Electronic Equipment firm from the Northeast will move out of this system next year: go out of business, move abroad, or move to another category of firm. There is a 0.021 probability that a firm in the *National Census of Manufacturers* will move into Electronics, or be created, or move in from abroad, into the Northeast. Finally, with probability 0.954 a firm out of the categories will stay out, according to this research.)

(a) Does the Markov model assumption of lack of history seem justified?

(b) Assume that the initial distribution is even, except that the value at Z is 0.9. Compute the vectors for $n = 1$ through $n = 4$.

(c) Suppose that the initial distribution is this.

NE	*NC*	*S*	*W*	*Z*
0.0000	0.6522	0.3478	0.0000	0.0000

Calculate the distributions for $n = 1$ through $n = 4$.

(d) Find the distribution for $n = 50$ and $n = 51$. Has the system settled down to an equilibrium?

4 [Wickens] Here is a model of some kinds of learning The learner starts in an undecided state s_U. Eventually the learner has to decide to do either response A (that is, end in state s_A) or response B (ending in s_B). However, the learner doesn't jump right from undecided to sure that A is the correct thing to do (or B). Instead, the learner spends some time in a "tentative-A" state, or a "tentative-B" state, trying the response out (denoted here t_A and t_B). Imagine that once the learner has decided, it is final, so once in s_A or s_B, the learner stays there. For the other state changes, we can posit transitions with probability p in either direction.

(a) Construct the transition matrix.

(b) Take $p = 0.25$ and take the initial vector to be 1 at s_U. Run this for five steps. What is the chance of ending up at s_A?

(c) Do the same for $p = 0.20$.

(d) Graph p versus the chance of ending at s_A. Is there a threshold value for p, above which the learner is almost sure not to take longer than five steps?

5 A certain town is in a certain country (this is a hypothetical problem). Each year ten percent of the town dwellers move to other parts of the country. Each year one percent of the people from elsewhere move to the town. Assume that there are two states s_T, living in town, and s_C, living elsewhere.

(a) Construct the transition matrix.

(b) Starting with an initial distribution $s_T = 0.3$ and $s_C = 0.7$, get the results for the first ten years.

(c) Do the same for $s_T = 0.2$.

(d) Are the two outcomes alike or different?

6 For the World Series application, use a computer to generate the seven vectors for $p = 0.55$ and $p = 0.6$.

(a) What is the chance of the National League team winning it all, even though they have only a probability of 0.45 or 0.40 of winning any one game?

(b) Graph the probability p against the chance that the American League team wins it all. Is there a threshold value—a p above which the better team is essentially ensured of winning?

7 Above we define a transition matrix to have each entry nonnegative and each column sum to 1.

(a) Check that the three transition matrices shown in this Topic meet these two conditions. Must any transition matrix do so?

(b) Observe that if $A\vec{v}_0 = \vec{v}_1$ and $A\vec{v}_1 = \vec{v}_2$ then A^2 is a transition matrix from $\vec{v}_0$ to $\vec{v}_2$. Show that a power of a transition matrix is also a transition matrix.

(c) Generalize the prior item by proving that the product of two appropriately-sized transition matrices is a transition matrix.

Topic

Orthonormal Matrices

In *The Elements*, Euclid considers two figures to be the same if they have the same size and shape. That is, while the triangles below are not equal because they are not the same set of points, they are, for Euclid's purposes, essentially indistinguishable because we can imagine picking the plane up, sliding it over and rotating it a bit, although not warping or stretching it, and then putting it back down, to superimpose the first figure on the second. (Euclid never explicitly states this principle but he uses it often [Casey].)

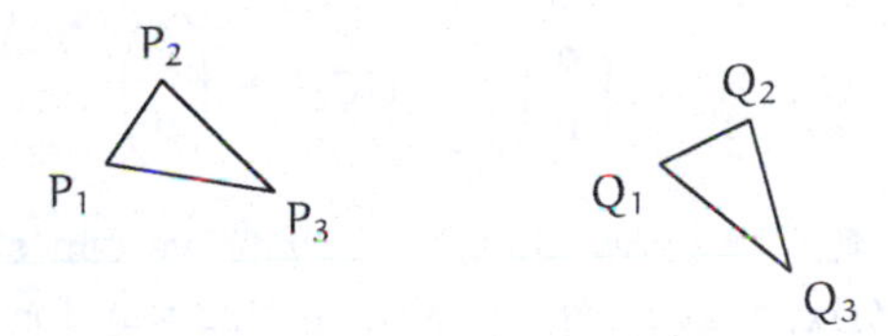

In modern terms "picking the plane up ..." is taking a map from the plane to itself. Euclid considers only transformations that may slide or turn the plane but not bend or stretch it. Accordingly, define a map $f\colon \mathbb{R}^2 \to \mathbb{R}^2$ to be *distance-preserving* or a *rigid motion* or an *isometry* if for all points $P_1, P_2 \in \mathbb{R}^2$, the distance from $f(P_1)$ to $f(P_2)$ equals the distance from P_1 to P_2. We also define a plane *figure* to be a set of points in the plane and we say that two figures are *congruent* if there is a distance-preserving map from the plane to itself that carries one figure onto the other.

Many statements from Euclidean geometry follow easily from these definitions. Some are: (i) collinearity is invariant under any distance-preserving map (that is, if P_1, P_2, and P_3 are collinear then so are $f(P_1)$, $f(P_2)$, and $f(P_3)$), (ii) betweeness is invariant under any distance-preserving map (if P_2 is between P_1 and P_3 then so is $f(P_2)$ between $f(P_1)$ and $f(P_3)$), (iii) the property of being a triangle is invariant under any distance-preserving map (if a figure is a triangle then the image of that figure is also a triangle), (iv) and the property of being a circle is invariant under any distance-preserving map. In 1872, F. Klein suggested that we can define Euclidean geometry as the study of properties that are invariant under these maps. (This forms part of Klein's Erlanger Program, which proposes the organizing principle that we can describe each kind of geometry — Euclidean, projective, etc. — as the study of the properties that are invariant under some group of transformations. The word 'group' here means more than just 'collection' but that lies outside of our scope.)

We can use linear algebra to characterize the distance-preserving maps of the plane.

To begin, observe that there are distance-preserving transformations of the plane that are not linear. The obvious example is this *translation*.

$$\begin{pmatrix} x \\ y \end{pmatrix} \quad \mapsto \quad \begin{pmatrix} x \\ y \end{pmatrix} + \begin{pmatrix} 1 \\ 0 \end{pmatrix} = \begin{pmatrix} x+1 \\ y \end{pmatrix}$$

However, this example turns out to be the only one, in that if f is distance-preserving and sends $\vec{0}$ to $\vec{v}_0$ then the map $\vec{v} \mapsto f(\vec{v}) - \vec{v}_0$ is linear. That will follow immediately from this statement: a map t that is distance-preserving and sends $\vec{0}$ to itself is linear. To prove this equivalent statement, consider the standard basis and suppose that

$$t(\vec{e}_1) = \begin{pmatrix} a \\ b \end{pmatrix} \qquad t(\vec{e}_2) = \begin{pmatrix} c \\ d \end{pmatrix}$$

for some $a, b, c, d \in \mathbb{R}$. To show that t is linear we can show that it can be represented by a matrix, that is, that t acts in this way for all $x, y \in \mathbb{R}$.

$$\vec{v} = \begin{pmatrix} x \\ y \end{pmatrix} \stackrel{t}{\longmapsto} \begin{pmatrix} ax+cy \\ bx+dy \end{pmatrix} \qquad (*)$$

Recall that if we fix three non-collinear points then we can determine any point by giving its distance from those three. So we can determine any point $\vec{v}$ in the domain by its distance from $\vec{0}$, $\vec{e}_1$, and $\vec{e}_2$. Similarly, we can determine any point $t(\vec{v})$ in the codomain by its distance from the three fixed points $t(\vec{0})$, $t(\vec{e}_1)$, and $t(\vec{e}_2)$ (these three are not collinear because, as mentioned above, collinearity is invariant and $\vec{0}$, $\vec{e}_1$, and $\vec{e}_2$ are not collinear). Because t is distance-preserving we can say more: for the point $\vec{v}$ in the plane that is determined by being the distance d_0 from $\vec{0}$, the distance d_1 from $\vec{e}_1$, and the distance d_2 from $\vec{e}_2$, its image $t(\vec{v})$ must be the unique point in the codomain that is determined by being d_0 from $t(\vec{0})$, d_1 from $t(\vec{e}_1)$, and d_2 from $t(\vec{e}_2)$. Because of the uniqueness, checking that the action in $(*)$ works in the d_0, d_1, and d_2 cases

$$\text{dist}(\begin{pmatrix} x \\ y \end{pmatrix}, \vec{0}) = \text{dist}(t(\begin{pmatrix} x \\ y \end{pmatrix}), t(\vec{0})) = \text{dist}(\begin{pmatrix} ax+cy \\ bx+dy \end{pmatrix}, \vec{0})$$

(we assumed that t maps $\vec{0}$ to itself)

$$\text{dist}(\begin{pmatrix} x \\ y \end{pmatrix}, \vec{e}_1) = \text{dist}(t(\begin{pmatrix} x \\ y \end{pmatrix}), t(\vec{e}_1)) = \text{dist}(\begin{pmatrix} ax + cy \\ bx + dy \end{pmatrix}, \begin{pmatrix} a \\ b \end{pmatrix})$$

and

$$\text{dist}(\begin{pmatrix} x \\ y \end{pmatrix}, \vec{e}_2) = \text{dist}(t(\begin{pmatrix} x \\ y \end{pmatrix}), t(\vec{e}_2)) = \text{dist}(\begin{pmatrix} ax + cy \\ bx + dy \end{pmatrix}, \begin{pmatrix} c \\ d \end{pmatrix})$$

suffices to show that $(*)$ describes t. Those checks are routine.

Thus any distance-preserving $f\colon \mathbb{R}^2 \to \mathbb{R}^2$ is a linear map plus a translation, $f(\vec{v}) = t(\vec{v}) + \vec{v}_0$ for some constant vector $\vec{v}_0$ and linear map t that is distance-preserving. So in order to understand distance-preserving maps what remains is to understand distance-preserving linear maps.

Not every linear map is distance-preserving. For example $\vec{v} \mapsto 2\vec{v}$ does not preserve distances.

But there is a neat characterization: a linear transformation t of the plane is distance-preserving if and only if both $\|t(\vec{e}_1)\| = \|t(\vec{e}_2)\| = 1$, and $t(\vec{e}_1)$ is orthogonal to $t(\vec{e}_2)$. The 'only if' half of that statement is easy — because t is distance-preserving it must preserve the lengths of vectors and because t is distance-preserving the Pythagorean theorem shows that it must preserve orthogonality. To show the 'if' half we can check that the map preserves lengths of vectors because then for all $\vec{p}$ and $\vec{q}$ the distance between the two is preserved $\|t(\vec{p} - \vec{q}\,)\| = \|t(\vec{p}) - t(\vec{q}\,)\| = \|\vec{p} - \vec{q}\,\|$. For that check let

$$\vec{v} = \begin{pmatrix} x \\ y \end{pmatrix} \qquad t(\vec{e}_1) = \begin{pmatrix} a \\ b \end{pmatrix} \qquad t(\vec{e}_2) = \begin{pmatrix} c \\ d \end{pmatrix}$$

and with the 'if' assumptions that $a^2 + b^2 = c^2 + d^2 = 1$ and $ac + bd = 0$ we have this.

$$\begin{aligned}
\|t(\vec{v}\,)\|^2 &= (ax + cy)^2 + (bx + dy)^2 \\
&= a^2x^2 + 2acxy + c^2y^2 + b^2x^2 + 2bdxy + d^2y^2 \\
&= x^2(a^2 + b^2) + y^2(c^2 + d^2) + 2xy(ac + bd) \\
&= x^2 + y^2 \\
&= \|\vec{v}\,\|^2
\end{aligned}$$

One thing that is neat about this characterization is that we can easily recognize matrices that represent such a map with respect to the standard bases: the columns are of length one and are mutually orthogonal. This is an *orthonormal matrix* (or, more informally, *orthogonal matrix* since people often use this term to mean not just that the columns are orthogonal but also that they have length one).

We can leverage this characterization to understand the geometric actions of distance-preserving maps. Because $\|t(\vec{v})\| = \|\vec{v}\|$, the map t sends any $\vec{v}$ somewhere on the circle about the origin that has radius equal to the length of $\vec{v}$. In particular, $\vec{e}_1$ and $\vec{e}_2$ map to the unit circle. What's more, once we fix the unit vector $\vec{e}_1$ as mapped to the vector with components a and b then there are only two places where $\vec{e}_2$ can go if its image is to be perpendicular to the first vector's image: it can map either to one where $\vec{e}_2$ maintains its position a quarter circle clockwise from $\vec{e}_1$

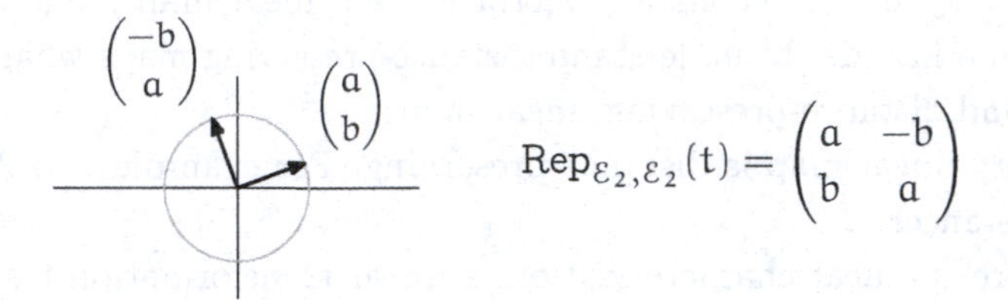

or to one where it goes a quarter circle counterclockwise.

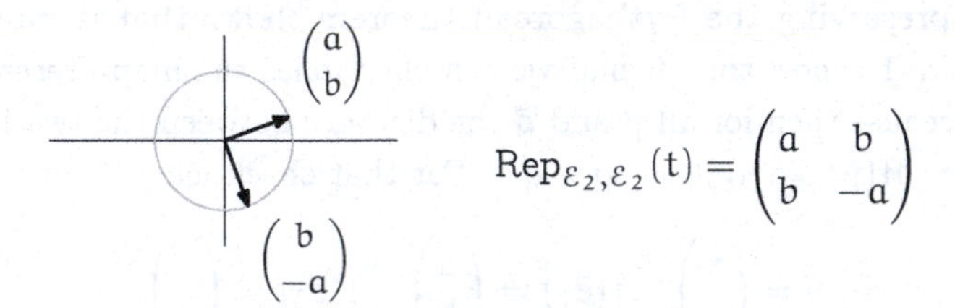

The geometric description of these two cases is easy. Let θ be the counterclockwise angle between the x-axis and the image of $\vec{e}_1$. The first matrix above represents, with respect to the standard bases, a *rotation* of the plane by θ radians.

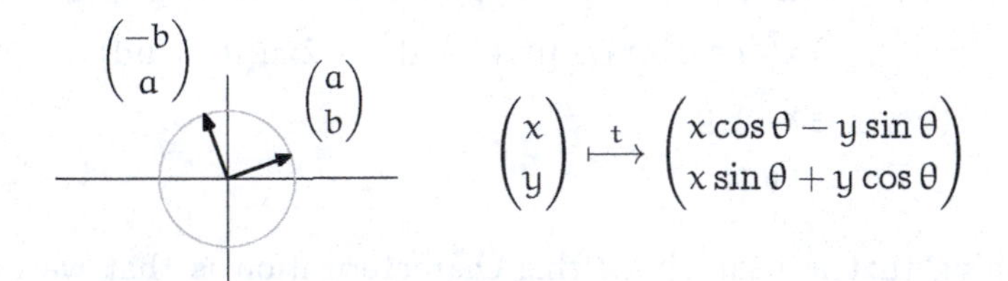

The second matrix above represents a *reflection* of the plane through the line bisecting the angle between $\vec{e}_1$ and $t(\vec{e}_1)$.

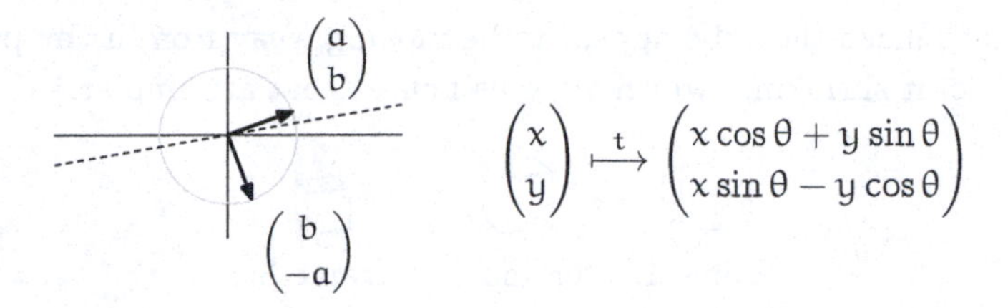

(This picture shows $\vec{e}_1$ reflected up into the first quadrant and $\vec{e}_2$ reflected down into the fourth quadrant.)

Note: in the domain the angle between $\vec{e}_1$ and $\vec{e}_2$ runs counterclockwise, and in the first map above the angle from $t(\vec{e}_1)$ to $t(\vec{e}_2)$ is also counterclockwise, so it preserves the orientation of the angle. But the second map reverses the orientation. A distance-preserving map is *direct* if it preserves orientations and *opposite* if it reverses orientation.

With that, we have characterized the Euclidean study of congruence. It considers, for plane figures, the properties that are invariant under combinations of (i) a rotation followed by a translation, or (ii) a reflection followed by a translation (a reflection followed by a non-trivial translation is a *glide reflection*).

Another idea encountered in elementary geometry, besides congruence of figures, is that figures are *similar* if they are congruent after a change of scale. The two triangles below are similar since the second is the same shape as the first but 3/2-ths the size.

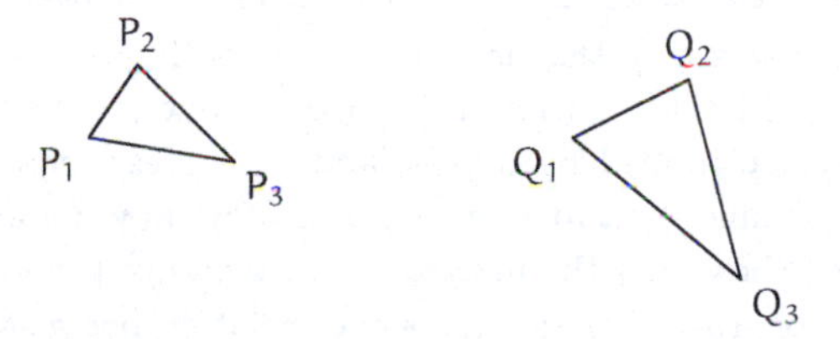

From the above work we have that figures are similar if there is an orthonormal matrix T such that the points $\vec{q}$ on one figure are the images of the points $\vec{p}$ on the other figure by $\vec{q} = (kT)\vec{v} + \vec{p}_0$ for some nonzero real number k and constant vector $\vec{p}_0$.

Although these ideas are from Euclid, mathematics is timeless and they are still in use today. One application of the maps studied above is in computer graphics. We can, for example, animate this top view of a cube by putting together film frames of it rotating; that's a rigid motion.

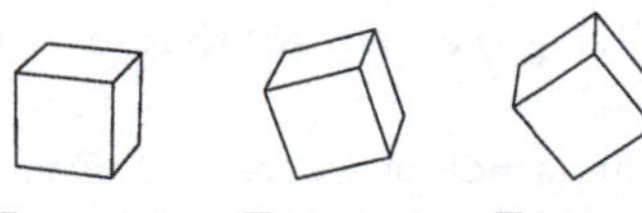

Frame 1 Frame 2 Frame 3

We could also make the cube appear to be moving away from us by producing film frames of it shrinking, which gives us figures that are similar.

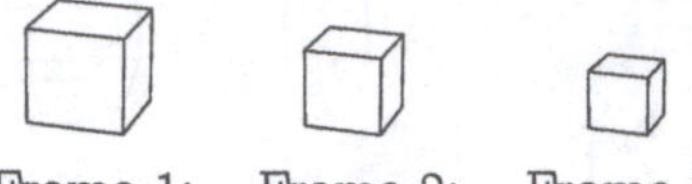

Frame 1: Frame 2: Frame 3:

Computer graphics incorporates techniques from linear algebra in many other ways (see Exercise 4).

A beautiful book that explores some of this area is [Weyl]. More on groups, of transformations and otherwise, is in any book on Modern Algebra, for instance [Birkhoff & MacLane]. More on Klein and the Erlanger Program is in [Yaglom].

Exercises

1 Decide if each of these is an orthonormal matrix.

(a) $\begin{pmatrix} 1/\sqrt{2} & -1/\sqrt{2} \\ -1/\sqrt{2} & -1/\sqrt{2} \end{pmatrix}$

(b) $\begin{pmatrix} 1/\sqrt{3} & -1/\sqrt{3} \\ -1/\sqrt{3} & -1/\sqrt{3} \end{pmatrix}$

(c) $\begin{pmatrix} 1/\sqrt{3} & -\sqrt{2}/\sqrt{3} \\ -\sqrt{2}/\sqrt{3} & -1/\sqrt{3} \end{pmatrix}$

2 Write down the formula for each of these distance-preserving maps.

(a) the map that rotates $\pi/6$ radians, and then translates by $\vec{e}_2$

(b) the map that reflects about the line $y = 2x$

(c) the map that reflects about $y = -2x$ and translates over 1 and up 1

3 (a) The proof that a map that is distance-preserving and sends the zero vector to itself incidentally shows that such a map is one-to-one and onto (the point in the domain determined by d_0, d_1, and d_2 corresponds to the point in the codomain determined by those three). Therefore any distance-preserving map has an inverse. Show that the inverse is also distance-preserving.

(b) Prove that congruence is an equivalence relation between plane figures.

4 In practice the matrix for the distance-preserving linear transformation and the translation are often combined into one. Check that these two computations yield the same first two components.

$$\begin{pmatrix} a & c \\ b & d \end{pmatrix}\begin{pmatrix} x \\ y \end{pmatrix} + \begin{pmatrix} e \\ f \end{pmatrix} \qquad \begin{pmatrix} a & c & e \\ b & d & f \\ 0 & 0 & 1 \end{pmatrix}\begin{pmatrix} x \\ y \\ 1 \end{pmatrix}$$

(These are *homogeneous coordinates*; see the Topic on Projective Geometry).

5 (a) Verify that the properties described in the second paragraph of this Topic as invariant under distance-preserving maps are indeed so.

(b) Give two more properties that are of interest in Euclidean geometry from your experience in studying that subject that are also invariant under distance-preserving maps.

(c) Give a property that is not of interest in Euclidean geometry and is not invariant under distance-preserving maps.

Chapter Four

Determinants

In the first chapter we highlighted the special case of linear systems with the same number of equations as unknowns, those of the form $T\vec{x} = \vec{b}$ where T is a square matrix. We noted that there are only two kinds of T's. If T is associated with a unique solution for any $\vec{b}$, such as for the homogeneous system $T\vec{x} = \vec{0}$, then T is associated with a unique solution for every such $\vec{b}$. We call such a matrix nonsingular. The other kind of T, where every linear system for which it is the matrix of coefficients has either no solution or infinitely many solutions, we call singular.

In our work since then this distinction has been a theme. For instance, we now know that an $n\times n$ matrix T is nonsingular if and only if each of these holds:

- any system $T\vec{x} = \vec{b}$ has a solution and that solution is unique;
- Gauss-Jordan reduction of T yields an identity matrix;
- the rows of T form a linearly independent set;
- the columns of T form a linearly independent set, a basis for $\mathbb{R}^n$;
- any map that T represents is an isomorphism;
- an inverse matrix T^{-1} exists.

So when we look at a square matrix, one of the first things that we ask is whether it is nonsingular.

This chapter develops a formula that determines whether T is nonsingular. More precisely, we will develop a formula for 1×1 matrices, one for 2×2 matrices, etc. These are naturally related; that is, we will develop a family of formulas, a scheme that describes the formula for each size.

Since we will restrict the discussion to square matrices, in this chapter we will often simply say 'matrix' in place of 'square matrix'.

I Definition

Determining nonsingularity is trivial for 1×1 matrices.

$$\begin{pmatrix} a \end{pmatrix} \quad \text{is nonsingular iff} \quad a \neq 0$$

Corollary Three.IV.4.11 gives the 2×2 formula.

$$\begin{pmatrix} a & b \\ c & d \end{pmatrix} \quad \text{is nonsingular iff} \quad ad - bc \neq 0$$

We can produce the 3×3 formula as we did the prior one, although the computation is intricate (see Exercise 9).

$$\begin{pmatrix} a & b & c \\ d & e & f \\ g & h & i \end{pmatrix} \quad \text{is nonsingular iff} \quad aei + bfg + cdh - hfa - idb - gec \neq 0$$

With these cases in mind, we posit a family of formulas: a, $ad-bc$, etc. For each n the formula defines a *determinant* function $\det_{n\times n} : \mathcal{M}_{n\times n} \to \mathbb{R}$ such that an $n\times n$ matrix T is nonsingular if and only if $\det_{n\times n}(T) \neq 0$. (We usually omit the subscript $n\times n$ because the size of T describes which determinant function we mean.)

I.1 Exploration

This subsection is an optional motivation and development of the general definition. The definition is in the next subsection.

Above, in each case the matrix is nonsingular if and only if some formula is nonzero. But the three formulas don't show an obvious pattern. We may spot that the 1×1 term a has one letter, that the 2×2 terms ad and bc have two letters, and that the 3×3 terms each have three letters. We may even spot that in those terms there is a letter from each row and column of the matrix, e.g., in the cdh term one letter comes from each row and from each column.

$$\begin{pmatrix} & & c \\ d & & \\ & h & \end{pmatrix}$$

But these observations are perhaps more puzzling than enlightening. For instance, we might wonder why some terms are added but some are subtracted.

A good strategy for solving problems is to explore which properties the solution must have, and then search for something with those properties. So we shall start by asking what properties we'd like the determinant formulas to have.

At this point, our main way to decide whether a matrix is singular or not is to do Gaussian reduction and then check whether the diagonal of the echelon form matrix has any zeroes, that is, whether the product down the diagonal is zero. So we could guess that whatever determinant formula we find, the proof that it is right may involve applying Gauss's Method to the matrix to show that in the end the product down the diagonal is zero if and only if our formula gives zero.

This suggests a plan: we will look for a family of determinant formulas that are unaffected by row operations and such that the determinant of an echelon form matrix is the product of its diagonal entries. In the rest of this subsection we will test this plan against the 2×2 and 3×3 formulas. In the end we will have to modify the "unaffected by row operations" part, but not by much.

First we check whether the 2×2 and 3×3 formulas are unaffected by the row operation of combining: if

$$T \xrightarrow{k\rho_i+\rho_j} \hat{T}$$

then is $\det(\hat{T}) = \det(T)$? This check of the 2×2 determinant after the $k\rho_1 + \rho_2$ operation

$$\det(\begin{pmatrix} a & b \\ ka+c & kb+d \end{pmatrix}) = a(kb+d) - (ka+c)b = ad - bc$$

shows that it is indeed unchanged, and the other 2×2 combination $k\rho_2+\rho_1$ gives the same result. Likewise, the 3×3 combination $k\rho_3 + \rho_2$ leaves the determinant unchanged

$$\begin{aligned}\det(\begin{pmatrix} a & b & c \\ kg+d & kh+e & ki+f \\ g & h & i \end{pmatrix}) &= a(kh+e)i + b(ki+f)g + c(kg+d)h \\ &\quad - h(ki+f)a - i(kg+d)b - g(kh+e)c \\ &= aei + bfg + cdh - hfa - idb - gec\end{aligned}$$

as do the other 3×3 row combination operations.

So there seems to be promise in the plan. Of course, perhaps if we had worked out the 4×4 determinant formula and tested it then we might have found that it is affected by row combinations. This is an exploration and we do not yet have all the facts. Nonetheless, so far, so good.

Next we compare $\det(\hat{T})$ with $\det(T)$ for row swaps. Here we hit a snag: the

2×2 row swap $\rho_1 \leftrightarrow \rho_2$ does not yield $ad - bc$.

$$\det(\begin{pmatrix} c & d \\ a & b \end{pmatrix}) = bc - ad$$

And this $\rho_1 \leftrightarrow \rho_3$ swap inside of a 3×3 matrix

$$\det(\begin{pmatrix} g & h & i \\ d & e & f \\ a & b & c \end{pmatrix}) = gec + hfa + idb - bfg - cdh - aei$$

also does not give the same determinant as before the swap since again there is a sign change. Trying a different 3×3 swap $\rho_1 \leftrightarrow \rho_2$

$$\det(\begin{pmatrix} d & e & f \\ a & b & c \\ g & h & i \end{pmatrix}) = dbi + ecg + fah - hcd - iae - gbf$$

also gives a change of sign.

So row swaps appear in this experiment to change the sign of a determinant. This does not wreck our plan entirely. We hope to decide nonsingularity by considering only whether the formula gives zero, not by considering its sign. Therefore, instead of expecting determinant formulas to be entirely unaffected by row operations we modify our plan so that on a swap they will change sign.

Obviously we finish by comparing $\det(\hat{T})$ with $\det(T)$ for the operation of multiplying a row by a scalar. This

$$\det(\begin{pmatrix} a & b \\ kc & kd \end{pmatrix}) = a(kd) - (kc)b = k\cdot(ad - bc)$$

ends with the entire determinant multiplied by k, and the other 2×2 case has the same result. This 3×3 case ends the same way

$$\begin{aligned}\det(\begin{pmatrix} a & b & c \\ d & e & f \\ kg & kh & ki \end{pmatrix}) &= ae(ki) + bf(kg) + cd(kh) \\ &\quad -(kh)fa - (ki)db - (kg)ec \\ &= k\cdot(aei + bfg + cdh - hfa - idb - gec)\end{aligned}$$

as do the other two 3×3 cases. These make us suspect that multiplying a row by k multiplies the determinant by k. As before, this modifies our plan but does not wreck it. We are asking only that the zero-ness of the determinant formula be unchanged, not focusing on the its sign or magnitude.

So in this exploration out plan got modified in some inessential ways and is now: we will look for $n\times n$ determinant functions that remain unchanged under the operation of row combination, that change sign on a row swap, that rescale on the rescaling of a row, and such that the determinant of an echelon form matrix is the product down the diagonal. In the next two subsections we will see that for each n there is one and only one such function.

Finally, for the next subsection note that factoring out scalars is a row-wise operation: here

$$\det(\begin{pmatrix} 3 & 3 & 9 \\ 2 & 1 & 1 \\ 5 & 11 & -5 \end{pmatrix}) = 3 \cdot \det(\begin{pmatrix} 1 & 1 & 3 \\ 2 & 1 & 1 \\ 5 & 11 & -5 \end{pmatrix})$$

the 3 comes only out of the top row only, leaving the other rows unchanged. Consequently in the definition of determinant we will write it as a function of the rows $\det(\vec{\rho}_1, \vec{\rho}_2, \ldots \vec{\rho}_n)$, rather than as $\det(T)$ or as a function of the entries $\det(t_{1,1}, \ldots, t_{n,n})$.

Exercises

✓ **1.1** Evaluate the determinant of each.

(a) $\begin{pmatrix} 3 & 1 \\ -1 & 1 \end{pmatrix}$ (b) $\begin{pmatrix} 2 & 0 & 1 \\ 3 & 1 & 1 \\ -1 & 0 & 1 \end{pmatrix}$ (c) $\begin{pmatrix} 4 & 0 & 1 \\ 0 & 0 & 1 \\ 1 & 3 & -1 \end{pmatrix}$

1.2 Evaluate the determinant of each.

(a) $\begin{pmatrix} 2 & 0 \\ -1 & 3 \end{pmatrix}$ (b) $\begin{pmatrix} 2 & 1 & 1 \\ 0 & 5 & -2 \\ 1 & -3 & 4 \end{pmatrix}$ (c) $\begin{pmatrix} 2 & 3 & 4 \\ 5 & 6 & 7 \\ 8 & 9 & 1 \end{pmatrix}$

✓ **1.3** Verify that the determinant of an upper-triangular 3×3 matrix is the product down the diagonal.

$$\det(\begin{pmatrix} a & b & c \\ 0 & e & f \\ 0 & 0 & i \end{pmatrix}) = aei$$

Do lower-triangular matrices work the same way?

✓ **1.4** Use the determinant to decide if each is singular or nonsingular.

(a) $\begin{pmatrix} 2 & 1 \\ 3 & 1 \end{pmatrix}$ (b) $\begin{pmatrix} 0 & 1 \\ 1 & -1 \end{pmatrix}$ (c) $\begin{pmatrix} 4 & 2 \\ 2 & 1 \end{pmatrix}$

1.5 Singular or nonsingular? Use the determinant to decide.

(a) $\begin{pmatrix} 2 & 1 & 1 \\ 3 & 2 & 2 \\ 0 & 1 & 4 \end{pmatrix}$ (b) $\begin{pmatrix} 1 & 0 & 1 \\ 2 & 1 & 1 \\ 4 & 1 & 3 \end{pmatrix}$ (c) $\begin{pmatrix} 2 & 1 & 0 \\ 3 & -2 & 0 \\ 1 & 0 & 0 \end{pmatrix}$

✓ **1.6** Each pair of matrices differ by one row operation. Use this operation to compare $\det(A)$ with $\det(B)$.

(a) $A = \begin{pmatrix} 1 & 2 \\ 2 & 3 \end{pmatrix}$ $B = \begin{pmatrix} 1 & 2 \\ 0 & -1 \end{pmatrix}$

(b) $A = \begin{pmatrix} 3 & 1 & 0 \\ 0 & 0 & 1 \\ 0 & 1 & 2 \end{pmatrix}$ $B = \begin{pmatrix} 3 & 1 & 0 \\ 0 & 1 & 2 \\ 0 & 0 & 1 \end{pmatrix}$

(c) $A = \begin{pmatrix} 1 & -1 & 3 \\ 2 & 2 & -6 \\ 1 & 0 & 4 \end{pmatrix}$ $B = \begin{pmatrix} 1 & -1 & 3 \\ 1 & 1 & -3 \\ 1 & 0 & 4 \end{pmatrix}$

1.7 Show this.

$$\det(\begin{pmatrix} 1 & 1 & 1 \\ a & b & c \\ a^2 & b^2 & c^2 \end{pmatrix}) = (b-a)(c-a)(c-b)$$

✓ 1.8 Which real numbers x make this matrix singular?

$$\begin{pmatrix} 12-x & 4 \\ 8 & 8-x \end{pmatrix}$$

1.9 Do the Gaussian reduction to check the formula for 3×3 matrices stated in the preamble to this section.

$$\begin{pmatrix} a & b & c \\ d & e & f \\ g & h & i \end{pmatrix} \text{ is nonsingular iff } aei + bfg + cdh - hfa - idb - gec \neq 0$$

1.10 Show that the equation of a line in $\mathbb{R}^2$ thru (x_1, y_1) and (x_2, y_2) is given by this determinant.

$$\det(\begin{pmatrix} x & y & 1 \\ x_1 & y_1 & 1 \\ x_2 & y_2 & 1 \end{pmatrix}) = 0 \qquad x_1 \neq x_2$$

✓ 1.11 Many people know this mnemonic for the determinant of a 3×3 matrix: first repeat the first two columns and then sum the products on the forward diagonals and subtract the products on the backward diagonals. That is, first write

$$\left(\begin{array}{ccc|cc} h_{1,1} & h_{1,2} & h_{1,3} & h_{1,1} & h_{1,2} \\ h_{2,1} & h_{2,2} & h_{2,3} & h_{2,1} & h_{2,2} \\ h_{3,1} & h_{3,2} & h_{3,3} & h_{3,1} & h_{3,2} \end{array}\right)$$

and then calculate this.

$$\begin{gathered} h_{1,1}h_{2,2}h_{3,3} + h_{1,2}h_{2,3}h_{3,1} + h_{1,3}h_{2,1}h_{3,2} \\ -h_{3,1}h_{2,2}h_{1,3} - h_{3,2}h_{2,3}h_{1,1} - h_{3,3}h_{2,1}h_{1,2} \end{gathered}$$

(a) Check that this agrees with the formula given in the preamble to this section.

(b) Does it extend to other-sized determinants?

1.12 The *cross product* of the vectors

$$\vec{x} = \begin{pmatrix} x_1 \\ x_2 \\ x_3 \end{pmatrix} \qquad \vec{y} = \begin{pmatrix} y_1 \\ y_2 \\ y_3 \end{pmatrix}$$

is the vector computed as this determinant.

$$\vec{x} \times \vec{y} = \det(\begin{pmatrix} \vec{e}_1 & \vec{e}_2 & \vec{e}_3 \\ x_1 & x_2 & x_3 \\ y_1 & y_2 & y_3 \end{pmatrix})$$

Note that the first row's entries are vectors, the vectors from the standard basis for $\mathbb{R}^3$. Show that the cross product of two vectors is perpendicular to each vector.

1.13 Prove that each statement holds for 2×2 matrices.

(a) The determinant of a product is the product of the determinants $\det(ST) = \det(S) \cdot \det(T)$.

(b) If T is invertible then the determinant of the inverse is the inverse of the determinant $\det(T^{-1}) = (\det(T))^{-1}$.

Matrices T and T′ are *similar* if there is a nonsingular matrix P such that $T' = PTP^{-1}$. (We shall look at this relationship in Chapter Five.) Show that similar 2×2 matrices have the same determinant.

✓ 1.14 Prove that the area of this region in the plane

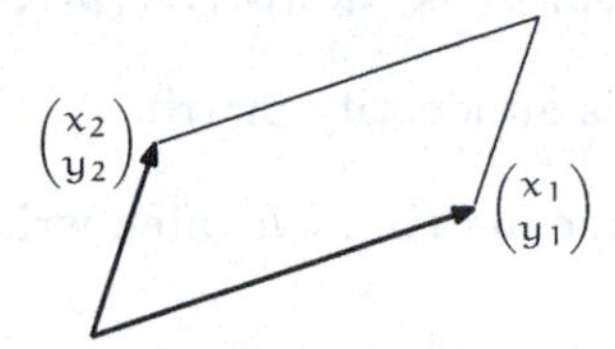

is equal to the value of this determinant.

$$\det(\begin{pmatrix} x_1 & x_2 \\ y_1 & y_2 \end{pmatrix})$$

Compare with this.

$$\det(\begin{pmatrix} x_2 & x_1 \\ y_2 & y_1 \end{pmatrix})$$

1.15 Prove that for 2×2 matrices, the determinant of a matrix equals the determinant of its transpose. Does that also hold for 3×3 matrices?

✓ 1.16 Is the determinant function linear — is $\det(x \cdot T + y \cdot S) = x \cdot \det(T) + y \cdot \det(S)$?

1.17 Show that if A is 3×3 then $\det(c \cdot A) = c^3 \cdot \det(A)$ for any scalar c.

1.18 Which real numbers θ make

$$\begin{pmatrix} \cos\theta & -\sin\theta \\ \sin\theta & \cos\theta \end{pmatrix}$$

singular? Explain geometrically.

? 1.19 [Am. Math. Mon., Apr. 1955] If a third order determinant has elements 1, 2, ..., 9, what is the maximum value it may have?

I.2 Properties of Determinants

We want a formula to determine whether an $n \times n$ matrix is nonsingular. We will not begin by stating such a formula. Instead we will begin by considering the function that such a formula calculates. We will define this function by its properties, then prove that the function with these properties exists and is unique, and also describe how to compute it. (Because we will eventually show that the function exists and is unique, from the start we will just say '$\det(T)$' instead of 'if there is a unique determinant function then $\det(T)$'.)

2.1 Definition A $n \times n$ *determinant* is a function $\det\colon \mathcal{M}_{n\times n} \to \mathbb{R}$ such that

(1) $\det(\vec{\rho}_1, \ldots, k\cdot\vec{\rho}_i + \vec{\rho}_j, \ldots, \vec{\rho}_n) = \det(\vec{\rho}_1, \ldots, \vec{\rho}_j, \ldots, \vec{\rho}_n)$ for $i \neq j$

(2) $\det(\vec{\rho}_1, \ldots, \vec{\rho}_j, \ldots, \vec{\rho}_i, \ldots, \vec{\rho}_n) = -\det(\vec{\rho}_1, \ldots, \vec{\rho}_i, \ldots, \vec{\rho}_j, \ldots, \vec{\rho}_n)$ for $i \neq j$

(3) $\det(\vec{\rho}_1, \ldots, k\vec{\rho}_i, \ldots, \vec{\rho}_n) = k\cdot\det(\vec{\rho}_1, \ldots, \vec{\rho}_i, \ldots, \vec{\rho}_n)$ for any scalar k

(4) $\det(I) = 1$ where I is an identity matrix

(the $\vec{\rho}$'s are the rows of the matrix). We often write $|T|$ for $\det(T)$.

2.2 Remark Condition (2) is redundant since

$$T \xrightarrow{\rho_i+\rho_j} \xrightarrow{-\rho_j+\rho_i} \xrightarrow{\rho_i+\rho_j} \xrightarrow{-\rho_i} \hat{T}$$

swaps rows i and j. We have listed it for consistency with the Gauss's Method presentation in earlier chapters.

2.3 Remark Condition (3) does not have a $k \neq 0$ restriction, although the Gauss's Method operation of multiplying a row by k does have it. The next result shows that we do not need that restriction here.

2.4 Lemma A matrix with two identical rows has a determinant of zero. A matrix with a zero row has a determinant of zero. A matrix is nonsingular if and only if its determinant is nonzero. The determinant of an echelon form matrix is the product down its diagonal.

Proof To verify the first sentence swap the two equal rows. The sign of the determinant changes but the matrix is the same and so its determinant is the same. Thus the determinant is zero.

For the second sentence multiply the zero row by two. That doubles the determinant but it also leaves the row unchanged, and hence leaves the determinant unchanged. Thus the determinant must be zero.

Do Gauss-Jordan reduction for the third sentence, $T \to \cdots \to \hat{T}$. By the first three properties the determinant of T is zero if and only if the determinant of $\hat{T}$ is zero (although the two could differ in sign or magnitude). A nonsingular matrix T Gauss-Jordan reduces to an identity matrix and so has a nonzero determinant. A singular T reduces to a $\hat{T}$ with a zero row; by the second sentence of this lemma its determinant is zero.

The fourth sentence has two cases. If the echelon form matrix is singular then it has a zero row. Thus it has a zero on its diagonal and the product down its diagonal is zero. By the third sentence of this result the determinant is zero and therefore this matrix's determinant equals the product down its diagonal.

If the echelon form matrix is nonsingular then none of its diagonal entries is zero. This means that we can divide by those entries and use condition (3) to get 1's on the diagonal.

$$\begin{vmatrix} t_{1,1} & t_{1,2} & & t_{1,n} \\ 0 & t_{2,2} & & t_{2,n} \\ & & \ddots & \\ 0 & & & t_{n,n} \end{vmatrix} = t_{1,1} \cdot t_{2,2} \cdots t_{n,n} \cdot \begin{vmatrix} 1 & t_{1,2}/t_{1,1} & & t_{1,n}/t_{1,1} \\ 0 & 1 & & t_{2,n}/t_{2,2} \\ & & \ddots & \\ 0 & & & 1 \end{vmatrix}$$

Then the Jordan half of Gauss-Jordan elimination leaves the identity matrix.

$$= t_{1,1} \cdot t_{2,2} \cdots t_{n,n} \cdot \begin{vmatrix} 1 & 0 & & 0 \\ 0 & 1 & & 0 \\ & & \ddots & \\ 0 & & & 1 \end{vmatrix} = t_{1,1} \cdot t_{2,2} \cdots t_{n,n} \cdot 1$$

So in this case also, the determinant is the product down the diagonal. QED

That gives us a way to compute the value of a determinant function on a matrix: do Gaussian reduction, keeping track of any changes of sign caused by row swaps and any scalars that we factor out, and finish by multiplying down the diagonal of the echelon form result. This algorithm is as fast as Gauss's Method and so is practical on all of the matrices that we will see.

2.5 Example Doing 2×2 determinants with Gauss's Method

$$\begin{vmatrix} 2 & 4 \\ -1 & 3 \end{vmatrix} = \begin{vmatrix} 2 & 4 \\ 0 & 5 \end{vmatrix} = 10$$

doesn't give a big time savings because the 2×2 determinant formula is easy. However, a 3×3 determinant is often easier to calculate with Gauss's Method than with its formula.

$$\begin{vmatrix} 2 & 2 & 6 \\ 4 & 4 & 3 \\ 0 & -3 & 5 \end{vmatrix} = \begin{vmatrix} 2 & 2 & 6 \\ 0 & 0 & -9 \\ 0 & -3 & 5 \end{vmatrix} = -\begin{vmatrix} 2 & 2 & 6 \\ 0 & -3 & 5 \\ 0 & 0 & -9 \end{vmatrix} = -54$$

2.6 Example Determinants bigger than 3×3 go quickly with the Gauss's Method procedure.

$$\begin{vmatrix} 1 & 0 & 1 & 3 \\ 0 & 1 & 1 & 4 \\ 0 & 0 & 0 & 5 \\ 0 & 1 & 0 & 1 \end{vmatrix} = \begin{vmatrix} 1 & 0 & 1 & 3 \\ 0 & 1 & 1 & 4 \\ 0 & 0 & 0 & 5 \\ 0 & 0 & -1 & -3 \end{vmatrix} = -\begin{vmatrix} 1 & 0 & 1 & 3 \\ 0 & 1 & 1 & 4 \\ 0 & 0 & -1 & -3 \\ 0 & 0 & 0 & 5 \end{vmatrix} = -(-5) = 5$$

The prior example illustrates an important point. Although we have not yet found a $4{\times}4$ determinant formula, if one exists then we know what value it gives to the matrix — if there is a function with properties (1)-(4) then on the above matrix the function must return 5.

2.7 Lemma For each n, if there is an $n{\times}n$ determinant function then it is unique.

PROOF Perform Gauss's Method on the matrix, keeping track of how the sign alternates on row swaps and any row-scaling factors, and then multiply down the diagonal of the echelon form result. By the definition and the lemma, all $n{\times}n$ determinant functions must return this value on the matrix. QED

The 'if there is an $n{\times}n$ determinant function' emphasizes that, although we can use Gauss's Method to compute the only value that a determinant function could possibly return, we haven't yet shown that such a function exists for all n. The rest of this section does that.

Exercises

For these, assume that an $n{\times}n$ determinant function exists for all n.

✓ **2.8** Use Gauss's Method to find each determinant.

(a) $\begin{vmatrix} 3 & 1 & 2 \\ 3 & 1 & 0 \\ 0 & 1 & 4 \end{vmatrix}$ **(b)** $\begin{vmatrix} 1 & 0 & 0 & 1 \\ 2 & 1 & 1 & 0 \\ -1 & 0 & 1 & 0 \\ 1 & 1 & 1 & 0 \end{vmatrix}$

2.9 Use Gauss's Method to find each.

(a) $\begin{vmatrix} 2 & -1 \\ -1 & -1 \end{vmatrix}$ **(b)** $\begin{vmatrix} 1 & 1 & 0 \\ 3 & 0 & 2 \\ 5 & 2 & 2 \end{vmatrix}$

2.10 For which values of k does this system have a unique solution?

$$\begin{aligned} x \quad + z - w &= 2 \\ y - 2z \quad &= 3 \\ x \quad + kz \quad &= 4 \\ z - w &= 2 \end{aligned}$$

✓ **2.11** Express each of these in terms of $|H|$.

(a) $\begin{vmatrix} h_{3,1} & h_{3,2} & h_{3,3} \\ h_{2,1} & h_{2,2} & h_{2,3} \\ h_{1,1} & h_{1,2} & h_{1,3} \end{vmatrix}$

(b) $\begin{vmatrix} -h_{1,1} & -h_{1,2} & -h_{1,3} \\ -2h_{2,1} & -2h_{2,2} & -2h_{2,3} \\ -3h_{3,1} & -3h_{3,2} & -3h_{3,3} \end{vmatrix}$

(c) $\begin{vmatrix} h_{1,1}+h_{3,1} & h_{1,2}+h_{3,2} & h_{1,3}+h_{3,3} \\ h_{2,1} & h_{2,2} & h_{2,3} \\ 5h_{3,1} & 5h_{3,2} & 5h_{3,3} \end{vmatrix}$

✓ **2.12** Find the determinant of a diagonal matrix.

2.13 Describe the solution set of a homogeneous linear system if the determinant of the matrix of coefficients is nonzero.

✓ **2.14** Show that this determinant is zero.

$$\begin{vmatrix} y+z & x+z & x+y \\ x & y & z \\ 1 & 1 & 1 \end{vmatrix}$$

2.15 (a) Find the 1×1, 2×2, and 3×3 matrices with i,j entry given by $(-1)^{i+j}$.
(b) Find the determinant of the square matrix with i,j entry $(-1)^{i+j}$.

2.16 (a) Find the 1×1, 2×2, and 3×3 matrices with i,j entry given by $i+j$.
(b) Find the determinant of the square matrix with i,j entry $i+j$.

✓ **2.17** Show that determinant functions are not linear by giving a case where $|A+B| \neq |A|+|B|$.

2.18 The second condition in the definition, that row swaps change the sign of a determinant, is somewhat annoying. It means we have to keep track of the number of swaps, to compute how the sign alternates. Can we get rid of it? Can we replace it with the condition that row swaps leave the determinant unchanged? (If so then we would need new 1×1, 2×2, and 3×3 formulas, but that would be a minor matter.)

2.19 Prove that the determinant of any triangular matrix, upper or lower, is the product down its diagonal.

2.20 Refer to the definition of elementary matrices in the Mechanics of Matrix Multiplication subsection.
(a) What is the determinant of each kind of elementary matrix?
(b) Prove that if E is any elementary matrix then $|ES|=|E||S|$ for any appropriately sized S.
(c) *(This question doesn't involve determinants.)* Prove that if T is singular then a product TS is also singular.
(d) Show that $|TS|=|T||S|$.
(e) Show that if T is nonsingular then $|T^{-1}|=|T|^{-1}$.

2.21 Prove that the determinant of a product is the product of the determinants $|TS| = |T|\,|S|$ in this way. Fix the $n\times n$ matrix S and consider the function $d\colon \mathcal{M}_{n\times n} \to \mathbb{R}$ given by $T \mapsto |TS|/|S|$.
(a) Check that d satisfies condition (1) in the definition of a determinant function.
(b) Check condition (2).
(c) Check condition (3).
(d) Check condition (4).
(e) Conclude the determinant of a product is the product of the determinants.

2.22 A *submatrix* of a given matrix A is one that we get by deleting some of the rows and columns of A. Thus, the first matrix here is a submatrix of the second.

$$\begin{pmatrix} 3 & 1 \\ 2 & 5 \end{pmatrix} \qquad \begin{pmatrix} 3 & 4 & 1 \\ 0 & 9 & -2 \\ 2 & -1 & 5 \end{pmatrix}$$

Prove that for any square matrix, the rank of the matrix is r if and only if r is the largest integer such that there is an $r\times r$ submatrix with a nonzero determinant.

✓ **2.23** Prove that a matrix with rational entries has a rational determinant.

? **2.24** [Am. Math. Mon., Feb. 1953] Find the element of likeness in (a) simplifying a fraction, (b) powdering the nose, (c) building new steps on the church, (d) keeping emeritus professors on campus, (e) putting B, C, D in the determinant

$$\begin{vmatrix} 1 & a & a^2 & a^3 \\ a^3 & 1 & a & a^2 \\ B & a^3 & 1 & a \\ C & D & a^3 & 1 \end{vmatrix}.$$

I.3 The Permutation Expansion

The prior subsection defines a function to be a determinant if it satisfies four conditions and shows that there is at most one $n\times n$ determinant function for each n. What is left is to show that for each n such a function exists.

But, we easily compute determinants: we use Gauss's Method, keeping track of the sign changes from row swaps, and end by multiplying down the diagonal. How could they not exist?

The difficulty is to show that the computation gives a well-defined—that is, unique—result. Consider these two Gauss's Method reductions of the same matrix, the first without any row swap

$$\begin{pmatrix} 1 & 2 \\ 3 & 4 \end{pmatrix} \xrightarrow{-3\rho_1+\rho_2} \begin{pmatrix} 1 & 2 \\ 0 & -2 \end{pmatrix}$$

and the second with one.

$$\begin{pmatrix} 1 & 2 \\ 3 & 4 \end{pmatrix} \xrightarrow{\rho_1\leftrightarrow\rho_2} \begin{pmatrix} 3 & 4 \\ 1 & 2 \end{pmatrix} \xrightarrow{-(1/3)\rho_1+\rho_2} \begin{pmatrix} 3 & 4 \\ 0 & 2/3 \end{pmatrix}$$

Both yield the determinant -2 since in the second one we note that the row swap changes the sign of the result we get by multiplying down the diagonal. The fact that we are able to proceed in two ways opens the possibility that the two give different answers. That is, the way that we have given to compute determinant values does not plainly eliminate the possibility that there might be, say, two reductions of some 7×7 matrix that lead to different determinant values. In that case we would not have a function, since the definition of a function is that for each input there must be exactly associated one output. The rest of this section shows that the definition Definition 2.1 never leads to a conflict.

To do this we will define an alternative way to find the value of a determinant. (This alternative is less useful in practice because it is slow. But it is very useful for theory.) The key idea is that condition (3) of Definition 2.1 shows that the determinant function is not linear.

3.1 Example With condition (3) scalars come out of each row separately,

$$\begin{vmatrix} 4 & 2 \\ -2 & 6 \end{vmatrix} = 2 \cdot \begin{vmatrix} 2 & 1 \\ -2 & 6 \end{vmatrix} = 4 \cdot \begin{vmatrix} 2 & 1 \\ -1 & 3 \end{vmatrix}$$

not from the entire matrix at once. So, where

$$A = \begin{pmatrix} 2 & 1 \\ -1 & 3 \end{pmatrix}$$

then $\det(2A) \neq 2 \cdot \det(A)$ (instead, $\det(2A) = 4 \cdot \det(A)$).

Since scalars come out a row at a time we might guess that determinants are linear a row at a time.

3.2 Definition Let V be a vector space. A map $f\colon V^n \to \mathbb{R}$ is *multilinear* if

(1) $f(\vec{\rho}_1, \ldots, \vec{v} + \vec{w}, \ldots, \vec{\rho}_n) = f(\vec{\rho}_1, \ldots, \vec{v}, \ldots, \vec{\rho}_n) + f(\vec{\rho}_1, \ldots, \vec{w}, \ldots, \vec{\rho}_n)$

(2) $f(\vec{\rho}_1, \ldots, k\vec{v}, \ldots, \vec{\rho}_n) = k \cdot f(\vec{\rho}_1, \ldots, \vec{v}, \ldots, \vec{\rho}_n)$

for $\vec{v}, \vec{w} \in V$ and $k \in \mathbb{R}$.

3.3 Lemma Determinants are multilinear.

Proof Property (2) here is just Definition 2.1's condition (3) so we need only verify property (1).

There are two cases. If the set of other rows $\{\vec{\rho}_1, \ldots, \vec{\rho}_{i-1}, \vec{\rho}_{i+1}, \ldots, \vec{\rho}_n\}$ is linearly dependent then all three matrices are singular and so all three determinants are zero and the equality is trivial.

Therefore assume that the set of other rows is linearly independent. We can make a basis by adding one more vector $\langle \vec{\rho}_1, \ldots, \vec{\rho}_{i-1}, \vec{\beta}, \vec{\rho}_{i+1}, \ldots, \vec{\rho}_n \rangle$. Express $\vec{v}$ and $\vec{w}$ with respect to this basis

$$\begin{aligned} \vec{v} &= v_1\vec{\rho}_1 + \cdots + v_{i-1}\vec{\rho}_{i-1} + v_i\vec{\beta} + v_{i+1}\vec{\rho}_{i+1} + \cdots + v_n\vec{\rho}_n \\ \vec{w} &= w_1\vec{\rho}_1 + \cdots + w_{i-1}\vec{\rho}_{i-1} + w_i\vec{\beta} + w_{i+1}\vec{\rho}_{i+1} + \cdots + w_n\vec{\rho}_n \end{aligned}$$

and add.

$$\vec{v} + \vec{w} = (v_1 + w_1)\vec{\rho}_1 + \cdots + (v_i + w_i)\vec{\beta} + \cdots + (v_n + w_n)\vec{\rho}_n$$

Consider the left side of (1) and expand $\vec{v} + \vec{w}$.

$$\det(\vec{\rho}_1, \ldots, \;(v_1 + w_1)\vec{\rho}_1 + \cdots + (v_i + w_i)\vec{\beta} + \cdots + (v_n + w_n)\vec{\rho}_n, \;\ldots, \vec{\rho}_n) \quad (*)$$

By the definition of determinant's condition (1), the value of $(*)$ is unchanged by the operation of adding $-(v_1+w_1)\vec{\rho}_1$ to the i-th row $\vec{v}+\vec{w}$. The i-th row becomes this.

$$\vec{v}+\vec{w}-(v_1+w_1)\vec{\rho}_1=(v_2+w_2)\vec{\rho}_2+\cdots+(v_i+w_i)\vec{\beta}+\cdots+(v_n+w_n)\vec{\rho}_n$$

Next add $-(v_2+w_2)\vec{\rho}_2$, etc., to eliminate all of the terms from the other rows. Apply condition (3) from the definition of determinant.

$$\begin{aligned}\det(\vec{\rho}_1,\ldots,\vec{v}+\vec{w},\ldots,\vec{\rho}_n)&\\ &=\det(\vec{\rho}_1,\ldots,(v_i+w_i)\cdot\vec{\beta},\ldots,\vec{\rho}_n)\\ &=(v_i+w_i)\cdot\det(\vec{\rho}_1,\ldots,\vec{\beta},\ldots,\vec{\rho}_n)\\ &=v_i\cdot\det(\vec{\rho}_1,\ldots,\vec{\beta},\ldots,\vec{\rho}_n)+w_i\cdot\det(\vec{\rho}_1,\ldots,\vec{\beta},\ldots,\vec{\rho}_n)\end{aligned}$$

Now this is a sum of two determinants. To finish, bring v_i and w_i back inside in front of the $\vec{\beta}$'s and use row combinations again, this time to reconstruct the expressions of $\vec{v}$ and $\vec{w}$ in terms of the basis. That is, start with the operations of adding $v_1\vec{\rho}_1$ to $v_i\vec{\beta}$ and $w_1\vec{\rho}_1$ to $w_i\vec{\rho}_1$, etc., to get the expansions of $\vec{v}$ and $\vec{w}$. QED

Multilinearity allows us to expand a determinant into a sum of determinants, each of which involves a simple matrix.

3.4 Example Use property (1) of multilinearity to break up the first row

$$\begin{vmatrix}2&1\\4&3\end{vmatrix}=\begin{vmatrix}2&0\\4&3\end{vmatrix}+\begin{vmatrix}0&1\\4&3\end{vmatrix}$$

and then use (1) again to break each along the second row.

$$=\begin{vmatrix}2&0\\4&0\end{vmatrix}+\begin{vmatrix}2&0\\0&3\end{vmatrix}+\begin{vmatrix}0&1\\4&0\end{vmatrix}+\begin{vmatrix}0&1\\0&3\end{vmatrix}$$

The result is four determinants. In each row of each of the four there is a single entry from the original matrix.

3.5 Example In the same way, a 3×3 determinant separates into a sum of many simpler determinants. Splitting along the first row produces three determinants (we have highlighted the zero in the $1,3$ position to set it off visually from the zeroes that appear as part of the splitting).

$$\begin{vmatrix}2&1&-1\\4&3&\boxed{0}\\2&1&5\end{vmatrix}=\begin{vmatrix}2&0&0\\4&3&\boxed{0}\\2&1&5\end{vmatrix}+\begin{vmatrix}0&1&0\\4&3&\boxed{0}\\2&1&5\end{vmatrix}+\begin{vmatrix}0&0&-1\\4&3&\boxed{0}\\2&1&5\end{vmatrix}$$

In turn, each of the above splits in three along the second row. Then each of the nine splits in three along the third row. The result is twenty seven determinants, such that each row contains a single entry from the starting matrix.

$$= \begin{vmatrix} 2 & 0 & 0 \\ 4 & 0 & 0 \\ 2 & 0 & 0 \end{vmatrix} + \begin{vmatrix} 2 & 0 & 0 \\ 4 & 0 & 0 \\ 0 & 1 & 0 \end{vmatrix} + \begin{vmatrix} 2 & 0 & 0 \\ 4 & 0 & 0 \\ 0 & 0 & 5 \end{vmatrix} + \begin{vmatrix} 2 & 0 & 0 \\ 0 & 3 & 0 \\ 2 & 0 & 0 \end{vmatrix} + \cdots + \begin{vmatrix} 0 & 0 & -1 \\ 0 & 0 & \boxed{0} \\ 0 & 0 & 5 \end{vmatrix}$$

So multilinearity will expand an $n \times n$ determinant into a sum of n^n-many determinants, where each row of each determinant contains a single entry from the starting matrix.

In this expansion, although there are lots of terms, most of them have a determinant of zero.

3.6 Example In each of these examples from the prior expansion, two of the entries from the original matrix are in the same column.

$$\begin{vmatrix} 2 & 0 & 0 \\ 4 & 0 & 0 \\ 0 & 1 & 0 \end{vmatrix} \qquad \begin{vmatrix} 0 & 0 & -1 \\ 0 & 3 & 0 \\ 0 & 0 & 5 \end{vmatrix} \qquad \begin{vmatrix} 0 & 1 & 0 \\ 0 & 0 & \boxed{0} \\ 0 & 0 & 5 \end{vmatrix}$$

For instance, in the first matrix the 2 and the 4 both come from the first column of the original matrix. In the second matrix the -1 and 5 both come from the third column. And in the third matrix the 0 and 5 both come from the third column. Any such matrix is singular because one row is a multiple of the other. Thus any such determinant is zero, by Lemma 2.4.

With that observation the above expansion of the 3×3 determinant into the sum of the twenty seven determinants simplifies to the sum of these six where the entries from the original matrix come one per row, and also one per column.

$$\begin{aligned} \begin{vmatrix} 2 & 1 & -1 \\ 4 & 3 & \boxed{0} \\ 2 & 1 & 5 \end{vmatrix} &= \begin{vmatrix} 2 & 0 & 0 \\ 0 & 3 & 0 \\ 0 & 0 & 5 \end{vmatrix} + \begin{vmatrix} 2 & 0 & 0 \\ 0 & 0 & \boxed{0} \\ 0 & 1 & 0 \end{vmatrix} \\ &\quad + \begin{vmatrix} 0 & 1 & 0 \\ 4 & 0 & 0 \\ 0 & 0 & 5 \end{vmatrix} + \begin{vmatrix} 0 & 1 & 0 \\ 0 & 0 & \boxed{0} \\ 2 & 0 & 0 \end{vmatrix} \\ &\quad + \begin{vmatrix} 0 & 0 & -1 \\ 4 & 0 & 0 \\ 0 & 1 & 0 \end{vmatrix} + \begin{vmatrix} 0 & 0 & -1 \\ 0 & 3 & 0 \\ 2 & 0 & 0 \end{vmatrix} \end{aligned}$$

In that expansion we can bring out the scalars.

$$= (2)(3)(5)\begin{vmatrix}1&0&0\\0&1&0\\0&0&1\end{vmatrix} + (2)(\boxed{0})(1)\begin{vmatrix}1&0&0\\0&0&1\\0&1&0\end{vmatrix}$$
$$+ (1)(4)(5)\begin{vmatrix}0&1&0\\1&0&0\\0&0&1\end{vmatrix} + (1)(\boxed{0})(2)\begin{vmatrix}0&1&0\\0&0&1\\1&0&0\end{vmatrix}$$
$$+ (-1)(4)(1)\begin{vmatrix}0&0&1\\1&0&0\\0&1&0\end{vmatrix} + (-1)(3)(2)\begin{vmatrix}0&0&1\\0&1&0\\1&0&0\end{vmatrix}$$

To finish, evaluate those six determinants by row-swapping them to the identity matrix, keeping track of the sign changes.

$$\begin{aligned}&= 30\cdot(+1) + 0\cdot(-1)\\ &\quad + 20\cdot(-1) + 0\cdot(+1)\\ &\quad - 4\cdot(+1) - 6\cdot(-1) = 12\end{aligned}$$

That example captures this subsection's new calculation scheme. Multilinearity expands a determinant into many separate determinants, each with one entry from the original matrix per row. Most of these have one row that is a multiple of another so we omit them. We are left with the determinants that have one entry per row and column from the original matrix. Factoring out the scalars further reduces the determinants that we must compute to the one-entry-per-row-and-column matrices where all entries are 1's.

Recall Definition Three.IV.3.14, that a *permutation matrix* is square, with entries 0's except for a single 1 in each row and column. We now introduce a notation for permutation matrices.

3.7 Definition An n-*permutation* is a function on the first n positive integers $\phi\colon \{1,\ldots,n\}\to\{1,\ldots,n\}$ that is one-to-one and onto.

In a permutation each number $1, \ldots, n$ appears as output for one and only one input. We can denote a permutation as a sequence $\phi = \langle\phi(1),\phi(2),\ldots,\phi(n)\rangle$.

3.8 Example The 2-permutations are the functions $\phi_1\colon\{1,2\}\to\{1,2\}$ given by $\phi_1(1)=1$, $\phi_1(2)=2$, and $\phi_2\colon\{1,2\}\to\{1,2\}$ given by $\phi_2(1)=2$, $\phi_2(2)=1$. The sequence notation is shorter: $\phi_1=\langle 1,2\rangle$ and $\phi_2=\langle 2,1\rangle$.

3.9 Example In the sequence notation the 3-permutations are $\phi_1=\langle 1,2,3\rangle$, $\phi_2=\langle 1,3,2\rangle$, $\phi_3=\langle 2,1,3\rangle$, $\phi_4=\langle 2,3,1\rangle$, $\phi_5=\langle 3,1,2\rangle$, and $\phi_6=\langle 3,2,1\rangle$.

We denote the row vector that is all 0's except for a 1 in entry j with ι_j so that the four-wide ι_2 is $(0\ 1\ 0\ 0)$. Now our notation for permutation matrices is: with any $\phi = \langle\phi(1),\ldots,\phi(n)\rangle$ associate the matrix whose rows are $\iota_{\phi(1)}$, $\ldots$, $\iota_{\phi(n)}$. For instance, associated with the 4-permutation $\phi = \langle 3,2,1,4\rangle$ is the matrix whose rows are the corresponding ι's.

$$P_\phi = \begin{pmatrix} \iota_3 \\ \iota_2 \\ \iota_1 \\ \iota_4 \end{pmatrix} = \begin{pmatrix} 0 & 0 & 1 & 0 \\ 0 & 1 & 0 & 0 \\ 1 & 0 & 0 & 0 \\ 0 & 0 & 0 & 1 \end{pmatrix}$$

3.10 Example These are the permutation matrices for the 2-permutations listed in Example 3.8.

$$P_{\phi_1} = \begin{pmatrix} \iota_1 \\ \iota_2 \end{pmatrix} = \begin{pmatrix} 1 & 0 \\ 0 & 1 \end{pmatrix} \qquad P_{\phi_2} = \begin{pmatrix} \iota_2 \\ \iota_1 \end{pmatrix} = \begin{pmatrix} 0 & 1 \\ 1 & 0 \end{pmatrix}$$

For instance, P_{ϕ_2}'s first row is $\iota_{\phi_2(1)} = \iota_2$ and its second is $\iota_{\phi_2(2)} = \iota_1$.

3.11 Example Consider the 3-permutation $\phi_5 = \langle 3,1,2\rangle$. The permutation matrix P_{ϕ_5} has rows $\iota_{\phi_5(1)} = \iota_3$, $\iota_{\phi_5(2)} = \iota_1$, and $\iota_{\phi_5(3)} = \iota_2$.

$$P_{\phi_5} = \begin{pmatrix} 0 & 0 & 1 \\ 1 & 0 & 0 \\ 0 & 1 & 0 \end{pmatrix}$$

3.12 Definition The *permutation expansion* for determinants is

$$\begin{vmatrix} t_{1,1} & t_{1,2} & \ldots & t_{1,n} \\ t_{2,1} & t_{2,2} & \ldots & t_{2,n} \\ & \vdots & & \\ t_{n,1} & t_{n,2} & \ldots & t_{n,n} \end{vmatrix} = \begin{aligned} & t_{1,\phi_1(1)}t_{2,\phi_1(2)}\cdots t_{n,\phi_1(n)}|P_{\phi_1}| \\ & + t_{1,\phi_2(1)}t_{2,\phi_2(2)}\cdots t_{n,\phi_2(n)}|P_{\phi_2}| \\ & \vdots \\ & + t_{1,\phi_k(1)}t_{2,\phi_k(2)}\cdots t_{n,\phi_k(n)}|P_{\phi_k}| \end{aligned}$$

where $\phi_1,\ldots,\phi_k$ are all of the n-permutations.

We can restate the formula in *summation notation*

$$|T| = \sum_{\text{permutations } \phi} t_{1,\phi(1)}t_{2,\phi(2)}\cdots t_{n,\phi(n)}\,|P_\phi|$$

read aloud as, "the sum, over all permutations ϕ, of terms having the form $t_{1,\phi(1)}t_{2,\phi(2)}\cdots t_{n,\phi(n)}|P_\phi|$."

3.13 Example The familiar 2×2 determinant formula follows from the above

$$\begin{aligned}\begin{vmatrix}t_{1,1} & t_{1,2}\\ t_{2,1} & t_{2,2}\end{vmatrix} &= t_{1,1}t_{2,2}\cdot|P_{\phi_1}|+t_{1,2}t_{2,1}\cdot|P_{\phi_2}|\\ &= t_{1,1}t_{2,2}\cdot\begin{vmatrix}1&0\\0&1\end{vmatrix}+t_{1,2}t_{2,1}\cdot\begin{vmatrix}0&1\\1&0\end{vmatrix}\\ &= t_{1,1}t_{2,2}-t_{1,2}t_{2,1}\end{aligned}$$

as does the 3×3 formula.

$$\begin{aligned}\begin{vmatrix}t_{1,1} & t_{1,2} & t_{1,3}\\ t_{2,1} & t_{2,2} & t_{2,3}\\ t_{3,1} & t_{3,2} & t_{3,3}\end{vmatrix} &= t_{1,1}t_{2,2}t_{3,3}\,|P_{\phi_1}|+t_{1,1}t_{2,3}t_{3,2}\,|P_{\phi_2}|+t_{1,2}t_{2,1}t_{3,3}\,|P_{\phi_3}|\\ &\quad+t_{1,2}t_{2,3}t_{3,1}\,|P_{\phi_4}|+t_{1,3}t_{2,1}t_{3,2}\,|P_{\phi_5}|+t_{1,3}t_{2,2}t_{3,1}\,|P_{\phi_6}|\\ &= t_{1,1}t_{2,2}t_{3,3}-t_{1,1}t_{2,3}t_{3,2}-t_{1,2}t_{2,1}t_{3,3}\\ &\quad+t_{1,2}t_{2,3}t_{3,1}+t_{1,3}t_{2,1}t_{3,2}-t_{1,3}t_{2,2}t_{3,1}\end{aligned}$$

Computing a determinant with the permutation expansion typically takes longer than with Gauss's Method. However, we will use it to prove that the determinant function is well-defined. We will just state the result here and defer its proof to the following subsection.

3.14 Theorem For each n there is an $n\times n$ determinant function.

Also in the next subsection is the proof of the result below (they are together because the two proofs overlap).

3.15 Theorem The determinant of a matrix equals the determinant of its transpose.

Because of this theorem, while we have so far stated determinant results in terms of rows, all of the results also hold in terms of columns.

3.16 Corollary A matrix with two equal columns is singular. Column swaps change the sign of a determinant. Determinants are multilinear in their columns.

Proof For the first statement, transposing the matrix results in a matrix with the same determinant, and with two equal rows, and hence a determinant of zero. Prove the other two in the same way. QED

We finish this subsection with a summary: determinant functions exist, are unique, and we know how to compute them. As for what determinants are about, perhaps these lines [Kemp] help make it memorable.

Determinant none,
 Solution: lots or none.
Determinant some,
 Solution: just one.

Exercises

This summarizes our notation for the 2- and 3- permutations.

i	1	2
$\phi_1(i)$	1	2
$\phi_2(i)$	2	1

i	1	2	3
$\phi_1(i)$	1	2	3
$\phi_2(i)$	1	3	2
$\phi_3(i)$	2	1	3
$\phi_4(i)$	2	3	1
$\phi_5(i)$	3	1	2
$\phi_6(i)$	3	2	1

✓ **3.17** Compute the determinant by using the permutation expansion.

(a) $\begin{vmatrix} 1 & 2 & 3 \\ 4 & 5 & 6 \\ 7 & 8 & 9 \end{vmatrix}$ (b) $\begin{vmatrix} 2 & 2 & 1 \\ 3 & -1 & 0 \\ -2 & 0 & 5 \end{vmatrix}$

✓ **3.18** Compute these both with Gauss's Method and the permutation expansion formula.

(a) $\begin{vmatrix} 2 & 1 \\ 3 & 1 \end{vmatrix}$ (b) $\begin{vmatrix} 0 & 1 & 4 \\ 0 & 2 & 3 \\ 1 & 5 & 1 \end{vmatrix}$

✓ **3.19** Use the permutation expansion formula to derive the formula for 3×3 determinants.

3.20 List all of the 4-permutations.

3.21 A permutation, regarded as a function from the set $\{1,..,n\}$ to itself, is one-to-one and onto. Therefore, each permutation has an inverse.

(a) Find the inverse of each 2-permutation.

(b) Find the inverse of each 3-permutation.

3.22 Prove that f is multilinear if and only if for all $\vec{v},\vec{w}\in V$ and $k_1,k_2\in\mathbb{R}$, this holds.

$$f(\vec{\rho}_1,\ldots,k_1\vec{v}_1+k_2\vec{v}_2,\ldots,\vec{\rho}_n)=k_1f(\vec{\rho}_1,\ldots,\vec{v}_1,\ldots,\vec{\rho}_n)+k_2f(\vec{\rho}_1,\ldots,\vec{v}_2,\ldots,\vec{\rho}_n)$$

3.23 How would determinants change if we changed property (4) of the definition to read that $|I|=2$?

3.24 Verify the second and third statements in Corollary 3.16.

✓ **3.25** Show that if an $n\times n$ matrix has a nonzero determinant then we can express any column vector $\vec{v}\in\mathbb{R}^n$ as a linear combination of the columns of the matrix.

3.26 [Strang 80] True or false: a matrix whose entries are only zeros or ones has a determinant equal to zero, one, or negative one.

3.27 (a) Show that there are 120 terms in the permutation expansion formula of a 5×5 matrix.

(b) How many are sure to be zero if the $1,2$ entry is zero?

3.28 How many n-permutations are there?

3.29 Show that the inverse of a permutation matrix is its transpose.

3.30 A matrix A is *skew-symmetric* if $A^{\mathsf{T}} = -A$, as in this matrix.

$$A = \begin{pmatrix} 0 & 3 \\ -3 & 0 \end{pmatrix}$$

Show that $n \times n$ skew-symmetric matrices with nonzero determinants exist only for even n.

✓ **3.31** What is the smallest number of zeros, and the placement of those zeros, needed to ensure that a 4×4 matrix has a determinant of zero?

✓ **3.32** If we have n data points $(x_1, y_1), (x_2, y_2), \ldots, (x_n, y_n)$ and want to find a polynomial $p(x) = a_{n-1}x^{n-1} + a_{n-2}x^{n-2} + \cdots + a_1 x + a_0$ passing through those points then we can plug in the points to get an n equation/n unknown linear system. The matrix of coefficients for that system is the *Vandermonde matrix*. Prove that the determinant of the transpose of that matrix of coefficients

$$\begin{vmatrix} 1 & 1 & \ldots & 1 \\ x_1 & x_2 & \ldots & x_n \\ {x_1}^2 & {x_2}^2 & \ldots & {x_n}^2 \\ & \vdots & & \\ {x_1}^{n-1} & {x_2}^{n-1} & \ldots & {x_n}^{n-1} \end{vmatrix}$$

equals the product, over all indices $i, j \in \{1, \ldots, n\}$ with $i < j$, of terms of the form $x_j - x_i$. (This shows that the determinant is zero, and the linear system has no solution, if and only if the x_i's in the data are not distinct.)

3.33 We can divide a matrix into *blocks*, as here,

$$\left(\begin{array}{cc|c} 1 & 2 & 0 \\ 3 & 4 & 0 \\ \hline 0 & 0 & -2 \end{array}\right)$$

which shows four blocks, the square 2×2 and 1×1 ones in the upper left and lower right, and the zero blocks in the upper right and lower left. Show that if a matrix is such that we can partition it as

$$T = \left(\begin{array}{c|c} J & Z_2 \\ \hline Z_1 & K \end{array}\right)$$

where J and K are square, and Z_1 and Z_2 are all zeroes, then $|T| = |J| \cdot |K|$.

✓ **3.34** Prove that for any $n \times n$ matrix T there are at most n distinct reals r such that the matrix $T - rI$ has determinant zero (we shall use this result in Chapter Five).

? **3.35** [Math. Mag., Jan. 1963, Q307] The nine positive digits can be arranged into 3×3 arrays in $9!$ ways. Find the sum of the determinants of these arrays.

3.36 [Math. Mag., Jan. 1963, Q237] Show that

$$\begin{vmatrix} x-2 & x-3 & x-4 \\ x+1 & x-1 & x-3 \\ x-4 & x-7 & x-10 \end{vmatrix} = 0.$$

? **3.37** [Am. Math. Mon., Jan. 1949] Let S be the sum of the integer elements of a magic square of order three and let D be the value of the square considered as a determinant. Show that D/S is an integer.

? **3.38** [Am. Math. Mon., Jun. 1931] Show that the determinant of the n^2 elements in the upper left corner of the Pascal triangle

$$\begin{matrix} 1 & 1 & 1 & 1 & . & . \\ 1 & 2 & 3 & . & . & \\ 1 & 3 & . & . & & \\ 1 & . & . & & & \\ . & & & & & \\ . & & & & & \end{matrix}$$

has the value unity.

I.4 Determinants Exist

This subsection contains proofs of two results from the prior subsection. It is optional. We will use the material developed here only in the Jordan Canonical Form subsection, which is also optional.

We wish to show that for any size n, the determinant function on $n{\times}n$ matrices is well-defined. The prior subsection develops the permutation expansion formula.

$$\begin{aligned}
\begin{vmatrix} t_{1,1} & t_{1,2} & \ldots & t_{1,n} \\ t_{2,1} & t_{2,2} & \ldots & t_{2,n} \\ & \vdots & & \\ t_{n,1} & t_{n,2} & \ldots & t_{n,n} \end{vmatrix}
&= t_{1,\phi_1(1)}t_{2,\phi_1(2)}\cdots t_{n,\phi_1(n)}|P_{\phi_1}| \\
&\quad + t_{1,\phi_2(1)}t_{2,\phi_2(2)}\cdots t_{n,\phi_2(n)}|P_{\phi_2}| \\
&\quad \vdots \\
&\quad + t_{1,\phi_k(1)}t_{2,\phi_k(2)}\cdots t_{n,\phi_k(n)}|P_{\phi_k}| \\
&= \sum_{\text{permutations }\phi} t_{1,\phi(1)}t_{2,\phi(2)}\cdots t_{n,\phi(n)}\,|P_\phi|
\end{aligned}$$

This reduces the problem of showing that the determinant is well-defined to only showing that the determinant is well-defined on the set of permutation matrices.

A permutation matrix can be row-swapped to the identity matrix. So one way that we can calculate its determinant is by keeping track of the number of swaps. However, we still must show that the result is well-defined. Recall what the difficulty is: the determinant of

$$P_\phi = \begin{pmatrix} 0 & 1 & 0 & 0 \\ 1 & 0 & 0 & 0 \\ 0 & 0 & 1 & 0 \\ 0 & 0 & 0 & 1 \end{pmatrix}$$

could be computed with one swap

$$P_\phi \xrightarrow{\rho_1\leftrightarrow\rho_2} \begin{pmatrix} 1&0&0&0\\ 0&1&0&0\\ 0&0&1&0\\ 0&0&0&1 \end{pmatrix}$$

or with three.

$$P_\phi \xrightarrow{\rho_3\leftrightarrow\rho_1} \xrightarrow{\rho_2\leftrightarrow\rho_3} \xrightarrow{\rho_1\leftrightarrow\rho_3} \begin{pmatrix} 1&0&0&0\\ 0&1&0&0\\ 0&0&1&0\\ 0&0&0&1 \end{pmatrix}$$

Both reductions have an odd number of swaps so in this case we figure that $|P_\phi| = -1$ but if there were some way to do it with an even number of swaps then we would have the determinant giving two different outputs from a single input. Below, Corollary 4.5 proves that this cannot happen—there is no permutation matrix that can be row-swapped to an identity matrix in two ways, one with an even number of swaps and the other with an odd number of swaps.

4.1 Definition In a permutation $\phi = \langle \ldots, k, \ldots, j, \ldots \rangle$, elements such that $k > j$ are in an *inversion* of their natural order. Similarly, in a permutation matrix two rows

$$P_\phi = \begin{pmatrix} \vdots \\ \iota_k \\ \vdots \\ \iota_j \\ \vdots \end{pmatrix}$$

such that $k > j$ are in an *inversion*.

4.2 Example This permutation matrix

$$\begin{pmatrix} 1&0&0&0\\ 0&0&1&0\\ 0&1&0&0\\ 0&0&0&1 \end{pmatrix} = \begin{pmatrix} \iota_1 \\ \iota_3 \\ \iota_2 \\ \iota_4 \end{pmatrix}$$

has a single inversion, that ι_3 precedes ι_2.

4.3 Example There are three inversions here:

$$\begin{pmatrix} 0&0&1\\ 0&1&0\\ 1&0&0 \end{pmatrix} = \begin{pmatrix} \iota_3 \\ \iota_2 \\ \iota_1 \end{pmatrix}$$

ι_3 precedes ι_1, ι_3 precedes ι_2, and ι_2 precedes ι_1.

4.4 Lemma A row-swap in a permutation matrix changes the number of inversions from even to odd, or from odd to even.

Proof Consider a swap of rows j and k, where $k > j$.

If the two rows are adjacent

$$P_\phi = \begin{pmatrix} \vdots \\ \iota_{\phi(j)} \\ \iota_{\phi(k)} \\ \vdots \end{pmatrix} \xrightarrow{\rho_k \leftrightarrow \rho_j} \begin{pmatrix} \vdots \\ \iota_{\phi(k)} \\ \iota_{\phi(j)} \\ \vdots \end{pmatrix}$$

then since inversions involving rows not in this pair are not affected, the swap changes the total number of inversions by one, either removing or producing one inversion depending on whether $\phi(j) > \phi(k)$ or not. Consequently, the total number of inversions changes from odd to even or from even to odd.

If the rows are not adjacent then we can swap them via a sequence of adjacent swaps, first bringing row k up

$$\begin{pmatrix} \vdots \\ \iota_{\phi(j)} \\ \iota_{\phi(j+1)} \\ \iota_{\phi(j+2)} \\ \vdots \\ \iota_{\phi(k)} \\ \vdots \end{pmatrix} \xrightarrow{\rho_k \leftrightarrow \rho_{k-1}} \xrightarrow{\rho_{k-1} \leftrightarrow \rho_{k-2}} \cdots \xrightarrow{\rho_{j+1} \leftrightarrow \rho_j} \begin{pmatrix} \vdots \\ \iota_{\phi(k)} \\ \iota_{\phi(j)} \\ \iota_{\phi(j+1)} \\ \vdots \\ \iota_{\phi(k-1)} \\ \vdots \end{pmatrix}$$

and then bringing row j down.

$$\xrightarrow{\rho_{j+1} \leftrightarrow \rho_{j+2}} \xrightarrow{\rho_{j+2} \leftrightarrow \rho_{j+3}} \cdots \xrightarrow{\rho_{k-1} \leftrightarrow \rho_k} \begin{pmatrix} \vdots \\ \iota_{\phi(k)} \\ \iota_{\phi(j+1)} \\ \iota_{\phi(j+2)} \\ \vdots \\ \iota_{\phi(j)} \\ \vdots \end{pmatrix}$$

Each of these adjacent swaps changes the number of inversions from odd to even or from even to odd. The total number of swaps $(k-j)+(k-j-1)$ is odd.

Thus, in aggregate, the number of inversions changes from even to odd, or from odd to even. QED

4.5 Corollary If a permutation matrix has an odd number of inversions then swapping it to the identity takes an odd number of swaps. If it has an even number of inversions then swapping to the identity takes an even number.

PROOF The identity matrix has zero inversions. To change an odd number to zero requires an odd number of swaps, and to change an even number to zero requires an even number of swaps. QED

4.6 Example The matrix in Example 4.3 can be brought to the identity with one swap $\rho_1 \leftrightarrow \rho_3$. (So the number of swaps needn't be the same as the number of inversions, but the oddness or evenness of the two numbers is the same.)

4.7 Definition The *signum* of a permutation $\operatorname{sgn}(\phi)$ is -1 if the number of inversions in ϕ is odd and is $+1$ if the number of inversions is even.

4.8 Example Using the notation for the 3-permutations from Example 3.8 we have

$$P_{\phi_1} = \begin{pmatrix} 1 & 0 & 0 \\ 0 & 1 & 0 \\ 0 & 0 & 1 \end{pmatrix} \qquad P_{\phi_2} = \begin{pmatrix} 1 & 0 & 0 \\ 0 & 0 & 1 \\ 0 & 1 & 0 \end{pmatrix}$$

so $\operatorname{sgn}(\phi_1) = 1$ because there are no inversions, while $\operatorname{sgn}(\phi_2) = -1$ because there is one.

We still have not shown that the determinant function is well-defined because we have not considered row operations on permutation matrices other than row swaps. We will finesse this issue. Define a function $d\colon \mathcal{M}_{n\times n} \to \mathbb{R}$ by altering the permutation expansion formula, replacing $|P_\phi|$ with $\operatorname{sgn}(\phi)$.

$$d(T) = \sum_{\text{permutations } \phi} t_{1,\phi(1)}t_{2,\phi(2)}\cdots t_{n,\phi(n)} \cdot \operatorname{sgn}(\phi)$$

The advantage of this formula is that the number of inversions is clearly well-defined—just count them. Therefore, we will be finished showing that an $n\times n$ determinant function exists when we show that this d satisfies the conditions required of a determinant.

4.9 Lemma The function d above is a determinant. Hence determinants exist for every n.

Proof We must check that it has the four conditions from the definition of determinant, Definition 2.1.

Condition (4) is easy: where I is the $n \times n$ identity, in

$$d(I) = \sum_{\text{perm } \phi} \iota_{1,\phi(1)} \iota_{2,\phi(2)} \cdots \iota_{n,\phi(n)} \operatorname{sgn}(\phi)$$

all of the terms in the summation are zero except for the one where the permutation ϕ is the identity, which gives the product down the diagonal, which is one.

For condition (3) suppose that $T \xrightarrow{k\rho_i} \hat{T}$ and consider $d(\hat{T})$.

$$\sum_{\text{perm } \phi} \hat{t}_{1,\phi(1)} \cdots \hat{t}_{i,\phi(i)} \cdots \hat{t}_{n,\phi(n)} \operatorname{sgn}(\phi)$$
$$= \sum_{\phi} t_{1,\phi(1)} \cdots k t_{i,\phi(i)} \cdots t_{n,\phi(n)} \operatorname{sgn}(\phi)$$

Factor out k to get the desired equality.

$$= k \cdot \sum_{\phi} t_{1,\phi(1)} \cdots t_{i,\phi(i)} \cdots t_{n,\phi(n)} \operatorname{sgn}(\phi) = k \cdot d(T)$$

For (2) suppose that $T \xrightarrow{\rho_i \leftrightarrow \rho_j} \hat{T}$. We must show that $d(\hat{T})$ is the negative of $d(T)$.

$$d(\hat{T}) = \sum_{\text{perm } \phi} \hat{t}_{1,\phi(1)} \cdots \hat{t}_{i,\phi(i)} \cdots \hat{t}_{j,\phi(j)} \cdots \hat{t}_{n,\phi(n)} \operatorname{sgn}(\phi) \qquad (*)$$

We will show that each term in $(*)$ is associated with a term in $d(T)$, and that the two terms are negatives of each other. Consider the matrix from the multilinear expansion of $d(\hat{T})$ giving the term $\hat{t}_{1,\phi(1)} \cdots \hat{t}_{i,\phi(i)} \cdots \hat{t}_{j,\phi(j)} \cdots \hat{t}_{n,\phi(n)} \operatorname{sgn}(\phi)$.

$$\begin{pmatrix} & \vdots & \\ \hat{t}_{i,\phi(i)} & & \\ & \vdots & \\ & & \hat{t}_{j,\phi(j)} \\ & \vdots & \end{pmatrix}$$

It is the result of the $\rho_i \leftrightarrow \rho_j$ operation performed on this matrix.

$$\begin{pmatrix} & \vdots & \\ & & t_{i,\phi(j)} \\ & \vdots & \\ t_{j,\phi(i)} & & \\ & \vdots & \end{pmatrix}$$

That is, the term with hatted t's is associated with this term from the $d(T)$ expansion: $t_{1,\sigma(1)} \cdots t_{j,\sigma(j)} \cdots t_{i,\sigma(i)} \cdots t_{n,\sigma(n)} \operatorname{sgn}(\sigma)$, where the permutation σ equals ϕ but with the i-th and j-th numbers interchanged, $\sigma(i) = \phi(j)$ and $\sigma(j) = \phi(i)$. The two terms have the same multiplicands $\hat{t}_{1,\phi(1)} = t_{1,\sigma(1)}$, ..., including the entries from the swapped rows $\hat{t}_{i,\phi(i)} = t_{j,\phi(i)} = t_{j,\sigma(j)}$ and $\hat{t}_{j,\phi(j)} = t_{i,\phi(j)} = t_{i,\sigma(i)}$. But the two terms are negatives of each other since $\operatorname{sgn}(\phi) = -\operatorname{sgn}(\sigma)$ by Lemma 4.4.

Now, any permutation ϕ can be derived from some other permutation σ by such a swap, in one and only one way. Therefore the summation in $(*)$ is in fact a sum over all permutations, taken once and only once.

$$\begin{aligned} d(\hat{T}) &= \sum_{\text{perm } \phi} \hat{t}_{1,\phi(1)} \cdots \hat{t}_{i,\phi(i)} \cdots \hat{t}_{j,\phi(j)} \cdots \hat{t}_{n,\phi(n)} \operatorname{sgn}(\phi) \\ &= \sum_{\text{perm } \sigma} t_{1,\sigma(1)} \cdots t_{j,\sigma(j)} \cdots t_{i,\sigma(i)} \cdots t_{n,\sigma(n)} \cdot \big(-\operatorname{sgn}(\sigma)\big) \end{aligned}$$

Thus $d(\hat{T}) = -d(T)$.

Finally, for condition (1) suppose that $T \xrightarrow{k\rho_i+\rho_j} \hat{T}$.

$$\begin{aligned} d(\hat{T}) &= \sum_{\text{perm } \phi} \hat{t}_{1,\phi(1)} \cdots \hat{t}_{i,\phi(i)} \cdots \hat{t}_{j,\phi(j)} \cdots \hat{t}_{n,\phi(n)} \operatorname{sgn}(\phi) \\ &= \sum_{\phi} t_{1,\phi(1)} \cdots t_{i,\phi(i)} \cdots (kt_{i,\phi(j)} + t_{j,\phi(j)}) \cdots t_{n,\phi(n)} \operatorname{sgn}(\phi) \end{aligned}$$

Distribute over the addition in $kt_{i,\phi(j)} + t_{j,\phi(j)}$.

$$\begin{aligned} &= \sum_{\phi} \big[t_{1,\phi(1)} \cdots t_{i,\phi(i)} \cdots kt_{i,\phi(j)} \cdots t_{n,\phi(n)} \operatorname{sgn}(\phi) \\ &\qquad + t_{1,\phi(1)} \cdots t_{i,\phi(i)} \cdots t_{j,\phi(j)} \cdots t_{n,\phi(n)} \operatorname{sgn}(\phi)\big] \end{aligned}$$

Break it into two summations.

$$\begin{aligned} &= \sum_{\phi} t_{1,\phi(1)} \cdots t_{i,\phi(i)} \cdots kt_{i,\phi(j)} \cdots t_{n,\phi(n)} \operatorname{sgn}(\phi) \\ &\qquad + \sum_{\phi} t_{1,\phi(1)} \cdots t_{i,\phi(i)} \cdots t_{j,\phi(j)} \cdots t_{n,\phi(n)} \operatorname{sgn}(\phi) \end{aligned}$$

Recognize the second one.

$$\begin{aligned} &= k \cdot \sum_{\phi} t_{1,\phi(1)} \cdots t_{i,\phi(i)} \cdots t_{i,\phi(j)} \cdots t_{n,\phi(n)} \operatorname{sgn}(\phi) \\ &\qquad + d(T) \end{aligned}$$

Consider the terms $t_{1,\phi(1)} \cdots t_{i,\phi(i)} \cdots t_{i,\phi(j)} \cdots t_{n,\phi(n)} \operatorname{sgn}(\phi)$. Notice the subscripts; the entry is $t_{i,\phi(j)}$, not $t_{j,\phi(j)}$. The sum of these terms is the determinant of a matrix S that is equal to T except that row j of S is a copy of row i of T, that is, S has two equal rows. In the same way that we proved Lemma 2.4 we can see that $d(S) = 0$: a swap of S's equal rows will change the sign of $d(S)$ but since the matrix is unchanged by that swap the value of $d(S)$ must also be unchanged, and so that value must be zero. QED

We have now proved that determinant functions exist for each size $n \times n$. We already know that for each size there is at most one determinant. Therefore, for each size there is one and only one determinant function.

We end this subsection by proving the other result remaining from the prior subsection.

4.10 Theorem The determinant of a matrix equals the determinant of its transpose.

PROOF The proof is best understood by doing the general 3×3 case. That the argument applies to the $n \times n$ case will be clear.

Compare the permutation expansion of the matrix T

$$\begin{aligned}\begin{vmatrix} t_{1,1} & t_{1,2} & t_{1,3} \\ t_{2,1} & t_{2,2} & t_{2,3} \\ t_{3,1} & t_{3,2} & t_{3,3} \end{vmatrix} &= t_{1,1}t_{2,2}t_{3,3}\begin{vmatrix} 1 & 0 & 0 \\ 0 & 1 & 0 \\ 0 & 0 & 1 \end{vmatrix} + t_{1,1}t_{2,3}t_{3,2}\begin{vmatrix} 1 & 0 & 0 \\ 0 & 0 & 1 \\ 0 & 1 & 0 \end{vmatrix} \\ &\quad + t_{1,2}t_{2,1}t_{3,3}\begin{vmatrix} 0 & 1 & 0 \\ 1 & 0 & 0 \\ 0 & 0 & 1 \end{vmatrix} + t_{1,2}t_{2,3}t_{3,1}\begin{vmatrix} 0 & 1 & 0 \\ 0 & 0 & 1 \\ 1 & 0 & 0 \end{vmatrix} \\ &\quad + t_{1,3}t_{2,1}t_{3,2}\begin{vmatrix} 0 & 0 & 1 \\ 1 & 0 & 0 \\ 0 & 1 & 0 \end{vmatrix} + t_{1,3}t_{2,2}t_{3,1}\begin{vmatrix} 0 & 0 & 1 \\ 0 & 1 & 0 \\ 1 & 0 & 0 \end{vmatrix}\end{aligned}$$

with the permutation expansion of its transpose.

$$\begin{aligned}\begin{vmatrix} t_{1,1} & t_{2,1} & t_{3,1} \\ t_{1,2} & t_{2,2} & t_{3,2} \\ t_{1,3} & t_{2,3} & t_{3,3} \end{vmatrix} &= t_{1,1}t_{2,2}t_{3,3}\begin{vmatrix} 1 & 0 & 0 \\ 0 & 1 & 0 \\ 0 & 0 & 1 \end{vmatrix} + t_{1,1}t_{3,2}t_{2,3}\begin{vmatrix} 1 & 0 & 0 \\ 0 & 0 & 1 \\ 0 & 1 & 0 \end{vmatrix} \\ &\quad + t_{2,1}t_{1,2}t_{3,3}\begin{vmatrix} 0 & 1 & 0 \\ 1 & 0 & 0 \\ 0 & 0 & 1 \end{vmatrix} + t_{2,1}t_{3,2}t_{1,3}\begin{vmatrix} 0 & 1 & 0 \\ 0 & 0 & 1 \\ 1 & 0 & 0 \end{vmatrix} \\ &\quad + t_{3,1}t_{1,2}t_{2,3}\begin{vmatrix} 0 & 0 & 1 \\ 1 & 0 & 0 \\ 0 & 1 & 0 \end{vmatrix} + t_{3,1}t_{2,2}t_{1,3}\begin{vmatrix} 0 & 0 & 1 \\ 0 & 1 & 0 \\ 1 & 0 & 0 \end{vmatrix}\end{aligned}$$

Compare first the six products of t's. The ones in the expansion of T are the same as the ones in the expansion of the transpose; for instance, $t_{1,2}t_{2,3}t_{3,1}$ is in the top and $t_{3,1}t_{1,2}t_{2,3}$ is in the bottom. That's perfectly sensible—the six in the top arise from all of the ways of picking one entry of T from each row and column while the six in the bottom are all of the ways of picking one entry of T from each column and row, so of course they are the same set.

Next observe that in the two expansions, each t-product expression is not necessarily associated with the same permutation matrix. For instance, on the top $t_{1,2}t_{2,3}t_{3,1}$ is associated with the matrix for the map $1 \mapsto 2$, $2 \mapsto 3$, $3 \mapsto 1$. On the bottom $t_{3,1}t_{1,2}t_{2,3}$ is associated with the matrix for the map $1 \mapsto 3$, $2 \mapsto 1$, $3 \mapsto 2$. The second map is inverse to the first. This is also perfectly sensible—both the matrix transpose and the map inverse flip the 1,2 to 2,1, flip the 2,3 to 3,2, and flip 3,1 to 1,3.

We finish by noting that the determinant of P_{ϕ} equals the determinant of $P_{\phi^{-1}}$, as Exercise 16shows. QED

Exercises

These summarize the notation used in this book for the 2- and 3- permutations.

i	1	2
$\phi_1(i)$	1	2
$\phi_2(i)$	2	1

i	1	2	3
$\phi_1(i)$	1	2	3
$\phi_2(i)$	1	3	2
$\phi_3(i)$	2	1	3
$\phi_4(i)$	2	3	1
$\phi_5(i)$	3	1	2
$\phi_6(i)$	3	2	1

4.11 Give the permutation expansion of a general 2×2 matrix and its transpose.

✓ **4.12** *This problem appears also in the prior subsection.*

(a) Find the inverse of each 2-permutation.

(b) Find the inverse of each 3-permutation.

✓ **4.13** (a) Find the signum of each 2-permutation.

(b) Find the signum of each 3-permutation.

4.14 Find the only nonzero term in the permutation expansion of this matrix.

$$\begin{vmatrix} 0 & 1 & 0 & 0 \\ 1 & 0 & 1 & 0 \\ 0 & 1 & 0 & 1 \\ 0 & 0 & 1 & 0 \end{vmatrix}$$

Compute that determinant by finding the signum of the associated permutation.

4.15 [Strang 80] What is the signum of the n-permutation $\phi = \langle n, n-1, \ldots, 2, 1 \rangle$?

4.16 Prove these.

(a) Every permutation has an inverse.

(b) $\operatorname{sgn}(\phi^{-1}) = \operatorname{sgn}(\phi)$

(c) Every permutation is the inverse of another.

4.17 Prove that the matrix of the permutation inverse is the transpose of the matrix of the permutation $P_{\phi^{-1}} = {P_{\phi}}^{\mathsf{T}}$, for any permutation ϕ.

✓ **4.18** Show that a permutation matrix with m inversions can be row swapped to the identity in m steps. Contrast this with Corollary 4.5.

✓ **4.19** For any permutation ϕ let $g(\phi)$ be the integer defined in this way.

$$g(\phi) = \prod_{i<j} [\phi(j) - \phi(i)]$$

(This is the product, over all indices i and j with $i < j$, of terms of the given form.)

(a) Compute the value of g on all 2-permutations.

(b) Compute the value of g on all 3-permutations.

(c) Prove that $g(\phi)$ is not 0.

(d) Prove this.

$$\operatorname{sgn}(\phi) = \frac{g(\phi)}{|g(\phi)|}$$

Many authors give this formula as the definition of the signum function.

II Geometry of Determinants

The prior section develops the determinant algebraically, by considering formulas satisfying certain conditions. This section complements that with a geometric approach. Beyond its intuitive appeal, an advantage of this approach is that while we have so far only considered whether or not a determinant is zero, here we shall give a meaning to the value of the determinant. (The prior section treats the determinant as a function of the rows but this section focuses on columns.)

II.1 Determinants as Size Functions

This parallelogram picture is familiar from the construction of the sum of the two vectors.

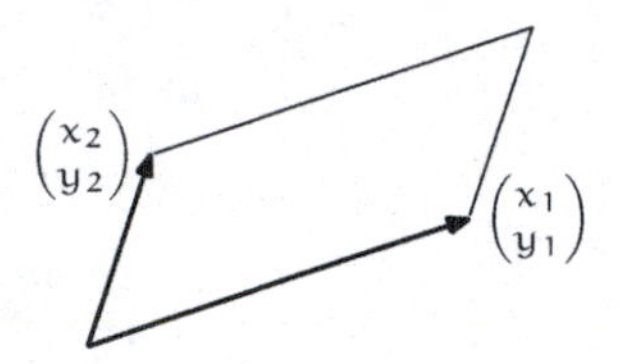

1.1 Definition In $\mathbb{R}^n$ the *box* (or *parallelepiped*) formed by $\langle\vec{v}_1,\ldots,\vec{v}_n\rangle$ is the set $\{t_1\vec{v}_1+\cdots+t_n\vec{v}_n \mid t_1,\ldots,t_n\in[0\,\ldots\,1]\}$.

Thus the parallelogram above is the box formed by $\langle\binom{x_1}{y_1},\binom{x_2}{y_2}\rangle$. A three-space box is shown in Example 1.4.

We can find the area of the above box by drawing an enclosing rectangle and subtracting away areas not in the box.

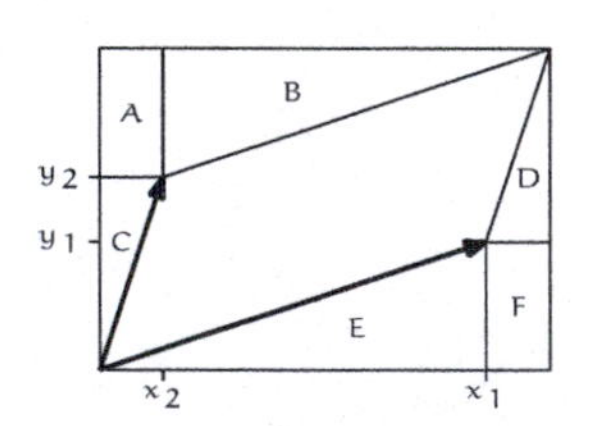

$$\begin{aligned}
&\text{area of parallelogram}\\
&\quad=\text{area of rectangle}-\text{area of }A-\text{area of }B\\
&\qquad-\cdots-\text{area of }F\\
&\quad=(x_1+x_2)(y_1+y_2)-x_2y_1-x_1y_1/2\\
&\qquad-x_2y_2/2-x_2y_2/2-x_1y_1/2-x_2y_1\\
&\quad=x_1y_2-x_2y_1
\end{aligned}$$

That the area equals the value of the determinant

$$\begin{vmatrix}x_1&x_2\\y_1&y_2\end{vmatrix}=x_1y_2-x_2y_1$$

is no coincidence. The definition of determinants contains four properties that we know lead to a unique function for each dimension n. We shall argue that these properties make good postulates for a function that measure the size of boxes in n-space.

For instance, a function that measures the size of the box should have the property that multiplying one of the box-defining vectors by a scalar will multiply the size by that scalar.

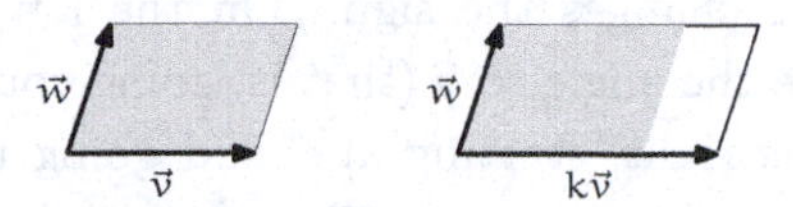

Shown here is $k = 1.4$. On the right the rescaled region is in solid lines with the original region shaded for comparison.

That is, we can reasonably expect that $\text{size}(\ldots, k\vec{v}, \ldots) = k \cdot \text{size}(\ldots, \vec{v}, \ldots)$. Of course, this condition is one of those in the definition of determinants.

Another property of determinants that should apply to any function measuring the size of a box is that it is unaffected by row combinations. Here are before-combining and after-combining boxes (the scalar shown is $k = -0.35$).

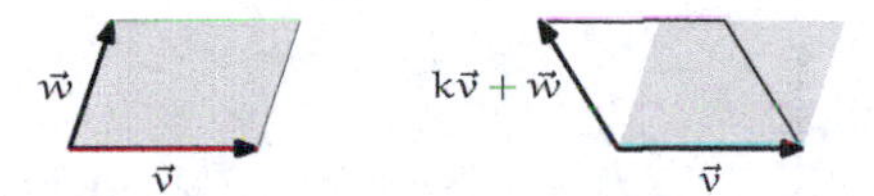

The box formed by v and $k\vec{v} + \vec{w}$ slants differently than the original one but the two have the same base and the same height, and hence the same area. So we expect that size is not affected by a shear operation $\text{size}(\ldots, \vec{v}, \ldots, \vec{w}, \ldots) = \text{size}(\ldots, \vec{v}, \ldots, k\vec{v} + \vec{w}, \ldots)$. Again, this is a determinant condition.

We expect that the box formed by unit vectors has unit size

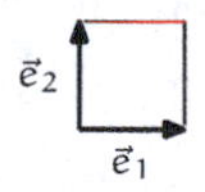

and we naturally extend that to any n-space $\text{size}(\vec{e}_1, \ldots, \vec{e}_n) = 1$.

Condition (2) of the definition of determinant is redundant, as remarked following the definition. We know from the prior section that for each n the determinant exists and is unique so we know that these postulates for size functions are consistent and that we do not need any more postulates. Therefore, we are justified in interpreting $\det(\vec{v}_1, \ldots, \vec{v}_n)$ as giving the size of the box formed by the vectors.

1.2 Remark Although condition (2) is redundant it raises an important point. Consider these two.

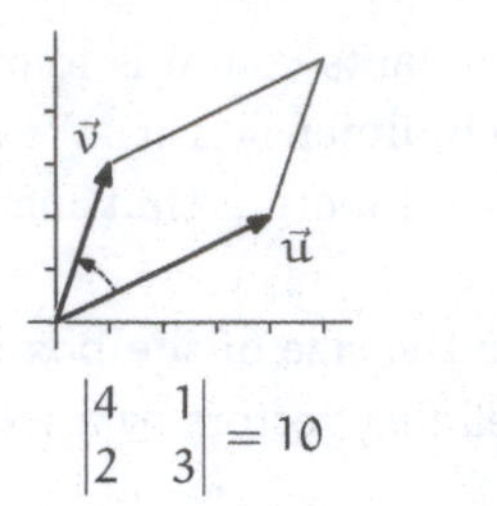

$$\begin{vmatrix} 4 & 1 \\ 2 & 3 \end{vmatrix} = 10$$

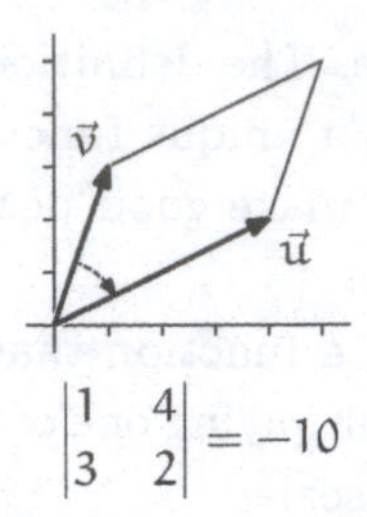

$$\begin{vmatrix} 1 & 4 \\ 3 & 2 \end{vmatrix} = -10$$

Swapping the columns changes the sign. On the left, starting with $\vec{u}$ and following the arc inside the angle to $\vec{v}$ (that is, going counterclockwise), we get a positive size. On the right, starting at $\vec{v}$ and going to $\vec{u}$, and so following the clockwise arc, gives a negative size. The sign returned by the size function reflects the *orientation* or *sense* of the box. (We see the same thing if we picture the effect of scalar multiplication by a negative scalar.)

1.3 Definition The *volume* of a box is the absolute value of the determinant of a matrix with those vectors as columns.

1.4 Example By the formula that takes the area of the base times the height, the volume of this parallelepiped is 12. That agrees with the determinant.

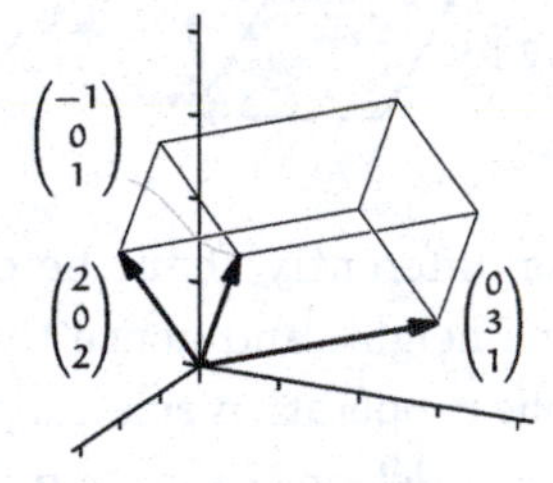

$$\begin{vmatrix} 2 & 0 & -1 \\ 0 & 3 & 0 \\ 2 & 1 & 1 \end{vmatrix} = 12$$

We can also compute the volume as the absolute value of this determinant.

$$\begin{vmatrix} 0 & 2 & 0 \\ 3 & 0 & 3 \\ 1 & 2 & 1 \end{vmatrix} = -12$$

1.5 Theorem A transformation $t\colon \mathbb{R}^n \to \mathbb{R}^n$ changes the size of all boxes by the same factor, namely, the size of the image of a box $|t(S)|$ is $|T|$ times the size of the box $|S|$, where T is the matrix representing t with respect to the standard basis.

That is, the determinant of a product is the product of the determinants $|TS| = |T| \cdot |S|$.

The two sentences say the same thing, first in map terms and then in matrix terms. This is because $|t(S)| = |TS|$, as both give the size of the box that is the image of the unit box $\mathcal{E}_n$ under the composition $t \circ s$, where the maps are represented with respect to the standard basis. We will prove the second sentence.

PROOF First consider the case that T is singular and thus does not have an inverse. Observe that if TS is invertible then there is an M such that $(TS)M = I$, so $T(SM) = I$, and so T is invertible. The contrapositive of that observation is that if T is not invertible then neither is TS—if $|T| = 0$ then $|TS| = 0$.

Now consider the case that T is nonsingular. Any nonsingular matrix factors into a product of elementary matrices $T = E_1E_2 \cdots E_r$. To finish this argument we will verify that $|ES| = |E| \cdot |S|$ for all matrices S and elementary matrices E. The result will then follow because $|TS| = |E_1 \cdots E_rS| = |E_1| \cdots |E_r| \cdot |S| = |E_1 \cdots E_r| \cdot |S| = |T| \cdot |S|$.

There are three types of elementary matrix. We will cover the $M_i(k)$ case; the $P_{i,j}$ and $C_{i,j}(k)$ checks are similar. The matrix $M_i(k)S$ equals S except that row i is multiplied by k. The third condition of determinant functions then gives that $|M_i(k)S| = k \cdot |S|$. But $|M_i(k)| = k$, again by the third condition because $M_i(k)$ is derived from the identity by multiplication of row i by k. Thus $|ES| = |E| \cdot |S|$ holds for $E = M_i(k)$. QED

1.6 Example Application of the map t represented with respect to the standard bases by

$$\begin{pmatrix} 1 & 1 \\ -2 & 0 \end{pmatrix}$$

will double sizes of boxes, e.g., from this

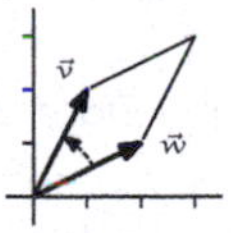

$$\begin{vmatrix} 2 & 1 \\ 1 & 2 \end{vmatrix} = 3$$

to this

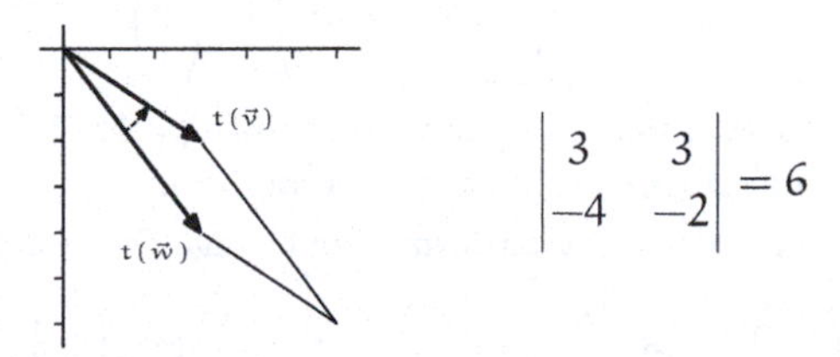

$$\begin{vmatrix} 3 & 3 \\ -4 & -2 \end{vmatrix} = 6$$

1.7 Corollary If a matrix is invertible then the determinant of its inverse is the inverse of its determinant $|T^{-1}| = 1/|T|$.

PROOF $1 = |I| = |TT^{-1}| = |T| \cdot |T^{-1}|$ QED

Exercises

1.8 Find the volume of the region defined by the vectors.

(a) $\langle\begin{pmatrix}1\\3\end{pmatrix},\begin{pmatrix}-1\\4\end{pmatrix}\rangle$

(b) $\langle\begin{pmatrix}2\\1\\0\end{pmatrix},\begin{pmatrix}3\\-2\\4\end{pmatrix},\begin{pmatrix}8\\-3\\8\end{pmatrix}\rangle$

(c) $\langle\begin{pmatrix}1\\2\\0\\1\end{pmatrix},\begin{pmatrix}2\\2\\2\\2\end{pmatrix},\begin{pmatrix}-1\\3\\0\\5\end{pmatrix},\begin{pmatrix}0\\1\\0\\7\end{pmatrix}\rangle$

✓ 1.9 Is

$$\begin{pmatrix}4\\1\\2\end{pmatrix}$$

inside of the box formed by these three?

$$\begin{pmatrix}3\\3\\1\end{pmatrix}\quad\begin{pmatrix}2\\6\\1\end{pmatrix}\quad\begin{pmatrix}1\\0\\5\end{pmatrix}$$

✓ 1.10 Find the volume of this region.

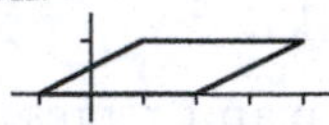

✓ 1.11 Suppose that $|A| = 3$. By what factor do these change volumes?

(a) A (b) A^2 (c) A^{-2}

✓ 1.12 Consider the linear transformation of $\mathbb{R}^3$ represented with respect to the standard bases by this matrix.

$$\begin{pmatrix}1&0&-1\\3&1&1\\-1&0&3\end{pmatrix}$$

(a) Compute the determinant of the matrix. Does the transformation preserve orientation or reverse it?

(b) Find the size of the box defined by these vectors. What is its orientation?

$$\begin{pmatrix}1\\-1\\2\end{pmatrix}\quad\begin{pmatrix}2\\0\\-1\end{pmatrix}\quad\begin{pmatrix}1\\1\\0\end{pmatrix}$$

(c) Find the images under t of the vectors in the prior item and find the size of the box that they define. What is the orientation?

1.13 By what factor does each transformation change the size of boxes?

(a) $\begin{pmatrix}x\\y\end{pmatrix}\mapsto\begin{pmatrix}2x\\3y\end{pmatrix}$ (b) $\begin{pmatrix}x\\y\end{pmatrix}\mapsto\begin{pmatrix}3x-y\\-2x+y\end{pmatrix}$ (c) $\begin{pmatrix}x\\y\\z\end{pmatrix}\mapsto\begin{pmatrix}x-y\\x+y+z\\y-2z\end{pmatrix}$

1.14 What is the area of the image of the rectangle $[2..4]\times[2..5]$ under the action of this matrix?

$$\begin{pmatrix}2&3\\4&-1\end{pmatrix}$$

1.15 If $t\colon \mathbb{R}^3 \to \mathbb{R}^3$ changes volumes by a factor of 7 and $s\colon \mathbb{R}^3 \to \mathbb{R}^3$ changes volumes by a factor of $3/2$ then by what factor will their composition changes volumes?

1.16 In what way does the definition of a box differ from the definition of a span?

✓ **1.17** Why doesn't this picture contradict Theorem 1.5?

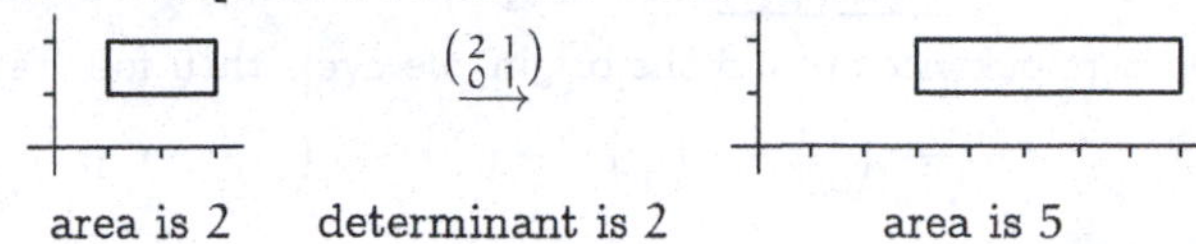

area is 2 determinant is 2 area is 5

✓ **1.18** Does $|TS| = |ST|$? $|T(SP)| = |(TS)P|$?

1.19 Show that there are no $2{\times}2$ matrices A and B satisfying these.

$$AB = \begin{pmatrix}1&-1\\2&0\end{pmatrix} \quad BA = \begin{pmatrix}2&1\\1&1\end{pmatrix}$$

1.20 (a) Suppose that $|A| = 3$ and that $|B| = 2$. Find $|A^2 \cdot B^T \cdot B^{-2} \cdot A^T|$.

(b) Assume that $|A| = 0$. Prove that $|6A^3 + 5A^2 + 2A| = 0$.

✓ **1.21** Let T be the matrix representing (with respect to the standard bases) the map that rotates plane vectors counterclockwise thru θ radians. By what factor does T change sizes?

✓ **1.22** Must a transformation $t\colon \mathbb{R}^2 \to \mathbb{R}^2$ that preserves areas also preserve lengths?

✓ **1.23** What is the volume of a parallelepiped in $\mathbb{R}^3$ bounded by a linearly dependent set?

✓ **1.24** Find the area of the triangle in $\mathbb{R}^3$ with endpoints $(1,2,1)$, $(3,-1,4)$, and $(2,2,2)$. (Area, not volume. The triangle defines a plane—what is the area of the triangle in that plane?)

✓ **1.25** An alternate proof of Theorem 1.5 uses the definition of determinant functions.

(a) Note that the vectors forming S make a linearly dependent set if and only if $|S| = 0$, and check that the result holds in this case.

(b) For the $|S| \neq 0$ case, to show that $|TS|/|S| = |T|$ for all transformations, consider the function $d\colon \mathcal{M}_{n\times n} \to \mathbb{R}$ given by $T \mapsto |TS|/|S|$. Show that d has the first property of a determinant.

(c) Show that d has the remaining three properties of a determinant function.

(d) Conclude that $|TS| = |T| \cdot |S|$.

1.26 Give a non-identity matrix with the property that $A^T = A^{-1}$. Show that if $A^T = A^{-1}$ then $|A| = \pm 1$. Does the converse hold?

1.27 The algebraic property of determinants that factoring a scalar out of a single row will multiply the determinant by that scalar shows that where H is $3{\times}3$, the determinant of cH is c^3 times the determinant of H. Explain this geometrically, that is, using Theorem 1.5. (The observation that increasing the linear size of a three-dimensional object by a factor of c will increase its volume by a factor of c^3 while only increasing its surface area by an amount proportional to a factor of c^2 is the *Square-cube law* [Wikipedia, Square-cube Law].)

✓ **1.28** We say that matrices H and G are *similar* if there is a nonsingular matrix P such that $H = P^{-1}GP$ (we will study this relation in Chapter Five). Show that similar matrices have the same determinant.

1.29 We usually represent vectors in $\mathbb{R}^2$ with respect to the standard basis so vectors in the first quadrant have both coordinates positive.

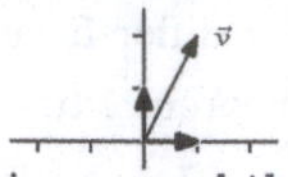

$$\mathrm{Rep}_{\mathcal{E}_2}(\vec{v}) = \begin{pmatrix} +3 \\ +2 \end{pmatrix}$$

Moving counterclockwise around the origin, we cycle thru four regions:

$$\cdots \longrightarrow \begin{pmatrix} + \\ + \end{pmatrix} \longrightarrow \begin{pmatrix} - \\ + \end{pmatrix} \longrightarrow \begin{pmatrix} - \\ - \end{pmatrix} \longrightarrow \begin{pmatrix} + \\ - \end{pmatrix} \longrightarrow \cdots.$$

Using this basis

$$B = \langle \begin{pmatrix} 0 \\ 1 \end{pmatrix}, \begin{pmatrix} -1 \\ 0 \end{pmatrix} \rangle$$

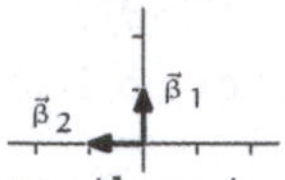

gives the same counterclockwise cycle. We say these two bases have the same *orientation*.

(a) Why do they give the same cycle?

(b) What other configurations of unit vectors on the axes give the same cycle?

(c) Find the determinants of the matrices formed from those (ordered) bases.

(d) What other counterclockwise cycles are possible, and what are the associated determinants?

(e) What happens in $\mathbb{R}^1$?

(f) What happens in $\mathbb{R}^3$?

A fascinating general-audience discussion of orientations is in [Gardner].

1.30 *This question uses material from the optional Determinant Functions Exist subsection.* Prove Theorem 1.5 by using the permutation expansion formula for the determinant.

✓ 1.31 (a) Show that this gives the equation of a line in $\mathbb{R}^2$ thru (x_2, y_2) and (x_3, y_3).

$$\begin{vmatrix} x & x_2 & x_3 \\ y & y_2 & y_3 \\ 1 & 1 & 1 \end{vmatrix} = 0$$

(b) [Petersen] Prove that the area of a triangle with vertices (x_1, y_1), (x_2, y_2), and (x_3, y_3) is

$$\frac{1}{2}\begin{vmatrix} x_1 & x_2 & x_3 \\ y_1 & y_2 & y_3 \\ 1 & 1 & 1 \end{vmatrix}.$$

(c) [Math. Mag., Jan. 1973] Prove that the area of a triangle with vertices at (x_1, y_1), (x_2, y_2), and (x_3, y_3) whose coordinates are integers has an area of N or $N/2$ for some positive integer N.

III Laplace's Formula

Determinants are a font of interesting and amusing formulas. Here is one that is often used to compute determinants by hand.

III.1 Laplace's Expansion

The example shows a 3×3 case but the approach works for any size $n > 1$.

1.1 Example Consider the permutation expansion.

$$\begin{aligned}
\begin{vmatrix} t_{1,1} & t_{1,2} & t_{1,3} \\ t_{2,1} & t_{2,2} & t_{2,3} \\ t_{3,1} & t_{3,2} & t_{3,3} \end{vmatrix} &= t_{1,1}t_{2,2}t_{3,3}\begin{vmatrix} 1 & 0 & 0 \\ 0 & 1 & 0 \\ 0 & 0 & 1 \end{vmatrix} + t_{1,1}t_{2,3}t_{3,2}\begin{vmatrix} 1 & 0 & 0 \\ 0 & 0 & 1 \\ 0 & 1 & 0 \end{vmatrix} \\
&\quad + t_{1,2}t_{2,1}t_{3,3}\begin{vmatrix} 0 & 1 & 0 \\ 1 & 0 & 0 \\ 0 & 0 & 1 \end{vmatrix} + t_{1,2}t_{2,3}t_{3,1}\begin{vmatrix} 0 & 1 & 0 \\ 0 & 0 & 1 \\ 1 & 0 & 0 \end{vmatrix} \\
&\quad + t_{1,3}t_{2,1}t_{3,2}\begin{vmatrix} 0 & 0 & 1 \\ 1 & 0 & 0 \\ 0 & 1 & 0 \end{vmatrix} + t_{1,3}t_{2,2}t_{3,1}\begin{vmatrix} 0 & 0 & 1 \\ 0 & 1 & 0 \\ 1 & 0 & 0 \end{vmatrix}
\end{aligned}$$

Pick a row or column and factor out its entries; here we do the entries in the first row.

$$\begin{aligned}
&= t_{1,1}\cdot\left[t_{2,2}t_{3,3}\begin{vmatrix} 1 & 0 & 0 \\ 0 & 1 & 0 \\ 0 & 0 & 1 \end{vmatrix} + t_{2,3}t_{3,2}\begin{vmatrix} 1 & 0 & 0 \\ 0 & 0 & 1 \\ 0 & 1 & 0 \end{vmatrix}\right] \\
&\quad + t_{1,2}\cdot\left[t_{2,1}t_{3,3}\begin{vmatrix} 0 & 1 & 0 \\ 1 & 0 & 0 \\ 0 & 0 & 1 \end{vmatrix} + t_{2,3}t_{3,1}\begin{vmatrix} 0 & 1 & 0 \\ 0 & 0 & 1 \\ 1 & 0 & 0 \end{vmatrix}\right] \\
&\quad + t_{1,3}\cdot\left[t_{2,1}t_{3,2}\begin{vmatrix} 0 & 0 & 1 \\ 1 & 0 & 0 \\ 0 & 1 & 0 \end{vmatrix} + t_{2,2}t_{3,1}\begin{vmatrix} 0 & 0 & 1 \\ 0 & 1 & 0 \\ 1 & 0 & 0 \end{vmatrix}\right]
\end{aligned}$$

In those permutation matrices, swap to get the first rows into place. This requires one swap to each of the permutation matrices on the second line, and two swaps to each on the third line. (Recall that row swaps change the sign of the determinant.)

$$
\begin{aligned}
&= t_{1,1} \cdot \left[t_{2,2}t_{3,3} \begin{vmatrix} 1 & 0 & 0 \\ 0 & 1 & 0 \\ 0 & 0 & 1 \end{vmatrix} + t_{2,3}t_{3,2} \begin{vmatrix} 1 & 0 & 0 \\ 0 & 0 & 1 \\ 0 & 1 & 0 \end{vmatrix} \right] \\
&\quad - t_{1,2} \cdot \left[t_{2,1}t_{3,3} \begin{vmatrix} 1 & 0 & 0 \\ 0 & 1 & 0 \\ 0 & 0 & 1 \end{vmatrix} + t_{2,3}t_{3,1} \begin{vmatrix} 1 & 0 & 0 \\ 0 & 0 & 1 \\ 0 & 1 & 0 \end{vmatrix} \right] \\
&\quad + t_{1,3} \cdot \left[t_{2,1}t_{3,2} \begin{vmatrix} 1 & 0 & 0 \\ 0 & 1 & 0 \\ 0 & 0 & 1 \end{vmatrix} + t_{2,2}t_{3,1} \begin{vmatrix} 1 & 0 & 0 \\ 0 & 0 & 1 \\ 0 & 1 & 0 \end{vmatrix} \right]
\end{aligned}
$$

On each line the terms in square brackets involve only the second and third row and column, and simplify to a 2×2 determinant.

$$
= t_{1,1} \cdot \begin{vmatrix} t_{2,2} & t_{2,3} \\ t_{3,2} & t_{3,3} \end{vmatrix} - t_{1,2} \cdot \begin{vmatrix} t_{2,1} & t_{2,3} \\ t_{3,1} & t_{3,3} \end{vmatrix} + t_{1,3} \cdot \begin{vmatrix} t_{2,1} & t_{2,2} \\ t_{3,1} & t_{3,2} \end{vmatrix}
$$

The formula given in Theorem 1.5, which generalizes this example, is a *recurrence* — the determinant is expressed as a combination of determinants. This formula isn't circular because it gives the $n\times n$ case in terms of smaller ones.

1.2 Definition For any $n\times n$ matrix T, the $(n-1)\times(n-1)$ matrix formed by deleting row i and column j of T is the i,j *minor* of T. The i,j *cofactor* $T_{i,j}$ of T is $(-1)^{i+j}$ times the determinant of the i,j minor of T.

1.3 Example The $1,2$ cofactor of the matrix from Example 1.1 is the negative of the second 2×2 determinant.

$$
T_{1,2} = -1 \cdot \begin{vmatrix} t_{2,1} & t_{2,3} \\ t_{3,1} & t_{3,3} \end{vmatrix}
$$

1.4 Example Where

$$
T = \begin{pmatrix} 1 & 2 & 3 \\ 4 & 5 & 6 \\ 7 & 8 & 9 \end{pmatrix}
$$

these are the $1,2$ and $2,2$ cofactors.

$$
T_{1,2} = (-1)^{1+2} \cdot \begin{vmatrix} 4 & 6 \\ 7 & 9 \end{vmatrix} = 6 \qquad T_{2,2} = (-1)^{2+2} \cdot \begin{vmatrix} 1 & 3 \\ 7 & 9 \end{vmatrix} = -12
$$

1.5 Theorem (Laplace Expansion of Determinants) Where T is an $n \times n$ matrix, we can find the determinant by expanding by cofactors on any row i or column j.

$$\begin{aligned}|T| &= t_{i,1} \cdot T_{i,1} + t_{i,2} \cdot T_{i,2} + \cdots + t_{i,n} \cdot T_{i,n} \\ &= t_{1,j} \cdot T_{1,j} + t_{2,j} \cdot T_{2,j} + \cdots + t_{n,j} \cdot T_{n,j}\end{aligned}$$

PROOF Exercise 25. QED

1.6 Example We can compute the determinant

$$|T| = \begin{vmatrix} 1 & 2 & 3 \\ 4 & 5 & 6 \\ 7 & 8 & 9 \end{vmatrix}$$

by expanding along the first row, as in Example 1.1.

$$|T| = 1 \cdot (+1) \begin{vmatrix} 5 & 6 \\ 8 & 9 \end{vmatrix} + 2 \cdot (-1) \begin{vmatrix} 4 & 6 \\ 7 & 9 \end{vmatrix} + 3 \cdot (+1) \begin{vmatrix} 4 & 5 \\ 7 & 8 \end{vmatrix} = -3 + 12 - 9 = 0$$

Or, we could expand down the second column.

$$|T| = 2 \cdot (-1) \begin{vmatrix} 4 & 6 \\ 7 & 9 \end{vmatrix} + 5 \cdot (+1) \begin{vmatrix} 1 & 3 \\ 7 & 9 \end{vmatrix} + 8 \cdot (-1) \begin{vmatrix} 1 & 3 \\ 4 & 6 \end{vmatrix} = 12 - 60 + 48 = 0$$

1.7 Example A row or column with many zeroes suggests a Laplace expansion.

$$\begin{vmatrix} 1 & 5 & 0 \\ 2 & 1 & 1 \\ 3 & -1 & 0 \end{vmatrix} = 0 \cdot (+1) \begin{vmatrix} 2 & 1 \\ 3 & -1 \end{vmatrix} + 1 \cdot (-1) \begin{vmatrix} 1 & 5 \\ 3 & -1 \end{vmatrix} + 0 \cdot (+1) \begin{vmatrix} 1 & 5 \\ 2 & 1 \end{vmatrix} = 16$$

We finish by applying Laplace's expansion to derive a new formula for the inverse of a matrix. With Theorem 1.5, we can calculate the determinant of a matrix by taking linear combinations of entries from a row with their associated cofactors.

$$t_{i,1} \cdot T_{i,1} + t_{i,2} \cdot T_{i,2} + \cdots + t_{i,n} \cdot T_{i,n} = |T| \qquad (*)$$

Recall that a matrix with two identical rows has a zero determinant. Thus, weighing the cofactors by entries from row k with $k \neq i$ gives zero

$$t_{i,1} \cdot T_{k,1} + t_{i,2} \cdot T_{k,2} + \cdots + t_{i,n} \cdot T_{k,n} = 0 \qquad (**)$$

because it represents the expansion along the row k of a matrix with row i equal to row k. This summarizes $(*)$ and $(**)$.

$$\begin{pmatrix} t_{1,1} & t_{1,2} & \ldots & t_{1,n} \\ t_{2,1} & t_{2,2} & \ldots & t_{2,n} \\ & \vdots & & \\ t_{n,1} & t_{n,2} & \ldots & t_{n,n} \end{pmatrix} \begin{pmatrix} T_{1,1} & T_{2,1} & \ldots & T_{n,1} \\ T_{1,2} & T_{2,2} & \ldots & T_{n,2} \\ & \vdots & & \\ T_{1,n} & T_{2,n} & \ldots & T_{n,n} \end{pmatrix} = \begin{pmatrix} |T| & 0 & \ldots & 0 \\ 0 & |T| & \ldots & 0 \\ & \vdots & & \\ 0 & 0 & \ldots & |T| \end{pmatrix}$$

Note that the order of the subscripts in the matrix of cofactors is opposite to the order of subscripts in the other matrix; e.g., along the first row of the matrix of cofactors the subscripts are $1,1$ then $2,1$, etc.

1.8 Definition The matrix *adjoint* to the square matrix T is

$$\operatorname{adj}(T)=\begin{pmatrix} T_{1,1} & T_{2,1} & \ldots & T_{n,1} \\ T_{1,2} & T_{2,2} & \ldots & T_{n,2} \\ & & \vdots & \\ T_{1,n} & T_{2,n} & \ldots & T_{n,n} \end{pmatrix}$$

where $T_{j,i}$ is the j,i cofactor.

1.9 Theorem Where T is a square matrix, $T\cdot\operatorname{adj}(T)=\operatorname{adj}(T)\cdot T=|T|\cdot I$. Thus if T has an inverse, if $|T|\neq 0$, then $T^{-1}=(1/|T|)\cdot\operatorname{adj}(T)$.

Proof Equations $(*)$ and $(**)$. QED

1.10 Example If

$$T=\begin{pmatrix} 1 & 0 & 4 \\ 2 & 1 & -1 \\ 1 & 0 & 1 \end{pmatrix}$$

then $\operatorname{adj}(T)$ is

$$\begin{pmatrix} T_{1,1} & T_{2,1} & T_{3,1} \\ T_{1,2} & T_{2,2} & T_{3,2} \\ T_{1,3} & T_{2,3} & T_{3,3} \end{pmatrix}=\begin{pmatrix} \begin{vmatrix} 1 & -1 \\ 0 & 1 \end{vmatrix} & -\begin{vmatrix} 0 & 4 \\ 0 & 1 \end{vmatrix} & \begin{vmatrix} 0 & 4 \\ 1 & -1 \end{vmatrix} \\ -\begin{vmatrix} 2 & -1 \\ 1 & 1 \end{vmatrix} & \begin{vmatrix} 1 & 4 \\ 1 & 1 \end{vmatrix} & -\begin{vmatrix} 1 & 4 \\ 2 & -1 \end{vmatrix} \\ \begin{vmatrix} 2 & 1 \\ 1 & 0 \end{vmatrix} & -\begin{vmatrix} 1 & 0 \\ 1 & 0 \end{vmatrix} & \begin{vmatrix} 1 & 0 \\ 2 & 1 \end{vmatrix} \end{pmatrix}=\begin{pmatrix} 1 & 0 & -4 \\ -3 & -3 & 9 \\ -1 & 0 & 1 \end{pmatrix}$$

and taking the product with T gives the diagonal matrix $|T|\cdot I$.

$$\begin{pmatrix} 1 & 0 & 4 \\ 2 & 1 & -1 \\ 1 & 0 & 1 \end{pmatrix}\begin{pmatrix} 1 & 0 & -4 \\ -3 & -3 & 9 \\ -1 & 0 & 1 \end{pmatrix}=\begin{pmatrix} -3 & 0 & 0 \\ 0 & -3 & 0 \\ 0 & 0 & -3 \end{pmatrix}$$

The inverse of T is $(1/-3)\cdot\operatorname{adj}(T)$.

$$T^{-1}=\begin{pmatrix} 1/-3 & 0/-3 & -4/-3 \\ -3/-3 & -3/-3 & 9/-3 \\ -1/-3 & 0/-3 & 1/-3 \end{pmatrix}=\begin{pmatrix} -1/3 & 0 & 4/3 \\ 1 & 1 & -3 \\ 1/3 & 0 & -1/3 \end{pmatrix}$$

The formulas from this subsection are often used for by-hand calculation and are sometimes useful with special types of matrices. However, for generic matrices they are not the best choice because they require more arithmetic than, for instance, the Gauss-Jordan method.

Exercises

✓ **1.11** Find the cofactor.

$$T = \begin{pmatrix} 1 & 0 & 2 \\ -1 & 1 & 3 \\ 0 & 2 & -1 \end{pmatrix}$$

(a) $T_{2,3}$ (b) $T_{3,2}$ (c) $T_{1,3}$

✓ **1.12** Find the determinant by expanding

$$\begin{vmatrix} 3 & 0 & 1 \\ 1 & 2 & 2 \\ -1 & 3 & 0 \end{vmatrix}$$

(a) on the first row (b) on the second row (c) on the third column.

1.13 Find the adjoint of the matrix in Example 1.6.

✓ **1.14** Find the matrix adjoint to each.

(a) $\begin{pmatrix} 2 & 1 & 4 \\ -1 & 0 & 2 \\ 1 & 0 & 1 \end{pmatrix}$ (b) $\begin{pmatrix} 3 & -1 \\ 2 & 4 \end{pmatrix}$ (c) $\begin{pmatrix} 1 & 1 \\ 5 & 0 \end{pmatrix}$ (d) $\begin{pmatrix} 1 & 4 & 3 \\ -1 & 0 & 3 \\ 1 & 8 & 9 \end{pmatrix}$

✓ **1.15** Find the inverse of each matrix in the prior question with Theorem 1.9.

1.16 Find the matrix adjoint to this one.

$$\begin{pmatrix} 2 & 1 & 0 & 0 \\ 1 & 2 & 1 & 0 \\ 0 & 1 & 2 & 1 \\ 0 & 0 & 1 & 2 \end{pmatrix}$$

✓ **1.17** Expand across the first row to derive the formula for the determinant of a 2×2 matrix.

✓ **1.18** Expand across the first row to derive the formula for the determinant of a 3×3 matrix.

✓ **1.19** (a) Give a formula for the adjoint of a 2×2 matrix.
(b) Use it to derive the formula for the inverse.

✓ **1.20** Can we compute a determinant by expanding down the diagonal?

1.21 Give a formula for the adjoint of a diagonal matrix.

✓ **1.22** Prove that the transpose of the adjoint is the adjoint of the transpose.

1.23 Prove or disprove: $\text{adj}(\text{adj}(T)) = T$.

1.24 A square matrix is *upper triangular* if each i, j entry is zero in the part above the diagonal, that is, when $i > j$.
(a) Must the adjoint of an upper triangular matrix be upper triangular? Lower triangular?

(b) Prove that the inverse of a upper triangular matrix is upper triangular, if an inverse exists.

1.25 *This question requires material from the optional Determinants Exist subsection.* Prove Theorem 1.5 by using the permutation expansion.

1.26 Prove that the determinant of a matrix equals the determinant of its transpose using Laplace's expansion and induction on the size of the matrix.

? **1.27** Show that

$$F_n = \begin{vmatrix} 1 & -1 & 1 & -1 & 1 & -1 & \ldots \\ 1 & 1 & 0 & 1 & 0 & 1 & \ldots \\ 0 & 1 & 1 & 0 & 1 & 0 & \ldots \\ 0 & 0 & 1 & 1 & 0 & 1 & \ldots \\ . & . & . & . & . & . & \ldots \end{vmatrix}$$

where F_n is the n-th term of $1, 1, 2, 3, 5, \ldots, x, y, x+y, \ldots$, the Fibonacci sequence, and the determinant is of order $n-1$. [Am. Math. Mon., Jun. 1949]

Topic

Cramer's Rule

A linear system is equivalent to a linear relationship among vectors.

$$\begin{aligned} x_1 + 2x_2 &= 6 \\ 3x_1 + x_2 &= 8 \end{aligned} \quad \iff \quad x_1 \cdot \begin{pmatrix} 1 \\ 3 \end{pmatrix} + x_2 \cdot \begin{pmatrix} 2 \\ 1 \end{pmatrix} = \begin{pmatrix} 6 \\ 8 \end{pmatrix}$$

In the picture below the small parallelogram is formed from sides that are the vectors $\binom{1}{3}$ and $\binom{2}{1}$. It is nested inside a parallelogram with sides $x_1\binom{1}{3}$ and $x_2\binom{2}{1}$. By the vector equation, the far corner of the larger parallelogram is $\binom{6}{8}$.

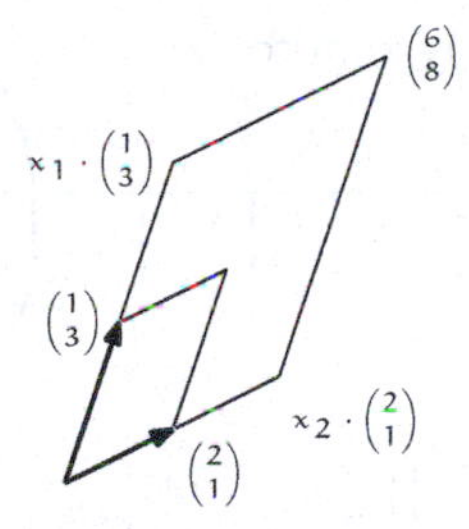

This drawing restates the algebraic question of finding the solution of a linear system into geometric terms: by what factors x_1 and x_2 must we dilate the sides of the starting parallelogram so that it will fill the other one?

We can use this picture, and our geometric understanding of determinants, to get a new formula for solving linear systems. Compare the sizes of these shaded boxes.

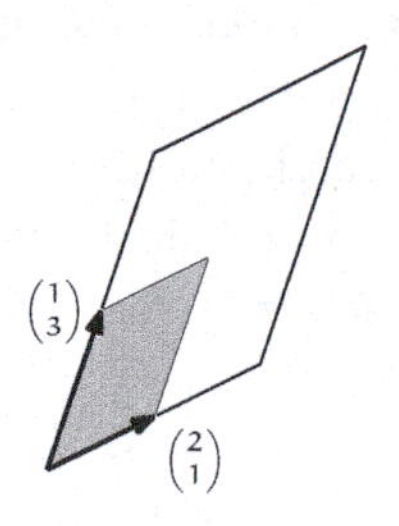

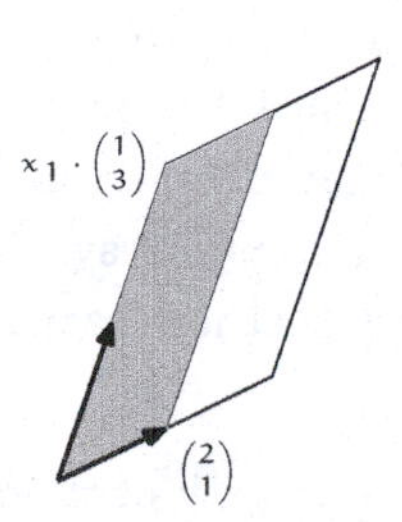

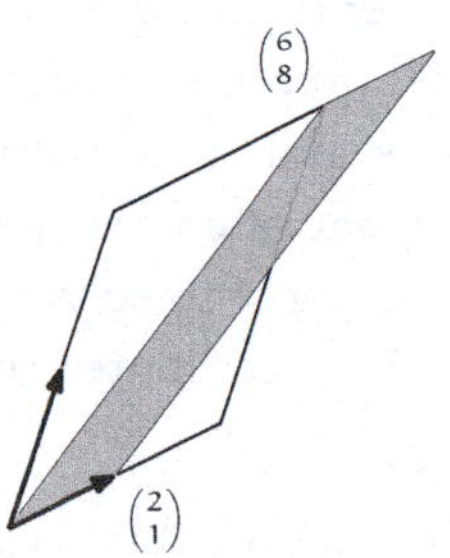

The second is defined by the vectors $x_1\binom{1}{3}$ and $\binom{2}{1}$ and one of the properties of the size function — the determinant — is that therefore the size of the second box is x_1 times the size of the first. The third box is derived from the second by shearing, adding $x_2\binom{2}{1}$ to $x_1\binom{1}{3}$ to get $x_1\binom{1}{3}+x_2\binom{2}{1}=\binom{6}{8}$, along with $\binom{2}{1}$. The determinant is not affected by shearing so the size of the third box equals that of the second.

Taken together we have this.

$$x_1\cdot\begin{vmatrix}1&2\\3&1\end{vmatrix}=\begin{vmatrix}x_1\cdot 1&2\\x_1\cdot 3&1\end{vmatrix}=\begin{vmatrix}x_1\cdot 1+x_2\cdot 2&2\\x_1\cdot 3+x_2\cdot 1&1\end{vmatrix}=\begin{vmatrix}6&2\\8&1\end{vmatrix}$$

Solving gives the value of one of the variables.

$$x_1=\frac{\begin{vmatrix}6&2\\8&1\end{vmatrix}}{\begin{vmatrix}1&2\\3&1\end{vmatrix}}=\frac{-10}{-5}=2$$

The generalization of this example is *Cramer's Rule*: if $|A|\neq 0$ then the system $A\vec{x}=\vec{b}$ has the unique solution $x_i=|B_i|/|A|$ where the matrix B_i is formed from A by replacing column i with the vector $\vec{b}$. The proof is Exercise 3.

For instance, to solve this system for x_2

$$\begin{pmatrix}1&0&4\\2&1&-1\\1&0&1\end{pmatrix}\begin{pmatrix}x_1\\x_2\\x_3\end{pmatrix}=\begin{pmatrix}2\\1\\-1\end{pmatrix}$$

we do this computation.

$$x_2=\frac{\begin{vmatrix}1&2&4\\2&1&-1\\1&-1&1\end{vmatrix}}{\begin{vmatrix}1&0&4\\2&1&-1\\1&0&1\end{vmatrix}}=\frac{-18}{-3}$$

Cramer's Rule lets us by-eye solve systems that are small and simple. For example, we can solve systems with two equations and two unknowns, or three equations and three unknowns, where the numbers are small integers. Such cases appear often enough that many people find this formula handy.

But using it to solving large or complex systems is not practical, either by hand or by a computer. A Gauss's Method-based approach is faster.

Exercises

1 Use Cramer's Rule to solve each for each of the variables.

(a) $\begin{aligned} x - y &= 4 \\ -x + 2y &= -7 \end{aligned}$ (b) $\begin{aligned} -2x + y &= -2 \\ x - 2y &= -2 \end{aligned}$

2 Use Cramer's Rule to solve this system for z.

$$\begin{aligned} 2x + y + z &= 1 \\ 3x \phantom{{}+y} + z &= 4 \\ x - y - z &= 2 \end{aligned}$$

3 Prove Cramer's Rule.

4 Here is an alternative proof of Cramer's Rule that doesn't overtly contain any geometry. Write X_i for the identity matrix with column i replaced by the vector $\vec{x}$ of unknowns $x_1, \ldots, x_n$.

(a) Observe that $AX_i = B_i$.

(b) Take the determinant of both sides.

5 Suppose that a linear system has as many equations as unknowns, that all of its coefficients and constants are integers, and that its matrix of coefficients has determinant 1. Prove that the entries in the solution are all integers. (*Remark.* This is often used to invent linear systems for exercises.)

6 Use Cramer's Rule to give a formula for the solution of a two equations/two unknowns linear system.

7 Can Cramer's Rule tell the difference between a system with no solutions and one with infinitely many?

8 The first picture in this Topic (the one that doesn't use determinants) shows a unique solution case. Produce a similar picture for the case of infinitely many solutions, and the case of no solutions.

Topic

Speed of Calculating Determinants

For large matrices, finding the determinant by using row operations is typically much faster than using the permutation expansion. We make this statement precise by finding how many operations each method performs.

To compare the speed of two algorithms, we find for each one how the time taken grows as the size of its input data set grows. For instance, if we increase the size of the input by a factor of ten does the time taken grow by a factor of ten, or by a factor of a hundred, or by a factor of a thousand? That is, is the time taken proportional to the size of the data set, or to the square of that size, or to the cube of that size, etc.? An algorithm whose time is proportional to the square is faster than one that takes time proportional to the cube.

First consider the permutation expansion formula.

$$\begin{vmatrix} t_{1,1} & t_{1,2} & \ldots & t_{1,n} \\ t_{2,1} & t_{2,2} & \ldots & t_{2,n} \\ & \vdots & & \\ t_{n,1} & t_{n,2} & \ldots & t_{n,n} \end{vmatrix} = \sum_{\text{permutations } \phi} t_{1,\phi(1)}t_{2,\phi(2)}\cdots t_{n,\phi(n)}\,|P_\phi|$$

There are $n! = n \cdot (n-1) \cdots 2 \cdot 1$ different n-permutations so for a matrix with n rows this sum has $n!$ terms (and inside each term is n-many multiplications). The factorial function grows quickly: when n is only 10 the expansion already has $10! = 3,628,800$ terms. Observe that growth proportional to the factorial is bigger than growth proportional to the square $n! > n^2$ because multiplying the first two factors in $n!$ gives $n \cdot (n-1)$, which for large n is approximately n^2 and then multiplying in more factors will make the factorial even larger. Similarly, the factorial function grows faster than n^3, etc. So an algorithm that uses the permutation expansion formula, and thus performs a number of operations at least as large as the factorial of the number of rows, would be very slow.

In contrast, the time taken by the row reduction method does not grow so fast. Below is a script for row reduction in the computer language Python. (*Note:* The code here is naive; for example it does not handle the case that the m(p_row, p_row) entry is zero. Analysis of a finished version that includes all of the tests and subcases is messier but would gives us roughly the same speed results.)

```
import random

def random_matrix(num_rows, num_cols):
    m = []
    for col in range(num_cols):
        new_row = []
        for row in range(num_rows):
            new_row.append(random.uniform(0,100))
        m.append(new_row)
    return m

def gauss_method(m):
    """Perform Gauss's Method on m.  This code is for illustration only
    and should not be used in practice.
      m  list of lists of numbers; each included list is a row
    """
    num_rows, num_cols = len(m), len(m[0])
    for p_row in range(num_rows):
        for row in range(p_row+1, num_rows):
            factor = -m[row][p_row] / float(m[p_row][p_row])
            new_row = []
            for col_num in range(num_cols):
                p_entry, entry = m[p_row][col_num], m[row][col_num]
                new_row.append(entry+factor*p_entry)
            m[row] = new_row
    return m

response = raw_input('number of rows? ')
num_rows = int(response)
m = random_matrix(num_rows, num_rows)
for row in m:
    print row
M = gauss_method(m)
print "-----"
for row in M:
    print row
```

Besides a routine to do Gauss's Method, this program also has a routine to generate a matrix filled with random numbers (the numbers are between 0 and 100, to make them readable below). This program prompts a user for the number of rows, generates a random square matrix of that size, and does row reduction on it.

```
$ python gauss_method.py
number of rows? 4
[69.48033741746909, 32.393754742132586, 91.35245787350696, 87.04557918402462]
[98.64189032145111, 28.58228108715638, 72.32273998878178, 26.310252241189257]
[85.22896214660841, 39.93894635139987, 4.061683241757219, 70.5925099861901]
[24.06322759315518, 26.699175587284373, 37.398583921673314, 87.42617087562161]
-----
[69.48033741746909, 32.393754742132586, 91.35245787350696, 87.04557918402462]
[0.0, -17.40743803545155, -57.37120602662462, -97.2691774792963]
[0.0, 0.0, -108.66513774392809, -37.31586824349682]
[0.0, 0.0, 0.0, -13.678536859817994]
```

Inside of the gauss_method routine, for each row prow, the routine performs $\text{factor} \cdot \rho_{\text{prow}} + \rho_{\text{row}}$ on the rows below. For each of these rows below, this involves operating on every entry in that row. That is a triply-nested loop. So this program has a running time that is something like the cube of the number of rows in the matrix. (*Comment.* We are glossing over many issues. For example, we may worry that the time taken by the program is dominated by the time to store and retrieve entries from memory, rather than by the row operations. However, development of a computation model is outside of our scope.)

If we add this code at the bottom,

```
def do_matrix(num_rows):
    gauss_method(random_matrix(num_rows, num_rows))

import timeit
for num_rows in [10,20,30,40,50,60,70,80,90,100]:
    s = "do_matrix("+str(num_rows)+")"
    t = timeit.timeit(stmt=s, setup="from __main__ import do_matrix",
                      number=100)
    print "num_rows=", num_rows, " seconds=", t
```

then Python will time the program. Here is the output from a timed test run.

```
num_rows= 10  seconds= 0.0162539482117
num_rows= 20  seconds= 0.0808238983154
num_rows= 30  seconds= 0.248152971268
num_rows= 40  seconds= 0.555531978607
num_rows= 50  seconds= 1.05453586578
num_rows= 60  seconds= 1.77881097794
num_rows= 70  seconds= 2.75969099998
num_rows= 80  seconds= 4.10647988319
num_rows= 90  seconds= 5.81125879288
num_rows= 100  seconds= 7.86893582344
```

Graphing that data gives part of the curve of a cubic.

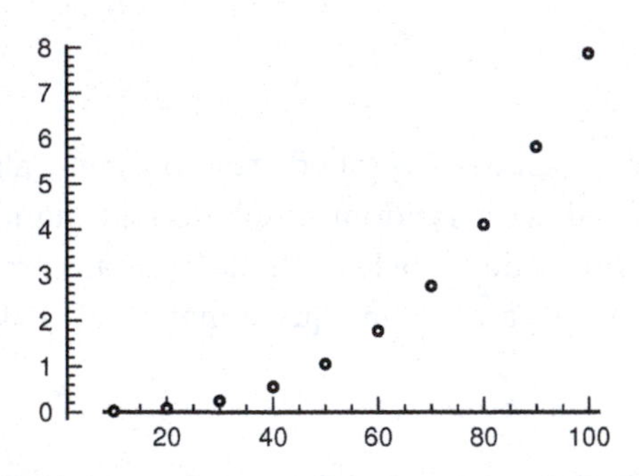

Finding the fastest algorithm to compute the determinant is a topic of current research. So far, researchers have found algorithms that run in time between the square and cube of the number of rows.

The contrast between the times taken by the two determinant computation methods of permutation expansion and row operations makes the point that although in principle they give the same answer, in practice we want the one with the best performance.

Exercises

1 To get an idea of what happens for typical matrices we can use the ability of computer systems to generate random numbers (of course, these are only pseudo-random in that they come from an algorithm but they pass a number of reasonable statistical tests for randomness).

(a) Fill a 5×5 array with random numbers say, in the range $[0\ldots1))$. See if it is singular. Repeat that experiment a few times. Are singular matrices frequent or rare in this sense?

(b) Time your computer algebra system at finding the determinant of ten 10×10 arrays of random numbers. Find the average time per array. Repeat the prior item for 20×20 arrays, 30×30 arrays, $\ldots$ 100×100 arrays, and compare to the numbers given above. (Notice that, when an array is singular, we can sometimes decide that quickly, for instance if the first row equals the second. In the light of your answer to the first part, do you expect that singular systems play a large role in your average?)

(c) Graph the input size versus the average time.

2 Compute the determinant of each of these by hand using the two methods discussed above.

(a) $\begin{vmatrix} 2 & 1 \\ 5 & -3 \end{vmatrix}$ (b) $\begin{vmatrix} 3 & 1 & 1 \\ -1 & 0 & 5 \\ -1 & 2 & -2 \end{vmatrix}$ (c) $\begin{vmatrix} 2 & 1 & 0 & 0 \\ 1 & 3 & 2 & 0 \\ 0 & -1 & -2 & 1 \\ 0 & 0 & -2 & 1 \end{vmatrix}$

Count the number of multiplications and divisions used in each case, for each of the methods.

3 The use by the timing routine of `do_matrix` has a bug. That routine does two things, generate a random matrix and then do `gauss_method` on it, and the timing number returned is for the combination. Produce code that times only the `gauss_method` routine.

4 What 10×10 array can you invent that takes your computer the longest time to reduce? The shortest?

5 Some computer language specifications requires that arrays be stored "by column," that is, the entire first column is stored contiguously, then the second column, etc. Does the code fragment given take advantage of this, or can it be rewritten to make it faster, by taking advantage of the fact that computer fetches are faster from contiguous locations?

Topic

Chiò's Method

When doing Gauss's Method on a matrix that contains only integers people often like to keep it that way. To avoid fractions in the reduction of this matrix

$$A=\begin{pmatrix}2&1&1\\3&4&-1\\1&5&1\end{pmatrix}$$

they may start by multiplying the lower rows by 2

$$\xrightarrow[2\rho_3]{2\rho_2}\begin{pmatrix}2&1&1\\6&8&-2\\2&10&2\end{pmatrix} \qquad (*)$$

so that elimination in the first column goes like this.

$$\xrightarrow[-\rho_1+\rho_3]{-3\rho_1+\rho_2}\begin{pmatrix}2&1&1\\0&5&-5\\0&8&0\end{pmatrix} \qquad (**)$$

This all-integer approach is easier for mental calculations. And, using integer arithmetic on a computer avoids some sticky issues involving floating point calculations [Kahan]. So there are sound reasons for this approach.

Another advantage of this approach is that we can easily apply Laplace's expansion to the first column of (**) and then get the determinant by remembering to divide by 4 because of (*).

Here is the general 3×3 case of this approach to finding the determinant. First, assuming $a_{1,1}\neq0$, we can rescale the lower rows.

$$A=\begin{pmatrix}a_{1,1}&a_{1,2}&a_{1,3}\\a_{2,1}&a_{2,2}&a_{2,3}\\a_{3,1}&a_{3,2}&a_{3,3}\end{pmatrix}\xrightarrow[a_{1,1}\rho_3]{a_{1,1}\rho_2}\begin{pmatrix}a_{1,1}&a_{1,2}&a_{1,3}\\a_{2,1}a_{1,1}&a_{2,2}a_{1,1}&a_{2,3}a_{1,1}\\a_{3,1}a_{1,1}&a_{3,2}a_{1,1}&a_{3,3}a_{1,1}\end{pmatrix}$$

This rescales the determinant by $a_{1,1}^2$. Now eliminate down the first column.

$$\xrightarrow[-a_{3,1}\rho_1+\rho_3]{-a_{2,1}\rho_1+\rho_2} \begin{pmatrix} a_{1,1} & a_{1,2} & a_{1,3} \\ 0 & a_{2,2}a_{1,1}-a_{2,1}a_{1,2} & a_{2,3}a_{1,1}-a_{2,1}a_{1,3} \\ 0 & a_{3,2}a_{1,1}-a_{3,1}a_{1,2} & a_{3,3}a_{1,1}-a_{3,1}a_{1,3} \end{pmatrix}$$

Let C be the $1,1$ minor. By Laplace the determinant of the above matrix is $a_{1,1}\det(C)$. We thus have $a_{1,1}^2\det(A) = a_{1,1}\det(C)$ and since $a_{1,1} \neq 0$ this gives $\det(A) = \det(C)/a_{1,1}$.

To do larger matrices we must see how to compute the minor's entries. The pattern above is that each element of the minor is a 2×2 determinant. For instance, the entry in the minor's upper left $a_{2,2}a_{1,1} - a_{2,1}a_{1,2}$, which is the $2,2$ entry in the above matrix, is the determinant of the matrix of these four elements of A.

$$\begin{pmatrix} \boxed{a_{1,1}} & \boxed{a_{1,2}} & a_{1,3} \\ \boxed{a_{2,1}} & \boxed{a_{2,2}} & a_{2,3} \\ a_{3,1} & a_{3,2} & a_{3,3} \end{pmatrix}$$

And the minor's lower left, the $3,2$ entry from above, is the determinant of the matrix of these four.

$$\begin{pmatrix} \boxed{a_{1,1}} & \boxed{a_{1,2}} & a_{1,3} \\ a_{2,1} & a_{2,2} & a_{2,3} \\ \boxed{a_{3,1}} & \boxed{a_{3,2}} & a_{3,3} \end{pmatrix}$$

So, where A is $n\times n$ for $n \geqslant 3$, we let Chiò's matrix C be the $(n-1)\times(n-1)$ matrix whose i,j entry is the determinant

$$\begin{vmatrix} a_{1,1} & a_{1,j+1} \\ a_{i+1,1} & a_{i+1,j+1} \end{vmatrix}$$

where $1 < i,j \leqslant n$. *Chiò's method* for finding the determinant of A is that if $a_{1,1} \neq 0$ then $\det(A) = \det(C)/a_{1,1}^{n-2}$. (By the way, nothing in Chiò's formula requires that the numbers be integers; it applies to reals as well.)

To illustrate we find the determinant of this 3×3 matrix.

$$A = \begin{pmatrix} 2 & 1 & 1 \\ 3 & 4 & -1 \\ 1 & 5 & 1 \end{pmatrix}$$

This is Chiò's matrix.

$$C = \begin{pmatrix} \begin{vmatrix} 2 & 1 \\ 3 & 4 \end{vmatrix} & \begin{vmatrix} 2 & 1 \\ 3 & -1 \end{vmatrix} \\ \begin{vmatrix} 2 & 1 \\ 1 & 5 \end{vmatrix} & \begin{vmatrix} 2 & 1 \\ 1 & 1 \end{vmatrix} \end{pmatrix} = \begin{pmatrix} 5 & -5 \\ 9 & 1 \end{pmatrix}$$

The formula for 3×3 matrices $\det(A) = \det(C)/a_{1,1}$ gives $\det(A) = (50/2) = 25$.

For a larger determinant we must do multiple steps but each involves only 2×2 determinants. So we can often calculate the determinant just by writing down a bit of intermediate information. For instance, with this 4×4 matrix

$$A = \begin{pmatrix} 3 & 0 & 1 & 1 \\ 1 & 2 & 0 & 1 \\ 2 & -1 & 0 & 3 \\ 1 & 0 & 0 & 1 \end{pmatrix}$$

we can mentally doing each of the 2×2 calculations and only write down the 3×3 result.

$$C_3 = \begin{pmatrix} \begin{vmatrix} 3 & 0 \\ 1 & 2 \end{vmatrix} & \begin{vmatrix} 3 & 1 \\ 1 & 0 \end{vmatrix} & \begin{vmatrix} 3 & 1 \\ 1 & 1 \end{vmatrix} \\ \begin{vmatrix} 3 & 0 \\ 2 & -1 \end{vmatrix} & \begin{vmatrix} 3 & 1 \\ 2 & 0 \end{vmatrix} & \begin{vmatrix} 3 & 1 \\ 2 & 3 \end{vmatrix} \\ \begin{vmatrix} 3 & 0 \\ 1 & 0 \end{vmatrix} & \begin{vmatrix} 3 & 1 \\ 1 & 0 \end{vmatrix} & \begin{vmatrix} 3 & 1 \\ 1 & 1 \end{vmatrix} \end{pmatrix} = \begin{pmatrix} 6 & -1 & 2 \\ -3 & -2 & 7 \\ 0 & -1 & 2 \end{pmatrix}$$

Note that the determinant of this is $a_{1,1}^{4-2} = 3^2$ times the determinant of A.

To finish, iterate. Here is Chiò's matrix of C_3.

$$C_2 = \begin{pmatrix} \begin{vmatrix} 6 & -1 \\ -3 & -2 \end{vmatrix} & \begin{vmatrix} 6 & 2 \\ -3 & 7 \end{vmatrix} \\ \begin{vmatrix} 6 & -1 \\ 0 & -1 \end{vmatrix} & \begin{vmatrix} 6 & 2 \\ 0 & 2 \end{vmatrix} \end{pmatrix} = \begin{pmatrix} -15 & 48 \\ -6 & 12 \end{pmatrix}$$

The determinant of this matrix is 6 times the determinant of C_3. The determinant of C_2 is 108. So $\det(A) = 108/(3^2 \cdot 6) = 2$.

Laplace's expansion formula reduces the calculation of an $n \times n$ determinant to the evaluation of a number of $(n-1) \times (n-1)$ ones. Chiò's formula is also recursive but it reduces an $n \times n$ determinant to a single $(n-1) \times (n-1)$ determinant, calculated from a number of 2×2 determinants. However, for large matrices Gauss's Method is better than either of these; for instance, it takes roughly half as many operations as Chiò's Method [Fuller & Logan].

Exercises

1 Use Chiò's Method to find each determinant.

(a) $\begin{vmatrix} 1 & 2 & 3 \\ 4 & 5 & 6 \\ 7 & 8 & 9 \end{vmatrix}$ (b) $\begin{vmatrix} 2 & 1 & 4 & 0 \\ 0 & 1 & 4 & 0 \\ 1 & 1 & 1 & 1 \\ 0 & 2 & 1 & 1 \end{vmatrix}$

2 What if $a_{1,1}$ is zero?

3 The *Rule of Sarrus* is a mnemonic that many people learn for the $3{\times}3$ determinant formula. To the right of the matrix, copy the first two columns.

$$\begin{array}{ccc|cc} a & b & c & a & b \\ d & e & f & d & e \\ g & h & i & g & h \end{array}$$

Then the determinant is the sum of the three upper-left to lower-right diagonals minus the three lower-left to upper-right diagonals $aei+bfg+cdh-gec-hfa-idb$. Count the operations involved in Sarrus's formula and in Chiò's.

4 Prove Chiò's formula.

Computer Code

This implements Chiò's Method. It is in the computer language Python.

```
#!/usr/bin/python
# chio.py
#  Calculate a determinant using Chio's method.
# Jim Hefferon; Public Domain
# For demonstration only; for instance, does not handle the M[0][0]=0 case

def det_two(a,b,c,d):
    """Return the determinant of the 2x2 matrix [[a,b], [c,d]]"""
    return a*d-b*c

def chio_mat(M):
    """Return the Chio matrix as a list of the rows
        M  nxn matrix, list of rows"""
    dim=len(M)
    C=[]
    for row in range(1,dim):
        C.append([])
        for col in range(1,dim):
            C[-1].append(det_two(M[0][0], M[0][col], M[row][0], M[row][col]))
    return C

def chio_det(M,show=None):
    """Find the determinant of M by Chio's method
        M  mxm matrix, list of rows"""
    dim=len(M)
    key_elet=M[0][0]
    if dim==1:
        return key_elet
    return chio_det(chio_mat(M))/(key_elet**(dim-2))


if __name__=='__main__':
    M=[[2,1,1], [3,4,-1], [1,5,1]]
    print "M=",M
    print "Det is", chio_det(M)
```

This is the result of calling the program from a command line.

```
$ python chio.py
M=[[2, 1, 1], [3, 4, -1], [1, 5, 1]]
Det is 25
```

Topic

Projective Geometry

There are geometries other than the familiar Euclidean one. One such geometry arose when artists observed that what a viewer sees is not necessarily what is there. As an example, here is Leonardo da Vinci's *The Last Supper*.

Look at where the ceiling meets the left and right walls. In the room those lines are parallel but da Vinci has painted lines that, if extended, would intersect. The intersection is the *vanishing point*. This aspect of perspective is familiar as an image of railroad tracks that appear to converge at the horizon.

Da Vinci has adopted a model of how we see. Imagine a person viewing a room. From the person's eye, in every direction, carry a ray outward until it intersects something, such as a point on the line where the wall meets the ceiling. This first intersection point is what the person sees in that direction. Overall what the person sees is the collection of three-dimensional intersection points projected to a common two dimensional image.

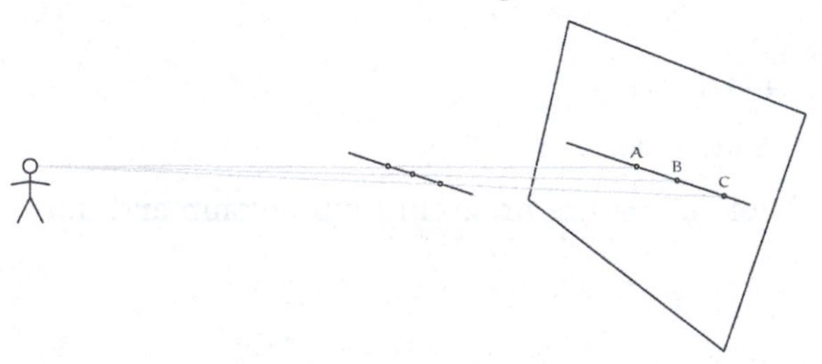

This is a *central projection* from a single point. As the sketch shows, this projection is not orthogonal like the ones we have seen earlier because the line from the viewer to C is not orthogonal to the image plane. (This model is only an approximation — it does not take into account such factors as that we have binocular vision or that our brain's processing greatly affects what we perceive. Nonetheless the model is interesting, both artistically and mathematically.)

The operation of central projection preserves some geometric properties, for instance lines project to lines. However, it fails to preserve some others. One example is that equal length segments can project to segments of unequal length (above, AB is longer than BC because the segment projected to AB is closer to the viewer and closer things look bigger). The study of the effects of central projections is projective geometry.

There are three cases of central projection. The first is the projection done by a movie projector.

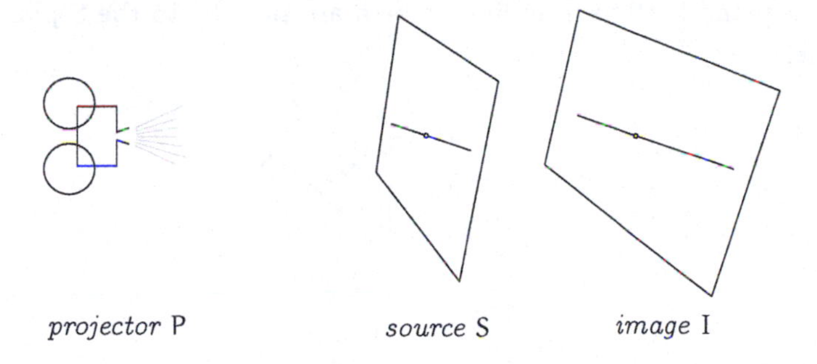

We can think that each source point is pushed from the domain plane S outward to the image plane I. The second case of projection is that of the artist pulling the source back to a canvas.

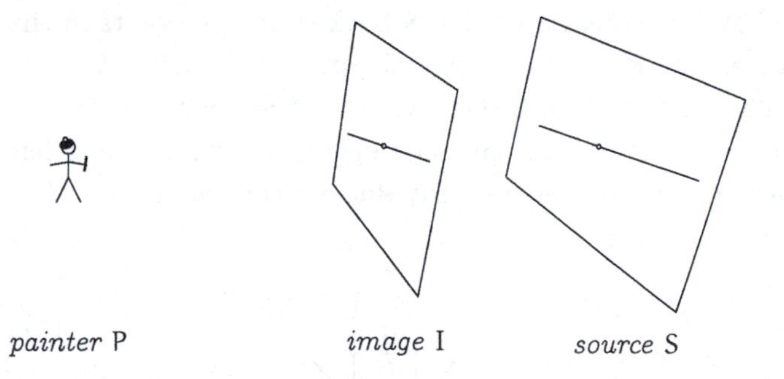

The two are different because first S is in the middle and then I. One more configuration can happen, with P in the middle. An example of this is when we use a pinhole to shine the image of a solar eclipse onto a paper.

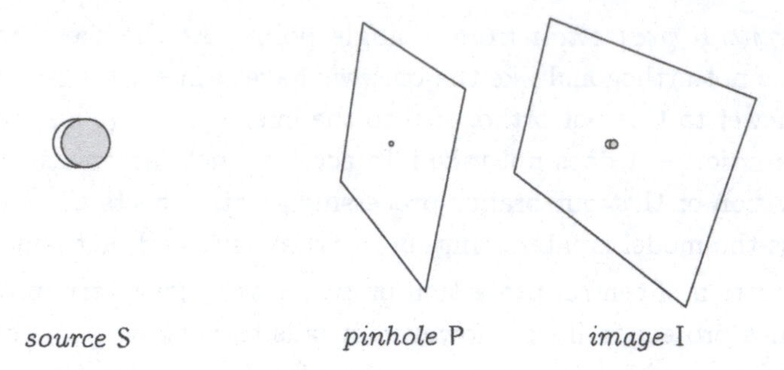

Although the three are not exactly the same, they are similar. We shall say that each is a central projection by P of S to I. We next look at three models of central projection, of increasing abstractness but also of increasing uniformity. The last model will bring out the linear algebra.

Consider again the effect of railroad tracks that appear to converge to a point. Model this with parallel lines in a domain plane S and a projection via a P to a codomain plane I. (The gray lines shown are parallel to the S plane and to the I plane.)

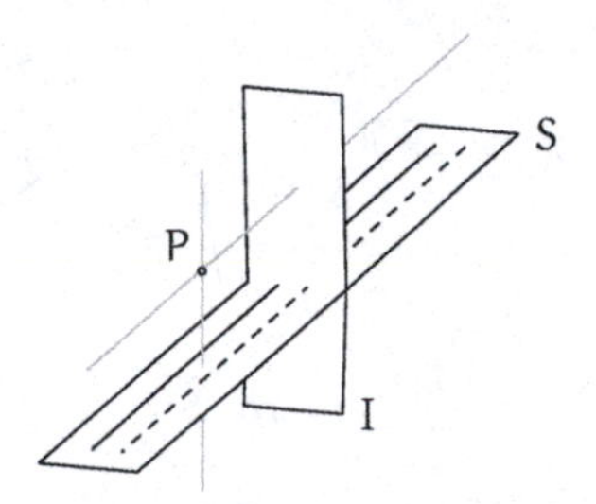

This single setting shows all three projection cases. The first picture below shows P acting as a movie projector by pushing points from part of S out to image points on the lower half of I. The middle picture shows P acting as the artist by pulling points from another part of S back to image points in the middle of I. In the third picture P acts as the pinhole, projecting points from S to the upper part of I. This third picture is the trickiest — the points that are projected near to the vanishing point are the ones that are far out on the lower left of S. Points in S that are near to the vertical gray line are sent high up on I.

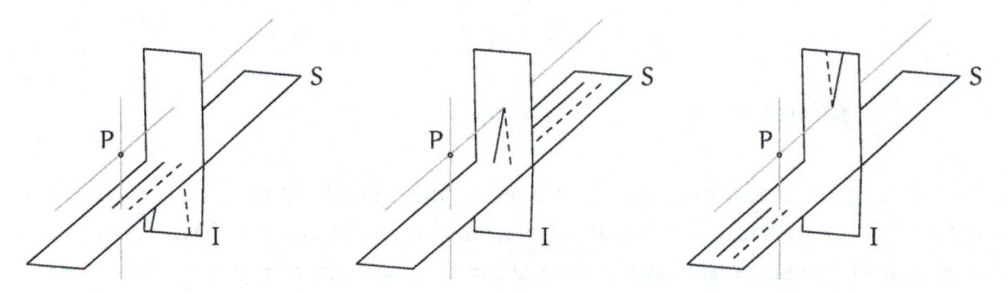

There are two awkward things here. First, neither of the two points in the domain nearest to the vertical gray line (see below) has an image because a projection from those two is along the gray line that is parallel to the codomain plane (we say that these two are projected to infinity). The second is that the vanishing point in I isn't the image of any point from S because a projection to this point would be along the gray line that is parallel to the domain plane (we say that the vanishing point is the image of a projection from infinity).

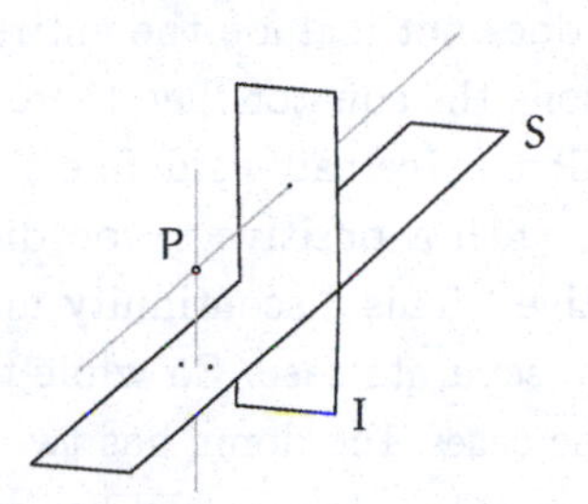

For a model that eliminates this awkwardness, cover the projector P with a hemispheric dome. In any direction, defined by a line through the origin, project anything in that direction to the single spot on the dome where the line intersects. This includes projecting things on the line between P and the dome, as with the movie projector. It includes projecting things on the line further from P than the dome, as with the painter. More subtly, it also includes things on the line that lie behind P, as with the pinhole case.

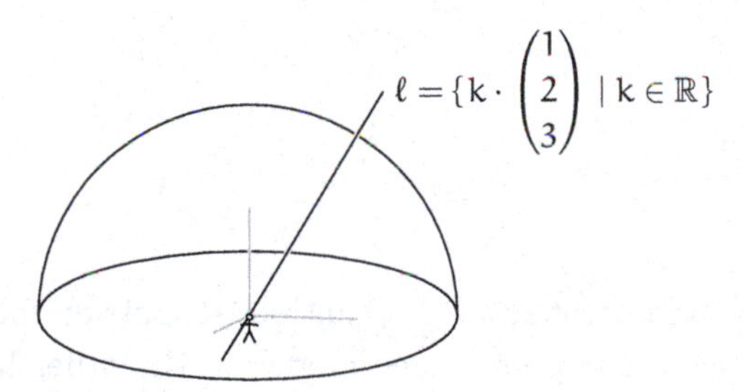

More formally, for any nonzero vector $\vec{v} \in \mathbb{R}^3$, let the associated *point* v *in the projective plane* be the set $\{k\vec{v} \mid k \in \mathbb{R} \text{ and } k \neq 0\}$ of nonzero vectors lying on the same line through the origin as $\vec{v}$. To describe a projective point we can give any representative member of the line, so that the projective point shown above can be represented in any of these three ways.

$$\begin{pmatrix} 1 \\ 2 \\ 3 \end{pmatrix} \quad \begin{pmatrix} 1/3 \\ 2/3 \\ 1 \end{pmatrix} \quad \begin{pmatrix} -2 \\ -4 \\ -6 \end{pmatrix}$$

Each of these is a *homogeneous coordinate vector* for the point ℓ.

This picture and definition clarifies central projection but there is still something ungainly about the dome model: what happens when P looks down? Consider, in the sketch above, the part of P's line of sight that comes up towards us, out of the page. Imagine that this part of the line falls, to the equator and below. Now the part of the line ℓ that intersects the dome lies behind the page.

That is, as the line of sight continues down past the equator, the projective point suddenly shifts from the front of the dome to the back of the dome. (This brings out that the dome does not include the entire equator or else when the viewer is looking exactly along the equator then there would be two points in the line that are both on the dome. Instead we define the dome so that it includes the points on the equator with a positive y coordinate, as well as the point where $y = 0$ and x is positive.) This discontinuity means that we often have to treat equatorial points as a separate case. So while the railroad track model of central projection has three cases, the dome has two.

We can do better, we can reduce to a model having a single case. Consider a sphere centered at the origin. Any line through the origin intersects the sphere in two spots, said to be antipodal. Because we associate each line through the origin with a point in the projective plane, we can draw such a point as a pair of antipodal spots on the sphere. Below, we show the two antipodal spots connected by a dashed line to emphasize that they are not two different points, the pair of spots together make one projective point.

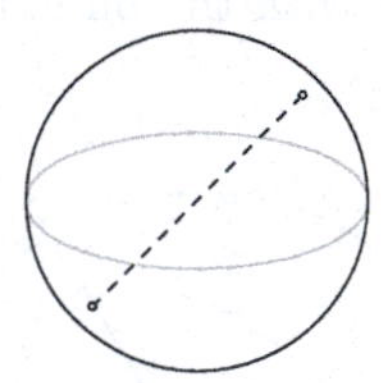

While drawing a point as a pair of antipodal spots on the sphere is not as intuitive as the one-spot-per-point dome mode, on the other hand the awkwardness of the dome model is gone in that as a line of view slides from north to south, no sudden changes happen. This central projection model is uniform.

So far we have described points in projective geometry. What about lines? What a viewer P at the origin sees as a line is shown below as a great circle, the intersection of the model sphere with a plane through the origin.

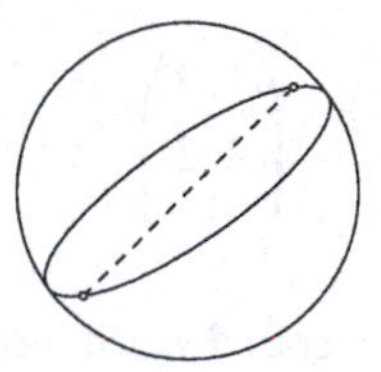

(We've included one of the projective points on this line to bring out a subtlety. Because two antipodal spots together make up a single projective point, the great circle's behind-the-paper part is the same set of projective points as its in-front-of-the-paper part.) Just as we did with each projective point, we can also describe a projective line with a triple of reals. For instance, the members of this plane through the origin in $\mathbb{R}^3$

$$\{\begin{pmatrix} x \\ y \\ z \end{pmatrix} \mid x + y - z = 0\}$$

project to a line that we can describe with $(1 \;\; 1 \;\; -1)$ (using a row vector for this typographically distinguishes lines from points). In general, for any nonzero three-wide row vector $\vec{L}$ we define the associated *line in the projective plane*, to be the set $L = \{k\vec{L} \mid k \in \mathbb{R} \text{ and } k \neq 0\}$.

The reason this description of a line as a triple is convenient is that in the projective plane a point v and a line L are *incident* — the point lies on the line, the line passes through the point — if and only if a dot product of their representatives $v_1 L_1 + v_2 L_2 + v_3 L_3$ is zero (Exercise 4 shows that this is independent of the choice of representatives $\vec{v}$ and $\vec{L}$). For instance, the projective point described above by the column vector with components 1, 2, and 3 lies in the projective line described by $(1 \;\; 1 \;\; -1)$, simply because any vector in $\mathbb{R}^3$ whose components are in ratio $1:2:3$ lies in the plane through the origin whose equation is of the form $k \cdot x + k \cdot y - k \cdot z = 0$ for any nonzero k. That is, the incidence formula is inherited from the three-space lines and planes of which v and L are projections.

With this, we can do analytic projective geometry. For instance, the projective line $L = (1 \;\; 1 \;\; -1)$ has the equation $1v_1 + 1v_2 - 1v_3 = 0$, meaning that for any projective point v incident with the line, any of v's representative homogeneous coordinate vectors will satisfy the equation. This is true simply because those vectors lie on the three space plane. One difference from Euclidean analytic geometry is that in projective geometry besides talking about the equation of a line, we also talk about the equation of a point. For the fixed point

$$v = \begin{pmatrix} 1 \\ 2 \\ 3 \end{pmatrix}$$

the property that characterizes lines incident on this point is that the components of any representatives satisfy $1L_1 + 2L_2 + 3L_3 = 0$ and so this is the equation of v.

This symmetry of the statements about lines and points is the *Duality Principle* of projective geometry: in any true statement, interchanging 'point' with 'line' results in another true statement. For example, just as two distinct points determine one and only one line, in the projective plane two distinct lines determine one and only one point. Here is a picture showing two projective lines that cross in antipodal spots and thus cross at one projective point.

$(*)$

Contrast this with Euclidean geometry, where two unequal lines may have a unique intersection or may be parallel. In this way, projective geometry is simpler, more uniform, than Euclidean geometry.

That simplicity is relevant because there is a relationship between the two spaces: we can view the projective plane as an extension of the Euclidean plane. Draw the sphere model of the projective plane as the unit sphere in $\mathbb{R}^3$. Take Euclidean 2-space to be the plane $z = 1$. As shown below, all of the points on the Euclidean plane are projections of antipodal spots from the sphere. Conversely, we can view some points in the projective plane as corresponding to points in Euclidean space. (Note that projective points on the equator don't correspond to points on the plane; instead we say these project out to infinity.)

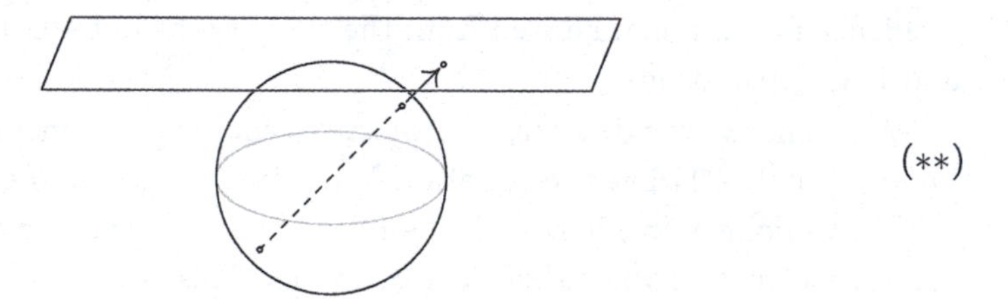

$(**)$

Thus we can think of projective space as consisting of the Euclidean plane with some extra points adjoined — the Euclidean plane is embedded in the projective plane. The extra points in projective space, the equatorial points, are called *ideal points* or *points at infinity* and the equator is called the *ideal line* or *line at infinity* (it is not a Euclidean line, it is a projective line).

The advantage of this extension from the Euclidean plane to the projective plane is that some of the nonuniformity of Euclidean geometry disappears. For instance, the projective lines shown above in $(*)$ cross at antipodal spots, a single projective point, on the sphere's equator. If we put those lines into $(**)$ then they correspond to Euclidean lines that are parallel. That is, in moving from the Euclidean plane to the projective plane, we move from having two cases, that distinct lines either intersect or are parallel, to having only one case, that distinct lines intersect (possibly at a point at infinity).

A disadvantage of the projective plane is that we don't have the same familiarity with it as we have with the Euclidean plane. Doing analytic geometry in the projective plane helps because the equations lead us to the right conclusions. Analytic projective geometry uses linear algebra. For instance, for three points of the projective plane t, u, and v, setting up the equations for those points by fixing vectors representing each shows that the three are collinear if and only if the resulting three-equation system has infinitely many row vector solutions representing their line. That in turn holds if and only if this determinant is zero.

$$\begin{vmatrix} t_1 & u_1 & v_1 \\ t_2 & u_2 & v_2 \\ t_3 & u_3 & v_3 \end{vmatrix}$$

Thus, three points in the projective plane are collinear if and only if any three representative column vectors are linearly dependent. Similarly, by duality, three lines in the projective plane are incident on a single point if and only if any three row vectors representing them are linearly dependent.

The following result is more evidence of the niceness of the geometry of the projective plane. These two triangles are *in perspective* from the point O because their corresponding vertices are collinear.

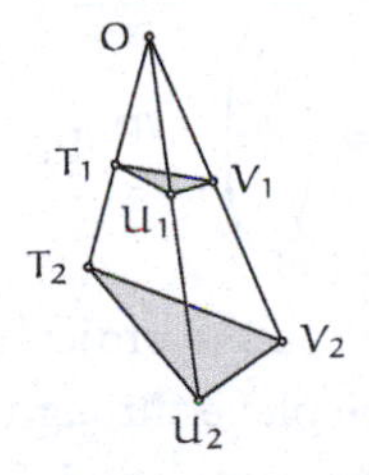

Consider the pairs of corresponding sides: the sides T_1U_1 and T_2U_2, the sides T_1V_1 and T_2V_2, and the sides U_1V_1 and U_2V_2. *Desargue's Theorem* is that when we extend the three pairs of corresponding sides, they intersect (shown here as the points TU, TV, and UV). What's more, those three intersection points are collinear.

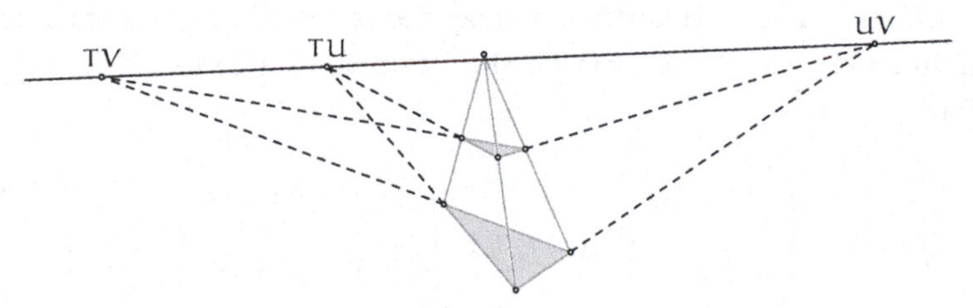

We will prove this using projective geometry. (We've drawn Euclidean figures because that is the more familiar image. To consider them as projective figures we can imagine that, although the line segments shown are parts of great circles and so are curved, the model has such a large radius compared to the size of the figures that the sides appear in our sketch to be straight.)

For the proof we need a preliminary lemma [Coxeter]: if W, X, Y, Z are four points in the projective plane, no three of which are collinear, then there are homogeneous coordinate vectors $\vec{w}$, $\vec{x}$, $\vec{y}$, and $\vec{z}$ for the projective points, and a basis B for $\mathbb{R}^3$, satisfying this.

$$\mathrm{Rep}_B(\vec{w}) = \begin{pmatrix} 1 \\ 0 \\ 0 \end{pmatrix} \quad \mathrm{Rep}_B(\vec{x}) = \begin{pmatrix} 0 \\ 1 \\ 0 \end{pmatrix} \quad \mathrm{Rep}_B(\vec{y}) = \begin{pmatrix} 0 \\ 0 \\ 1 \end{pmatrix} \quad \mathrm{Rep}_B(\vec{z}) = \begin{pmatrix} 1 \\ 1 \\ 1 \end{pmatrix}$$

To prove the lemma, because W, X, and Y are not on the same projective line, any homogeneous coordinate vectors $\vec{w}_0$, $\vec{x}_0$, and $\vec{y}_0$ do not line on the same plane through the origin in $\mathbb{R}^3$ and so form a spanning set for $\mathbb{R}^3$. Thus any homogeneous coordinate vector for Z is a combination $\vec{z}_0 = a \cdot \vec{w}_0 + b \cdot \vec{x}_0 + c \cdot \vec{y}_0$. Then let the basis be $B = \langle \vec{w}, \vec{x}, \vec{y} \rangle$ and take $\vec{w} = a \cdot \vec{w}_0$, $\vec{x} = b \cdot \vec{x}_0$, $\vec{y} = c \cdot \vec{y}_0$, and $\vec{z} = \vec{z}_0$.

To prove Desargue's Theorem use the lemma to fix homogeneous coordinate vectors and a basis.

$$\mathrm{Rep}_B(\vec{t}_1) = \begin{pmatrix} 1 \\ 0 \\ 0 \end{pmatrix} \quad \mathrm{Rep}_B(\vec{u}_1) = \begin{pmatrix} 0 \\ 1 \\ 0 \end{pmatrix} \quad \mathrm{Rep}_B(\vec{v}_1) = \begin{pmatrix} 0 \\ 0 \\ 1 \end{pmatrix} \quad \mathrm{Rep}_B(\vec{o}) = \begin{pmatrix} 1 \\ 1 \\ 1 \end{pmatrix}$$

The projective point T_2 is incident on the projective line OT_1 so any homogeneous coordinate vector for T_2 lies in the plane through the origin in $\mathbb{R}^3$ that is spanned by homogeneous coordinate vectors of O and T_1:

$$\mathrm{Rep}_B(\vec{t}_2) = a \begin{pmatrix} 1 \\ 1 \\ 1 \end{pmatrix} + b \begin{pmatrix} 1 \\ 0 \\ 0 \end{pmatrix}$$

for some scalars a and b. Hence the homogeneous coordinate vectors of members T_2 of the line OT_1 are of the form on the left below. The forms for U_2 and V_2 are similar.

$$\mathrm{Rep}_B(\vec{t}_2) = \begin{pmatrix} t_2 \\ 1 \\ 1 \end{pmatrix} \quad \mathrm{Rep}_B(\vec{u}_2) = \begin{pmatrix} 1 \\ u_2 \\ 1 \end{pmatrix} \quad \mathrm{Rep}_B(\vec{v}_2) = \begin{pmatrix} 1 \\ 1 \\ v_2 \end{pmatrix}$$

The projective line T_1U_1 is the projection of a plane through the origin in $\mathbb{R}^3$. One way to get its equation is to note that any vector in it is linearly dependent on the vectors for T_1 and U_1 and so this determinant is zero.

$$\begin{vmatrix} 1 & 0 & x \\ 0 & 1 & y \\ 0 & 0 & z \end{vmatrix} = 0 \qquad \Longrightarrow \qquad z = 0$$

The equation of the plane in $\mathbb{R}^3$ whose image is the projective line T_2U_2 is this.

$$\begin{vmatrix} t_2 & 1 & x \\ 1 & u_2 & y \\ 1 & 1 & z \end{vmatrix} = 0 \qquad \Longrightarrow \qquad (1-u_2)\cdot x + (1-t_2)\cdot y + (t_2u_2 - 1)\cdot z = 0$$

Finding the intersection of the two is routine.

$$T_1U_1 \cap T_2U_2 = \begin{pmatrix} t_2 - 1 \\ 1 - u_2 \\ 0 \end{pmatrix}$$

(This is, of course, a homogeneous coordinate vector of a projective point.) The other two intersections are similar.

$$T_1V_1 \cap T_2V_2 = \begin{pmatrix} 1 - t_2 \\ 0 \\ v_2 - 1 \end{pmatrix} \qquad U_1V_1 \cap U_2V_2 = \begin{pmatrix} 0 \\ u_2 - 1 \\ 1 - v_2 \end{pmatrix}$$

Finish the proof by noting that these projective points are on one projective line because the sum of the three homogeneous coordinate vectors is zero.

Every projective theorem has a translation to a Euclidean version, although the Euclidean result may be messier to state and prove. Desargue's theorem illustrates this. In the translation to Euclidean space, we must treat separately the case where O lies on the ideal line, for then the lines T_1T_2, U_1U_2, and V_1V_2 are parallel.

The remark following the statement of Desargue's Theorem suggests thinking of the Euclidean pictures as figures from projective geometry for a sphere model with very large radius. That is, just as a small area of the world seems to people living there to be flat, the projective plane is locally Euclidean.

We finish by pointing out one more thing about the projective plane. Although its local properties are familiar, the projective plane has a perhaps unfamiliar global property. The picture below shows a projective point. At that point we have drawn Cartesian axes, xy-axes. Of course, the axes appear in the picture at both antipodal spots, one in the northern hemisphere (that is, shown on the right) and the other in the south. Observe that in the northern hemisphere a person who puts their right hand on the sphere, palm down, with their thumb on the y axis will have their fingers pointing along the x-axis in the positive direction.

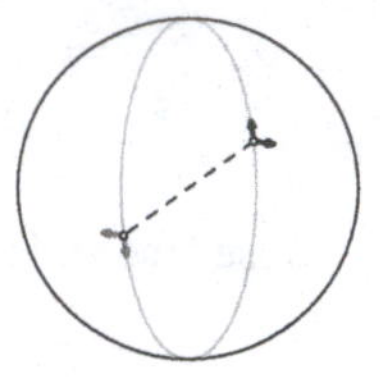

The sequence of pictures below show a trip around this space: the antipodal spots rotate around the sphere with the spot in the northern hemisphere moving up and over the north pole, ending on the far side of the sphere, and its companion coming to the front. (Be careful: the trip shown is not halfway around the projective plane. It is a full circuit. The spots at either end of the dashed line are the same projective point. So by the third sphere below the trip has pretty much returned to the same projective point where we drew it starting above.)

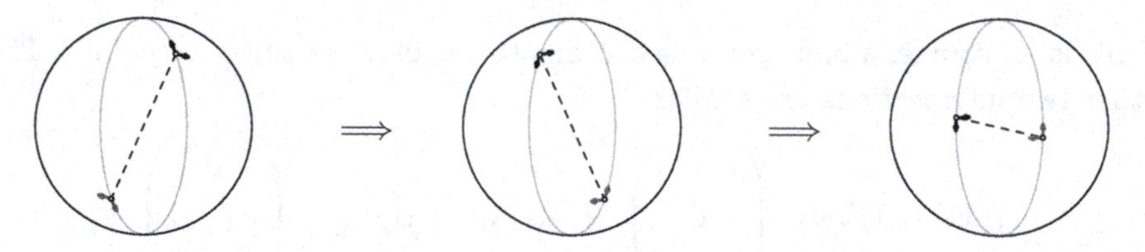

At the end of the circuit, the x part of the xy-axes sticks out in the other direction. That is, for a person to put their thumb on the y-axis and have their fingers point positively on the x-axis, they must use their left hand. The projective plane is not orientable — in this geometry, left and right handedness are not fixed properties of figures (said another way, we cannot describe a spiral as clockwise or counterclockwise).

This exhibition of the existence of a non-orientable space raises the question of whether our universe orientable. Could an astronaut leave earth right-handed and return left-handed? [Gardner] is a nontechnical reference. [Clarke] is a classic science fiction story about orientation reversal.

For an overview of projective geometry see [Courant & Robbins]. The approach we've taken here, the analytic approach, leads to quick theorems and illustrates the power of linear algebra; see [Hanes], [Ryan], and [Eggar]. But another approach, the synthetic approach of deriving the results from an axiom system, is both extraordinarily beautiful and is also the historical route of development. Two fine sources for this approach are [Coxeter] or [Seidenberg]. An easy and interesting application is in [Davies].

Exercises

1 What is the equation of this point?

$$\begin{pmatrix}1\\0\\0\end{pmatrix}$$

2 (a) Find the line incident on these points in the projective plane.

$$\begin{pmatrix}1\\2\\3\end{pmatrix},\ \begin{pmatrix}4\\5\\6\end{pmatrix}$$

(b) Find the point incident on both of these projective lines.

$$(1\ \ 2\ \ 3),\ (4\ \ 5\ \ 6)$$

3 Find the formula for the line incident on two projective points. Find the formula for the point incident on two projective lines.

4 Prove that the definition of incidence is independent of the choice of the representatives of p and L. That is, if p_1, p_2, p_3, and q_1, q_2, q_3 are two triples of homogeneous coordinates for p, and L_1, L_2, L_3, and M_1, M_2, M_3 are two triples of homogeneous coordinates for L, prove that $p_1L_1 + p_2L_2 + p_3L_3 = 0$ if and only if $q_1M_1 + q_2M_2 + q_3M_3 = 0$.

5 Give a drawing to show that central projection does not preserve circles, that a circle may project to an ellipse. Can a (non-circular) ellipse project to a circle?

6 Give the formula for the correspondence between the non-equatorial part of the antipodal modal of the projective plane, and the plane $z = 1$.

7 (Pappus's Theorem) Assume that T_0, U_0, and V_0 are collinear and that T_1, U_1, and V_1 are collinear. Consider these three points: (i) the intersection V_2 of the lines T_0U_1 and T_1U_0, (ii) the intersection U_2 of the lines T_0V_1 and T_1V_0, and (iii) the intersection T_2 of U_0V_1 and U_1V_0.

(a) Draw a (Euclidean) picture.

(b) Apply the lemma used in Desargue's Theorem to get simple homogeneous coordinate vectors for the T's and V_0.

(c) Find the resulting homogeneous coordinate vectors for U's (these must each involve a parameter as, e.g., U_0 could be anywhere on the T_0V_0 line).

(d) Find the resulting homogeneous coordinate vectors for V_1. (*Hint:* it involves two parameters.)

(e) Find the resulting homogeneous coordinate vectors for V_2. (It also involves two parameters.)

(f) Show that the product of the three parameters is 1.

(g) Verify that V_2 is on the T_2U_2 line.

Chapter Five

Similarity

We have shown that for any homomorphism there are bases B and D such that the matrix representing the map has a block partial-identity form.

$$\text{Rep}_{B,D}(h) = \left(\begin{array}{c|c} \textit{Identity} & \textit{Zero} \\ \hline \textit{Zero} & \textit{Zero} \end{array}\right)$$

This representation describes the map as sending $c_1\vec{\beta}_1 + \cdots + c_n\vec{\beta}_n$ to $c_1\vec{\delta}_1 + \cdots + c_k\vec{\delta}_k + \vec{0} + \cdots + \vec{0}$, where n is the dimension of the domain and k is the dimension of the range. Under this representation the action of the map is easy to understand because most of the matrix entries are zero.

This chapter considers the special case where the domain and codomain are the same. Here we naturally ask for the domain basis and codomain basis to be the same. That is, we want a basis B so that $\text{Rep}_{B,B}(t)$ is as simple as possible, where we take 'simple' to mean that it has many zeroes. We will find that we cannot always get a matrix having the above block partial-identity form but we will develop a form that comes close, a representation that is nearly diagonal.

I Complex Vector Spaces

This chapter requires that we factor polynomials. But many polynomials do not factor over the real numbers; for instance, $x^2 + 1$ does not factor into a product of two linear polynomials with real coefficients; instead it requires complex numbers $x^2 + 1 = (x - i)(x + i)$.

Consequently in this chapter we shall use complex numbers for our scalars, including entries in vectors and matrices. That is, we shift from studying vector spaces over the real numbers to vector spaces over the complex numbers. Any real number is a complex number and in this chapter most of the examples use only real numbers but nonetheless, the critical theorems require that the scalars be complex. So this first section is a review of complex numbers.

In this book our approach is to shift to this more general context of taking scalars to be complex for the pragmatic reason that we must do so in order to move forward. However, the idea of doing vector spaces by taking scalars from a structure other than the real numbers is an interesting and useful one. Delightful presentations that take this approach from the start are in [Halmos] and [Hoffman & Kunze].

I.1 Polynomial Factoring and Complex Numbers

This subsection is a review only. For a full development, including proofs, see [Ebbinghaus].

Consider a polynomial $p(x) = c_n x^n + \cdots + c_1 x + c_0$ with leading coefficient $c_n \neq 0$. The degree of the polynomial is n. If $n = 0$ then p is a constant polynomial $p(x) = c_0$. Constant polynomials that are not the zero polynomial, $c_0 \neq 0$, have degree zero. We define the zero polynomial to have degree $-\infty$.

1.1 Remark Defining the degree of the zero polynomial to be $-\infty$ allows the equation $\text{degree}(fg) = \text{degree}(f) + \text{degree}(g)$ to hold for all polynomials.

Just as integers have a division operation—e.g., '4 goes 5 times into 21 with remainder 1'—so do polynomials.

1.2 Theorem (Division Theorem for Polynomials) Let $p(x)$ be a polynomial. If $d(x)$ is a non-zero polynomial then there are *quotient* and *remainder* polynomials $q(x)$ and $r(x)$ such that

$$p(x) = d(x) \cdot q(x) + r(x)$$

where the degree of $r(x)$ is strictly less than the degree of $d(x)$.

The point of the integer statement '4 goes 5 times into 21 with remainder 1' is that the remainder is less than 4—while 4 goes 5 times, it does not go 6 times. Similarly, the final clause of the polynomial division statement is crucial.

1.3 Example If $p(x) = 2x^3 - 3x^2 + 4x$ and $d(x) = x^2 + 1$ then $q(x) = 2x - 3$ and

$r(x) = 2x + 3$. Note that $r(x)$ has a lower degree than does $d(x)$.

1.4 Corollary The remainder when $p(x)$ is divided by $x - \lambda$ is the constant polynomial $r(x) = p(\lambda)$.

PROOF The remainder must be a constant polynomial because it is of degree less than the divisor $x - \lambda$. To determine the constant, take the theorem's divisor $d(x)$ to be $x - \lambda$ and substitute λ for x. QED

If a divisor $d(x)$ goes into a dividend $p(x)$ evenly, meaning that $r(x)$ is the zero polynomial, then $d(x)$ is a called a factor of $p(x)$. Any root of the factor, any $\lambda \in \mathbb{R}$ such that $d(\lambda) = 0$, is a root of $p(x)$ since $p(\lambda) = d(\lambda) \cdot q(\lambda) = 0$.

1.5 Corollary If λ is a root of the polynomial $p(x)$ then $x - \lambda$ divides $p(x)$ evenly, that is, $x - \lambda$ is a factor of $p(x)$.

PROOF By the above corollary $p(x) = (x - \lambda) \cdot q(x) + p(\lambda)$. Since λ is a root, $p(\lambda) = 0$ so $x - \lambda$ is a factor. QED

A repeated root of a polynomial is a number λ such that the polynomial is evenly divisible by $(x - \lambda)^n$ for some power larger than one. The largest such power is called the multiplicity of λ.

Finding the roots and factors of a high-degree polynomial can be hard. But for second-degree polynomials we have the quadratic formula: the roots of $ax^2 + bx + c$ are these

$$\lambda_1 = \frac{-b + \sqrt{b^2 - 4ac}}{2a} \qquad \lambda_2 = \frac{-b - \sqrt{b^2 - 4ac}}{2a}$$

(if the discriminant $b^2 - 4ac$ is negative then the polynomial has no real number roots). A polynomial that cannot be factored into two lower-degree polynomials with real number coefficients is said to be irreducible over the reals.

1.6 Theorem Any constant or linear polynomial is irreducible over the reals. A quadratic polynomial is irreducible over the reals if and only if its discriminant is negative. No cubic or higher-degree polynomial is irreducible over the reals.

1.7 Corollary Any polynomial with real coefficients can be factored into linear and irreducible quadratic polynomials. That factorization is unique; any two factorizations have the same powers of the same factors.

Note the analogy with the prime factorization of integers. In both cases the uniqueness clause is very useful.

1.8 Example Because of uniqueness we know, without multiplying them out, that $(x+3)^2(x^2+1)^3$ does not equal $(x+3)^4(x^2+x+1)^2$.

1.9 Example By uniqueness, if $c(x) = m(x) \cdot q(x)$ then where $c(x) = (x-3)^2(x+2)^3$ and $m(x) = (x-3)(x+2)^2$, we know that $q(x) = (x-3)(x+2)$.

While x^2+1 has no real roots and so doesn't factor over the real numbers, if we imagine a root—traditionally denoted i, so that $i^2+1=0$—then x^2+1 factors into a product of linears $(x-i)(x+i)$. When we adjoin this root i to the reals and close the new system with respect to addition and multiplication then we have the complex numbers $\mathbb{C} = \{a+bi \mid a, b \in \mathbb{R} \text{ and } i^2 = -1\}$. (These are often pictured on a plane with a plotted on the horizontal axis and b on the vertical; note that the distance of the point from the origin is $|a+bi| = \sqrt{a^2+b^2}$.)

In $\mathbb{C}$ all quadratics factor. That is, in contrast with the reals, $\mathbb{C}$ has no irreducible quadratics.

$$ax^2+bx+c = a \cdot \big(x - \frac{-b+\sqrt{b^2-4ac}}{2a}\big) \cdot \big(x - \frac{-b-\sqrt{b^2-4ac}}{2a}\big)$$

1.10 Example The second degree polynomial x^2+x+1 factors over the complex numbers into the product of two first degree polynomials.

$$\big(x - \frac{-1+\sqrt{-3}}{2}\big)\big(x - \frac{-1-\sqrt{-3}}{2}\big) = \big(x - (-\frac{1}{2} + \frac{\sqrt{3}}{2}i)\big)\big(x - (-\frac{1}{2} - \frac{\sqrt{3}}{2}i)\big)$$

1.11 Theorem (Fundamental Theorem of Algebra) Polynomials with complex coefficients factor into linear polynomials with complex coefficients. The factorization is unique.

I.2 Complex Representations

Recall the definitions of the complex number addition

$$(a+bi)+(c+di) = (a+c)+(b+d)i$$

and multiplication.

$$\begin{aligned}(a+bi)(c+di) &= ac+adi+bci+bd(-1)\\ &= (ac-bd)+(ad+bc)i\end{aligned}$$

2.1 Example For instance, $(1-2i)+(5+4i) = 6+2i$ and $(2-3i)(4-0.5i) = 6.5-13i$.

With these rules, all of the operations that we've used for real vector spaces carry over unchanged to vector spaces with complex scalars.

2.2 Example Matrix multiplication is the same, although the scalar arithmetic involves more bookkeeping.

$$\begin{pmatrix} 1+1i & 2-0i \\ i & -2+3i \end{pmatrix} \begin{pmatrix} 1+0i & 1-0i \\ 3i & -i \end{pmatrix}$$
$$= \begin{pmatrix} (1+1i)\cdot(1+0i)+(2-0i)\cdot(3i) & (1+1i)\cdot(1-0i)+(2-0i)\cdot(-i) \\ (i)\cdot(1+0i)+(-2+3i)\cdot(3i) & (i)\cdot(1-0i)+(-2+3i)\cdot(-i) \end{pmatrix}$$
$$= \begin{pmatrix} 1+7i & 1-1i \\ -9-5i & 3+3i \end{pmatrix}$$

We shall carry over unchanged from the previous chapters everything that we can. For instance, we shall call this

$$\langle \begin{pmatrix} 1+0i \\ 0+0i \\ \vdots \\ 0+0i \end{pmatrix}, \dots, \begin{pmatrix} 0+0i \\ 0+0i \\ \vdots \\ 1+0i \end{pmatrix} \rangle$$

the *standard basis* for $\mathbb{C}^n$ as a vector space over $\mathbb{C}$ and again denote it $\mathcal{E}_n$. Another example is that $\mathcal{P}_n$ will be the vector space of degree n polynomials with coefficients that are complex.

II Similarity

We've defined two matrices H and $\hat{H}$ to be matrix equivalent if there are nonsingular P and Q such that $\hat{H} = PHQ$. We were motivated by this diagram showing H and $\hat{H}$ both representing a map h, but with respect to different pairs of bases, B, D and $\hat{B}, \hat{D}$.

$$\begin{array}{ccc} V_{\mathit{wrt}\ B} & \xrightarrow[H]{h} & W_{\mathit{wrt}\ D} \\ \mathrm{id}\downarrow & & \mathrm{id}\downarrow \\ V_{\mathit{wrt}\ \hat{B}} & \xrightarrow[\hat{H}]{h} & W_{\mathit{wrt}\ \hat{D}} \end{array}$$

We now consider the special case of transformations, where the codomain equals the domain, and we add the requirement that the codomain's basis equals the domain's basis. So, we are considering representations with respect to B, B and D, D.

$$\begin{array}{ccc} V_{\mathit{wrt}\ B} & \xrightarrow[T]{t} & V_{\mathit{wrt}\ B} \\ \mathrm{id}\downarrow & & \mathrm{id}\downarrow \\ V_{\mathit{wrt}\ D} & \xrightarrow[\hat{T}]{t} & V_{\mathit{wrt}\ D} \end{array}$$

In matrix terms, $\mathrm{Rep}_{D,D}(t) = \mathrm{Rep}_{B,D}(\mathrm{id})\,\mathrm{Rep}_{B,B}(t)\,\bigl(\mathrm{Rep}_{B,D}(\mathrm{id})\bigr)^{-1}$.

II.1 Definition and Examples

1.1 Example Consider the derivative transformation $d/dx\colon \mathcal{P}_2 \to \mathcal{P}_2$, and two bases for that space $B = \langle x^2, x, 1\rangle$ and $D = \langle 1, 1+x, 1+x^2\rangle$ We will compute the four sides of the arrow square.

$$\begin{array}{ccc} \mathcal{P}_{2\ \mathit{wrt}\ B} & \xrightarrow[T]{d/dx} & \mathcal{P}_{2\ \mathit{wrt}\ B} \\ \mathrm{id}\downarrow & & \mathrm{id}\downarrow \\ \mathcal{P}_{2\ \mathit{wrt}\ D} & \xrightarrow[\hat{T}]{d/dx} & \mathcal{P}_{2\ \mathit{wrt}\ D} \end{array}$$

The top is first. The effect of the transformation on the starting basis B

$$x^2 \overset{d/dx}{\longmapsto} 2x \quad x \overset{d/dx}{\longmapsto} 1 \quad 1 \overset{d/dx}{\longmapsto} 0$$

represented with respect to the ending basis (also B)

$$\text{Rep}_B(2x) = \begin{pmatrix} 0 \\ 2 \\ 0 \end{pmatrix} \qquad \text{Rep}_B(1) = \begin{pmatrix} 0 \\ 0 \\ 1 \end{pmatrix} \qquad \text{Rep}_B(0) = \begin{pmatrix} 0 \\ 0 \\ 0 \end{pmatrix}$$

gives the representation of the map.

$$T = \text{Rep}_{B,B}(d/dx) = \begin{pmatrix} 0 & 0 & 0 \\ 2 & 0 & 0 \\ 0 & 1 & 0 \end{pmatrix}$$

Next, the bottom. The effect of the transformation on elements of D

$$1 \overset{d/dx}{\longmapsto} 0 \qquad 1+x \overset{d/dx}{\longmapsto} 1 \qquad 1+x^2 \overset{d/dx}{\longmapsto} 2x$$

represented with respect to D gives the matrix $\hat{T}$.

$$\hat{T} = \text{Rep}_{D,D}(d/dx) = \begin{pmatrix} 0 & 1 & -2 \\ 0 & 0 & 2 \\ 0 & 0 & 0 \end{pmatrix}$$

Third, computing the matrix for the right-hand side involves finding the effect of the identity map on the elements of B. Of course, the identity map does not transform them at all so to find the matrix we represent B's elements with respect to D.

$$\text{Rep}_D(x^2) = \begin{pmatrix} -1 \\ 0 \\ 1 \end{pmatrix} \quad \text{Rep}_D(x) = \begin{pmatrix} -1 \\ 1 \\ 0 \end{pmatrix} \quad \text{Rep}_D(1) = \begin{pmatrix} 1 \\ 0 \\ 0 \end{pmatrix}$$

So the matrix for going down the right side is the concatenation of those.

$$P = \text{Rep}_{B,D}(\text{id}) = \begin{pmatrix} -1 & -1 & 1 \\ 0 & 1 & 0 \\ 1 & 0 & 0 \end{pmatrix}$$

With that, we have two options to compute the matrix for going up on left side. The direct computation represents elements of D with respect to B

$$\text{Rep}_B(1) = \begin{pmatrix} 0 \\ 0 \\ 1 \end{pmatrix} \quad \text{Rep}_B(1+x) = \begin{pmatrix} 0 \\ 1 \\ 1 \end{pmatrix} \quad \text{Rep}_B(1+x^2) = \begin{pmatrix} 1 \\ 0 \\ 1 \end{pmatrix}$$

and concatenates to make the matrix.

$$\begin{pmatrix} 0 & 0 & 1 \\ 0 & 1 & 0 \\ 1 & 1 & 1 \end{pmatrix}$$

The other option to compute the matrix for going up on the left is to take the inverse of the matrix P for going down on the right.

$$\left(\begin{array}{ccc|ccc} -1 & -1 & 1 & 1 & 0 & 0 \\ 0 & 1 & 0 & 0 & 1 & 0 \\ 1 & 0 & 0 & 0 & 0 & 1 \end{array}\right) \longrightarrow \cdots \longrightarrow \left(\begin{array}{ccc|ccc} 1 & 0 & 0 & 0 & 0 & 1 \\ 0 & 1 & 0 & 0 & 1 & 0 \\ 0 & 0 & 1 & 1 & 1 & 1 \end{array}\right)$$

1.2 Definition The matrices T and $\hat{T}$ are *similar* if there is a nonsingular P such that $\hat{T} = PTP^{-1}$.

Since nonsingular matrices are square, T and $\hat{T}$ must be square and of the same size. Exercise 15 checks that similarity is an equivalence relation.

1.3 Example The definition does not require that we consider a map. Calculation with these two

$$P = \begin{pmatrix} 2 & 1 \\ 1 & 1 \end{pmatrix} \qquad T = \begin{pmatrix} 2 & -3 \\ 1 & -1 \end{pmatrix}$$

gives that T is similar to this matrix.

$$\hat{T} = \begin{pmatrix} 12 & -19 \\ 7 & -11 \end{pmatrix}$$

1.4 Example The only matrix similar to the zero matrix is itself: $PZP^{-1} = PZ = Z$. The identity matrix has the same property: $PIP^{-1} = PP^{-1} = I$.

Matrix similarity is a special case of matrix equivalence so if two matrices are similar then they are matrix equivalent. What about the converse: if they are square, must any two matrix equivalent matrices be similar? No; the matrix equivalence class of an identity matrix consists of all nonsingular matrices of that size while the prior example shows that the only member of the similarity class of an identity matrix is itself. Thus these two are matrix equivalent but not similar.

$$T = \begin{pmatrix} 1 & 0 \\ 0 & 1 \end{pmatrix} \qquad S = \begin{pmatrix} 1 & 2 \\ 0 & 3 \end{pmatrix}$$

So some matrix equivalence classes split into two or more similarity classes—similarity gives a finer partition than does matrix equivalence. This shows some matrix equivalence classes subdivided into similarity classes.

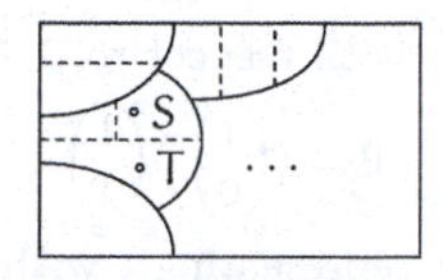

To understand the similarity relation we shall study the similarity classes. We approach this question in the same way that we've studied both the row equivalence and matrix equivalence relations, by finding a canonical form for representatives of the similarity classes, called Jordan form. With this canonical form, we can decide if two matrices are similar by checking whether they are in a class with the same representative. We've also seen with both row equivalence and matrix equivalence that a canonical form gives us insight into the ways in which members of the same class are alike (e.g., two identically-sized matrices are matrix equivalent if and only if they have the same rank).

Exercises

1.5 For

$$T=\begin{pmatrix}1&3\\-2&-6\end{pmatrix}\quad \hat{T}=\begin{pmatrix}0&0\\-11/2&-5\end{pmatrix}\quad P=\begin{pmatrix}4&2\\-3&2\end{pmatrix}$$

check that $\hat{T}=PTP^{-1}$.

1.6 Example 1.4 shows that the only matrix similar to a zero matrix is itself and that the only matrix similar to the identity is itself.

(a) Show that the 1×1 matrix whose single entry is 2 is also similar only to itself.

(b) Is a matrix of the form cI for some scalar c similar only to itself?

(c) Is a diagonal matrix similar only to itself?

✓ **1.7** Consider this transformation of $\mathbb{C}^3$

$$t(\begin{pmatrix}x\\y\\z\end{pmatrix})=\begin{pmatrix}x-z\\z\\2y\end{pmatrix}$$

and these bases.

$$B=\langle\begin{pmatrix}1\\2\\3\end{pmatrix},\begin{pmatrix}0\\1\\0\end{pmatrix},\begin{pmatrix}0\\0\\1\end{pmatrix}\rangle\qquad D=\langle\begin{pmatrix}1\\0\\0\end{pmatrix},\begin{pmatrix}1\\1\\0\end{pmatrix},\begin{pmatrix}1\\0\\1\end{pmatrix}\rangle$$

We will compute the parts of the arrow diagram to represent the transformation using two similar matrices.

(a) Draw the arrow diagram, specialized for this case.

(b) Compute $T=\text{Rep}_{B,B}(t)$.

(c) Compute $\hat{T}=\text{Rep}_{D,D}(t)$.

(d) Compute the matrices for other the two sides of the arrow square.

1.8 Consider the transformation $t\colon \mathcal{P}_2\to\mathcal{P}_2$ described by $x^2\mapsto x+1$, $x\mapsto x^2-1$, and $1\mapsto 3$.

(a) Find $T=\text{Rep}_{B,B}(t)$ where $B=\langle x^2,x,1\rangle$.

(b) Find $\hat{T}=\text{Rep}_{D,D}(t)$ where $D=\langle 1,1+x,1+x+x^2\rangle$.

(c) Find the matrix P such that $\hat{T}=PTP^{-1}$.

1.9 Let T represent $t\colon \mathbb{C}^2 \to \mathbb{C}^2$ with respect to B, B.

$$T = \begin{pmatrix} 1 & -1 \\ 2 & 1 \end{pmatrix} \qquad B = \langle \begin{pmatrix} 1 \\ 0 \end{pmatrix}, \begin{pmatrix} 1 \\ 1 \end{pmatrix} \rangle, \quad D = \langle \begin{pmatrix} 2 \\ 0 \end{pmatrix}, \begin{pmatrix} 0 \\ -2 \end{pmatrix} \rangle$$

We will convert to the matrix representing t with resepct to D, D.

(a) Draw the arrow diagram.

(b) Give the matrix that represents the left and right sides of that diagram, in the direction that we traverse the diagram to make the conversion.

(c) Find $\text{Rep}_{D,D}(t)$.

✓ 1.10 Exhibit an nontrivial similarity relationship by letting $t\colon \mathbb{C}^2 \to \mathbb{C}^2$ act in this way,

$$\begin{pmatrix} 1 \\ 2 \end{pmatrix} \mapsto \begin{pmatrix} 3 \\ 0 \end{pmatrix} \qquad \begin{pmatrix} -1 \\ 1 \end{pmatrix} \mapsto \begin{pmatrix} -1 \\ 2 \end{pmatrix}$$

picking two bases B, D, and representing t with respect to them, $\hat{T} = \text{Rep}_{B,B}(t)$ and $T = \text{Rep}_{D,D}(t)$. Then compute the P and P^{-1} to change bases from B to D and back again.

✓ 1.11 Show that these matrices are not similar.

$$\begin{pmatrix} 1 & 0 & 4 \\ 1 & 1 & 3 \\ 2 & 1 & 7 \end{pmatrix} \qquad \begin{pmatrix} 1 & 0 & 1 \\ 0 & 1 & 1 \\ 3 & 1 & 2 \end{pmatrix}$$

1.12 Explain Example 1.4 in terms of maps.

✓ 1.13 [Halmos] Are there two matrices A and B that are similar while A^2 and B^2 are not similar?

✓ 1.14 Prove that if two matrices are similar and one is invertible then so is the other.

✓ 1.15 Show that similarity is an equivalence relation. (The definition given earlier already reflects this, so instead start here with the definition that $\hat{T}$ is similar to T if $\hat{T} = PTP^{-1}$.)

1.16 Consider a matrix representing, with respect to some B, B, reflection across the x-axis in $\mathbb{R}^2$. Consider also a matrix representing, with respect to some D, D, reflection across the y-axis. Must they be similar?

1.17 Prove that similarity preserves determinants and rank. Does the converse hold?

1.18 Is there a matrix equivalence class with only one matrix similarity class inside? One with infinitely many similarity classes?

1.19 Can two different diagonal matrices be in the same similarity class?

✓ 1.20 Prove that if two matrices are similar then their k-th powers are similar when $k > 0$. What if $k \leqslant 0$?

✓ 1.21 Let $p(x)$ be the polynomial $c_n x^n + \cdots + c_1 x + c_0$. Show that if T is similar to S then $p(T) = c_n T^n + \cdots + c_1 T + c_0 I$ is similar to $p(S) = c_n S^n + \cdots + c_1 S + c_0 I$.

1.22 List all of the matrix equivalence classes of $1{\times}1$ matrices. Also list the similarity classes, and describe which similarity classes are contained inside of each matrix equivalence class.

1.23 Does similarity preserve sums?

1.24 Show that if $T - \lambda I$ and N are similar matrices then T and $N + \lambda I$ are also similar.

II.2 Diagonalizability

The prior subsection shows that although similar matrices are necessarily matrix equivalent, the converse does not hold. Some matrix equivalence classes break into two or more similarity classes; for instance, the nonsingular 2×2 matrices form one matrix equivalence class but more than one similarity class.

Thus we cannot use the canonical form for matrix equivalence, a block partial-identity matrix, as a canonical form for matrix similarity. The diagram below illustrates. The stars are similarity class representatives. Each dashed-line similarity class subdivision has one star but each solid-curve matrix equivalence class division has only one partial identity matrix.

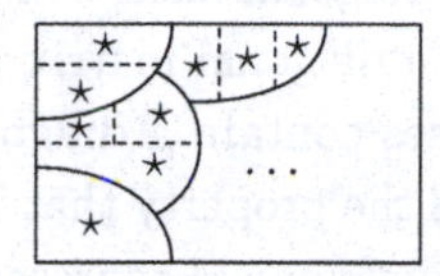

To develop a canonical form for representatives of the similarity classes we naturally build on previous work. This means first that the partial identity matrices should represent the similarity classes into which they fall. Beyond that, the representatives should be as simple as possible. The simplest extension of the partial identity form is the diagonal form.

2.1 Definition A transformation is *diagonalizable* if it has a diagonal representation with respect to the same basis for the codomain as for the domain. A *diagonalizable matrix* is one that is similar to a diagonal matrix: T is diagonalizable if there is a nonsingular P such that PTP^{-1} is diagonal.

2.2 Example The matrix

$$\begin{pmatrix} 4 & -2 \\ 1 & 1 \end{pmatrix}$$

is diagonalizable.

$$\begin{pmatrix} 2 & 0 \\ 0 & 3 \end{pmatrix} = \begin{pmatrix} -1 & 2 \\ 1 & -1 \end{pmatrix} \begin{pmatrix} 4 & -2 \\ 1 & 1 \end{pmatrix} \begin{pmatrix} -1 & 2 \\ 1 & -1 \end{pmatrix}^{-1}$$

2.3 Example We will show that this matrix is not diagonalizable.

$$N = \begin{pmatrix} 0 & 0 \\ 1 & 0 \end{pmatrix}$$

The fact that N is not the zero matrix means that it cannot be similar to the zero matrix, because the zero matrix is similar only to itself. Thus if N were to be similar to a diagonal matrix then that matrix would have have at least one nonzero entry on its diagonal.

The square of N is the zero matrix. This imples that for any map n represented by N (with respect to some B, B) the composition $n \circ n$ is the zero map. This in turn implies that for any matrix representing n (with respect to some $\hat{B}, \hat{B}$), its square is the zero matrix. But the square of a nonzero diagonal matrix cannot be the zero matrix, because the square of a diagonal matrix is the diagonal matrix whose entries are the squares of the entries from the starting matrix. Thus there is no $\hat{B}, \hat{B}$ such that n is represented by a diagonal matrix—the matrix N is not diagonalizable.

That example shows that a diagonal form will not suffice as a canonical form for similarity—we cannot find a diagonal matrix in each matrix similarity class. However, some similarity classes contain a diagonal matrix and the canonical form that we are developing has the property that if a matrix can be diagonalized then the diagonal matrix is the canonical representative of its similarity class.

2.4 Lemma A transformation t is diagonalizable if and only if there is a basis $B = \langle \vec{\beta}_1, \ldots, \vec{\beta}_n \rangle$ and scalars $\lambda_1, \ldots, \lambda_n$ such that $t(\vec{\beta}_i) = \lambda_i \vec{\beta}_i$ for each i.

Proof Consider a diagonal representation matrix.

$$\text{Rep}_{B,B}(t) = \begin{pmatrix} \vdots & & \vdots \\ \text{Rep}_B(t(\vec{\beta}_1)) & \cdots & \text{Rep}_B(t(\vec{\beta}_n)) \\ \vdots & & \vdots \end{pmatrix} = \begin{pmatrix} \lambda_1 & & 0 \\ \vdots & \ddots & \vdots \\ 0 & & \lambda_n \end{pmatrix}$$

Consider the representation of a member of this basis with respect to the basis $\text{Rep}_B(\vec{\beta}_i)$. The product of the diagonal matrix and the representation vector

$$\text{Rep}_B(t(\vec{\beta}_i)) = \begin{pmatrix} \lambda_1 & & 0 \\ \vdots & \ddots & \vdots \\ 0 & & \lambda_n \end{pmatrix} \begin{pmatrix} 0 \\ \vdots \\ 1 \\ \vdots \\ 0 \end{pmatrix} = \begin{pmatrix} 0 \\ \vdots \\ \lambda_i \\ \vdots \\ 0 \end{pmatrix}$$

has the stated action. QED

2.5 Example To diagonalize

$$T = \begin{pmatrix} 3 & 2 \\ 0 & 1 \end{pmatrix}$$

we take T as the representation of a transformation with respect to the standard basis $\mathrm{Rep}_{\mathcal{E}_2,\mathcal{E}_2}(t)$ and look for a basis $B = \langle\vec{\beta}_1, \vec{\beta}_2\rangle$ such that

$$\mathrm{Rep}_{B,B}(t) = \begin{pmatrix} \lambda_1 & 0 \\ 0 & \lambda_2 \end{pmatrix}$$

that is, such that $t(\vec{\beta}_1) = \lambda_1\vec{\beta}_1$ and $t(\vec{\beta}_2) = \lambda_2\vec{\beta}_2$.

$$\begin{pmatrix} 3 & 2 \\ 0 & 1 \end{pmatrix}\vec{\beta}_1 = \lambda_1 \cdot \vec{\beta}_1 \qquad \begin{pmatrix} 3 & 2 \\ 0 & 1 \end{pmatrix}\vec{\beta}_2 = \lambda_2 \cdot \vec{\beta}_2$$

We are looking for scalars x such that this equation

$$\begin{pmatrix} 3 & 2 \\ 0 & 1 \end{pmatrix}\begin{pmatrix} b_1 \\ b_2 \end{pmatrix} = x \cdot \begin{pmatrix} b_1 \\ b_2 \end{pmatrix}$$

has solutions b_1 and b_2 that are not both 0 (the zero vector is not the member of any basis). That's a linear system.

$$\begin{aligned} (3-x) \cdot b_1 + \quad\quad 2 \cdot b_2 &= 0 \\ (1-x) \cdot b_2 &= 0 \end{aligned} \qquad (*)$$

Focus first on the bottom equation. There are two cases: either $b_2 = 0$ or $x = 1$.

In the $b_2 = 0$ case the first equation gives that either $b_1 = 0$ or $x = 3$. Since we've disallowed the possibility that both $b_2 = 0$ and $b_1 = 0$, we are left with the first diagonal entry $\lambda_1 = 3$. With that, $(*)$'s first equation is $0 \cdot b_1 + 2 \cdot b_2 = 0$ and so associated with $\lambda_1 = 3$ are vectors having a second component of zero while the first component is free.

$$\begin{pmatrix} 3 & 2 \\ 0 & 1 \end{pmatrix}\begin{pmatrix} b_1 \\ 0 \end{pmatrix} = 3 \cdot \begin{pmatrix} b_1 \\ 0 \end{pmatrix}$$

To get a first basis vector choose any nonzero b_1.

$$\vec{\beta}_1 = \begin{pmatrix} 1 \\ 0 \end{pmatrix}$$

The other case for the bottom equation of $(*)$ is $\lambda_2 = 1$. Then $(*)$'s first equation is $2 \cdot b_1 + 2 \cdot b_2 = 0$ and so associated with this case are vectors whose second component is the negative of the first.

$$\begin{pmatrix} 3 & 2 \\ 0 & 1 \end{pmatrix}\begin{pmatrix} b_1 \\ -b_1 \end{pmatrix} = 1 \cdot \begin{pmatrix} b_1 \\ -b_1 \end{pmatrix}$$

Get the second basis vector by choosing a nonzero one of these.

$$\vec{\beta}_2 = \begin{pmatrix} 1 \\ -1 \end{pmatrix}$$

Now draw the similarity diagram

$$\begin{array}{ccc} \mathbb{R}^2_{wrt\ \mathcal{E}_2} & \xrightarrow[T]{t} & \mathbb{R}^2_{wrt\ \mathcal{E}_2} \\ {\scriptstyle id}\big\downarrow & & {\scriptstyle id}\big\downarrow \\ \mathbb{R}^2_{wrt\ B} & \xrightarrow[D]{t} & \mathbb{R}^2_{wrt\ B} \end{array}$$

and note that the matrix $\text{Rep}_{B,\mathcal{E}_2}(\text{id})$ is easy, giving this diagonalization.

$$\begin{pmatrix} 3 & 0 \\ 0 & 1 \end{pmatrix} = \begin{pmatrix} 1 & 1 \\ 0 & -1 \end{pmatrix}^{-1} \begin{pmatrix} 3 & 2 \\ 0 & 1 \end{pmatrix} \begin{pmatrix} 1 & 1 \\ 0 & -1 \end{pmatrix}$$

In the next subsection we will expand on that example by considering more closely the property of Lemma 2.4. This includes seeing a streamlined way to find the λ's.

Exercises

✓ **2.6** Repeat Example 2.5 for the matrix from Example 2.2.

2.7 Diagonalize these upper triangular matrices.

(a) $\begin{pmatrix} -2 & 1 \\ 0 & 2 \end{pmatrix}$ (b) $\begin{pmatrix} 5 & 4 \\ 0 & 1 \end{pmatrix}$

✓ **2.8** What form do the powers of a diagonal matrix have?

2.9 Give two same-sized diagonal matrices that are not similar. Must any two different diagonal matrices come from different similarity classes?

2.10 Give a nonsingular diagonal matrix. Can a diagonal matrix ever be singular?

✓ **2.11** Show that the inverse of a diagonal matrix is the diagonal of the the inverses, if no element on that diagonal is zero. What happens when a diagonal entry is zero?

2.12 The equation ending Example 2.5

$$\begin{pmatrix} 1 & 1 \\ 0 & -1 \end{pmatrix}^{-1} \begin{pmatrix} 3 & 2 \\ 0 & 1 \end{pmatrix} \begin{pmatrix} 1 & 1 \\ 0 & -1 \end{pmatrix} = \begin{pmatrix} 3 & 0 \\ 0 & 1 \end{pmatrix}$$

is a bit jarring because for P we must take the first matrix, which is shown as an inverse, and for P^{-1} we take the inverse of the first matrix, so that the two -1 powers cancel and this matrix is shown without a superscript -1.

(a) Check that this nicer-appearing equation holds.

$$\begin{pmatrix} 3 & 0 \\ 0 & 1 \end{pmatrix} = \begin{pmatrix} 1 & 1 \\ 0 & -1 \end{pmatrix} \begin{pmatrix} 3 & 2 \\ 0 & 1 \end{pmatrix} \begin{pmatrix} 1 & 1 \\ 0 & -1 \end{pmatrix}^{-1}$$

(b) Is the previous item a coincidence? Or can we always switch the P and the P^{-1}?

2.13 Show that the P used to diagonalize in Example 2.5 is not unique.

2.14 Find a formula for the powers of this matrix *Hint*: see Exercise 8.

$$\begin{pmatrix} -3 & 1 \\ -4 & 2 \end{pmatrix}$$

✓ **2.15** Diagonalize these.

(a) $\begin{pmatrix}1&1\\0&0\end{pmatrix}$ (b) $\begin{pmatrix}0&1\\1&0\end{pmatrix}$

2.16 We can ask how diagonalization interacts with the matrix operations. Assume that $t, s\colon V \to V$ are each diagonalizable. Is ct diagonalizable for all scalars c? What about $t+s$? $t \circ s$?

✓ **2.17** Show that matrices of this form are not diagonalizable.

$$\begin{pmatrix}1&c\\0&1\end{pmatrix} \qquad c \neq 0$$

2.18 Show that each of these is diagonalizable.

(a) $\begin{pmatrix}1&2\\2&1\end{pmatrix}$ (b) $\begin{pmatrix}x&y\\y&z\end{pmatrix}$ $\quad x, y, z$ scalars

II.3 Eigenvalues and Eigenvectors

We will next focus on the property of Lemma 2.4.

3.1 Definition A transformation $t\colon V \to V$ has a scalar *eigenvalue* λ if there is a nonzero *eigenvector* $\vec{\zeta} \in V$ such that $t(\vec{\zeta}) = \lambda \cdot \vec{\zeta}$.

("Eigen" is German for "characteristic of" or "peculiar to." Some authors call these *characteristic* values and vectors. No authors call them "peculiar.")

3.2 Example The projection map

$$\begin{pmatrix}x\\y\\z\end{pmatrix} \stackrel{\pi}{\longmapsto} \begin{pmatrix}x\\y\\0\end{pmatrix} \qquad x, y, z \in \mathbb{C}$$

has an eigenvalue of 1 associated with any eigenvector

$$\begin{pmatrix}x\\y\\0\end{pmatrix}$$

where x and y are scalars that are not both zero.

In contrast, a number that is not an eigenvalue of of this map is 2, since assuming that π doubles a vector leads to the equations $x = 2x$, $y = 2y$, and $0 = 2z$, and thus no non-$\vec{0}$ vector is doubled.

Note that the definition requires that the eigenvector be non-$\vec{0}$. Some authors allow $\vec{0}$ as an eigenvector for λ as long as there are also non-$\vec{0}$ vectors associated with λ. The key point is to disallow the trivial case where λ is such that $t(\vec{v}) = \lambda\vec{v}$ for only the single vector $\vec{v} = \vec{0}$.

Also, note that the eigenvalue λ could be 0. The issue is whether $\vec{\zeta}$ equals $\vec{0}$.

3.3 Example The only transformation on the trivial space $\{\vec{0}\}$ is $\vec{0} \mapsto \vec{0}$. This map has no eigenvalues because there are no non-$\vec{0}$ vectors $\vec{v}$ mapped to a scalar multiple $\lambda \cdot \vec{v}$ of themselves.

3.4 Example Consider the homomorphism $t\colon \mathcal{P}_1 \to \mathcal{P}_1$ given by $c_0 + c_1 x \mapsto (c_0 + c_1) + (c_0 + c_1)x$. While the codomain $\mathcal{P}_1$ of t is two-dimensional, its range is one-dimensional $\mathscr{R}(t) = \{c + cx \mid c \in \mathbb{C}\}$. Application of t to a vector in that range will simply rescale the vector $c + cx \mapsto (2c) + (2c)x$. That is, t has an eigenvalue of 2 associated with eigenvectors of the form $c + cx$ where $c \neq 0$.

This map also has an eigenvalue of 0 associated with eigenvectors of the form $c - cx$ where $c \neq 0$.

The definition above is for maps. We can give a matrix version.

3.5 Definition A square matrix T has a scalar *eigenvalue* λ associated with the nonzero *eigenvector* $\vec{\zeta}$ if $T\vec{\zeta} = \lambda \cdot \vec{\zeta}$.

This extension of the definition for maps to a definition for matrices is natural but there is a point on which we must take care. The eigenvalues of a map are also the eigenvalues of matrices r epresenting that map, and so similar matrices have the same eigenvalues. However, the eigenvectors can differ — similar matrices need not have the same eigenvectors. The next example explains.

3.6 Example These matrices are similar

$$T = \begin{pmatrix} 2 & 0 \\ 0 & 0 \end{pmatrix} \quad \hat{T} = \begin{pmatrix} 4 & -2 \\ 4 & -2 \end{pmatrix}$$

since $\hat{T} = PTP^{-1}$ for this P.

$$P = \begin{pmatrix} 1 & 1 \\ 1 & 2 \end{pmatrix} \quad P^{-1} = \begin{pmatrix} 2 & -1 \\ -1 & 1 \end{pmatrix}$$

The matrix T has two eigenvalues, $\lambda_1 = 2$ and $\lambda_2 = 0$. The first one is associated with this eigenvector.

$$T\vec{e}_1 = \begin{pmatrix} 2 & 0 \\ 0 & 0 \end{pmatrix} \begin{pmatrix} 1 \\ 0 \end{pmatrix} = \begin{pmatrix} 2 \\ 0 \end{pmatrix} = 2\vec{e}_1$$

Suppose that T represents a transformation $t\colon \mathbb{C}^2 \to \mathbb{C}^2$ with respect to the standard basis. Then the action of this transformation t is simple.

$$\begin{pmatrix} x \\ y \end{pmatrix} \stackrel{t}{\longmapsto} \begin{pmatrix} 2x \\ 0 \end{pmatrix}$$

Of course, $\hat{T}$ represents the same transformation but with respect to a different basis B. We can easily find this basis. The arrow diagram

$$\begin{array}{ccc} V_{wrt\,\mathcal{E}_3} & \xrightarrow[T]{t} & V_{wrt\,\mathcal{E}_3} \\ {\scriptstyle id}\big\downarrow & & {\scriptstyle id}\big\downarrow \\ V_{wrt\,B} & \xrightarrow[\hat{T}]{t} & V_{wrt\,B} \end{array}$$

shows that $P^{-1} = \text{Rep}_{B,\mathcal{E}_3}(\text{id})$. By the definition of the matrix representation of a map, its first column is $\text{Rep}_{\mathcal{E}_3}(\text{id}(\vec{\beta}_1)) = \text{Rep}_{\mathcal{E}_3}(\vec{\beta}_1)$. With respect to the standard basis any vector is represented by itself, so the first basis element $\vec{\beta}_1$ is the first column of P^{-1}. The same goes for the other one.

$$B = \langle \begin{pmatrix} 2 \\ -1 \end{pmatrix}, \begin{pmatrix} -1 \\ 1 \end{pmatrix} \rangle$$

Since the matrices T and $\hat{T}$ both represent the transformation t, both reflect the action $t(\vec{e}_1) = 2\vec{e}_1$.

$$\begin{aligned} \text{Rep}_{\mathcal{E}_2,\mathcal{E}_2}(t) \cdot \text{Rep}_{\mathcal{E}_2}(\vec{e}_1) &= T \cdot \text{Rep}_{\mathcal{E}_2}(\vec{e}_1) = 2 \cdot \text{Rep}_{\mathcal{E}_2}(\vec{e}_1) \\ \text{Rep}_{B,B}(t) \cdot \text{Rep}_{B}(\vec{e}_1) &= \hat{T} \cdot \text{Rep}_{B}(\vec{e}_1) = 2 \cdot \text{Rep}_{B}(\vec{e}_1) \end{aligned}$$

But while in those two equations the eigenvalue 2's are the same, the vector representations differ.

$$\begin{aligned} T \cdot \text{Rep}_{\mathcal{E}_2}(\vec{e}_1) &= T \begin{pmatrix} 1 \\ 0 \end{pmatrix} = 2 \cdot \begin{pmatrix} 1 \\ 0 \end{pmatrix} \\ \hat{T} \cdot \text{Rep}_{B}(\vec{e}_1) &= \hat{T} \cdot \begin{pmatrix} 1 \\ 1 \end{pmatrix} = 2 \cdot \begin{pmatrix} 1 \\ 1 \end{pmatrix} \end{aligned}$$

That is, when the matrix representing the transformation is $T = \text{Rep}_{\mathcal{E}_2,\mathcal{E}_2}(t)$ then it "assumes" that column vectors are representations with respect to $\mathcal{E}_2$. However $\hat{T} = \text{Rep}_{B,B}(t)$ "assumes" that column vectors are representations with respect to B, and so the column vectors that get doubled are different.

We next see the basic tool for finding eigenvectors and eigenvalues.

3.7 Example If

$$T = \begin{pmatrix} 1 & 2 & 1 \\ 2 & 0 & -2 \\ -1 & 2 & 3 \end{pmatrix}$$

then to find the scalars x such that $T\vec{\zeta} = x\vec{\zeta}$ for nonzero eigenvectors $\vec{\zeta}$, bring everything to the left-hand side

$$\begin{pmatrix} 1 & 2 & 1 \\ 2 & 0 & -2 \\ -1 & 2 & 3 \end{pmatrix}\begin{pmatrix} z_1 \\ z_2 \\ z_3 \end{pmatrix} - x\begin{pmatrix} z_1 \\ z_2 \\ z_3 \end{pmatrix} = \vec{0}$$

and factor $(T - xI)\vec{\zeta} = \vec{0}$. (Note that it says $T - xI$. The expression $T - x$ doesn't make sense because T is a matrix while x is a scalar.) This homogeneous linear system

$$\begin{pmatrix} 1-x & 2 & 1 \\ 2 & 0-x & -2 \\ -1 & 2 & 3-x \end{pmatrix}\begin{pmatrix} z_1 \\ z_2 \\ z_3 \end{pmatrix} = \begin{pmatrix} 0 \\ 0 \\ 0 \end{pmatrix}$$

has a nonzero solution $\vec{z}$ if and only if the matrix is singular. We can determine when that happens.

$$\begin{aligned} 0 &= |T - xI| \\ &= \begin{vmatrix} 1-x & 2 & 1 \\ 2 & 0-x & -2 \\ -1 & 2 & 3-x \end{vmatrix} \\ &= x^3 - 4x^2 + 4x \\ &= x(x-2)^2 \end{aligned}$$

The eigenvalues are $\lambda_1 = 0$ and $\lambda_2 = 2$. To find the associated eigenvectors plug in each eigenvalue. Plugging in $\lambda_1 = 0$ gives

$$\begin{pmatrix} 1-0 & 2 & 1 \\ 2 & 0-0 & -2 \\ -1 & 2 & 3-0 \end{pmatrix}\begin{pmatrix} z_1 \\ z_2 \\ z_3 \end{pmatrix} = \begin{pmatrix} 0 \\ 0 \\ 0 \end{pmatrix} \quad \Longrightarrow \quad \begin{pmatrix} z_1 \\ z_2 \\ z_3 \end{pmatrix} = \begin{pmatrix} a \\ -a \\ a \end{pmatrix}$$

for $a \neq 0$ (a must be non-0 because eigenvectors are defined to be non-$\vec{0}$). Plugging in $\lambda_2 = 2$ gives

$$\begin{pmatrix} 1-2 & 2 & 1 \\ 2 & 0-2 & -2 \\ -1 & 2 & 3-2 \end{pmatrix}\begin{pmatrix} z_1 \\ z_2 \\ z_3 \end{pmatrix} = \begin{pmatrix} 0 \\ 0 \\ 0 \end{pmatrix} \quad \Longrightarrow \quad \begin{pmatrix} z_1 \\ z_2 \\ z_3 \end{pmatrix} = \begin{pmatrix} b \\ 0 \\ b \end{pmatrix}$$

with $b \neq 0$.

3.8 Example If

$$S = \begin{pmatrix} \pi & 1 \\ 0 & 3 \end{pmatrix}$$

(here π is not a projection map, it is the number 3.14...) then

$$\begin{vmatrix} \pi - x & 1 \\ 0 & 3 - x \end{vmatrix} = (x - \pi)(x - 3)$$

so S has eigenvalues of $\lambda_1 = \pi$ and $\lambda_2 = 3$. To find associated eigenvectors, first plug in λ_1 for x

$$\begin{pmatrix} \pi - \pi & 1 \\ 0 & 3 - \pi \end{pmatrix} \begin{pmatrix} z_1 \\ z_2 \end{pmatrix} = \begin{pmatrix} 0 \\ 0 \end{pmatrix} \quad \Longrightarrow \quad \begin{pmatrix} z_1 \\ z_2 \end{pmatrix} = \begin{pmatrix} a \\ 0 \end{pmatrix}$$

for a scalar $a \neq 0$. Then plug in λ_2

$$\begin{pmatrix} \pi - 3 & 1 \\ 0 & 3 - 3 \end{pmatrix} \begin{pmatrix} z_1 \\ z_2 \end{pmatrix} = \begin{pmatrix} 0 \\ 0 \end{pmatrix} \quad \Longrightarrow \quad \begin{pmatrix} z_1 \\ z_2 \end{pmatrix} = \begin{pmatrix} -b/(\pi - 3) \\ b \end{pmatrix}$$

where $b \neq 0$.

3.9 Definition The *characteristic polynomial of a square matrix* T is the determinant $|T - xI|$ where x is a variable. The *characteristic equation* is $|T - xI| = 0$. The *characteristic polynomial of a transformation* t is the characteristic polynomial of any matrix representation $\mathrm{Rep}_{B,B}(t)$.

Exercise 30 checks that the characteristic polynomial of a transformation is well-defined, that is, that the characteristic polynomial is the same no matter which basis we use for the representation.

3.10 Lemma A linear transformation on a nontrivial vector space has at least one eigenvalue.

PROOF Any root of the characteristic polynomial is an eigenvalue. Over the complex numbers, any polynomial of degree one or greater has a root. QED

3.11 Remark That result is the reason that in this chapter we use scalars that are complex numbers.

3.12 Definition The *eigenspace of a transformation* t *associated with the eigenvalue* λ is $V_\lambda = \{\vec{\zeta} \mid t(\vec{\zeta}) = \lambda\vec{\zeta}\}$. The eigenspace of a matrix is analogous.

3.13 Lemma An eigenspace is a subspace.

PROOF Fix an eigenvalue λ. Notice first that V_λ contains the zero vector since $t(\vec{0}) = \vec{0}$, which equals $\lambda\vec{0}$. So the eigenspace is a nonempty subset of the space. What remains is to check closure of this set under linear combinations. Take $\vec{\zeta}_1, \ldots, \vec{\zeta}_n \in V_\lambda$ and then verify

$$\begin{aligned} t(c_1\vec{\zeta}_1 + c_2\vec{\zeta}_2 + \cdots + c_n\vec{\zeta}_n) &= c_1 t(\vec{\zeta}_1) + \cdots + c_n t(\vec{\zeta}_n) \\ &= c_1\lambda\vec{\zeta}_1 + \cdots + c_n\lambda\vec{\zeta}_n \\ &= \lambda(c_1\vec{\zeta}_1 + \cdots + c_n\vec{\zeta}_n) \end{aligned}$$

that the combination is also an element of V_λ. QED

3.14 Example In Example 3.7 these are the eigenspaces associated with the eigenvalues 0 and 2.

$$V_0 = \{\begin{pmatrix} a \\ -a \\ a \end{pmatrix} \mid a \in \mathbb{C}\}, \qquad V_2 = \{\begin{pmatrix} b \\ 0 \\ b \end{pmatrix} \mid b \in \mathbb{C}\}.$$

3.15 Example In Example 3.8 these are the eigenspaces associated with the eigenvalues π and 3.

$$V_\pi = \{\begin{pmatrix} a \\ 0 \end{pmatrix} \mid a \in \mathbb{C}\} \qquad V_3 = \{\begin{pmatrix} -b/(\pi - 3) \\ b \end{pmatrix} \mid b \in \mathbb{C}\}$$

The characteristic equation in Example 3.7 is $0 = x(x-2)^2$ so in some sense 2 is an eigenvalue twice. However there are not twice as many eigenvectors in that the dimension of the associated eigenspace V_2 is one, not two. The next example is a case where a number is a double root of the characteristic equation and the dimension of the associated eigenspace is two.

3.16 Example With respect to the standard bases, this matrix

$$\begin{pmatrix} 1 & 0 & 0 \\ 0 & 1 & 0 \\ 0 & 0 & 0 \end{pmatrix}$$

represents projection.

$$\begin{pmatrix} x \\ y \\ z \end{pmatrix} \overset{\pi}{\longmapsto} \begin{pmatrix} x \\ y \\ 0 \end{pmatrix} \qquad x, y, z \in \mathbb{C}$$

Its characteristic equation

$$\begin{aligned} 0 &= |T - xI| \\ &= \begin{vmatrix} 1-x & 0 & 0 \\ 0 & 1-x & 0 \\ 0 & 0 & 0-x \end{vmatrix} \\ &= (1-x)^2(0-x) \end{aligned}$$

has the double root $x = 1$ along with the single root $x = 0$. Its eigenspace associated with the eigenvalue 0 and its eigenspace associated with the eigenvalue 1 are easy to find.

$$V_0 = \{ \begin{pmatrix} 0 \\ 0 \\ c_3 \end{pmatrix} \mid c_3 \in \mathbb{C} \} \qquad V_1 = \{ \begin{pmatrix} c_1 \\ c_2 \\ 0 \end{pmatrix} \mid c_1, c_2 \in \mathbb{C} \}$$

Note that V_1 has dimension two.

By Lemma 3.13 if two eigenvectors $\vec{v}_1$ and $\vec{v}_2$ are associated with the same eigenvalue then a linear combination of those two is also an eigenvector, associated with the same eigenvalue. As an illustration, referring to the prior example, this sum of two members of V_1

$$\begin{pmatrix} 1 \\ 0 \\ 0 \end{pmatrix} + \begin{pmatrix} 0 \\ 1 \\ 0 \end{pmatrix}$$

yields another member of V_1.

The next result speaks to the situation where the vectors come from different eigenspaces.

3.17 Theorem For any set of distinct eigenvalues of a map or matrix, a set of associated eigenvectors, one per eigenvalue, is linearly independent.

Proof We will use induction on the number of eigenvalues. The base step is that there are zero eigenvalues. Then the set of associated vectors is empty and so is linearly independent.

For the inductive step assume that the statement is true for any set of $k \geqslant 0$ distinct eigenvalues. Consider distinct eigenvalues $\lambda_1, \ldots, \lambda_{k+1}$ and let $\vec{v}_1, \ldots, \vec{v}_{k+1}$ be associated eigenvectors. Suppose that $\vec{0} = c_1\vec{v}_1 + \cdots + c_k\vec{v}_k + c_{k+1}\vec{v}_{k+1}$. Derive two equations from that, the first by multiplying by λ_{k+1} on both sides $\vec{0} = c_1\lambda_{k+1}\vec{v}_1 + \cdots + c_{k+1}\lambda_{k+1}\vec{v}_{k+1}$ and the second by applying the map to both sides $\vec{0} = c_1 t(\vec{v}_1) + \cdots + c_{k+1} t(\vec{v}_{k+1}) = c_1\lambda_1\vec{v}_1 + \cdots + c_{k+1}\lambda_{k+1}\vec{v}_{k+1}$ (applying the matrix gives the same result). Subtract the second from the first.

$$\vec{0} = c_1(\lambda_{k+1} - \lambda_1)\vec{v}_1 + \cdots + c_k(\lambda_{k+1} - \lambda_k)\vec{v}_k + c_{k+1}(\lambda_{k+1} - \lambda_{k+1})\vec{v}_{k+1}$$

The $\vec{v}_{k+1}$ term vanishes. Then the induction hypothesis gives that $c_1(\lambda_{k+1} - \lambda_1) = 0$, ..., $c_k(\lambda_{k+1} - \lambda_k) = 0$. The eigenvalues are distinct so the coefficients c_1, ..., c_k are all 0. With that we are left with the equation $\vec{0} = c_{k+1}\vec{v}_{k+1}$ so c_{k+1} is also 0. QED

3.18 Example The eigenvalues of

$$\begin{pmatrix} 2 & -2 & 2 \\ 0 & 1 & 1 \\ -4 & 8 & 3 \end{pmatrix}$$

are distinct: $\lambda_1 = 1$, $\lambda_2 = 2$, and $\lambda_3 = 3$. A set of associated eigenvectors

$$\{ \begin{pmatrix} 2 \\ 1 \\ 0 \end{pmatrix}, \begin{pmatrix} 9 \\ 4 \\ 4 \end{pmatrix}, \begin{pmatrix} 2 \\ 1 \\ 2 \end{pmatrix} \}$$

is linearly independent.

3.19 Corollary An $n \times n$ matrix with n distinct eigenvalues is diagonalizable.

Proof Form a basis of eigenvectors. Apply Lemma 2.4. QED

This section observes that some matrices are similar to a diagonal matrix. The idea of eigenvalues arose as the entries of that diagonal matrix, although the definition applies more broadly than just to diagonalizable matrices. To find eigenvalues we defined the characteristic equation and that led to the final result, a criteria for diagonalizability. (While it is useful for the theory, note that in applications finding eigenvalues this way is typically impractical; for one thing the matrix may be large and finding roots of large-degree polynomials is hard.)

In the next section we study matrices that cannot be diagonalized.

Exercises

3.20 For each, find the characteristic polynomial and the eigenvalues.

(a) $\begin{pmatrix} 10 & -9 \\ 4 & -2 \end{pmatrix}$ (b) $\begin{pmatrix} 1 & 2 \\ 4 & 3 \end{pmatrix}$ (c) $\begin{pmatrix} 0 & 3 \\ 7 & 0 \end{pmatrix}$ (d) $\begin{pmatrix} 0 & 0 \\ 0 & 0 \end{pmatrix}$

(e) $\begin{pmatrix} 1 & 0 \\ 0 & 1 \end{pmatrix}$

✓ **3.21** For each matrix, find the characteristic equation, and the eigenvalues and associated eigenvectors.

(a) $\begin{pmatrix} 3 & 0 \\ 8 & -1 \end{pmatrix}$

(b) $\begin{pmatrix} 3 & 2 \\ -1 & 0 \end{pmatrix}$

3.22 Find the characteristic equation, and the eigenvalues and associated eigenvectors for this matrix. *Hint.* The eigenvalues are complex.

$$\begin{pmatrix} -2 & -1 \\ 5 & 2 \end{pmatrix}$$

3.23 Find the characteristic polynomial, the eigenvalues, and the associated eigenvectors of this matrix.

$$\begin{pmatrix} 1 & 1 & 1 \\ 0 & 0 & 1 \\ 0 & 0 & 1 \end{pmatrix}$$

✓ **3.24** For each matrix, find the characteristic equation, and the eigenvalues and associated eigenvectors.

(a) $\begin{pmatrix} 3 & -2 & 0 \\ -2 & 3 & 0 \\ 0 & 0 & 5 \end{pmatrix}$ **(b)** $\begin{pmatrix} 0 & 1 & 0 \\ 0 & 0 & 1 \\ 4 & -17 & 8 \end{pmatrix}$

✓ **3.25** Let $t\colon \mathcal{P}_2 \to \mathcal{P}_2$ be

$$a_0 + a_1x + a_2x^2 \mapsto (5a_0 + 6a_1 + 2a_2) - (a_1 + 8a_2)x + (a_0 - 2a_2)x^2.$$

Find its eigenvalues and the associated eigenvectors.

3.26 Find the eigenvalues and eigenvectors of this map $t\colon \mathcal{M}_2 \to \mathcal{M}_2$.

$$\begin{pmatrix} a & b \\ c & d \end{pmatrix} \mapsto \begin{pmatrix} 2c & a+c \\ b-2c & d \end{pmatrix}$$

✓ **3.27** Find the eigenvalues and associated eigenvectors of the differentiation operator $d/dx\colon \mathcal{P}_3 \to \mathcal{P}_3$.

3.28 Prove that the eigenvalues of a triangular matrix (upper or lower triangular) are the entries on the diagonal.

✓ **3.29** Find the formula for the characteristic polynomial of a 2×2 matrix.

3.30 Prove that the characteristic polynomial of a transformation is well-defined.

3.31 Prove or disprove: if all the eigenvalues of a matrix are 0 then it must be the zero matrix.

✓ **3.32** **(a)** Show that any non-$\vec{0}$ vector in any nontrivial vector space can be a eigenvector. That is, given a $\vec{v} \neq \vec{0}$ from a nontrivial V, show that there is a transformation $t\colon V \to V$ having a scalar eigenvalue $\lambda \in \mathbb{R}$ such that $\vec{v} \in V_\lambda$.

(b) What if we are given a scalar λ? Can any non-$\vec{0}$ member of any nontrivial vector space be an eigenvector associated with λ?

✓ **3.33** Suppose that $t\colon V \to V$ and $T = \text{Rep}_{B,B}(t)$. Prove that the eigenvectors of T associated with λ are the non-$\vec{0}$ vectors in the kernel of the map represented (with respect to the same bases) by $T - \lambda I$.

3.34 Prove that if $a, \ldots, d$ are all integers and $a + b = c + d$ then

$$\begin{pmatrix} a & b \\ c & d \end{pmatrix}$$

has integral eigenvalues, namely $a + b$ and $a - c$.

✓ **3.35** Prove that if T is nonsingular and has eigenvalues $\lambda_1, \ldots, \lambda_n$ then T^{-1} has eigenvalues $1/\lambda_1, \ldots, 1/\lambda_n$. Is the converse true?

✓ **3.36** Suppose that T is $n \times n$ and c, d are scalars.

(a) Prove that if T has the eigenvalue λ with an associated eigenvector $\vec{v}$ then $\vec{v}$ is an eigenvector of $cT + dI$ associated with eigenvalue $c\lambda + d$.

(b) Prove that if T is diagonalizable then so is $cT + dI$.

✓ **3.37** Show that λ is an eigenvalue of T if and only if the map represented by $T - \lambda I$ is not an isomorphism.

3.38 [Strang 80]

(a) Show that if λ is an eigenvalue of A then λ^k is an eigenvalue of A^k.

(b) What is wrong with this proof generalizing that? "If λ is an eigenvalue of A and μ is an eigenvalue for B, then $\lambda\mu$ is an eigenvalue for AB, for, if $A\vec{x} = \lambda\vec{x}$ and $B\vec{x} = \mu\vec{x}$ then $AB\vec{x} = A\mu\vec{x} = \mu A\vec{x} = \mu\lambda\vec{x}$"?

3.39 Do matrix equivalent matrices have the same eigenvalues?

3.40 Show that a square matrix with real entries and an odd number of rows has at least one real eigenvalue.

3.41 Diagonalize.

$$\begin{pmatrix} -1 & 2 & 2 \\ 2 & 2 & 2 \\ -3 & -6 & -6 \end{pmatrix}$$

3.42 Suppose that P is a nonsingular $n \times n$ matrix. Show that the *similarity transformation* map $t_P \colon \mathcal{M}_{n \times n} \to \mathcal{M}_{n \times n}$ sending $T \mapsto PTP^{-1}$ is an isomorphism.

? **3.43** [Math. Mag., Nov. 1967] Show that if A is an n square matrix and each row (column) sums to c then c is a characteristic root of A. ("Characteristic root" is a synonym for eigenvalue.)

III Nilpotence

This chapter shows that every square matrix is similar to one that is a sum of two kinds of simple matrices. The prior section focused on the first simple kind, diagonal matrices. We now consider the other kind.

III.1 Self-Composition

Because a linear transformation $t\colon V \to V$ has the same domain as codomain, we can compose t with itself $t^2 = t \circ t$, and $t^3 = t \circ t \circ t$, etc.*

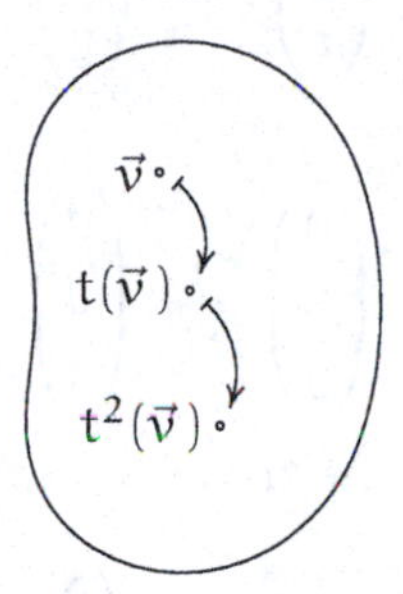

Note that the superscript power notation t^j for iterates of the transformations fits with the notation that we've used for their square matrix representations because if $\text{Rep}_{B,B}(t) = T$ then $\text{Rep}_{B,B}(t^j) = T^j$.

1.1 Example For the derivative map $d/dx\colon \mathcal{P}_3 \to \mathcal{P}_3$ given by

$$a + bx + cx^2 + dx^3 \overset{d/dx}{\longmapsto} b + 2cx + 3dx^2$$

the second power is the second derivative

$$a + bx + cx^2 + dx^3 \overset{d^2/dx^2}{\longmapsto} 2c + 6dx$$

the third power is the third derivative

$$a + bx + cx^2 + dx^3 \overset{d^3/dx^3}{\longmapsto} 6d$$

and any higher power is the zero map.

1.2 Example This transformation of the space $\mathcal{M}_{2\times 2}$ of 2×2 matrices

$$\begin{pmatrix} a & b \\ c & d \end{pmatrix} \overset{t}{\longmapsto} \begin{pmatrix} b & a \\ d & 0 \end{pmatrix}$$

* More information on function iteration is in the appendix.

has this second power

$$\begin{pmatrix} a & b \\ c & d \end{pmatrix} \stackrel{t^2}{\longmapsto} \begin{pmatrix} a & b \\ 0 & 0 \end{pmatrix}$$

and this third power.

$$\begin{pmatrix} a & b \\ c & d \end{pmatrix} \stackrel{t^3}{\longmapsto} \begin{pmatrix} b & a \\ 0 & 0 \end{pmatrix}$$

After that, $t^4 = t^2$ and $t^5 = t^3$, etc.

1.3 Example Consider the shift transformation $t\colon \mathbb{C}^3 \to \mathbb{C}^3$.

$$\begin{pmatrix} x \\ y \\ z \end{pmatrix} \stackrel{t}{\longmapsto} \begin{pmatrix} 0 \\ x \\ y \end{pmatrix}$$

We have that

$$\begin{pmatrix} x \\ y \\ z \end{pmatrix} \stackrel{t}{\longmapsto} \begin{pmatrix} 0 \\ x \\ y \end{pmatrix} \stackrel{t}{\longmapsto} \begin{pmatrix} 0 \\ 0 \\ x \end{pmatrix} \stackrel{t}{\longmapsto} \begin{pmatrix} 0 \\ 0 \\ 0 \end{pmatrix}$$

so the range spaces descend to the trivial subspace.

$$\mathscr{R}(t) = \{\begin{pmatrix} 0 \\ a \\ b \end{pmatrix} \mid a, b \in \mathbb{C}\} \qquad \mathscr{R}(t^2) = \{\begin{pmatrix} 0 \\ 0 \\ c \end{pmatrix} \mid c \in \mathbb{C}\} \qquad \mathscr{R}(t^3) = \{\begin{pmatrix} 0 \\ 0 \\ 0 \end{pmatrix}\}$$

These examples suggest that after some number of iterations the map settles down.

1.4 Lemma For any transformation $t\colon V \to V$, the range spaces of the powers form a descending chain

$$V \supseteq \mathscr{R}(t) \supseteq \mathscr{R}(t^2) \supseteq \cdots$$

and the null spaces form an ascending chain.

$$\{\vec{0}\} \subseteq \mathscr{N}(t) \subseteq \mathscr{N}(t^2) \subseteq \cdots$$

Further, there is a k such that for powers less than k the subsets are proper so that if $j < k$ then $\mathscr{R}(t^j) \supset \mathscr{R}(t^{j+1})$ and $\mathscr{N}(t^j) \subset \mathscr{N}(t^{j+1})$, while for higher powers the sets are equal so that if $j \geqslant k$ then $\mathscr{R}(t^j) = \mathscr{R}(t^{j+1})$ and $\mathscr{N}(t^j) = \mathscr{N}(t^{j+1})$).

Proof First recall that for any map the dimension of its range space plus the dimension of its null space equals the dimension of its domain. So if the dimensions of the range spaces shrink then the dimensions of the null spaces must rise. We will do the range space half here and leave the rest for Exercise 14.

We start by showing that the range spaces form a chain. If $\vec{w} \in \mathscr{R}(t^{j+1})$, so that $\vec{w} = t^{j+1}(\vec{v})$ for some $\vec{v}$, then $\vec{w} = t^{j}(\,t(\vec{v})\,)$. Thus $\vec{w} \in \mathscr{R}(t^{j})$.

Next we verify the "further" property: in the chain the subsets containments are proper initially, and then from some power k onward the range spaces are equal. We first show that if any pair of adjacent range spaces in the chain are equal $\mathscr{R}(t^{k}) = \mathscr{R}(t^{k+1})$ then all subsequent ones are also equal $\mathscr{R}(t^{k+1}) = \mathscr{R}(t^{k+2})$, etc. This holds because $t\colon \mathscr{R}(t^{k+1}) \to \mathscr{R}(t^{k+2})$ is the same map, with the same domain, as $t\colon \mathscr{R}(t^{k}) \to \mathscr{R}(t^{k+1})$ and it therefore has the same range $\mathscr{R}(t^{k+1}) = \mathscr{R}(t^{k+2})$ (it holds for all higher powers by induction). So if the chain of range spaces ever stops strictly decreasing then from that point onward it is stable.

We end by showing that the chain must eventually stop decreasing. Each range space is a subspace of the one before it. For it to be a proper subspace it must be of strictly lower dimension (see Exercise 12). These spaces are finite-dimensional and so the chain can fall for only finitely many steps. That is, the power k is at most the dimension of V. QED

1.5 Example The derivative map $a + bx + cx^2 + dx^3 \overset{d/dx}{\longmapsto} b + 2cx + 3dx^2$ on $\mathcal{P}_3$ has this chain of range spaces

$$\mathscr{R}(t^0) = \mathcal{P}_3 \supset \mathscr{R}(t^1) = \mathcal{P}_2 \supset \mathscr{R}(t^2) = \mathcal{P}_1 \supset \mathscr{R}(t^3) = \mathcal{P}_0 \supset \mathscr{R}(t^4) = \{\vec{0}\}$$

(all later elements of the chain are the trivial space). And it has this chain of null spaces

$$\mathscr{N}(t^0) = \{\vec{0}\} \subset \mathscr{N}(t^1) = \mathcal{P}_0 \subset \mathscr{N}(t^2) = \mathcal{P}_1 \subset \mathscr{N}(t^3) = \mathcal{P}_2 \subset \mathscr{N}(t^4) = \mathcal{P}_3$$

(later elements are the entire space).

1.6 Example Let $t\colon \mathcal{P}_2 \to \mathcal{P}_2$ be the map $c_0 + c_1x + c_2x^2 \mapsto 2c_0 + c_2x$. As the lemma describes, on iteration the range space shrinks

$$\mathscr{R}(t^0) = \mathcal{P}_2 \quad \mathscr{R}(t) = \{a + bx \mid a, b \in \mathbb{C}\} \quad \mathscr{R}(t^2) = \{a \mid a \in \mathbb{C}\}$$

and then stabilizes $\mathscr{R}(t^2) = \mathscr{R}(t^3) = \cdots$ while the null space grows

$$\mathscr{N}(t^0) = \{0\} \quad \mathscr{N}(t) = \{cx \mid c \in \mathbb{C}\} \quad \mathscr{N}(t^2) = \{cx + d \mid c, d \in \mathbb{C}\}$$

and then stabilizes $\mathscr{N}(t^2) = \mathscr{N}(t^3) = \cdots$.

1.7 Example The transformation $\pi\colon \mathbb{C}^3 \to \mathbb{C}^3$ projecting onto the first two coordinates

$$\begin{pmatrix} c_1 \\ c_2 \\ c_3 \end{pmatrix} \overset{\pi}{\longmapsto} \begin{pmatrix} c_1 \\ c_2 \\ 0 \end{pmatrix}$$

has $\mathbb{C}^3 \supset \mathscr{R}(\pi) = \mathscr{R}(\pi^2) = \cdots$ and $\{\vec{0}\} \subset \mathscr{N}(\pi) = \mathscr{N}(\pi^2) = \cdots$ where this is the range space and the null space.

$$\mathscr{R}(\pi) = \{\begin{pmatrix} a \\ b \\ 0 \end{pmatrix} \mid a, b \in \mathbb{C}\} \qquad \mathscr{N}(\pi) = \{\begin{pmatrix} 0 \\ 0 \\ c \end{pmatrix} \mid c \in \mathbb{C}\}$$

1.8 Definition Let t be a transformation on an n-dimensional space. The *generalized range space* (or *closure of the range space*) is $\mathscr{R}_\infty(t) = \mathscr{R}(t^n)$. The *generalized null space* (or *closure of the null space*) is $\mathscr{N}_\infty(t) = \mathscr{N}(t^n)$.

This graph illustrates. The horizontal axis gives the power j of a transformation. The vertical axis gives the dimension of the range space of t^j as the distance above zero, and thus also shows the dimension of the null space because the two add to the dimension n of the domain.

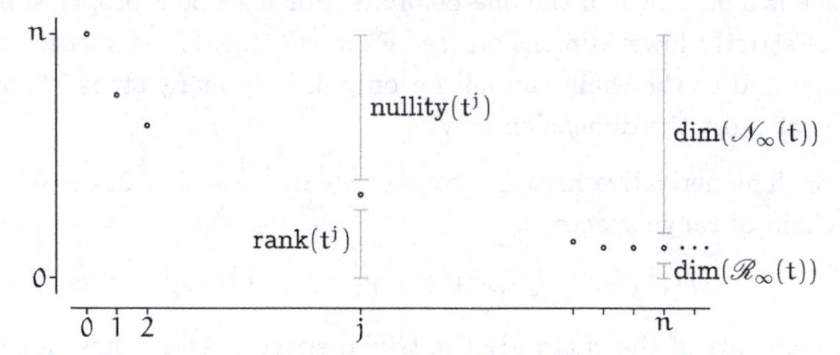

On iteration the rank falls and the nullity rises until there is some k such that the map reaches a steady state $\mathscr{R}(t^k) = \mathscr{R}(t^{k+1}) = \mathscr{R}_\infty(t)$ and $\mathscr{N}(t^k) = \mathscr{N}(t^{k+1}) = \mathscr{N}_\infty(t)$. This must happen by the n-th iterate.

Exercises

✓ **1.9** Give the chains of range spaces and null spaces for the zero and identity transformations.

✓ **1.10** For each map, give the chain of range spaces and the chain of null spaces, and the generalized range space and the generalized null space.

(a) $t_0 \colon \mathcal{P}_2 \to \mathcal{P}_2$, $a + bx + cx^2 \mapsto b + cx^2$

(b) $t_1 \colon \mathbb{R}^2 \to \mathbb{R}^2$,

$$\begin{pmatrix} a \\ b \end{pmatrix} \mapsto \begin{pmatrix} 0 \\ a \end{pmatrix}$$

(c) $t_2 \colon \mathcal{P}_2 \to \mathcal{P}_2$, $a + bx + cx^2 \mapsto b + cx + ax^2$

(d) $t_3 \colon \mathbb{R}^3 \to \mathbb{R}^3$,

$$\begin{pmatrix} a \\ b \\ c \end{pmatrix} \mapsto \begin{pmatrix} a \\ a \\ b \end{pmatrix}$$

1.11 Prove that function composition is associative $(t \circ t) \circ t = t \circ (t \circ t)$ and so we can write t^3 without specifying a grouping.

1.12 Check that a subspace must be of dimension less than or equal to the dimension of its superspace. Check that if the subspace is proper (the subspace does not equal the superspace) then the dimension is strictly less. *(This is used in the proof of Lemma 1.4.)*

✓ **1.13** Prove that the generalized range space $\mathscr{R}_\infty(t)$ is the entire space, and the generalized null space $\mathscr{N}_\infty(t)$ is trivial, if the transformation t is nonsingular. Is this 'only if' also?

1.14 Verify the null space half of Lemma 1.4.

✓ **1.15** Give an example of a transformation on a three dimensional space whose range has dimension two. What is its null space? Iterate your example until the range space and null space stabilize.

1.16 Show that the range space and null space of a linear transformation need not be disjoint. Are they ever disjoint?

III.2 Strings

This requires material from the optional Combining Subspaces subsection.

The prior subsection shows that as j increases the dimensions of the $\mathscr{R}(t^j)$'s fall while the dimensions of the $\mathscr{N}(t^j)$'s rise, in such a way that this rank and nullity split between them the dimension of V. Can we say more; do the two split a basis—is $V = \mathscr{R}(t^j) \oplus \mathscr{N}(t^j)$?

The answer is yes for the smallest power $j = 0$ since $V = \mathscr{R}(t^0) \oplus \mathscr{N}(t^0) = V \oplus \{\vec{0}\}$. The answer is also yes at the other extreme.

2.1 Lemma For any linear $t\colon V \to V$ the function $t\colon \mathscr{R}_\infty(t) \to \mathscr{R}_\infty(t)$ is one-to-one.

Proof Let the dimension of V be n. Because $\mathscr{R}(t^n) = \mathscr{R}(t^{n+1})$, the map $t\colon \mathscr{R}_\infty(t) \to \mathscr{R}_\infty(t)$ is a dimension-preserving homomorphism. Therefore, by Theorem Two.II.2.20 it is one-to-one. QED

2.2 Corollary Where $t\colon V \to V$ is a linear transformation, the space is the direct sum $V = \mathscr{R}_\infty(t) \oplus \mathscr{N}_\infty(t)$. That is, both (1) $\dim(V) = \dim(\mathscr{R}_\infty(t)) + \dim(\mathscr{N}_\infty(t))$ and (2) $\mathscr{R}_\infty(t) \cap \mathscr{N}_\infty(t) = \{\vec{0}\}$.

Proof Let the dimension of V be n. We will verify the second sentence, which is equivalent to the first. Clause (1) is true because any transformation satisfies that its rank plus its nullity equals the dimension of the space, and in particular this holds for the transformation t^n.

For clause (2), assume that $\vec{v} \in \mathscr{R}_\infty(t) \cap \mathscr{N}_\infty(t)$ to prove that $\vec{v} = \vec{0}$. Because $\vec{v}$ is in the generalized null space, $t^n(\vec{v}) = \vec{0}$. On the other hand, by the lemma $t\colon \mathscr{R}_\infty(t) \to \mathscr{R}_\infty(t)$ is one-to-one and a composition of one-to-one maps is one-to-one, so $t^n\colon \mathscr{R}_\infty(t) \to \mathscr{R}_\infty(t)$ is one-to-one. Only $\vec{0}$ is sent by a one-to-one linear map to $\vec{0}$ so the fact that $t^n(\vec{v}) = \vec{0}$ implies that $\vec{v} = \vec{0}$. QED

2.3 Remark Technically there is a difference between the map $t\colon V \to V$ and the map on the subspace $t\colon \mathscr{R}_\infty(t) \to \mathscr{R}_\infty(t)$ if the generalized range space is not equal to V, because the domains are different. But the difference is small because the second is the restriction of the first to $\mathscr{R}_\infty(t)$.

For powers between $j = 0$ and $j = n$, the space V might not be the direct sum of $\mathscr{R}(t^j)$ and $\mathscr{N}(t^j)$. The next example shows that the two can have a nontrivial intersection.

2.4 Example Consider the transformation of $\mathbb{C}^2$ defined by this action on the elements of the standard basis.

$$\begin{pmatrix}1\\0\end{pmatrix} \overset{n}{\longmapsto} \begin{pmatrix}0\\1\end{pmatrix} \quad \begin{pmatrix}0\\1\end{pmatrix} \overset{n}{\longmapsto} \begin{pmatrix}0\\0\end{pmatrix} \qquad N = \mathrm{Rep}_{\mathcal{E}_2,\mathcal{E}_2}(n) = \begin{pmatrix}0&0\\1&0\end{pmatrix}$$

This is a *shift map* and is clearly nilpotent of index two.

$$\begin{pmatrix}x\\y\end{pmatrix} \mapsto \begin{pmatrix}0\\x\end{pmatrix}$$

Another way to depict this map's action is with a *string*.

$$\vec{e}_1 \mapsto \vec{e}_2 \mapsto \vec{0}$$

The vector

$$\vec{e}_2 = \begin{pmatrix}0\\1\end{pmatrix}$$

is in both the range space and null space.

2.5 Example A map $\hat{n}\colon \mathbb{C}^4 \to \mathbb{C}^4$ whose action on $\mathcal{E}_4$ is given by the string

$$\vec{e}_1 \mapsto \vec{e}_2 \mapsto \vec{e}_3 \mapsto \vec{e}_4 \mapsto \vec{0}$$

has $\mathscr{R}(\hat{n}) \cap \mathscr{N}(\hat{n})$ equal to the span $[\{\vec{e}_4\}]$, has $\mathscr{R}(\hat{n}^2) \cap \mathscr{N}(\hat{n}^2) = [\{\vec{e}_3, \vec{e}_4\}]$, and has $\mathscr{R}(\hat{n}^3) \cap \mathscr{N}(\hat{n}^3) = [\{\vec{e}_4\}]$. It is nilpotent of index four. The matrix representation is all zeros except for some subdiagonal ones.

$$\hat{N} = \mathrm{Rep}_{\mathcal{E}_4,\mathcal{E}_4}(\hat{n}) = \begin{pmatrix}0&0&0&0\\1&0&0&0\\0&1&0&0\\0&0&1&0\end{pmatrix}$$

2.6 Example Transformations can act via more than one string. A transformation t acting on a basis $B = \langle \vec{\beta}_1, \ldots, \vec{\beta}_5 \rangle$ by

$$\begin{array}{l} \vec{\beta}_1 \mapsto \vec{\beta}_2 \mapsto \vec{\beta}_3 \mapsto \vec{0} \\ \vec{\beta}_4 \mapsto \vec{\beta}_5 \mapsto \vec{0} \end{array}$$

is represented by a matrix that is all zeros except for blocks of subdiagonal ones

$$\text{Rep}_{B,B}(t) = \left(\begin{array}{ccc|cc} 0 & 0 & 0 & 0 & 0 \\ 1 & 0 & 0 & 0 & 0 \\ 0 & 1 & 0 & 0 & 0 \\ \hline 0 & 0 & 0 & 0 & 0 \\ 0 & 0 & 0 & 1 & 0 \end{array}\right)$$

(the lines just visually organize the blocks).

In those examples all vectors are eventually transformed to zero.

2.7 Definition A *nilpotent* transformation is one with a power that is the zero map. A *nilpotent matrix* is one with a power that is the zero matrix. In either case, the least such power is the *index of nilpotency*.

2.8 Example In Example 2.4 the index of nilpotency is two. In Example 2.5 it is four. In Example 2.6 it is three.

2.9 Example The differentiation map $d/dx\colon \mathcal{P}_2 \to \mathcal{P}_2$ is nilpotent of index three since the third derivative of any quadratic polynomial is zero. This map's action is described by the string $x^2 \mapsto 2x \mapsto 2 \mapsto 0$ and taking the basis $B = \langle x^2, 2x, 2 \rangle$ gives this representation.

$$\text{Rep}_{B,B}(d/dx) = \begin{pmatrix} 0 & 0 & 0 \\ 1 & 0 & 0 \\ 0 & 1 & 0 \end{pmatrix}$$

Not all nilpotent matrices are all zeros except for blocks of subdiagonal ones.

2.10 Example With the matrix $\hat{N}$ from Example 2.5, and this four-vector basis

$$D = \langle \begin{pmatrix} 1 \\ 0 \\ 1 \\ 0 \end{pmatrix}, \begin{pmatrix} 0 \\ 2 \\ 1 \\ 0 \end{pmatrix}, \begin{pmatrix} 1 \\ 1 \\ 1 \\ 0 \end{pmatrix}, \begin{pmatrix} 0 \\ 0 \\ 0 \\ 1 \end{pmatrix} \rangle$$

a change of basis operation produces this representation with respect to D, D.

$$\begin{pmatrix} 1 & 0 & 1 & 0 \\ 0 & 2 & 1 & 0 \\ 1 & 1 & 1 & 0 \\ 0 & 0 & 0 & 1 \end{pmatrix} \begin{pmatrix} 0 & 0 & 0 & 0 \\ 1 & 0 & 0 & 0 \\ 0 & 1 & 0 & 0 \\ 0 & 0 & 1 & 0 \end{pmatrix} \begin{pmatrix} 1 & 0 & 1 & 0 \\ 0 & 2 & 1 & 0 \\ 1 & 1 & 1 & 0 \\ 0 & 0 & 0 & 1 \end{pmatrix}^{-1} = \begin{pmatrix} -1 & 0 & 1 & 0 \\ -3 & -2 & 5 & 0 \\ -2 & -1 & 3 & 0 \\ 2 & 1 & -2 & 0 \end{pmatrix}$$

The new matrix is nilpotent; its fourth power is the zero matrix. We could verify this with a tedious computation or we can instead just observe that it is nilpotent since its fourth power is similar to $\hat{N}^4$, the zero matrix, and the only matrix similar to the zero matrix is itself.

$$(P\hat{N}P^{-1})^4 = P\hat{N}P^{-1} \cdot P\hat{N}P^{-1} \cdot P\hat{N}P^{-1} \cdot P\hat{N}P^{-1} = P\hat{N}^4P^{-1}$$

The goal of this subsection is to show that the prior example is prototypical in that every nilpotent matrix is similar to one that is all zeros except for blocks of subdiagonal ones.

2.11 Definition Let t be a nilpotent transformation on V. A t-*string of length* k *generated by* $\vec{v} \in V$ is a sequence $\langle \vec{v}, t(\vec{v}), \ldots, t^{k-1}(\vec{v}) \rangle$. A t-*string basis* is a basis that is a concatenation of t-strings.

2.12 Example Consider differentiation $d/dx\colon \mathcal{P}_2 \to \mathcal{P}_2$. The sequence $\langle x^2, 2x, 2, 0 \rangle$ is a d/dx-string of length 4. The sequence $\langle x^2, 2x, 2 \rangle$ is a d/dx-string of length 3 that is a basis for $\mathcal{P}_2$.

Note that the strings cannot form a basis under concatenation if they are not disjoint because a basis cannot have a repeated vector.

2.13 Example In Example 2.6, we can concatenate the t-strings $\langle \vec{\beta}_1, \vec{\beta}_2, \vec{\beta}_3 \rangle$ and $\langle \vec{\beta}_4, \vec{\beta}_5 \rangle$ to make a basis for the domain of t.

2.14 Lemma If a space has a t-string basis then the index of nilpotency of t is the length of the longest string in that basis.

PROOF Let the space have a basis of t-strings and let t's index of nilpotency be k. We cannot have that the longest string in that basis is longer than t's index of nilpotency because t^k sends any vector, including the vector starting the longest string, to $\vec{0}$. Therefore instead suppose that the space has a t-string basis B where all of the strings are shorter than length k. Because t has index k, there is a vector $\vec{v}$ such that $t^{k-1}(\vec{v}) \neq \vec{0}$. Represent $\vec{v}$ as a linear combination of elements from B and apply t^{k-1}. We are supposing that t^{k-1} maps each element of B to $\vec{0}$, and therefore maps each term in the linear combination to $\vec{0}$, but also that it does not map $\vec{v}$ to $\vec{0}$. That is a contradiction. QED

We shall show that each nilpotent map has an associated string basis, a basis of disjoint strings.

To see the main idea of the argument, imagine that we want to construct a counterexample, a map that is nilpotent but without an associated disjoint string basis. We might think to make something like the map $t\colon \mathbb{C}^5 \to \mathbb{C}^5$ with this action.

$$\begin{array}{l} \vec{e}_1 \\ \quad\searrow \vec{e}_3 \mapsto \vec{0} \\ \quad\nearrow \\ \vec{e}_2 \\ \\ \vec{e}_4 \mapsto \vec{e}_5 \mapsto \vec{0} \end{array} \qquad \mathrm{Rep}_{\mathcal{E}_5,\mathcal{E}_5}(t) = \begin{pmatrix} 0&0&0&0&0 \\ 0&0&0&0&0 \\ 1&1&0&0&0 \\ 0&0&0&0&0 \\ 0&0&0&1&0 \end{pmatrix}$$

But, the fact that the shown basis isn't disjoint doesn't mean that there isn't another basis that consists of disjoint strings.

To produce such a basis for this map we will first find the number and lengths of its strings. Observer that t's index of nilpotency is two. Lemma 2.14 says that at least one string in a disjoint string basis has length two. There are five basis elements so if there is a disjoint string basis then the map must act in one of these ways.

$$\begin{array}{lll} \vec{\beta}_1 \mapsto \vec{\beta}_2 \mapsto \vec{0} \\ \vec{\beta}_3 \mapsto \vec{\beta}_4 \mapsto \vec{0} \\ \vec{\beta}_5 \mapsto \vec{0} \end{array} \qquad \begin{array}{l} \vec{\beta}_1 \mapsto \vec{\beta}_2 \mapsto \vec{0} \\ \vec{\beta}_3 \mapsto \vec{0} \\ \vec{\beta}_4 \mapsto \vec{0} \\ \vec{\beta}_5 \mapsto \vec{0} \end{array}$$

Now, the key point. A transformation with the left-hand action has a null space of dimension three since that's how many basis vectors are mapped to zero. A transformation with the right-hand action has a null space of dimension four. Wit the matrix representation above we can determine which of the two possible shapes is right.

$$\mathscr{N}(t) = \{ \begin{pmatrix} x \\ -x \\ z \\ 0 \\ r \end{pmatrix} \mid x, z, r \in \mathbb{C} \}$$

This is three-dimensional, meaning that of the two disjoint string basis forms above, t's basis has the left-hand one.

To produce a string basis for t, first pick $\vec{\beta}_2$ and $\vec{\beta}_4$ from $\mathscr{R}(t) \cap \mathscr{N}(t)$.

$$\vec{\beta}_2 = \begin{pmatrix} 0 \\ 0 \\ 1 \\ 0 \\ 0 \end{pmatrix} \qquad \vec{\beta}_4 = \begin{pmatrix} 0 \\ 0 \\ 0 \\ 0 \\ 1 \end{pmatrix}$$

(Other choices are possible, just be sure that the set $\{\vec{\beta}_2, \vec{\beta}_4\}$ is linearly independent.) For $\vec{\beta}_5$ pick a vector from $\mathscr{N}(t)$ that is not in the span of $\{\vec{\beta}_2, \vec{\beta}_4\}$.

$$\vec{\beta}_5 = \begin{pmatrix} 1 \\ -1 \\ 0 \\ 0 \\ 0 \end{pmatrix}$$

Finally, take $\vec{\beta}_1$ and $\vec{\beta}_3$ such that $t(\vec{\beta}_1) = \vec{\beta}_2$ and $t(\vec{\beta}_3) = \vec{\beta}_4$.

$$\vec{\beta}_1 = \begin{pmatrix} 0 \\ 1 \\ 0 \\ 0 \\ 0 \end{pmatrix} \qquad \vec{\beta}_3 = \begin{pmatrix} 0 \\ 0 \\ 0 \\ 1 \\ 0 \end{pmatrix}$$

Therefore, we have a string basis $B = \langle \vec{\beta}_1, \dots, \vec{\beta}_5 \rangle$ and with respect to that basis the matrix of t has blocks of subdiagonal 1's.

$$\text{Rep}_{B,B}(t) = \left(\begin{array}{cc|cc|c} 0 & 0 & 0 & 0 & 0 \\ 1 & 0 & 0 & 0 & 0 \\ \cline{1-4} 0 & 0 & 0 & 0 & 0 \\ 0 & 0 & 1 & 0 & 0 \\ \cline{3-5} 0 & 0 & 0 & 0 & 0 \end{array}\right)$$

2.15 Theorem Any nilpotent transformation t is associated with a t-string basis. While the basis is not unique, the number and the length of the strings is determined by t.

This illustrates the proof, which describes three kinds of basis vectors (shown in squares if they are in the null space and in circles if they are not).

$$\begin{array}{llllllllll}
\textcircled{3} & \mapsto & \textcircled{1} & \mapsto \cdots & & \cdots \mapsto & \textcircled{1} & \mapsto & \boxed{1} & \mapsto \vec{0} \\
\textcircled{3} & \mapsto & \textcircled{1} & \mapsto \cdots & & \cdots \mapsto & \textcircled{1} & \mapsto & \boxed{1} & \mapsto \vec{0} \\
 & \vdots & & & & & & & & \\
\textcircled{3} & \mapsto & \textcircled{1} & \mapsto \cdots & \mapsto \textcircled{1} & \mapsto \boxed{1} & \mapsto \vec{0} & & & \\
\boxed{2} & \mapsto & \vec{0} & & & & & & & \\
 & \vdots & & & & & & & & \\
\boxed{2} & \mapsto & \vec{0} & & & & & & &
\end{array}$$

Proof Fix a vector space V. We will argue by induction on the index of nilpotency. If the map $t\colon V \to V$ has index of nilpotency 1 then it is the zero map and any basis is a string basis $\vec{\beta}_1 \mapsto \vec{0}, \dots, \vec{\beta}_n \mapsto \vec{0}$.

For the inductive step, assume that the theorem holds for any transformation $t\colon V \to V$ with an index of nilpotency between 1 and $k-1$ (with $k > 1$) and consider the index k case.

Observe that the restriction of t to the range space $t\colon \mathscr{R}(t) \to \mathscr{R}(t)$ is also nilpotent, of index $k-1$. Apply the inductive hypothesis to get a string basis for $\mathscr{R}(t)$, where the number and length of the strings is determined by t.

$$B = \langle \vec{\beta}_1, t(\vec{\beta}_1), \ldots, t^{h_1}(\vec{\beta}_1)\rangle^\frown\langle \vec{\beta}_2, \ldots, t^{h_2}(\vec{\beta}_2)\rangle^\frown \cdots ^\frown\langle \vec{\beta}_i, \ldots, t^{h_i}(\vec{\beta}_i)\rangle$$

(In the illustration above these are the vectors of kind 1.)

Note that taking the final nonzero vector in each of these strings gives a basis $C = \langle t^{h_1}(\vec{\beta}_1), \ldots, t^{h_i}(\vec{\beta}_i)\rangle$ for the intersection $\mathscr{R}(t) \cap \mathscr{N}(t)$. This is because a member of $\mathscr{R}(t)$ maps to zero if and only if it is a linear combination of those basis vectors that map to zero. (The illustration shows these as 1's in squares.)

Now extend C to a basis for all of $\mathscr{N}(t)$.

$$\hat{C} = C^\frown\langle \vec{\xi}_1, \ldots, \vec{\xi}_p\rangle$$

(In the illustration the $\vec{\xi}$'s are the vectors of kind 2 and so the set $\hat{C}$ is the set of vectors in squares.) While the vectors $\vec{\xi}$ we choose aren't uniquely determined by t, what is uniquely determined is the number of them: it is the dimension of $\mathscr{N}(t)$ minus the dimension of $\mathscr{R}(t) \cap \mathscr{N}(t)$.

Finally, $B^\frown\hat{C}$ is a basis for $\mathscr{R}(t) + \mathscr{N}(t)$ because any sum of something in the range space with something in the null space can be represented using elements of B for the range space part and elements of $\hat{C}$ for the part from the null space. Note that

$$\begin{aligned}
\dim\big(\mathscr{R}(t) + \mathscr{N}(t)\big) &= \dim(\mathscr{R}(t)) + \dim(\mathscr{N}(t)) - \dim(\mathscr{R}(t) \cap \mathscr{N}(t)) \\
&= \operatorname{rank}(t) + \operatorname{nullity}(t) - i \\
&= \dim(V) - i
\end{aligned}$$

and so we can extend $B^\frown\hat{C}$ to a basis for all of V by the addition of i more vectors, provided that they are not linearly dependent on what we have already. Recall that each of $\vec{\beta}_1, \ldots, \vec{\beta}_i$ is in $\mathscr{R}(t)$, and extend $B^\frown\hat{C}$ with vectors $\vec{v}_1, \ldots, \vec{v}_i$ such that $t(\vec{v}_1) = \vec{\beta}_1, \ldots, t(\vec{v}_i) = \vec{\beta}_i$. (In the illustration these are the 3's.) The check that this extension preserves linear independence is Exercise 31. QED

2.16 Corollary Every nilpotent matrix is similar to a matrix that is all zeros except for blocks of subdiagonal ones. That is, every nilpotent map is represented with respect to some basis by such a matrix.

This form is unique in the sense that if a nilpotent matrix is similar to two such matrices then those two simply have their blocks ordered differently. Thus this is a canonical form for the similarity classes of nilpotent matrices provided that we order the blocks, say, from longest to shortest.

2.17 Example The matrix

$$M = \begin{pmatrix} 1 & -1 \\ 1 & -1 \end{pmatrix}$$

has an index of nilpotency of two, as this calculation shows.

power p	M^p	$\mathscr{N}(M^p)$
1	$M = \begin{pmatrix} 1 & -1 \\ 1 & -1 \end{pmatrix}$	$\{\begin{pmatrix} x \\ x \end{pmatrix} \mid x \in \mathbb{C}\}$
2	$M^2 = \begin{pmatrix} 0 & 0 \\ 0 & 0 \end{pmatrix}$	$\mathbb{C}^2$

Because the matrix is 2×2, any transformation that it represents is on a space of dimension two. The nullspace of one application of the map $\mathscr{N}(m)$ has dimension one, and the nullspace of two applications $\mathscr{N}(m^2)$ has dimension two. Thus the action of m on a string basis is $\vec{\beta}_1 \mapsto \vec{\beta}_2 \mapsto \vec{0}$ and the canonical form of the matrix is this.

$$N = \begin{pmatrix} 0 & 0 \\ 1 & 0 \end{pmatrix}$$

We can exhibit such a string basis, and also the change of basis matrices witnessing the matrix similarity between M and N. Suppose that $m\colon \mathbb{C}^2 \to \mathbb{C}^2$ is such that M represents it with respect to the standard bases. (We could take M to be a representation with respect to some other basis but the standard one is convenient.) Pick $\vec{\beta}_2 \in \mathscr{N}(m)$. Also pick $\vec{\beta}_1$ so that $m(\vec{\beta}_1) = \vec{\beta}_2$.

$$\vec{\beta}_2 = \begin{pmatrix} 1 \\ 1 \end{pmatrix} \qquad \vec{\beta}_1 = \begin{pmatrix} 1 \\ 0 \end{pmatrix}$$

For the change of basis matrices, recall the similarity diagram.

$$\begin{array}{ccc} \mathbb{C}^2_{wrt\ \mathcal{E}_2} & \xrightarrow[M]{m} & \mathbb{C}^2_{wrt\ \mathcal{E}_2} \\ \text{id}\big\downarrow P & & \text{id}\big\downarrow P \\ \mathbb{C}^2_{wrt\ B} & \xrightarrow[N]{m} & \mathbb{C}^2_{wrt\ B} \end{array}$$

The canonical form equals $\text{Rep}_{B,B}(m) = PMP^{-1}$, where

$$P^{-1} = \text{Rep}_{B,\mathcal{E}_2}(\text{id}) = \begin{pmatrix} 1 & 1 \\ 0 & 1 \end{pmatrix} \qquad P = (P^{-1})^{-1} = \begin{pmatrix} 1 & -1 \\ 0 & 1 \end{pmatrix}$$

and the verification of the matrix calculation is routine.

$$\begin{pmatrix}1&-1\\0&1\end{pmatrix}\begin{pmatrix}1&-1\\1&-1\end{pmatrix}\begin{pmatrix}1&1\\0&1\end{pmatrix}=\begin{pmatrix}0&0\\1&0\end{pmatrix}$$

2.18 Example This matrix

$$\begin{pmatrix}0&0&0&0&0\\1&0&0&0&0\\-1&1&1&-1&1\\0&1&0&0&0\\1&0&-1&1&-1\end{pmatrix}$$

is nilpotent, of index 3.

power p	N^p	$\mathscr{N}(N^p)$
1	$\begin{pmatrix}0&0&0&0&0\\1&0&0&0&0\\-1&1&1&-1&1\\0&1&0&0&0\\1&0&-1&1&-1\end{pmatrix}$	$\{\begin{pmatrix}0\\0\\u-v\\u\\v\end{pmatrix}\mid u,v\in\mathbb{C}\}$
2	$\begin{pmatrix}0&0&0&0&0\\0&0&0&0&0\\1&0&0&0&0\\1&0&0&0&0\\0&0&0&0&0\end{pmatrix}$	$\{\begin{pmatrix}0\\y\\z\\u\\v\end{pmatrix}\mid y,z,u,v\in\mathbb{C}\}$
3	*–zero matrix–*	$\mathbb{C}^5$

The table tells us this about any string basis: the null space after one map application has dimension two so two basis vectors map directly to zero, the null space after the second application has dimension four so two additional basis vectors map to zero by the second iteration, and the null space after three applications is of dimension five so the remaining one basis vector maps to zero in three hops.

$$\begin{array}{l}\vec\beta_1\mapsto\vec\beta_2\mapsto\vec\beta_3\mapsto\vec 0\\ \vec\beta_4\mapsto\vec\beta_5\mapsto\vec 0\end{array}$$

To produce such a basis, first pick two vectors from $\mathscr{N}(n)$ that form a linearly independent set.

$$\vec\beta_3=\begin{pmatrix}0\\0\\1\\1\\0\end{pmatrix}\quad\vec\beta_5=\begin{pmatrix}0\\0\\0\\1\\1\end{pmatrix}$$

Then add $\vec{\beta}_2, \vec{\beta}_4 \in \mathscr{N}(n^2)$ such that $n(\vec{\beta}_2) = \vec{\beta}_3$ and $n(\vec{\beta}_4) = \vec{\beta}_5$.

$$\vec{\beta}_2 = \begin{pmatrix} 0 \\ 1 \\ 0 \\ 0 \\ 0 \end{pmatrix} \quad \vec{\beta}_4 = \begin{pmatrix} 0 \\ 1 \\ 0 \\ 1 \\ 0 \end{pmatrix}$$

Finish by adding $\vec{\beta}_1$ such that $n(\vec{\beta}_1) = \vec{\beta}_2$.

$$\vec{\beta}_1 = \begin{pmatrix} 1 \\ 0 \\ 1 \\ 0 \\ 0 \end{pmatrix}$$

Exercises

✓ **2.19** What is the index of nilpotency of the *right-shift* operator, here acting on the space of triples of reals?

$$(x, y, z) \mapsto (0, x, y)$$

✓ **2.20** For each string basis state the index of nilpotency and give the dimension of the range space and null space of each iteration of the nilpotent map.

(a) $\vec{\beta}_1 \mapsto \vec{\beta}_2 \mapsto \vec{0}$
$\vec{\beta}_3 \mapsto \vec{\beta}_4 \mapsto \vec{0}$

(b) $\vec{\beta}_1 \mapsto \vec{\beta}_2 \mapsto \vec{\beta}_3 \mapsto \vec{0}$
$\vec{\beta}_4 \mapsto \vec{0}$
$\vec{\beta}_5 \mapsto \vec{0}$
$\vec{\beta}_6 \mapsto \vec{0}$

(c) $\vec{\beta}_1 \mapsto \vec{\beta}_2 \mapsto \vec{\beta}_3 \mapsto \vec{0}$

Also give the canonical form of the matrix.

2.21 Decide which of these matrices are nilpotent.

(a) $\begin{pmatrix} -2 & 4 \\ -1 & 2 \end{pmatrix}$ **(b)** $\begin{pmatrix} 3 & 1 \\ 1 & 3 \end{pmatrix}$ **(c)** $\begin{pmatrix} -3 & 2 & 1 \\ -3 & 2 & 1 \\ -3 & 2 & 1 \end{pmatrix}$ **(d)** $\begin{pmatrix} 1 & 1 & 4 \\ 3 & 0 & -1 \\ 5 & 2 & 7 \end{pmatrix}$

(e) $\begin{pmatrix} 45 & -22 & -19 \\ 33 & -16 & -14 \\ 69 & -34 & -29 \end{pmatrix}$

✓ **2.22** Find the canonical form of this matrix.

$$\begin{pmatrix} 0 & 1 & 1 & 0 & 1 \\ 0 & 0 & 1 & 1 & 1 \\ 0 & 0 & 0 & 0 & 0 \\ 0 & 0 & 0 & 0 & 0 \\ 0 & 0 & 0 & 0 & 0 \end{pmatrix}$$

✓ 2.23 Consider the matrix from Example 2.18.

(a) Use the action of the map on the string basis to give the canonical form.

(b) Find the change of basis matrices that bring the matrix to canonical form.

(c) Use the answer in the prior item to check the answer in the first item.

✓ 2.24 Each of these matrices is nilpotent.

(a) $\begin{pmatrix} 1/2 & -1/2 \\ 1/2 & -1/2 \end{pmatrix}$ (b) $\begin{pmatrix} 0 & 0 & 0 \\ 0 & -1 & 1 \\ 0 & -1 & 1 \end{pmatrix}$ (c) $\begin{pmatrix} -1 & 1 & -1 \\ 1 & 0 & 1 \\ 1 & -1 & 1 \end{pmatrix}$

Put each in canonical form.

2.25 Describe the effect of left or right multiplication by a matrix that is in the canonical form for nilpotent matrices.

2.26 Is nilpotence invariant under similarity? That is, must a matrix similar to a nilpotent matrix also be nilpotent? If so, with the same index?

✓ 2.27 Show that the only eigenvalue of a nilpotent matrix is zero.

2.28 Is there a nilpotent transformation of index three on a two-dimensional space?

2.29 In the proof of Theorem 2.15, why isn't the proof's base case that the index of nilpotency is zero?

✓ 2.30 Let $t\colon V \to V$ be a linear transformation and suppose $\vec{v} \in V$ is such that $t^k(\vec{v}) = \vec{0}$ but $t^{k-1}(\vec{v}) \neq \vec{0}$. Consider the t-string $\langle \vec{v}, t(\vec{v}), \ldots, t^{k-1}(\vec{v}) \rangle$.

(a) Prove that t is a transformation on the span of the set of vectors in the string, that is, prove that t restricted to the span has a range that is a subset of the span. We say that the span is a t-*invariant* subspace.

(b) Prove that the restriction is nilpotent.

(c) Prove that the t-string is linearly independent and so is a basis for its span.

(d) Represent the restriction map with respect to the t-string basis.

2.31 Finish the proof of Theorem 2.15.

2.32 Show that the terms 'nilpotent transformation' and 'nilpotent matrix', as given in Definition 2.7, fit with each other: a map is nilpotent if and only if it is represented by a nilpotent matrix. (Is it that a transformation is nilpotent if an only if there is a basis such that the map's representation with respect to that basis is a nilpotent matrix, or that any representation is a nilpotent matrix?)

2.33 Let T be nilpotent of index four. How big can the range space of T^3 be?

2.34 Recall that similar matrices have the same eigenvalues. Show that the converse does not hold.

2.35 Lemma 2.1 shows that any for any linear transformation $t\colon V \to V$ the restriction $t\colon \mathscr{R}_\infty(t) \to \mathscr{R}_\infty(t)$ is one-to-one. Show that it is also onto, so it is an automorphism. Must it be the identity map?

2.36 Prove that a nilpotent matrix is similar to one that is all zeros except for blocks of super-diagonal ones.

✓ 2.37 Prove that if a transformation has the same range space as null space. then the dimension of its domain is even.

2.38 Prove that if two nilpotent matrices commute then their product and sum are also nilpotent.

IV Jordan Form

This section uses material from three optional subsections: Combining Subspaces, Determinants Exist, and Laplace's Expansion.

We began this chapter by remembering that every linear map $h\colon V \to W$ can be represented by a partial identity matrix with respect to some bases $B \subset V$ and $D \subset W$. That is, the partial identity form is a canonical form for matrix equivalence. This chapter considers transformations, where the codomain equals the domain, so we naturally ask what is possible when the two bases are equal $\mathrm{Rep}_{B,B}(t)$. In short, we want a canonical form for matrix similarity.

We noted that in the B, B case a partial identity matrix is not always possible. We therefore extended the matrix forms of interest to the natural generalization, diagonal matrices, and showed that a transformation or square matrix can be diagonalized if its eigenvalues are distinct. But at the same time we gave an example of a square matrix that cannot be diagonalized (because it is nilpotent) and thus diagonal form won't suffice as the canonical form for matrix similarity.

The prior section developed that example to get a canonical form, subdiagonal ones, for nilpotent matrices.

This section finishes our program by showing that for any linear transformation there is a basis such that the matrix representation $\mathrm{Rep}_{B,B}(t)$ is the sum of a diagonal matrix and a nilpotent matrix. This is Jordan canonical form.

IV.1 Polynomials of Maps and Matrices

Recall that the set of square matrices $\mathcal{M}_{n\times n}$ is a vector space under entry-by-entry addition and scalar multiplication, and that this space has dimension n^2. Thus, for any $n\times n$ matrix T the n^2+1-member set $\{I, T, T^2, \dots, T^{n^2}\}$ is linearly dependent and so there are scalars $c_0, \dots, c_{n^2}$, not all zero, such that

$$c_{n^2}T^{n^2} + \cdots + c_1 T + c_0 I$$

is the zero matrix. Therefore every transformation has a kind of generalized nilpotency: the powers of a square matrix cannot climb forever without a "repeat."

1.1 Example Rotation of plane vectors $\pi/6$ radians counterclockwise is represented with respect to the standard basis by

$$T = \begin{pmatrix} \sqrt{3}/2 & -1/2 \\ 1/2 & \sqrt{3}/2 \end{pmatrix}$$

and verifying that $0T^4 + 0T^3 + 1T^2 - 2T - 1I$ equals the zero matrix is easy.

1.2 Definition Let t be a linear transformation of a vector space V. Where $f(x) = c_n x^n + \cdots + c_1 x + c_0$ is a polynomial, $f(t)$ is the transformation $c_n t^n + \cdots + c_1 t + c_0(\text{id})$ on V. In the same way, if T is a square matrix then $f(T)$ is the matrix $c_n T^n + \cdots + c_1 T + c_0 I$.

The polynomial of the matrix represents the polynomial of the map: if $T = \text{Rep}_{B,B}(t)$ then $f(T) = \text{Rep}_{B,B}(f(t))$. This is because $T^j = \text{Rep}_{B,B}(t^j)$, and $cT = \text{Rep}_{B,B}(ct)$, and $T_1 + T_2 = \text{Rep}_{B,B}(t_1 + t_2)$.

1.3 Remark Most authors write the matrix polynomial slightly differently than the map polynomial. For instance, if $f(x) = x - 3$ then most authors explicitly write the identity matrix $f(T) = T - 3I$ but don't write the identity map $f(t) = t - 3$. We shall follow this convention.

Consider again Example 1.1. The space $\mathcal{M}_{2\times 2}$ has dimension four so we know that for any 2×2 matrix there is a fourth degree polynomial f such that $f(T)$ equals the zero matrix. But for the T in that example we exhibited a polynomial of degree less than four that gives the zero matrix. So while degree n^2 always suffices, in some cases a smaller-degree polynomial works.

1.4 Definition The *minimal polynomial* $m(x)$ of a transformation t or a square matrix T is the polynomial of least degree and with leading coefficient one such that $m(t)$ is the zero map or $m(T)$ is the zero matrix.

A minimal polynomial cannot be the zero polynomial because of the restriction on the leading coefficient. Obviously no other constant polynomial would do, so a minimal polynomial must have degree at least one. Thus, the zero matrix has minimal polynomial $p(x) = x$ while the identity matrix has minimal polynomial $\hat{p}(x) = x - 1$.

1.5 Lemma Any transformation or square matrix has a unique minimal polynomial.

Proof We first prove existence. By the earlier observation that degree n^2 suffices, there is at least one polynomial $p(x) = c_k x^k + \cdots + c_0$ that takes the map or matrix to zero, and it is not the zero polynomial by the prior paragraph. From among all such polynomials there must be at least one with minimal degree. Divide this polynomial by its leading coefficient c_k to get a leading 1. Hence any map or matrix has a minimal polynomial.

Now for uniqueness. Suppose that $m(x)$ and $\hat{m}(x)$ both take the map or matrix to zero, are both of minimal degree and are thus of equal degree, and both have a leading 1. Subtract: $d(x) = m(x) - \hat{m}(x)$. This polynomial takes the map or matrix to zero and since the leading terms of m and $\hat{m}$ cancel, d is of smaller degree than the other two. If d were to have a nonzero leading coefficient then we could divide by it to get a polynomial that takes the map or matrix to zero and has leading coefficient 1. This would contradict the minimality of the degree of m and $\hat{m}$. Thus the leading coefficient of d is zero, so $m(x) - \hat{m}(x)$ is the zero polynomial, and so the two are equal. QED

1.6 Example We can compute that $m(x) = x^2 - 2x - 1$ is minimal for the matrix of Example 1.1 by finding the powers of T up to $n^2 = 4$.

$$T^2 = \begin{pmatrix} 1/2 & -\sqrt{3}/2 \\ \sqrt{3}/2 & 1/2 \end{pmatrix} \quad T^3 = \begin{pmatrix} 0 & -1 \\ 1 & 0 \end{pmatrix} \quad T^4 = \begin{pmatrix} -1/2 & -\sqrt{3}/2 \\ \sqrt{3}/2 & -1/2 \end{pmatrix}$$

Put $c_4T^4 + c_3T^3 + c_2T^2 + c_1T + c_0I$ equal to the zero matrix

$$\begin{array}{rl} -(1/2)c_4 \qquad\quad + \quad (1/2)c_2 + (\sqrt{3}/2)c_1 + c_0 &= 0 \\ -(\sqrt{3}/2)c_4 - c_3 - (\sqrt{3}/2)c_2 - \quad (1/2)c_1 \qquad\quad &= 0 \\ (\sqrt{3}/2)c_4 + c_3 + (\sqrt{3}/2)c_2 + \quad (1/2)c_1 \qquad\quad &= 0 \\ -(1/2)c_4 \qquad\quad + \quad (1/2)c_2 + (\sqrt{3}/2)c_1 + c_0 &= 0 \end{array}$$

and use Gauss' Method.

$$\begin{array}{rl} c_4 \qquad - \quad c_2 - \sqrt{3}c_1 - \quad 2c_0 &= 0 \\ c_3 + \sqrt{3}c_2 + \quad 2c_1 + \sqrt{3}c_0 &= 0 \end{array}$$

Setting c_4, c_3, and c_2 to zero forces c_1 and c_0 to also come out as zero. To get a leading one, the most we can do is to set c_4 and c_3 to zero. Thus the minimal polynomial is quadratic.

Using the method of that example to find the minimal polynomial of a 3×3 matrix would mean doing Gaussian reduction on a system with nine equations in ten unknowns. We shall develop an alternative.

1.7 Lemma Suppose that the polynomial $f(x) = c_nx^n + \cdots + c_1x + c_0$ factors as $k(x - \lambda_1)^{q_1} \cdots (x - \lambda_z)^{q_z}$. If t is a linear transformation then these two are equal maps.

$$c_nt^n + \cdots + c_1t + c_0 = k \cdot (t - \lambda_1)^{q_1} \circ \cdots \circ (t - \lambda_z)^{q_z}$$

Consequently, if T is a square matrix then $f(T)$ and $k \cdot (T - \lambda_1 I)^{q_1} \cdots (T - \lambda_z I)^{q_z}$ are equal matrices.

Proof We use induction on the degree of the polynomial. The cases where the polynomial is of degree zero and degree one are clear. The full induction argument is Exercise 1.7 but we will give its sense with the degree two case.

A quadratic polynomial factors into two linear terms $f(x) = k(x - \lambda_1) \cdot (x - \lambda_2) = k(x^2 + (-\lambda_1 - \lambda_2)x + \lambda_1\lambda_2)$ (the roots λ_1 and λ_2 could be equal). We can check that substituting t for x in the factored and unfactored versions gives the same map.

$$\begin{aligned}
\big(k \cdot (t - \lambda_1) \circ (t - \lambda_2)\big)\,(\vec{v}) &= \big(k \cdot (t - \lambda_1)\big)\,(t(\vec{v}) - \lambda_2\vec{v}) \\
&= k \cdot \big(t(t(\vec{v})) - t(\lambda_2\vec{v}) - \lambda_1 t(\vec{v}) - \lambda_1\lambda_2\vec{v}\big) \\
&= k \cdot \big(t \circ t\,(\vec{v}) - (\lambda_1 + \lambda_2)t(\vec{v}) + \lambda_1\lambda_2\vec{v}\big) \\
&= k \cdot (t^2 - (\lambda_1 + \lambda_2)t + \lambda_1\lambda_2)\,(\vec{v})
\end{aligned}$$

The third equality holds because the scalar λ_2 comes out of the second term, since t is linear. QED

In particular, if a minimal polynomial $m(x)$ for a transformation t factors as $m(x) = (x - \lambda_1)^{q_1} \cdots (x - \lambda_z)^{q_z}$ then $m(t) = (t - \lambda_1)^{q_1} \circ \cdots \circ (t - \lambda_z)^{q_z}$ is the zero map. Since $m(t)$ sends every vector to zero, at least one of the maps $t - \lambda_i$ sends some nonzero vectors to zero. Exactly the same holds in the matrix case—if m is minimal for T then $m(T) = (T - \lambda_1 I)^{q_1} \cdots (T - \lambda_z I)^{q_z}$ is the zero matrix and at least one of the matrices $T - \lambda_i I$ sends some nonzero vectors to zero. That is, in both cases at least some of the λ_i are eigenvalues. (Exercise 29 expands on this.)

The next result is that every root of the minimal polynomial is an eigenvalue, and further that every eigenvalue is a root of the minimal polynomial (i.e, below it says '$1 \leqslant q_i$' and not just '$0 \leqslant q_i$'). For that result, recall that to find eigenvalues we solve $|T - xI| = 0$ and this determinant gives a polynomial in x, called the characteristic polynomial, whose roots are the eigenvalues.

1.8 Theorem (Cayley-Hamilton) If the characteristic polynomial of a transformation or square matrix factors into

$$k \cdot (x - \lambda_1)^{p_1}(x - \lambda_2)^{p_2} \cdots (x - \lambda_z)^{p_z}$$

then its minimal polynomial factors into

$$(x - \lambda_1)^{q_1}(x - \lambda_2)^{q_2} \cdots (x - \lambda_z)^{q_z}$$

where $1 \leqslant q_i \leqslant p_i$ for each i between 1 and z.

The proof takes up the next three lemmas. We will state them in matrix terms but they apply equally well to maps. (The matrix version is convenient for the first proof.)

The first result is the key. For the proof, observe that we can view a matrix of polynomials as a polynomial with matrix coefficients.

$$\begin{pmatrix} 2x^2+3x-1 & x^2+2 \\ 3x^2+4x+1 & 4x^2+x+1 \end{pmatrix} = \begin{pmatrix} 2 & 1 \\ 3 & 4 \end{pmatrix} x^2 + \begin{pmatrix} 3 & 0 \\ 4 & 1 \end{pmatrix} x + \begin{pmatrix} -1 & 2 \\ 1 & 1 \end{pmatrix}$$

1.9 Lemma If T is a square matrix with characteristic polynomial $c(x)$ then $c(T)$ is the zero matrix.

Proof Let C be $T - xI$, the matrix whose determinant is the characteristic polynomial $c(x) = c_n x^n + \cdots + c_1 x + c_0$.

$$C = \begin{pmatrix} t_{1,1}-x & t_{1,2} & \ldots & \\ t_{2,1} & t_{2,2}-x & & \\ \vdots & & \ddots & \\ & & & t_{n,n}-x \end{pmatrix}$$

Recall Theorem Four.III.1.9, that the product of a matrix with its adjoint equals the determinant of the matrix times the identity.

$$c(x) \cdot I = \operatorname{adj}(C)C = \operatorname{adj}(C)(T - xI) = \operatorname{adj}(C)T - \operatorname{adj}(C) \cdot x \qquad (*)$$

The left side of $(*)$ is $c_n I x^n + c_{n-1} I x^{n-1} + \cdots + c_1 I x + c_0 I$. For the right side, the entries of $\operatorname{adj}(C)$ are polynomials, each of degree at most $n-1$ since the minors of a matrix drop a row and column. As suggested before the proof, rewrite it as a polynomial with matrix coefficients: $\operatorname{adj}(C) = C_{n-1}x^{n-1} + \cdots + C_1 x + C_0$ where each C_i is a matrix of scalars. Now this is the right side of $(*)$.

$$[(C_{n-1}T)x^{n-1} + \cdots + (C_1 T)x + C_0 T] - [C_{n-1}x^n - C_{n-2}x^{n-1} - \cdots - C_0 x]$$

Equate the left and right side of $(*)$'s coefficients of x^n, of x^{n-1}, etc.

$$\begin{aligned} c_n I &= -C_{n-1} \\ c_{n-1} I &= -C_{n-2} + C_{n-1}T \\ &\vdots \\ c_1 I &= -C_0 + C_1 T \\ c_0 I &= C_0 T \end{aligned}$$

Multiply, from the right, both sides of the first equation by T^n, both sides of the second equation by T^{n-1}, etc.

$$\begin{aligned}
c_n T^n &= -C_{n-1} T^n \\
c_{n-1} T^{n-1} &= -C_{n-2} T^{n-1} + C_{n-1} T^n \\
&\vdots \\
c_1 T &= -C_0 T + C_1 T^2 \\
c_0 I &= C_0 T
\end{aligned}$$

Add. The left is $c_n T^n + c_{n-1} T^{n-1} + \cdots + c_0 I$. The right telescopes; for instance $-C_{n-1} T^n$ from the first line combines with the $C_{n-1} T^n$ half of the second line. The total on the right is the zero matrix. QED

We refer to that result by saying that a matrix or map *satisfies* its characteristic polynomial.

1.10 Lemma Where $f(x)$ is a polynomial, if $f(T)$ is the zero matrix then $f(x)$ is divisible by the minimal polynomial of T. That is, any polynomial that is satisfied by T is divisible by T's minimal polynomial.

PROOF Let $m(x)$ be minimal for T. The Division Theorem for Polynomials gives $f(x) = q(x)m(x) + r(x)$ where the degree of r is strictly less than the degree of m. Because T satisfies both f and m, plugging T into that equation gives that $r(T)$ is the zero matrix. That contradicts the minimality of m unless r is the zero polynomial. QED

Combining the prior two lemmas shows that the minimal polynomial divides the characteristic polynomial. Thus any root of the minimal polynomial is also a root of the characteristic polynomial. That is, so far we have that if $m(x) = (x-\lambda_1)^{q_1} \cdots (x-\lambda_i)^{q_i}$ then $c(x)$ has the form $(x-\lambda_1)^{p_1} \cdots (x-\lambda_i)^{p_i}(x-\lambda_{i+1})^{p_{i+1}} \cdots (x-\lambda_z)^{p_z}$ where each q_j is less than or equal to p_j. We finish the proof of the Cayley-Hamilton Theorem by showing that the characteristic polynomial has no additional roots, that is, there are no λ_{i+1}, λ_{i+2}, etc.

1.11 Lemma Each linear factor of the characteristic polynomial of a square matrix is also a linear factor of the minimal polynomial.

PROOF Let T be a square matrix with minimal polynomial $m(x)$ and assume that $x-\lambda$ is a factor of the characteristic polynomial of T, that λ is an eigenvalue of T. We must show that $x-\lambda$ is a factor of m, i.e., that $m(\lambda) = 0$.

Suppose that λ is an eigenvalue of T with associated eigenvector $\vec{v}$. Then $T \cdot T\vec{v} = T \cdot \lambda\vec{v} = \lambda T\vec{v} = \lambda^2\vec{v}$. Similarly, $T^n\vec{v} = \lambda^n\vec{v}$. With that, we have that for any polynomial function $p(x)$, application of the matrix $p(T)$ to $\vec{v}$ equals the result of multiplying $\vec{v}$ by the scalar $p(\lambda)$.

$$\begin{aligned} p(T)\cdot\vec{v} &= (c_kT^k+\cdots+c_1T+c_0I)\cdot\vec{v} = c_kT^k\vec{v}+\cdots+c_1T\vec{v}+c_0\vec{v} \\ &= c_k\lambda^k\vec{v}+\cdots+c_1\lambda\vec{v}+c_0\vec{v} = p(\lambda)\cdot\vec{v} \end{aligned}$$

Since $m(T)$ is the zero matrix, $\vec{0} = m(T)(\vec{v}) = m(\lambda)\cdot\vec{v}$ for all $\vec{v}$, and hence $m(\lambda) = 0$. QED

That concludes the proof of the Cayley-Hamilton Theorem.

1.12 Example We can use the Cayley-Hamilton Theorem to find the minimal polynomial of this matrix.

$$T = \begin{pmatrix} 2 & 0 & 0 & 1 \\ 1 & 2 & 0 & 2 \\ 0 & 0 & 2 & -1 \\ 0 & 0 & 0 & 1 \end{pmatrix}$$

First we find its characteristic polynomial $c(x) = (x-1)(x-2)^3$ with the usual determinant. Now, the Cayley-Hamilton Theorem says that T's minimal polynomial is either $(x-1)(x-2)$ or $(x-1)(x-2)^2$ or $(x-1)(x-2)^3$. We can decide among the choices just by computing

$$(T-1I)(T-2I) = \begin{pmatrix} 1 & 0 & 0 & 1 \\ 1 & 1 & 0 & 2 \\ 0 & 0 & 1 & -1 \\ 0 & 0 & 0 & 0 \end{pmatrix}\begin{pmatrix} 0 & 0 & 0 & 1 \\ 1 & 0 & 0 & 2 \\ 0 & 0 & 0 & -1 \\ 0 & 0 & 0 & -1 \end{pmatrix} = \begin{pmatrix} 0 & 0 & 0 & 0 \\ 1 & 0 & 0 & 1 \\ 0 & 0 & 0 & 0 \\ 0 & 0 & 0 & 0 \end{pmatrix}$$

and

$$(T-1I)(T-2I)^2 = \begin{pmatrix} 0 & 0 & 0 & 0 \\ 1 & 0 & 0 & 1 \\ 0 & 0 & 0 & 0 \\ 0 & 0 & 0 & 0 \end{pmatrix}\begin{pmatrix} 0 & 0 & 0 & 1 \\ 1 & 0 & 0 & 2 \\ 0 & 0 & 0 & -1 \\ 0 & 0 & 0 & -1 \end{pmatrix} = \begin{pmatrix} 0 & 0 & 0 & 0 \\ 0 & 0 & 0 & 0 \\ 0 & 0 & 0 & 0 \\ 0 & 0 & 0 & 0 \end{pmatrix}$$

and so $m(x) = (x-1)(x-2)^2$.

Exercises

✓ **1.13** What are the possible minimal polynomials if a matrix has the given characteristic polynomial?

(a) $(x-3)^4$ **(b)** $(x+1)^3(x-4)$ **(c)** $(x-2)^2(x-5)^2$
(d) $(x+3)^2(x-1)(x-2)^2$

What is the degree of each possibility?

✓ **1.14** Find the minimal polynomial of each matrix.

(a) $\begin{pmatrix}3&0&0\\1&3&0\\0&0&4\end{pmatrix}$ (b) $\begin{pmatrix}3&0&0\\1&3&0\\0&0&3\end{pmatrix}$ (c) $\begin{pmatrix}3&0&0\\1&3&0\\0&1&3\end{pmatrix}$ (d) $\begin{pmatrix}2&0&1\\0&6&2\\0&0&2\end{pmatrix}$

(e) $\begin{pmatrix}2&2&1\\0&6&2\\0&0&2\end{pmatrix}$ (f) $\begin{pmatrix}-1&4&0&0&0\\0&3&0&0&0\\0&-4&-1&0&0\\3&-9&-4&2&-1\\1&5&4&1&4\end{pmatrix}$

1.15 Find the minimal polynomial of this matrix.

$$\begin{pmatrix}0&1&0\\0&0&1\\1&0&0\end{pmatrix}$$

✓ 1.16 What is the minimal polynomial of the differentiation operator d/dx on $\mathcal{P}_n$?

✓ 1.17 Find the minimal polynomial of matrices of this form

$$\begin{pmatrix}\lambda&0&0&\ldots&&0\\1&\lambda&0&&&0\\0&1&\lambda&&&\\&&&\ddots&&\\&&&&\lambda&0\\0&0&\ldots&&1&\lambda\end{pmatrix}$$

where the scalar λ is fixed (i.e., is not a variable).

1.18 What is the minimal polynomial of the transformation of $\mathcal{P}_n$ that sends $p(x)$ to $p(x+1)$?

1.19 What is the minimal polynomial of the map $\pi\colon \mathbb{C}^3 \to \mathbb{C}^3$ projecting onto the first two coordinates?

1.20 Find a 3×3 matrix whose minimal polynomial is x^2.

1.21 What is wrong with this claimed proof of Lemma 1.9: "if $c(x) = |T - xI|$ then $c(T) = |T - TI| = 0$"? [Cullen]

1.22 Verify Lemma 1.9 for 2×2 matrices by direct calculation.

✓ 1.23 Prove that the minimal polynomial of an $n\times n$ matrix has degree at most n (not n^2 as a person might guess from this subsection's opening). Verify that this maximum, n, can happen.

✓ 1.24 Show that, on a nontrivial vector space, a linear transformation is nilpotent if and only if its only eigenvalue is zero.

1.25 What is the minimal polynomial of a zero map or matrix? Of an identity map or matrix?

✓ 1.26 Interpret the minimal polynomial of Example 1.1 geometrically.

1.27 What is the minimal polynomial of a diagonal matrix?

✓ 1.28 A *projection* is any transformation t such that $t^2 = t$. (For instance, consider the transformation of the plane $\mathbb{R}^2$ projecting each vector onto its first coordinate. If we project twice then we get the same result as if we project just once.) What is the minimal polynomial of a projection?

1.29 *The first two items of this question are review.*

(a) Prove that the composition of one-to-one maps is one-to-one.

(b) Prove that if a linear map is not one-to-one then at least one nonzero vector from the domain maps to the zero vector in the codomain.

(c) Verify the statement, excerpted here, that precedes Theorem 1.8.

> ... if a minimal polynomial $m(x)$ for a transformation t factors as $m(x) = (x-\lambda_1)^{q_1} \cdots (x-\lambda_z)^{q_z}$ then $m(t) = (t-\lambda_1)^{q_1} \circ \cdots \circ (t-\lambda_z)^{q_z}$ is the zero map. Since $m(t)$ sends every vector to zero, at least one of the maps $t-\lambda_i$ sends some nonzero vectors to zero. ... That is, ... at least some of the λ_i are eigenvalues.

1.30 True or false: for a transformation on an n dimensional space, if the minimal polynomial has degree n then the map is diagonalizable.

1.31 Let $f(x)$ be a polynomial. Prove that if A and B are similar matrices then $f(A)$ is similar to $f(B)$.

(a) Now show that similar matrices have the same characteristic polynomial.

(b) Show that similar matrices have the same minimal polynomial.

(c) Decide if these are similar.

$$\begin{pmatrix} 1 & 3 \\ 2 & 3 \end{pmatrix} \qquad \begin{pmatrix} 4 & -1 \\ 1 & 1 \end{pmatrix}$$

1.32 (a) Show that a matrix is invertible if and only if the constant term in its minimal polynomial is not 0.

(b) Show that if a square matrix T is not invertible then there is a nonzero matrix S such that ST and TS both equal the zero matrix.

✓ 1.33 (a) Finish the proof of Lemma 1.7.

(b) Give an example to show that the result does not hold if t is not linear.

1.34 Any transformation or square matrix has a minimal polynomial. Does the converse hold?

IV.2 Jordan Canonical Form

We are looking for a canonical form for matrix similarity. This subsection completes this program by moving from the canonical form for the classes of nilpotent matrices to the canonical form for all classes.

2.1 Lemma A linear transformation on a nontrivial vector space is nilpotent if and only if its only eigenvalue is zero.

Proof Let the linear transformation be $t\colon V \to V$. If t is nilpotent then there is an n such that t^n is the zero map, so t satisfies the polynomial $p(x) = x^n = (x-0)^n$. By Lemma 1.10 the minimal polynomial of t divides p, so the minimal polynomial has only zero for a root. By Cayley-Hamilton, Theorem 1.8, the characteristic polynomial has only zero for a root. Thus the only eigenvalue of t is zero.

Conversely, if a transformation t on an n-dimensional space has only the single eigenvalue of zero then its characteristic polynomial is x^n. Lemma 1.9 says that a map satisfies its characteristic polynomial so t^n is the zero map. Thus t is nilpotent. QED

The 'nontrivial vector space' is in the statement of that lemma because on a trivial space $\{\vec{0}\}$ the only transformation is the zero map, which has no eigenvalues because there are no associated nonzero eigenvectors.

2.2 Corollary The transformation $t-\lambda$ is nilpotent if and only if t's only eigenvalue is λ.

PROOF The transformation $t-\lambda$ is nilpotent if and only if $t-\lambda$'s only eigenvalue is 0. That holds if and only if t's only eigenvalue is λ, because $t(\vec{v}) = \lambda\vec{v}$ if and only if $(t-\lambda)(\vec{v}) = 0 \cdot \vec{v}$. QED

We already have the canonical form that we want for the case of nilpotent matrices, that is, for each matrix whose only eigenvalue is zero. Corollary III.2.16 says that each such matrix is similar to one that is all zeroes except for blocks of subdiagonal ones.

2.3 Lemma If the matrices $T-\lambda I$ and N are similar then T and $N+\lambda I$ are also similar, via the same change of basis matrices.

PROOF With $N = P(T-\lambda I)P^{-1} = PTP^{-1} - P(\lambda I)P^{-1}$ we have $N = PTP^{-1} - PP^{-1}(\lambda I)$ since the diagonal matrix λI commutes with anything, and so $N = PTP^{-1} - \lambda I$. Therefore $N + \lambda I = PTP^{-1}$. QED

2.4 Example The characteristic polynomial of

$$T = \begin{pmatrix} 2 & -1 \\ 1 & 4 \end{pmatrix}$$

is $(x-3)^2$ and so T has only the single eigenvalue 3. Thus for

$$T - 3I = \begin{pmatrix} -1 & -1 \\ 1 & 1 \end{pmatrix}$$

the only eigenvalue is 0 and $T-3I$ is nilpotent. Finding the null spaces is routine; to ease this computation we take T to represent a transformation $t\colon \mathbb{C}^2 \to \mathbb{C}^2$ with respect to the standard basis (we shall do this for the rest of the chapter).

$$\mathscr{N}(t-3) = \{\begin{pmatrix} -y \\ y \end{pmatrix} \mid y \in \mathbb{C}\} \qquad \mathscr{N}((t-3)^2) = \mathbb{C}^2$$

The dimension of each null space shows that the action of the map $t-3$ on a string basis is $\vec{\beta}_1 \mapsto \vec{\beta}_2 \mapsto \vec{0}$. Thus, here is the canonical form for $t-3$ with one choice for a string basis.

$$\mathrm{Rep}_{B,B}(t-3) = N = \begin{pmatrix} 0 & 0 \\ 1 & 0 \end{pmatrix} \qquad B = \langle \begin{pmatrix} 1 \\ 1 \end{pmatrix}, \begin{pmatrix} -2 \\ 2 \end{pmatrix} \rangle$$

By Lemma 2.3, T is similar to this matrix.

$$\mathrm{Rep}_{B,B}(t) = N + 3I = \begin{pmatrix} 3 & 0 \\ 1 & 3 \end{pmatrix}$$

We can produce the similarity computation. Recall how to find the change of basis matrices P and P^{-1} to express N as $P(T-3I)P^{-1}$. The similarity diagram

$$\begin{array}{ccc} \mathbb{C}^2_{wrt\ \mathcal{E}_2} & \xrightarrow[T-3I]{t-3} & \mathbb{C}^2_{wrt\ \mathcal{E}_2} \\ \mathrm{id}\big\downarrow P & & \mathrm{id}\big\downarrow P \\ \mathbb{C}^2_{wrt\ B} & \xrightarrow[N]{t-3} & \mathbb{C}^2_{wrt\ B} \end{array}$$

describes that to move from the lower left to the upper left we multiply by

$$P^{-1} = \left(\mathrm{Rep}_{\mathcal{E}_2,B}(\mathrm{id})\right)^{-1} = \mathrm{Rep}_{B,\mathcal{E}_2}(\mathrm{id}) = \begin{pmatrix} 1 & -2 \\ 1 & 2 \end{pmatrix}$$

and to move from the upper right to the lower right we multiply by this matrix.

$$P = \begin{pmatrix} 1 & -2 \\ 1 & 2 \end{pmatrix}^{-1} = \begin{pmatrix} 1/2 & 1/2 \\ -1/4 & 1/4 \end{pmatrix}$$

So this equation expresses the similarity.

$$\begin{pmatrix} 3 & 0 \\ 1 & 3 \end{pmatrix} = \begin{pmatrix} 1/2 & 1/2 \\ -1/4 & 1/4 \end{pmatrix} \begin{pmatrix} 2 & -1 \\ 1 & 4 \end{pmatrix} \begin{pmatrix} 1 & -2 \\ 1 & 2 \end{pmatrix}$$

2.5 Example This matrix has characteristic polynomial $(x-4)^4$

$$T = \begin{pmatrix} 4 & 1 & 0 & -1 \\ 0 & 3 & 0 & 1 \\ 0 & 0 & 4 & 0 \\ 1 & 0 & 0 & 5 \end{pmatrix}$$

and so has the single eigenvalue 4. The null space of $t-4$ has dimension two, the null space of $(t-4)^2$ has dimension three, and the null space of $(t-4)^3$ has dimension four. Thus, $t-4$ has the action on a string basis of $\vec{\beta}_1 \mapsto \vec{\beta}_2 \mapsto \vec{\beta}_3 \mapsto \vec{0}$ and $\vec{\beta}_4 \mapsto \vec{0}$. This gives the canonical form N for $t-4$, which in turn gives the form for t.

$$N + 4I = \begin{pmatrix} 4 & 0 & 0 & 0 \\ 1 & 4 & 0 & 0 \\ 0 & 1 & 4 & 0 \\ 0 & 0 & 0 & 4 \end{pmatrix}$$

An array that is all zeroes, except for some number λ down the diagonal and blocks of subdiagonal ones, is a *Jordan block*. We have shown that Jordan block matrices are canonical representatives of the similarity classes of single-eigenvalue matrices.

2.6 Example The 3×3 matrices whose only eigenvalue is $1/2$ separate into three similarity classes. The three classes have these canonical representatives.

$$\begin{pmatrix} 1/2 & 0 & 0 \\ 0 & 1/2 & 0 \\ 0 & 0 & 1/2 \end{pmatrix} \qquad \begin{pmatrix} 1/2 & 0 & 0 \\ 1 & 1/2 & 0 \\ 0 & 0 & 1/2 \end{pmatrix} \qquad \begin{pmatrix} 1/2 & 0 & 0 \\ 1 & 1/2 & 0 \\ 0 & 1 & 1/2 \end{pmatrix}$$

In particular, this matrix

$$\begin{pmatrix} 1/2 & 0 & 0 \\ 0 & 1/2 & 0 \\ 0 & 1 & 1/2 \end{pmatrix}$$

belongs to the similarity class represented by the middle one, because we have adopted the convention of ordering the blocks of subdiagonal ones from the longest block to the shortest.

We will finish the program of this chapter by extending this work to cover maps and matrices with multiple eigenvalues. The best possibility for general maps and matrices would be if we could break them into a part involving their first eigenvalue λ_1 (which we represent using its Jordan block), a part with λ_2, etc.

This best possibility is what happens. For any transformation $t\colon V \to V$, we shall break the space V into the direct sum of a part on which $t-\lambda_1$ is nilpotent, a part on which $t-\lambda_2$ is nilpotent, etc.

Suppose that $t\colon V \to V$ is a linear transformation. The restriction of t to a subspace M need not be a linear transformation on M because there may be an $\vec{m} \in M$ with $t(\vec{m}) \notin M$ (for instance, the transformation that rotates the plane by a quarter turn does not map most members of the $x = y$ line subspace back within that subspace). To ensure that the restriction of a transformation to a part of a space is a transformation on the part we need the next condition.

2.7 Definition Let $t\colon V \to V$ be a transformation. A subspace M is t *invariant* if whenever $\vec{m} \in M$ then $t(\vec{m}) \in M$ (shorter: $t(M) \subseteq M$).

Recall that Lemma III.1.4 shows that for any transformation t on an n dimensional space the range spaces of iterates are stable

$$\mathscr{R}(t^n) = \mathscr{R}(t^{n+1}) = \cdots = \mathscr{R}_\infty(t)$$

as are the null spaces.

$$\mathscr{N}(t^n) = \mathscr{N}(t^{n+1}) = \cdots = \mathscr{N}_\infty(t)$$

Thus, the generalized null space $\mathscr{N}_\infty(t)$ and the generalized range space $\mathscr{R}_\infty(t)$ are t invariant. In particular, $\mathscr{N}_\infty(t-\lambda_i)$ and $\mathscr{R}_\infty(t-\lambda_i)$ are $t-\lambda_i$ invariant.

The action of the transformation $t-\lambda_i$ on $\mathscr{N}_\infty(t-\lambda_i)$ is especially easy to understand. Observe that any transformation t is nilpotent on $\mathscr{N}_\infty(t)$, because if $\vec{v} \in \mathscr{N}_\infty(t)$ then by definition $t^n(\vec{v}) = \vec{0}$. Thus $t-\lambda_i$ is nilpotent on $\mathscr{N}_\infty(t-\lambda_i)$.

We shall take three steps to prove this section's major result. The next result is the first.

2.8 Lemma A subspace is t invariant if and only if it is $t-\lambda$ invariant for all scalars λ. In particular, if λ_i is an eigenvalue of a linear transformation t then for any other eigenvalue λ_j the spaces $\mathscr{N}_\infty(t-\lambda_i)$ and $\mathscr{R}_\infty(t-\lambda_i)$ are $t-\lambda_j$ invariant.

Proof For the first sentence we check the two implications separately. The 'if' half is easy: if the subspace is $t-\lambda$ invariant for all scalars λ then using $\lambda = 0$ shows that it is t invariant. For 'only if' suppose that the subspace is t invariant, so that if $\vec{m} \in M$ then $t(\vec{m}) \in M$, and let λ be a scalar. The subspace M is closed under linear combinations and so if $t(\vec{m}) \in M$ then $t(\vec{m}) - \lambda\vec{m} \in M$. Thus if $\vec{m} \in M$ then $(t-\lambda)(\vec{m}) \in M$.

The lemma's second sentence follows from its first. The two spaces are $t-\lambda_i$ invariant so they are t invariant. Apply the first sentence again to conclude that they are also $t-\lambda_j$ invariant. QED

The second step of the three that we will take to prove this section's major result makes use of an additional property of $\mathscr{N}_\infty(t-\lambda_i)$ and $\mathscr{R}_\infty(t-\lambda_i)$, that they are complementary. Recall that if a space is the direct sum of two others $V = \mathscr{N} \oplus \mathscr{R}$ then any vector $\vec{v}$ in the space breaks into two parts $\vec{v} = \vec{n} + \vec{r}$ where $\vec{n} \in \mathscr{N}$ and $\vec{r} \in \mathscr{R}$, and recall also that if $B_{\mathscr{N}}$ and $B_{\mathscr{R}}$ are bases for $\mathscr{N}$ and $\mathscr{R}$ then the concatenation $B_{\mathscr{N}} \frown B_{\mathscr{R}}$ is linearly independent. The next result says that for any subspaces $\mathscr{N}$ and $\mathscr{R}$ that are complementary as well as t invariant, the action of t on $\vec{v}$ breaks into the actions of t on $\vec{n}$ and on $\vec{r}$.

2.9 Lemma Let $t\colon V \to V$ be a transformation and let $\mathscr{N}$ and $\mathscr{R}$ be t invariant complementary subspaces of V. Then we can represent t by a matrix with blocks of square submatrices T_1 and T_2

$$\left(\begin{array}{c|c} T_1 & Z_2 \\ \hline Z_1 & T_2 \end{array}\right) \begin{array}{l} \}\ \dim(\mathscr{N})\text{-many rows} \\ \}\ \dim(\mathscr{R})\text{-many rows} \end{array}$$

where Z_1 and Z_2 are blocks of zeroes.

Proof Since the two subspaces are complementary, the concatenation of a basis for $\mathscr{N}$ with a basis for $\mathscr{R}$ makes a basis $B = \langle \vec{\nu}_1, \ldots, \vec{\nu}_p, \vec{\mu}_1, \ldots, \vec{\mu}_q \rangle$ for V. We shall show that the matrix

$$\text{Rep}_{B,B}(t) = \begin{pmatrix} \vdots & & \vdots \\ \text{Rep}_B(t(\vec{\nu}_1)) & \cdots & \text{Rep}_B(t(\vec{\mu}_q)) \\ \vdots & & \vdots \end{pmatrix}$$

has the desired form.

Any vector $\vec{v} \in V$ is a member of $\mathscr{N}$ if and only if when it is represented with respect to B the final q coefficients are zero. As $\mathscr{N}$ is t invariant, each of the vectors $\text{Rep}_B(t(\vec{\nu}_1))$, ..., $\text{Rep}_B(t(\vec{\nu}_p))$ has this form. Hence the lower left of $\text{Rep}_{B,B}(t)$ is all zeroes. The argument for the upper right is similar. QED

To see that we have decomposed t into its action on the parts, let $B_{\mathscr{N}} = \langle \vec{\nu}_1, \ldots, \vec{\nu}_p \rangle$ and $B_{\mathscr{R}} = \langle \vec{\mu}_1, \ldots, \vec{\mu}_q \rangle$. The restrictions of t to the subspaces $\mathscr{N}$ and $\mathscr{R}$ are represented with respect to the bases $B_{\mathscr{N}}, B_{\mathscr{N}}$ and $B_{\mathscr{R}}, B_{\mathscr{R}}$ by the matrices T_1 and T_2. So with subspaces that are invariant and complementary we can split the problem of examining a linear transformation into two lower-dimensional subproblems. The next result illustrates this decomposition into blocks.

2.10 Lemma If T is a matrix with square submatrices T_1 and T_2

$$T = \left(\begin{array}{c|c} T_1 & Z_2 \\ \hline Z_1 & T_2 \end{array}\right)$$

where the Z's are blocks of zeroes, then $|T| = |T_1| \cdot |T_2|$.

Proof Suppose that T is $n \times n$, that T_1 is $p \times p$, and that T_2 is $q \times q$. In the permutation formula for the determinant

$$|T| = \sum_{\text{permutations } \phi} t_{1,\phi(1)} t_{2,\phi(2)} \cdots t_{n,\phi(n)} \operatorname{sgn}(\phi)$$

each term comes from a rearrangement of the column numbers $1, \ldots, n$ into a new order $\phi(1), \ldots, \phi(n)$. The upper right block Z_2 is all zeroes, so if a ϕ has at least one of $p+1, \ldots, n$ among its first p column numbers $\phi(1), \ldots, \phi(p)$ then the term arising from ϕ does not contribute to the sum because it is zero, e.g., if $\phi(1) = n$ then $t_{1,\phi(1)} t_{2,\phi(2)} \ldots t_{n,\phi(n)} = 0 \cdot t_{2,\phi(2)} \ldots t_{n,\phi(n)} = 0$.

So the above formula reduces to a sum over all permutations with two halves: any contributing ϕ is the composition of a ϕ_1 that rearranges only $1, \ldots, p$ and a ϕ_2 that rearranges only $p+1, \ldots, p+q$. Now, the distributive law and the fact that the signum of a composition is the product of the signums gives that this

$$|T_1| \cdot |T_2| = \Big(\sum_{\substack{\text{perms } \phi_1 \\ \text{of } 1,\ldots,p}} t_{1,\phi_1(1)} \cdots t_{p,\phi_1(p)} \operatorname{sgn}(\phi_1) \Big) \cdot \Big(\sum_{\substack{\text{perms } \phi_2 \\ \text{of } p+1,\ldots,p+q}} t_{p+1,\phi_2(p+1)} \cdots t_{p+q,\phi_2(p+q)} \operatorname{sgn}(\phi_2) \Big)$$

equals $|T| = \sum_{\text{contributing } \phi} t_{1,\phi(1)} t_{2,\phi(2)} \cdots t_{n,\phi(n)} \operatorname{sgn}(\phi)$. QED

2.11 Example

$$\begin{vmatrix} 2 & 0 & 0 & 0 \\ 1 & 2 & 0 & 0 \\ 0 & 0 & 3 & 0 \\ 0 & 0 & 0 & 3 \end{vmatrix} = \begin{vmatrix} 2 & 0 \\ 1 & 2 \end{vmatrix} \cdot \begin{vmatrix} 3 & 0 \\ 0 & 3 \end{vmatrix} = 36$$

From Lemma 2.10 we conclude that if two subspaces are complementary and t invariant then t is one-to-one if and only if its restriction to each subspace is nonsingular.

Now for the promised third, and final, step to the main result.

2.12 Lemma If a linear transformation $t \colon V \to V$ has the characteristic polynomial $(x - \lambda_1)^{p_1} \ldots (x - \lambda_k)^{p_k}$ then (1) $V = \mathscr{N}_\infty(t - \lambda_1) \oplus \cdots \oplus \mathscr{N}_\infty(t - \lambda_k)$ and (2) $\dim(\mathscr{N}_\infty(t - \lambda_i)) = p_i$.

Proof This argument consists of proving two preliminary claims, followed by proofs of clauses (1) and (2).

The first claim is that $\mathscr{N}_\infty(t - \lambda_i) \cap \mathscr{N}_\infty(t - \lambda_j) = \{\vec{0}\}$ when $i \neq j$. By Lemma 2.8 both $\mathscr{N}_\infty(t - \lambda_i)$ and $\mathscr{N}_\infty(t - \lambda_j)$ are t invariant. The intersection of t invariant subspaces is t invariant and so the restriction of t to $\mathscr{N}_\infty(t - \lambda_i) \cap \mathscr{N}_\infty(t - \lambda_j)$ is a linear transformation. Now, $t - \lambda_i$ is nilpotent on $\mathscr{N}_\infty(t - \lambda_i)$ and $t - \lambda_j$ is nilpotent on $\mathscr{N}_\infty(t - \lambda_j)$, so both $t - \lambda_i$ and $t - \lambda_j$ are nilpotent on the intersection. Therefore by Lemma 2.1 and the observation following it, if t has any eigenvalues on the intersection then the "only" eigenvalue is both λ_i and λ_j. This cannot be, so the restriction has no eigenvalues: $\mathscr{N}_\infty(t-\lambda_i) \cap \mathscr{N}_\infty(t-\lambda_j)$ is the trivial space (Lemma 3.10 shows that the only transformation that is without any eigenvalues is the transformation on the trivial space).

The second claim is that $\mathscr{N}_\infty(t-\lambda_i) \subseteq \mathscr{R}_\infty(t-\lambda_j)$, where $i \neq j$. To verify it we will show that $t-\lambda_j$ is one-to-one on $\mathscr{N}_\infty(t-\lambda_i)$ so that, since $\mathscr{N}_\infty(t-\lambda_i)$ is $t-\lambda_j$ invariant by Lemma 2.8, the map $t-\lambda_j$ is an automorphism of the subspace $\mathscr{N}_\infty(t-\lambda_i)$ and therefore that $\mathscr{N}_\infty(t-\lambda_i)$ is a subset of each $\mathscr{R}(t-\lambda_j)$, $\mathscr{R}((t-\lambda_j)^2)$, etc. For the verification that the map is one-to-one suppose that $\vec{v} \in \mathscr{N}_\infty(t-\lambda_i)$ is in the null space of $t-\lambda_j$, aiming to show that $\vec{v} = \vec{0}$. Consider the map $[(t-\lambda_i)-(t-\lambda_j)]^n$. On the one hand, the only vector that $(t-\lambda_i)-(t-\lambda_j) = \lambda_i - \lambda_j$ maps to zero is the zero vector. On the other hand, as in the proof of Lemma 1.7 we can apply the binomial expansion to get this.

$$(t-\lambda_i)^n(\vec{v}) + \binom{n}{1}(t-\lambda_i)^{n-1}(t-\lambda_j)^1(\vec{v}) + \binom{n}{2}(t-\lambda_i)^{n-2}(t-\lambda_j)^2(\vec{v}) + \cdots$$

The first term is zero because $\vec{v} \in \mathscr{N}_\infty(t-\lambda_i)$ while the remaining terms are zero because $\vec{v}$ is in the null space of $t-\lambda_j$. Therefore $\vec{v} = \vec{0}$.

With those two preliminary claims done we can prove clause (1), that the space is the direct sum of the generalized null spaces. By Corollary III.2.2 the space is the direct sum $V = \mathscr{N}_\infty(t-\lambda_1) \oplus \mathscr{R}_\infty(t-\lambda_1)$. By the second claim $\mathscr{N}_\infty(t-\lambda_2) \subseteq \mathscr{R}_\infty(t-\lambda_1)$ and so we can get a basis for $\mathscr{R}_\infty(t-\lambda_1)$ by starting with a basis for $\mathscr{N}_\infty(t-\lambda_2)$ and adding extra basis elements taken from $\mathscr{R}_\infty(t-\lambda_1) \cap \mathscr{R}_\infty(t-\lambda_2)$. Thus $V = \mathscr{N}_\infty(t-\lambda_1) \oplus \mathscr{N}_\infty(t-\lambda_2) \oplus (\mathscr{R}_\infty(t-\lambda_1) \cap \mathscr{R}_\infty(t-\lambda_2))$. Continuing in this way we get this.

$$V = \mathscr{N}_\infty(t-\lambda_1) \oplus \cdots \oplus \mathscr{R}_\infty(t-\lambda_k) \oplus (\mathscr{R}_\infty(t-\lambda_1) \cap \cdots \cap \mathscr{R}_\infty(t-\lambda_k))$$

The first claim above shows that the final space is trivial.

We finish by verifying clause (2). Decompose V as $\mathscr{N}_\infty(t-\lambda_i) \oplus \mathscr{R}_\infty(t-\lambda_i)$ and apply Lemma 2.9.

$$T = \left(\begin{array}{c|c} T_1 & Z_2 \\ \hline Z_1 & T_2 \end{array}\right) \begin{array}{l} \}\ \dim(\,\mathscr{N}_\infty(t-\lambda_i)\,)\text{-many rows} \\ \}\ \dim(\,\mathscr{R}_\infty(t-\lambda_i)\,)\text{-many rows} \end{array}$$

Lemma 2.10 says that $|T - xI| = |T_1 - xI| \cdot |T_2 - xI|$. By the uniqueness clause of the Fundamental Theorem of Algebra, Theorem I.1.11, the determinants of the blocks have the same factors as the characteristic polynomial $|T_1 - xI| = (x-\lambda_1)^{q_1} \cdots (x-\lambda_z)^{q_k}$ and $|T_2 - xI| = (x-\lambda_1)^{r_1} \cdots (x-\lambda_z)^{r_k}$, where $q_1 + r_1 = p_1$, $\ldots$, $q_k + r_k = p_k$. We will finish by establishing that (i) $q_j = 0$ for all $j \neq i$, and (ii) $q_i = p_i$. Together these prove clause (2) because they show that the degree of the polynomial $|T_1 - xI|$ is q_i and the degree of that polynomial equals the dimension of the generalized null space $\mathscr{N}_\infty(t-\lambda_i)$.

For (i), because the restriction of $t-\lambda_i$ to $\mathscr{N}_\infty(t-\lambda_i)$ is nilpotent on that space, t's only eigenvalue on that space is λ_i, by Lemma 2.2. So $q_j = 0$ for $j \neq i$.

For (ii), consider the restriction of t to $\mathscr{R}_\infty(t-\lambda_i)$. By Lemma III.2.1, the map $t-\lambda_i$ is one-to-one on $\mathscr{R}_\infty(t-\lambda_i)$ and so λ_i is not an eigenvalue of t on that subspace. Therefore $x-\lambda_i$ is not a factor of $|T_2 - xI|$, so $r_i = 0$, and so $q_i = p_i$. QED

Recall the goal of this chapter, to give a canonical form for matrix similarity. That result is next. It translates the above steps into matrix terms.

2.13 Theorem Any square matrix is similar to one in *Jordan form*

$$\begin{pmatrix} J_{\lambda_1} & & \textit{–zeroes–} & & \\ & J_{\lambda_2} & & & \\ & & \ddots & & \\ & & & J_{\lambda_{k-1}} & \\ & & \textit{–zeroes–} & & J_{\lambda_k} \end{pmatrix}$$

where each J_λ is the Jordan block associated with an eigenvalue λ of the original matrix (that is, each J_λ is all zeroes except for λ's down the diagonal and some subdiagonal ones).

Proof Given an $n \times n$ matrix T, consider the linear map $t\colon \mathbb{C}^n \to \mathbb{C}^n$ that it represents with respect to the standard bases. Use the prior lemma to write $\mathbb{C}^n = \mathscr{N}_\infty(t-\lambda_1) \oplus \cdots \oplus \mathscr{N}_\infty(t-\lambda_k)$ where $\lambda_1, \ldots, \lambda_k$ are the eigenvalues of t. Because each $\mathscr{N}_\infty(t-\lambda_i)$ is t invariant, Lemma 2.9 and the prior lemma show that t is represented by a matrix that is all zeroes except for square blocks along the diagonal. To make those blocks into Jordan blocks, pick each B_{λ_i} to be a string basis for the action of $t-\lambda_i$ on $\mathscr{N}_\infty(t-\lambda_i)$. QED

2.14 Corollary Every square matrix is similar to the sum of a diagonal matrix and a nilpotent matrix.

For Jordan form a canonical form for matrix similarity, strictly speaking it must be unique. That is, for any square matrix there needs to be one and only one matrix J similar to it and of the specified form. As stated the theorem allows us to rearrange the Jordan blocks. We could make this form unique, say by arranging the Jordan blocks so the eigenvalues are in order, and then arranging the blocks of subdiagonal ones from longest to shortest. Below, we won't bother with that.

2.15 Example This matrix has the characteristic polynomial $(x-2)^2(x-6)$.

$$T = \begin{pmatrix} 2 & 0 & 1 \\ 0 & 6 & 2 \\ 0 & 0 & 2 \end{pmatrix}$$

First we do the eigenvalue 2. Computation of the powers of $T-2I$, and of the null spaces and nullities, is routine. (Recall from Example 2.4 our convention of taking T to represent a transformation $t\colon \mathbb{C}^3 \to \mathbb{C}^3$ with respect to the standard basis.)

p	$(T-2I)^p$	$\mathscr{N}((t-2)^p)$	*nullity*
1	$\begin{pmatrix} 0 & 0 & 1 \\ 0 & 4 & 2 \\ 0 & 0 & 0 \end{pmatrix}$	$\{\begin{pmatrix} x \\ 0 \\ 0 \end{pmatrix} \mid x \in \mathbb{C}\}$	1
2	$\begin{pmatrix} 0 & 0 & 0 \\ 0 & 16 & 8 \\ 0 & 0 & 0 \end{pmatrix}$	$\{\begin{pmatrix} x \\ -z/2 \\ z \end{pmatrix} \mid x, z \in \mathbb{C}\}$	2
3	$\begin{pmatrix} 0 & 0 & 0 \\ 0 & 64 & 32 \\ 0 & 0 & 0 \end{pmatrix}$	*–same–*	*–same–*

So the generalized null space $\mathscr{N}_\infty(t-2)$ has dimension two. We know that the restriction of $t-2$ is nilpotent on this subspace. From the way that the nullities grow we know that the action of $t-2$ on a string basis is $\vec{\beta}_1 \mapsto \vec{\beta}_2 \mapsto \vec{0}$. Thus we can represent the restriction in the canonical form

$$N_2 = \begin{pmatrix} 0 & 0 \\ 1 & 0 \end{pmatrix} = \text{Rep}_{B,B}(t-2) \qquad B_2 = \langle \begin{pmatrix} 1 \\ 1 \\ -2 \end{pmatrix}, \begin{pmatrix} -2 \\ 0 \\ 0 \end{pmatrix} \rangle$$

(other choices of basis are possible). Consequently, the action of the restriction of t to $\mathscr{N}_\infty(t-2)$ is represented by this matrix.

$$J_2 = N_2 + 2I = \text{Rep}_{B_2,B_2}(t) = \begin{pmatrix} 2 & 0 \\ 1 & 2 \end{pmatrix}$$

The second eigenvalue is 6. Its computations are easier. Because the power of $x-6$ in the characteristic polynomial is one, the restriction of $t-6$ to $\mathscr{N}_\infty(t-6)$

must be nilpotent, of index one (it can't be of index less than one and since $x-6$ is a factor of the characteristic polynomial with the exponent one it can't be of index more than one either). Its action on a string basis must be $\vec{\beta}_3 \mapsto \vec{0}$ and since it is the zero map, its canonical form N_6 is the 1×1 zero matrix. Consequently, the canonical form J_6 for the action of t on $\mathscr{N}_\infty(t-6)$ is the 1×1 matrix with the single entry 6. For the basis we can use any nonzero vector from the generalized null space.

$$B_6 = \langle \begin{pmatrix} 0 \\ 1 \\ 0 \end{pmatrix} \rangle$$

Taken together, these two give that the Jordan form of T is

$$\text{Rep}_{B,B}(t) = \begin{pmatrix} 2 & 0 & 0 \\ 1 & 2 & 0 \\ 0 & 0 & 6 \end{pmatrix}$$

where B is the concatenation of B_2 and B_6.

2.16 Example As a contrast with the prior example, this matrix

$$T = \begin{pmatrix} 2 & 2 & 1 \\ 0 & 6 & 2 \\ 0 & 0 & 2 \end{pmatrix}$$

has the same characteristic polynomial $(x-2)^2(x-6)$, but here

p	$(T-6I)^p$	$\mathscr{N}((t-6)^p)$	*nullity*
1	$\begin{pmatrix} -4 & 3 & 1 \\ 0 & 0 & 2 \\ 0 & 0 & -4 \end{pmatrix}$	$\{\begin{pmatrix} x \\ (4/3)x \\ 0 \end{pmatrix} \mid x \in \mathbb{C}\}$	1
2	$\begin{pmatrix} 16 & -12 & -2 \\ 0 & 0 & -8 \\ 0 & 0 & 16 \end{pmatrix}$	*–same–*	—

the action of $t-2$ is stable after only one application—the restriction of $t-2$ to $\mathscr{N}_\infty(t-2)$ is nilpotent of index one. The restriction of $t-2$ to the generalized null space acts on a string basis via the two strings $\vec{\beta}_1 \mapsto \vec{0}$ and $\vec{\beta}_2 \mapsto \vec{0}$. We have this Jordan block associated with the eigenvalue 2.

$$J_2 = \begin{pmatrix} 2 & 0 \\ 0 & 2 \end{pmatrix}$$

So the contrast with the prior example is that while the characteristic polynomial tells us to look at the action of $t-2$ on its generalized null space, the characteristic polynomial does not completely describe $t-2$'s action. We must do some computations to find that the minimal polynomial is $(x-2)(x-6)$.

For the eigenvalue 6 the arguments for the second eigenvalue of the prior example apply again. The restriction of $t-6$ to $\mathscr{N}_\infty(t-6)$ is nilpotent of index one. Thus $t-6$'s canonical form N_6 is the 1×1 zero matrix, and the associated Jordan block J_6 is the 1×1 matrix with entry 6.

Therefore the Jordan form for T is a diagonal matrix.

$$\text{Rep}_{B,B}(t)=\begin{pmatrix}2&0&0\\0&2&0\\0&0&6\end{pmatrix}\qquad B=B_2{}^\frown B_6=\langle\begin{pmatrix}1\\0\\0\end{pmatrix},\begin{pmatrix}0\\1\\-2\end{pmatrix},\begin{pmatrix}2\\4\\0\end{pmatrix}\rangle$$

(Checking that the third vector in B is in the null space of $t-6$ is routine.)

2.17 Example A bit of computing with

$$T=\begin{pmatrix}-1&4&0&0&0\\0&3&0&0&0\\0&-4&-1&0&0\\3&-9&-4&2&-1\\1&5&4&1&4\end{pmatrix}$$

shows that its characteristic polynomial is $(x-3)^3(x+1)^2$. This table

p	$(T-3I)^p$	$\mathscr{N}((t-3)^p)$	*nullity*
1	$\begin{pmatrix}-4&4&0&0&0\\0&0&0&0&0\\0&-4&-4&0&0\\3&-9&-4&-1&-1\\1&5&4&1&1\end{pmatrix}$	$\{\begin{pmatrix}-(u+v)/2\\-(u+v)/2\\(u+v)/2\\u\\v\end{pmatrix}\mid u,v\in\mathbb{C}\}$	2
2	$\begin{pmatrix}16&-16&0&0&0\\0&0&0&0&0\\0&16&16&0&0\\-16&32&16&0&0\\0&-16&-16&0&0\end{pmatrix}$	$\{\begin{pmatrix}-z\\-z\\z\\u\\v\end{pmatrix}\mid z,u,v\in\mathbb{C}\}$	3
3	$\begin{pmatrix}-64&64&0&0&0\\0&0&0&0&0\\0&-64&-64&0&0\\64&-128&-64&0&0\\0&64&64&0&0\end{pmatrix}$	*–same–*	*–same–*

shows that the restriction of $t-3$ to $\mathscr{N}_\infty(t-3)$ acts on a string basis via the two strings $\vec{\beta}_1 \mapsto \vec{\beta}_2 \mapsto \vec{0}$ and $\vec{\beta}_3 \mapsto \vec{0}$.

A similar calculation for the other eigenvalue

p	$(T+1I)^p$	$\mathscr{N}((t+1)^p)$	*nullity*
1	$\begin{pmatrix} 0 & 4 & 0 & 0 & 0 \\ 0 & 4 & 0 & 0 & 0 \\ 0 & -4 & 0 & 0 & 0 \\ 3 & -9 & -4 & 3 & -1 \\ 1 & 5 & 4 & 1 & 5 \end{pmatrix}$	$\{\begin{pmatrix} -(u+v) \\ 0 \\ -v \\ u \\ v \end{pmatrix} \mid u, v \in \mathbb{C}\}$	2
2	$\begin{pmatrix} 0 & 16 & 0 & 0 & 0 \\ 0 & 16 & 0 & 0 & 0 \\ 0 & -16 & 0 & 0 & 0 \\ 8 & -40 & -16 & 8 & -8 \\ 8 & 24 & 16 & 8 & 24 \end{pmatrix}$	*–same–*	*–same–*

gives that the restriction of $t+1$ to its generalized null space acts on a string basis via the two separate strings $\vec{\beta}_4 \mapsto \vec{0}$ and $\vec{\beta}_5 \mapsto \vec{0}$.

Therefore T is similar to this Jordan form matrix.

$$\begin{pmatrix} -1 & 0 & 0 & 0 & 0 \\ 0 & -1 & 0 & 0 & 0 \\ 0 & 0 & 3 & 0 & 0 \\ 0 & 0 & 1 & 3 & 0 \\ 0 & 0 & 0 & 0 & 3 \end{pmatrix}$$

Exercises

2.18 Do the check for Example 2.4.

2.19 Each matrix is in Jordan form. State its characteristic polynomial and its minimal polynomial.

(a) $\begin{pmatrix} 3 & 0 \\ 1 & 3 \end{pmatrix}$ (b) $\begin{pmatrix} -1 & 0 \\ 0 & -1 \end{pmatrix}$ (c) $\begin{pmatrix} 2 & 0 & 0 \\ 1 & 2 & 0 \\ 0 & 0 & -1/2 \end{pmatrix}$ (d) $\begin{pmatrix} 3 & 0 & 0 \\ 1 & 3 & 0 \\ 0 & 1 & 3 \end{pmatrix}$

(e) $\begin{pmatrix} 3 & 0 & 0 & 0 \\ 1 & 3 & 0 & 0 \\ 0 & 0 & 3 & 0 \\ 0 & 0 & 1 & 3 \end{pmatrix}$ (f) $\begin{pmatrix} 4 & 0 & 0 & 0 \\ 1 & 4 & 0 & 0 \\ 0 & 0 & -4 & 0 \\ 0 & 0 & 1 & -4 \end{pmatrix}$ (g) $\begin{pmatrix} 5 & 0 & 0 \\ 0 & 2 & 0 \\ 0 & 0 & 3 \end{pmatrix}$

(h) $\begin{pmatrix} 5 & 0 & 0 & 0 \\ 0 & 2 & 0 & 0 \\ 0 & 0 & 2 & 0 \\ 0 & 0 & 0 & 3 \end{pmatrix}$ (i) $\begin{pmatrix} 5 & 0 & 0 & 0 \\ 0 & 2 & 0 & 0 \\ 0 & 1 & 2 & 0 \\ 0 & 0 & 0 & 3 \end{pmatrix}$

✓ **2.20** Find the Jordan form from the given data.

(a) The matrix T is 5×5 with the single eigenvalue 3. The nullities of the powers are: $T-3I$ has nullity two, $(T-3I)^2$ has nullity three, $(T-3I)^3$ has nullity four, and $(T-3I)^4$ has nullity five.

(b) The matrix S is 5×5 with two eigenvalues. For the eigenvalue 2 the nullities are: $S-2I$ has nullity two, and $(S-2I)^2$ has nullity four. For the eigenvalue -1 the nullities are: $S+1I$ has nullity one.

2.21 Find the change of basis matrices for each example.
(a) Example 2.15 (b) Example 2.16 (c) Example 2.17

✓ **2.22** Find the Jordan form and a Jordan basis for each matrix.

(a) $\begin{pmatrix}-10&4\\-25&10\end{pmatrix}$ (b) $\begin{pmatrix}5&-4\\9&-7\end{pmatrix}$ (c) $\begin{pmatrix}4&0&0\\2&1&3\\5&0&4\end{pmatrix}$ (d) $\begin{pmatrix}5&4&3\\-1&0&-3\\1&-2&1\end{pmatrix}$

(e) $\begin{pmatrix}9&7&3\\-9&-7&-4\\4&4&4\end{pmatrix}$ (f) $\begin{pmatrix}2&2&-1\\-1&-1&1\\-1&-2&2\end{pmatrix}$ (g) $\begin{pmatrix}7&1&2&2\\1&4&-1&-1\\-2&1&5&-1\\1&1&2&8\end{pmatrix}$

✓ **2.23** Find all possible Jordan forms of a transformation with characteristic polynomial $(x-1)^2(x+2)^2$.

2.24 Find all possible Jordan forms of a transformation with characteristic polynomial $(x-1)^3(x+2)$.

✓ **2.25** Find all possible Jordan forms of a transformation with characteristic polynomial $(x-2)^3(x+1)$ and minimal polynomial $(x-2)^2(x+1)$.

2.26 Find all possible Jordan forms of a transformation with characteristic polynomial $(x-2)^4(x+1)$ and minimal polynomial $(x-2)^2(x+1)$.

✓ **2.27** Diagonalize these.

(a) $\begin{pmatrix}1&1\\0&0\end{pmatrix}$ (b) $\begin{pmatrix}0&1\\1&0\end{pmatrix}$

✓ **2.28** Find the Jordan matrix representing the differentiation operator on $\mathcal{P}_3$.

✓ **2.29** Decide if these two are similar.

$$\begin{pmatrix}1&-1\\4&-3\end{pmatrix}\quad\begin{pmatrix}-1&0\\1&-1\end{pmatrix}$$

2.30 Find the Jordan form of this matrix.

$$\begin{pmatrix}0&-1\\1&0\end{pmatrix}$$

Also give a Jordan basis.

2.31 How many similarity classes are there for 3×3 matrices whose only eigenvalues are -3 and 4?

✓ **2.32** Prove that a matrix is diagonalizable if and only if its minimal polynomial has only linear factors.

2.33 Give an example of a linear transformation on a vector space that has no non-trivial invariant subspaces.

2.34 Show that a subspace is $t-\lambda_1$ invariant if and only if it is $t-\lambda_2$ invariant.

2.35 Prove or disprove: two $n\times n$ matrices are similar if and only if they have the same characteristic and minimal polynomials.

2.36 The *trace* of a square matrix is the sum of its diagonal entries.

(a) Find the formula for the characteristic polynomial of a 2×2 matrix.

(b) Show that trace is invariant under similarity, and so we can sensibly speak of the 'trace of a map'. (*Hint:* see the prior item.)

(c) Is trace invariant under matrix equivalence?

(d) Show that the trace of a map is the sum of its eigenvalues (counting multiplicities).

(e) Show that the trace of a nilpotent map is zero. Does the converse hold?

2.37 To use Definition 2.7 to check whether a subspace is t invariant, we seemingly have to check all of the infinitely many vectors in a (nontrivial) subspace to see if they satisfy the condition. Prove that a subspace is t invariant if and only if its subbasis has the property that for all of its elements, $t(\vec{\beta})$ is in the subspace.

✓ 2.38 Is t invariance preserved under intersection? Under union? Complementation? Sums of subspaces?

2.39 Give a way to order the Jordan blocks if some of the eigenvalues are complex numbers. That is, suggest a reasonable ordering for the complex numbers.

2.40 Let $\mathcal{P}_j(\mathbb{R})$ be the vector space over the reals of degree j polynomials. Show that if $j\leqslant k$ then $\mathcal{P}_j(\mathbb{R})$ is an invariant subspace of $\mathcal{P}_k(\mathbb{R})$ under the differentiation operator. In $\mathcal{P}_7(\mathbb{R})$, does any of $\mathcal{P}_0(\mathbb{R})$, ..., $\mathcal{P}_6(\mathbb{R})$ have an invariant complement?

2.41 In $\mathcal{P}_n(\mathbb{R})$, the vector space (over the reals) of degree n polynomials,

$$\mathcal{E}=\{p(x)\in\mathcal{P}_n(\mathbb{R})\mid p(-x)=p(x)\text{ for all }x\}$$

and

$$\mathcal{O}=\{p(x)\in\mathcal{P}_n(\mathbb{R})\mid p(-x)=-p(x)\text{ for all }x\}$$

are the *even* and the *odd* polynomials; $p(x)=x^2$ is even while $p(x)=x^3$ is odd. Show that they are subspaces. Are they complementary? Are they invariant under the differentiation transformation?

2.42 Lemma 2.9 says that if M and N are invariant complements then t has a representation in the given block form (with respect to the same ending as starting basis, of course). Does the implication reverse?

2.43 A matrix S is the *square root* of another T if $S^2=T$. Show that any nonsingular matrix has a square root.

Topic

Method of Powers

In applications matrices can be large. Calculating eigenvalues and eigenvectors by finding and solving the characteristic polynomial is impractical, too slow and too error-prone. Some techniques avoid the characteristic polynomial. Here we shall see a method that is suitable for large matrices that are *sparse*, meaning that the great majority of the entries are zero.

Suppose that the $n \times n$ matrix T has n distinct eigenvalues $\lambda_1, \lambda_2, \ldots, \lambda_n$. Then $\mathbb{C}^n$ has a basis made of the associated eigenvectors $\langle \vec{\zeta}_1, \ldots, \vec{\zeta}_n \rangle$. For any $\vec{v} \in \mathbb{C}^n$, writing $\vec{v} = c_1\vec{\zeta}_1 + \cdots + c_n\vec{\zeta}_n$ and iterating T on $\vec{v}$ gives these.

$$\begin{aligned}
T\vec{v} &= c_1\lambda_1\vec{\zeta}_1 + c_2\lambda_2\vec{\zeta}_2 + \cdots + c_n\lambda_n\vec{\zeta}_n \\
T^2\vec{v} &= c_1\lambda_1^2\vec{\zeta}_1 + c_2\lambda_2^2\vec{\zeta}_2 + \cdots + c_n\lambda_n^2\vec{\zeta}_n \\
T^3\vec{v} &= c_1\lambda_1^3\vec{\zeta}_1 + c_2\lambda_2^3\vec{\zeta}_2 + \cdots + c_n\lambda_n^3\vec{\zeta}_n \\
&\vdots \\
T^k\vec{v} &= c_1\lambda_1^k\vec{\zeta}_1 + c_2\lambda_2^k\vec{\zeta}_2 + \cdots + c_n\lambda_n^k\vec{\zeta}_n
\end{aligned}$$

Assuming that $|\lambda_1|$ is the largest and dividing through

$$\frac{T^k\vec{v}}{\lambda_1^k} = c_1\vec{\zeta}_1 + c_2\frac{\lambda_2^k}{\lambda_1^k}\vec{\zeta}_2 + \cdots + c_n\frac{\lambda_n^k}{\lambda_1^k}\vec{\zeta}_n$$

shows that as k gets larger the fractions go to zero and so λ_1's term will dominate the expression and that expression has a limit of $c_1\vec{\zeta}_1$.

Thus if $c_1 \neq 0$, as k increases the vectors $T^k\vec{v}$ will tend toward the direction of the eigenvectors associated with the dominant eigenvalue. Consequently, the ratios of the vector lengths $|T^k\vec{v}|/|T^{k-1}\vec{v}|$ tend to that dominant eigenvalue.

For example, the eigenvalues of the matrix

$$T = \begin{pmatrix} 3 & 0 \\ 8 & -1 \end{pmatrix}$$

are 3 and -1. If $\vec{v}$ has the components 1 and 1 then iterating gives this.

$\vec{v}$	$T\vec{v}$	$T^2\vec{v}$	$\cdots$	$T^9\vec{v}$	$T^{10}\vec{v}$
$\begin{pmatrix}1\\1\end{pmatrix}$	$\begin{pmatrix}3\\7\end{pmatrix}$	$\begin{pmatrix}9\\17\end{pmatrix}$	$\cdots$	$\begin{pmatrix}19\,683\\39\,367\end{pmatrix}$	$\begin{pmatrix}59\,049\\118\,097\end{pmatrix}$

The ratio between the lengths of the last two is $2.999\,9$.

We note two implementation issues. First, instead of finding the powers of T and applying them to $\vec{v}$, we will compute $\vec{v}_1$ as $T\vec{v}$ and then compute $\vec{v}_2$ as $T\vec{v}_1$, etc. (that is, we do not separately calculate T^2, T^3, ...). We can quickly do these matrix-vector products even if T is large, provided that it is sparse. The second issue is that to avoid generating numbers that are so large that they overflow our computer's capability, we can normalize the $\vec{v}_i$'s at each step. For instance, we can divide each $\vec{v}_i$ by its length (other possibilities are to divide it by its largest component, or simply by its first component). We thus implement this method by generating

$$\begin{aligned}
\vec{w}_0 &= \vec{v}_0/|\vec{v}_0| \\
\vec{v}_1 &= T\vec{w}_0 \\
\vec{w}_1 &= \vec{v}_1/|\vec{v}_1| \\
\vec{v}_2 &= T\vec{w}_2 \\
&\vdots \\
\vec{w}_{k-1} &= \vec{v}_{k-1}/|\vec{v}_{k-1}| \\
\vec{v}_k &= T\vec{w}_k
\end{aligned}$$

until we are satisfied. Then $\vec{v}_k$ is an approximation of an eigenvector, and the approximation of the dominant eigenvalue is the ratio $(T \cdot \vec{v}_k)/(\vec{v}_k \cdot \vec{v}_k) \approx (\lambda_1 \vec{v}_k \cdot \vec{v}_k)/(\vec{v}_k \cdot \vec{v}_k) = \lambda_1$.

One way that we could be 'satisfied' is to iterate until our approximation of the eigenvalue settles down. We could decide for instance to stop the iteration process not after some fixed number of steps, but instead when $|\vec{v}_k|$ differs from $|\vec{v}_{k-1}|$ by less than one percent, or when they agree up to the second significant digit.

The rate of convergence is determined by the rate at which the powers of $|\lambda_2/\lambda_1|$ go to zero, where λ_2 is the eigenvalue of second largest length. If that ratio is much less than one then convergence is fast but if it is only slightly less than one then convergence can be quite slow. Consequently, the method of powers is not the most commonly used way of finding eigenvalues (although it is the simplest one, which is why it is here). Instead, there are a variety of methods that generally work by first replacing the given matrix T with another that is similar to it and so has the same eigenvalues, but is in some reduced form such as tridiagonal form, where the only nonzero entries are on the diagonal, or just above or below it. Then special case techniques can find the eigenvalues. Once we know the eigenvalues then we can easily compute the eigenvectors of T. These other methods are outside of our scope. A good reference is [Goult, *et al.*]

Exercises

1 Use ten iterations to estimate the largest eigenvalue of these matrices, starting from the vector with components 1 and 2. Compare the answer with the one obtained by solving the characteristic equation.

(a) $\begin{pmatrix}1&5\\0&4\end{pmatrix}$ (b) $\begin{pmatrix}3&2\\-1&0\end{pmatrix}$

2 Redo the prior exercise by iterating until $|\vec{v}_k|-|\vec{v}_{k-1}|$ has absolute value less than 0.01 At each step, normalize by dividing each vector by its length. How many iterations does it take? Are the answers significantly different?

3 Use ten iterations to estimate the largest eigenvalue of these matrices, starting from the vector with components 1, 2, and 3. Compare the answer with the one obtained by solving the characteristic equation.

(a) $\begin{pmatrix}4&0&1\\-2&1&0\\-2&0&1\end{pmatrix}$ (b) $\begin{pmatrix}-1&2&2\\2&2&2\\-3&-6&-6\end{pmatrix}$

4 Redo the prior exercise by iterating until $|\vec{v}_k|-|\vec{v}_{k-1}|$ has absolute value less than 0.01. At each step, normalize by dividing each vector by its length. How many iterations does it take? Are the answers significantly different?

5 What happens if $c_1=0$? That is, what happens if the initial vector does not to have any component in the direction of the relevant eigenvector?

6 How can we adapt the method of powers to find the smallest eigenvalue?

Computer Code

This is the code for the computer algebra system Octave that did the calculation above. (It has been lightly edited to remove blank lines, etc.)

```
>T=[3, 0;
    8, -1]
T=
   3   0
   8  -1
>v0=[1; 2]
v0=
   1
   1
>v1=T*v0
v1=
   3
   7
>v2=T*v1
v2=
   9
   17
>T9=T**9
T9=
  19683   0
```

```
  39368 -1
>T10=T**10
T10=
  59049  0
 118096  1
>v9=T9*v0
v9=
  19683
  39367
>v10=T10*v0
v10=
  59049
  118096
>norm(v10)/norm(v9)
 ans=2.9999
```

Remark. This does not use the full power of Octave; it has built-in functions to automatically apply sophisticated methods to find eigenvalues and eigenvectors.

Topic

Stable Populations

Imagine a reserve park with animals from a species that we are protecting. The park doesn't have a fence so animals cross the boundary, both from the inside out and from the outside in. Every year, 10% of the animals from inside of the park leave and 1% of the animals from the outside find their way in. Can we reach a stable level; are there populations for the park and the rest of the world that will stay constant over time, with the number of animals leaving equal to the number of animals entering?

Let p_n be the year n population in the park and let r_n be the population in the rest of the world.

$$\begin{aligned} p_{n+1} &= .90p_n + .01r_n \\ r_{n+1} &= .10p_n + .99r_n \end{aligned}$$

We have this matrix equation.

$$\begin{pmatrix} p_{n+1} \\ r_{n+1} \end{pmatrix} = \begin{pmatrix} .90 & .01 \\ .10 & .99 \end{pmatrix} \begin{pmatrix} p_n \\ r_n \end{pmatrix}$$

The population will be stable if $p_{n+1} = p_n$ and $r_{n+1} = r_n$ so that the matrix equation $\vec{v}_{n+1} = T\vec{v}_n$ becomes $\vec{v} = T\vec{v}$. We are therefore looking for eigenvectors for T that are associated with the eigenvalue $\lambda = 1$. The equation $\vec{0} = (\lambda I - T)\vec{v} = (I - T)\vec{v}$ is

$$\begin{pmatrix} 0.10 & -0.01 \\ -0.10 & 0.01 \end{pmatrix} \begin{pmatrix} p \\ r \end{pmatrix} = \begin{pmatrix} 0 \\ 0 \end{pmatrix}$$

and gives the eigenspace of vectors with the restriction that $p = .1r$. For example, if we start with a park population $p = 10\,000$ animals and a rest of the world population of $r = 100\,000$ animals then every year ten percent of those inside leave the park (this is a thousand animals), and every year one percent of those from the rest of the world enter the park (also a thousand animals). The population is stable, self-sustaining.

Now imagine that we are trying to raise the total world population of this species. We are trying to have the world population grow at 1% per year. This makes the population level stable in some sense, although it is a dynamic stability, in contrast to the static population level of the $\lambda = 1$ case. The equation $\vec{v}_{n+1} = 1.01 \cdot \vec{v}_n = T\vec{v}_n$ leads to $((1.01I - T)\vec{v} = \vec{0}$, which gives this system.

$$\begin{pmatrix} 0.11 & -0.01 \\ -0.10 & 0.02 \end{pmatrix} \begin{pmatrix} p \\ r \end{pmatrix} = \begin{pmatrix} 0 \\ 0 \end{pmatrix}$$

This matrix is nonsingular and so the only solution is $p = 0$, $r = 0$. Thus there is no nontrivial initial population that would lead to a regular annual one percent growth rate in p and r.

We can look for the rates that allow an initial population for the park that results in a steady growth behavior. We consider $\lambda\vec{v} = T\vec{v}$ and solve for λ.

$$0 = \begin{vmatrix} \lambda - .9 & .01 \\ .10 & \lambda - .99 \end{vmatrix} = (\lambda - .9)(\lambda - .99) - (.10)(.01) = \lambda^2 - 1.89\lambda + .89$$

We already know that $\lambda = 1$ is one solution of this characteristic equation. The other is 0.89. Thus there are two ways to have a dynamically stable p and r, where the two grow at the same rate despite the leaky park boundaries: have a world population that is does not grow or shrink, and have a world population that shrinks by 11% every year.

So one way to look at eigenvalues and eigenvectors is that they give a stable state for a system. If the eigenvalue is one then the system is static and if the eigenvalue isn't one then it is a dynamic stability.

Exercises

1 For the park discussed above, what should be the initial park population in the case where the populations decline by 11% every year?

2 What will happen to the population of the park in the event of a growth in world population of 1% per year? Will it lag the world growth, or lead it? Assume that the initial park population is ten thousand, and the world population is one hundred thousand, and calculate over a ten year span.

3 The park discussed above is partially fenced so that now, every year, only 5% of the animals from inside of the park leave (still, about 1% of the animals from the outside find their way in). Under what conditions can the park maintain a stable population now?

4 Suppose that a species of bird only lives in Canada, the United States, or in Mexico. Every year, 4% of the Canadian birds travel to the US, and 1% of them travel to Mexico. Every year, 6% of the US birds travel to Canada, and 4% go to Mexico. From Mexico, every year 10% travel to the US, and 0% go to Canada.

(a) Give the transition matrix.

(b) Is there a way for the three countries to have constant populations?

Topic

Page Ranking

Imagine that you are looking for the best book on Linear Algebra. You probably would try a web search engine such as Google. These lists pages ranked by importance. The ranking is defined, as Google's founders have said in [Brin & Page], that a page is important if other important pages link to it: "a page can have a high PageRank if there are many pages that point to it, or if there are some pages that point to it and have a high PageRank." But isn't that circular— how can they tell whether a page is important without first deciding on the important pages? With eigenvalues and eigenvectors.

We will present a simplified version of the Page Rank algorithm. For that we will model the World Wide Web as a collection of pages connected by links. This diagram, from [Wills], shows the pages as circles, and the links as arrows; for instance, page p_1 has a link to page p_2.

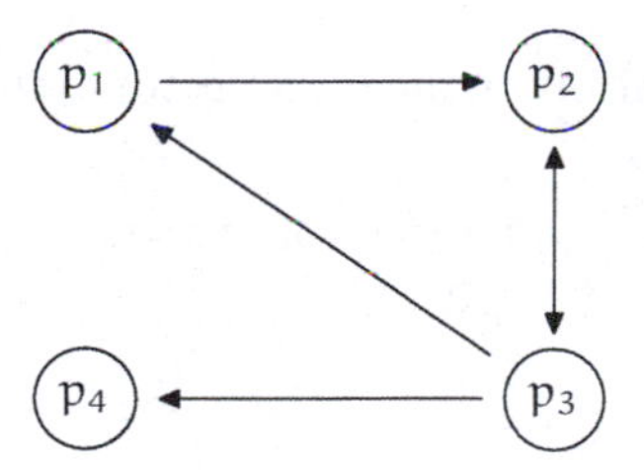

The key idea is that pages that should be highly ranked if they are cited often by other pages. That is, we raise the importance of a page p_i if it is linked-to from page p_j. The increment depends on the importance of the linking page p_j divided by how many out-links a_j are on that page.

$$\mathcal{I}(p_i) = \sum_{\text{in-linking pages } p_j} \frac{\mathcal{I}(p_j)}{a_j}$$

This matrix stores the information.

$$\begin{pmatrix} 0 & 0 & 1/3 & 0 \\ 1 & 0 & 1/3 & 0 \\ 0 & 1 & 0 & 0 \\ 0 & 0 & 1/3 & 0 \end{pmatrix}$$

The algorithm's inventors describe a way to think about that matrix.

> PageRank can be thought of as a model of user behavior. We assume there is a 'random surfer' who is given a web page at random and keeps clicking on links, never hitting "back" ... The probability that the random surfer visits a page is its PageRank. [Brin & Page]

In the diagram, a surfer on page p_3 has a probability $1/3$ of going next to each of the other pages.

That leads us to the problem of page p_4. Many pages are *dangling* or *sink links*, without any outbound links. The simplest model of what happens here is to imagine that the surfer goes to a next page entirely at random.

$$H = \begin{pmatrix} 0 & 0 & 1/3 & 1/4 \\ 1 & 0 & 1/3 & 1/4 \\ 0 & 1 & 0 & 1/4 \\ 0 & 0 & 1/3 & 1/4 \end{pmatrix}$$

We will find vector $\vec{\mathcal{I}}$ whose components are the importance rankings of each page $\mathcal{I}(p_i)$. With this notation, our requirements for the page rank are that $H\vec{\mathcal{I}} = \vec{\mathcal{I}}$. That is, we want an eigenvector of the matrix associated with the eigenvalue $\lambda = 1$.

Here is *Sage*'s calculation of the eigenvectors (slightly edited to fit on the page).

```
sage: H=matrix([[0,0,1/3,1/4], [1,0,1/3,1/4], [0,1,0,1/4], [0,0,1/3,1/4]])
sage: H.eigenvectors_right()
[(1, [
(1, 2, 9/4, 1)
], 1), (0, [
(0, 1, 3, -4)
], 1), (-0.3750000000000000? - 0.4389855730355308?*I,
          [(1, -0.1250000000000000? + 1.316956719106593?*I,
           -1.875000000000000? - 1.316956719106593?*I, 1)], 1),
        (-0.3750000000000000? + 0.4389855730355308?*I,
          [(1, -0.1250000000000000? - 1.316956719106593?*I,
            -1.875000000000000? + 1.316956719106593?*I, 1)], 1)]
```

The eigenvector that *Sage* gives associated with the eigenvalue $\lambda = 1$ is this.

$$\begin{pmatrix} 1 \\ 2 \\ 9/4 \\ 1 \end{pmatrix}$$

Of course, there are many vectors in that eigenspace. To get a page rank number we normalize to length one.

```
sage: v=vector([1, 2, 9/4, 1])
sage: v/v.norm()
(4/177*sqrt(177), 8/177*sqrt(177), 3/59*sqrt(177), 4/177*sqrt(177))
sage: w=v/v.norm()
sage: w.n()
(0.300658411201132, 0.601316822402263, 0.676481425202546, 0.300658411201132)
```

So we rank the first and fourth pages as of equal importance. We rank the second and third pages as much more important than those, and about equal in importance as each other.

We'll add one more refinement. We will allow the surfer to pick a new page at random even if they are not on a dangling page. Let this happen with probability α.

$$G = \alpha \cdot \begin{pmatrix} 0 & 0 & 1/3 & 1/4 \\ 1 & 0 & 1/3 & 1/4 \\ 0 & 1 & 0 & 1/4 \\ 0 & 0 & 1/3 & 1/4 \end{pmatrix} + (1-\alpha) \cdot \begin{pmatrix} 1/4 & 1/4 & 1/4 & 1/4 \\ 1/4 & 1/4 & 1/4 & 1/4 \\ 1/4 & 1/4 & 1/4 & 1/4 \\ 1/4 & 1/4 & 1/4 & 1/4 \end{pmatrix}$$

This is the *Google matrix*.

In practice α is typically between 0.85 and 0.99. Here are the ranks for the four pages with various α's.

α	0.85	0.90	0.95	0.99
p_1	0.325	0.317	0.309	0.302
p_2	0.602	0.602	0.602	0.601
p_3	0.652	0.661	0.669	0.675
p_4	0.325	0.317	0.309	0.302

The details of the algorithms used by commercial search engines are secret, no doubt have many refinements, and also change frequently. But the inventors of Google were gracious enough to outline the basis for their work in [Brin & Page]. A more current source is [Wikipedia, Google Page Rank]. Two additional excellent expositions are [Wills] and [Austin].

Exercises

1 A square matrix is *stochastic* if the sum of the entries in each column is one. The Google matrix is computed by taking a combination $G = \alpha * H + (1-\alpha) * S$ of two stochastic matrices. Show that G must be stochastic.

2 For this web of pages, the importance of each page should be equal. Verify it for $\alpha = 0.85$.

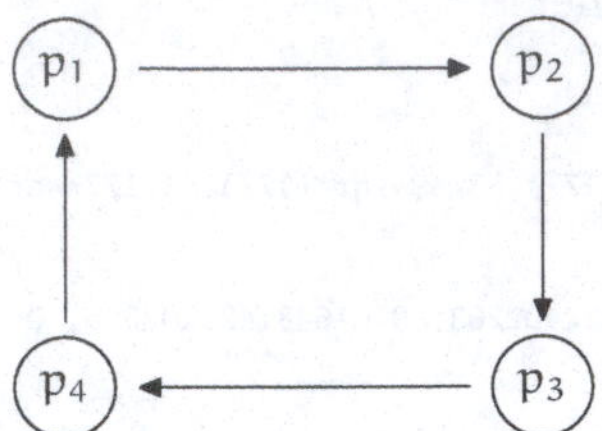

3 [Bryan & Leise] Give the importance ranking for this web of pages.

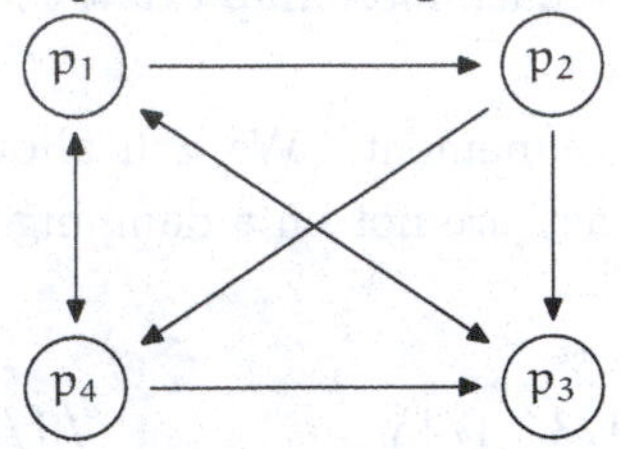

(a) Use $\alpha = 0.85$.
(b) Use $\alpha = 0.95$.
(c) Observe that while p_3 is linked-to from all other pages, and therefore seems important, it is not the highest ranked page. What is the highest ranked page? Explain.

Topic

Linear Recurrences

In 1202 Leonardo of Pisa, known as Fibonacci, posed this problem.

> A certain man put a pair of rabbits in a place surrounded on all sides by a wall. How many pairs of rabbits can be produced from that pair in a year if it is supposed that every month each pair begets a new pair which from the second month on becomes productive?

This moves past an elementary exponential growth model for populations to include that newborns are not fertile for some period, here a month. However, it retains other simplifying assumptions such as that there is an age after which the rabbits are infertile.

To get next month's total number of pairs we add the number of pairs alive going into next month to the number of pairs that will be newly born next month. The latter equals the number of pairs that will be productive going into next month, which is the number that next month will have been alive for at least two months.

$$F(n) = F(n-1) + F(n-2) \qquad \text{where } F(0) = 0,\ F(1) = 1 \qquad (*)$$

On the left is a *recurrence relation.* It gets that name because F recurs in its own defining equation. On the right are the initial conditions. From $(*)$ we can compute $F(2)$, $F(3)$, etc., to work up to the answer for Fibonacci's question.

month n	0	1	2	3	4	5	6	7	8	9	10	11	12
pairs $F(n)$	0	1	1	2	3	5	8	13	21	34	55	89	144

We will use linear algebra to get a formula that calculates $F(n)$ without having to first calculate the intermediate values $F(2)$, $F(3)$, etc.

We start by giving $(*)$ a matrix formulation.

$$\begin{pmatrix} F(n) \\ F(n-1) \end{pmatrix} = \begin{pmatrix} 1 & 1 \\ 1 & 0 \end{pmatrix} \begin{pmatrix} F(n-1) \\ F(n-2) \end{pmatrix} \qquad \text{where } \begin{pmatrix} F(1) \\ F(0) \end{pmatrix} = \begin{pmatrix} 1 \\ 0 \end{pmatrix}$$

Write T for the matrix and $\vec{v}_n$ for the vector with components $F(n)$ and $F(n-1)$ so that $\vec{v}_n = T^{n-1}\vec{v}_1$ for $n \geqslant 1$. If we diagonalize T then we have a fast way to compute its powers: where $T = PDP^{-1}$ then $T^n = PD^nP^{-1}$ and the n-th power of the diagonal matrix D is the diagonal matrix whose entries are the n-th powers of the entries of D.

The characteristic equation of T is $\lambda^2 - \lambda - 1 = 0$. The quadratic formula gives its roots as $(1+\sqrt{5})/2$ and $(1-\sqrt{5})/2$. (These are sometimes called "golden ratios;" see [Falbo].) Diagonalizing gives this.

$$\begin{pmatrix} 1 & 1 \\ 1 & 0 \end{pmatrix} = \begin{pmatrix} \frac{1+\sqrt{5}}{2} & \frac{1-\sqrt{5}}{2} \\ 1 & 1 \end{pmatrix} \begin{pmatrix} \frac{1+\sqrt{5}}{2} & 0 \\ 0 & \frac{1-\sqrt{5}}{2} \end{pmatrix} \begin{pmatrix} \frac{1}{\sqrt{5}} & -(\frac{1-\sqrt{5}}{2\sqrt{5}}) \\ \frac{-1}{\sqrt{5}} & \frac{1+\sqrt{5}}{2\sqrt{5}} \end{pmatrix}$$

Introducing the vectors and taking the n-th power, we have

$$\begin{aligned} \begin{pmatrix} F(n) \\ F(n-1) \end{pmatrix} &= \begin{pmatrix} 1 & 1 \\ 1 & 0 \end{pmatrix}^{n-1} \begin{pmatrix} f(1) \\ f(0) \end{pmatrix} \\ &= \begin{pmatrix} \frac{1+\sqrt{5}}{2} & \frac{1-\sqrt{5}}{2} \\ 1 & 1 \end{pmatrix} \begin{pmatrix} \left(\frac{1+\sqrt{5}}{2}\right)^{n-1} & 0 \\ 0 & \left(\frac{1-\sqrt{5}}{2}\right)^{n-1} \end{pmatrix} \begin{pmatrix} \frac{1}{\sqrt{5}} & -(\frac{1-\sqrt{5}}{2\sqrt{5}}) \\ \frac{-1}{\sqrt{5}} & \frac{1+\sqrt{5}}{2\sqrt{5}} \end{pmatrix} \begin{pmatrix} 1 \\ 0 \end{pmatrix} \end{aligned}$$

The calculation is ugly but not hard.

$$\begin{aligned} \begin{pmatrix} F(n) \\ F(n-1) \end{pmatrix} &= \begin{pmatrix} \frac{1+\sqrt{5}}{2} & \frac{1-\sqrt{5}}{2} \\ 1 & 1 \end{pmatrix} \begin{pmatrix} \left(\frac{1+\sqrt{5}}{2}\right)^{n-1} & 0 \\ 0 & \left(\frac{1-\sqrt{5}}{2}\right)^{n-1} \end{pmatrix} \begin{pmatrix} \frac{1}{\sqrt{5}} \\ -\frac{1}{\sqrt{5}} \end{pmatrix} \\ &= \frac{1}{\sqrt{5}} \begin{pmatrix} \frac{1+\sqrt{5}}{2} & \frac{1-\sqrt{5}}{2} \\ 1 & 1 \end{pmatrix} \begin{pmatrix} \left(\frac{1+\sqrt{5}}{2}\right)^{n-1} \\ -\left(\frac{1-\sqrt{5}}{2}\right)^{n-1} \end{pmatrix} \\ &= \frac{1}{\sqrt{5}} \begin{pmatrix} \left(\frac{1+\sqrt{5}}{2}\right)^{n} - \left(\frac{1-\sqrt{5}}{2}\right)^{n} \\ \left(\frac{1+\sqrt{5}}{2}\right)^{n-1} - \left(\frac{1-\sqrt{5}}{2}\right)^{n-1} \end{pmatrix} \end{aligned}$$

We want the first component.

$$F(n) = \frac{1}{\sqrt{5}} \left[\left(\frac{1+\sqrt{5}}{2}\right)^{n} - \left(\frac{1-\sqrt{5}}{2}\right)^{n} \right]$$

This formula gives the value of any member of the sequence without having to first find the intermediate values.

Because $(1-\sqrt{5})/2 \approx -0.618$ has absolute value less than one, its powers go to zero and so the $F(n)$ formula is dominated by its first term. Although we have extended the elementary model of population growth by adding a delay period before the onset of fertility, we nonetheless still get a function that is asymptotically exponential.

In general, a *homogeneous linear recurrence relation of order* k has this form.

$$f(n) = a_{n-1}f(n-1) + a_{n-2}f(n-2) + \cdots + a_{n-k}f(n-k)$$

This recurrence relation is homogeneous because it has no constant term, i.e, we can rewrite it as $0 = -f(n) + a_{n-1}f(n-1) + a_{n-2}f(n-2) + \cdots + a_{n-k}f(n-k)$. It is of order k because it uses k-many prior terms to calculate $f(n)$. The relation, cimbined with initial conditions giving values for $f(0)$, ..., $f(k-1)$, completely determines a sequence, simply because we can compute $f(n)$ by first computing $f(k)$, $f(k+1)$, etc. As with the Fibonacci case we will find a formula that solves the recurrence, that directly gives $f(n)$

Let V be the set of functions with domain $\mathbb{N} = \{0, 1, 2, \ldots\}$ and codomain $\mathbb{C}$. (Where convenient we sometimes use the domain $\mathbb{Z}^+ = \{1, 2, \ldots\}$.) This is a vector space under the usual meaning for addition and scalar multiplication, that $f + g$ is the map $x \mapsto f(x) + g(x)$ and cf is the map $x \mapsto c \cdot f(x)$.

If we put aside any initial conditions and look only at the recurrence, then there may be many functions satisfying the relation. For example, the Fibonacci recurrence that each value beyond the initial ones is the sum of the prior two is satisfied by the function L whose first few values are $L(0) = 2$, $L(1) = 1$, $L(2) = 3$, $L(3) = 4$, and $L(4) = 7$.

Fix a homogeneous linear recurrence relation of order k and consider the subset S of functions satisfying the relation (without initial conditions). This S is a subspace of V. It is nonempty because the zero function is a solution, by homogeneity. It is closed under addition because if f_1 and f_2 are solutions then this holds.

$$\begin{aligned}
-(f_1+f_2)(n) + a_{n-1}(f_1+f_2)(n-1) + \cdots + a_{n-k}(f_1+f_2)(n-k)& \\
= (-f_1(n) + \cdots + a_{n-k}f_1(n-k))& \\
+ (-f_2(n) + \cdots + a_{n-k}f_2(n-k))& \\
= 0 + 0 = 0&
\end{aligned}$$

It is also closed under scalar multiplication.

$$\begin{aligned}
-(rf_1)(n) + a_{n-1}(rf_1)(n-1) + \cdots + a_{n-k}(rf_1)(n-k)& \\
= r \cdot (-f_1(n) + \cdots + a_{n-k}f_1(n-k))& \\
= r \cdot 0& \\
= 0&
\end{aligned}$$

We can find the dimension of S. Where k is the order of the recurrence, consider this map from the set of functions S to the set of k-tall vectors.

$$f \mapsto \begin{pmatrix} f(0) \\ f(1) \\ \vdots \\ f(k-1) \end{pmatrix}$$

Exercise 4 shows that this is linear. Any solution of the recurrence is uniquely determined by the k-many initial conditions so this map is one-to-one and onto. Thus it is an isomorphism, and S has dimension k.

So we can describe the set of solutions of our linear homogeneous recurrence relation of order k by finding a basis consisting of k-many linearly independent functions. To produce those we give the recurrence a matrix formulation.

$$\begin{pmatrix} f(n) \\ f(n-1) \\ \vdots \\ f(n-k+1) \end{pmatrix} = \begin{pmatrix} a_{n-1} & a_{n-2} & a_{n-3} & \ldots & a_{n-k+1} & a_{n-k} \\ 1 & 0 & 0 & \ldots & 0 & 0 \\ 0 & 1 & 0 & & & \\ 0 & 0 & 1 & & & \\ \vdots & \vdots & & \ddots & & \vdots \\ 0 & 0 & 0 & \ldots & 1 & 0 \end{pmatrix} \begin{pmatrix} f(n-1) \\ f(n-2) \\ \vdots \\ f(n-k) \end{pmatrix}$$

Call the matrix A. We want its characteristic function, the determinant of $A - \lambda I$. The pattern in the 2×2 case

$$\begin{pmatrix} a_{n-1}-\lambda & a_{n-2} \\ 1 & -\lambda \end{pmatrix} = \lambda^2 - a_{n-1}\lambda - a_{n-2}$$

and the 3×3 case

$$\begin{pmatrix} a_{n-1}-\lambda & a_{n-2} & a_{n-3} \\ 1 & -\lambda & 0 \\ 0 & 1 & -\lambda \end{pmatrix} = -\lambda^3 + a_{n-1}\lambda^2 + a_{n-2}\lambda + a_{n-3}$$

leads us to expect, and Exercise 5 verifies, that this is the characteristic equation.

$$0 = \begin{vmatrix} a_{n-1}-\lambda & a_{n-2} & a_{n-3} & \ldots & a_{n-k+1} & a_{n-k} \\ 1 & -\lambda & 0 & \ldots & 0 & 0 \\ 0 & 1 & -\lambda & & & \\ 0 & 0 & 1 & & & \\ \vdots & \vdots & & \ddots & & \vdots \\ 0 & 0 & 0 & \ldots & 1 & -\lambda \end{vmatrix}$$
$$= \pm(-\lambda^k + a_{n-1}\lambda^{k-1} + a_{n-2}\lambda^{k-2} + \cdots + a_{n-k+1}\lambda + a_{n-k})$$

The $\pm$ is not relevant to find the roots so we drop it. We say that the polynomial $-\lambda^k + a_{n-1}\lambda^{k-1} + a_{n-2}\lambda^{k-2} + \cdots + a_{n-k+1}\lambda + a_{n-k}$ is associated with the recurrence relation.

If the characteristic equation has no repeated roots then the matrix is diagonalizable and we can, in theory, get a formula for $f(n)$, as in the Fibonacci case. But because we know that the subspace of solutions has dimension k we do not need to do the diagonalization calculation, provided that we can exhibit k different linearly independent functions satisfying the relation.

Where $r_1, r_2, \ldots, r_k$ are the distinct roots, consider the functions of powers of those roots, $f_{r_1}(n) = r_1^n$ through $f_{r_k}(n) = r_k^n$. Exercise 6 shows that each is a solution of the recurrence and that they form a linearly independent set. So, if the roots of the associated polynomial are distinct, any solution of the relation has the form $f(n) = c_1 r_1^n + c_2 r_2^n + \cdots + c_k r_k^n$ for some scalars $c_1, \ldots, c_n$. (The case of repeated roots is similar but we won't cover it here; see any text on Discrete Mathematics.)

Now we bring in the initial conditions. Use them to solve for $c_1, \ldots, c_n$. For instance, the polynomial associated with the Fibonacci relation is $-\lambda^2 + \lambda + 1$, whose roots are $r_1 = (1+\sqrt{5})/2$ and $r_2 = (1-\sqrt{5})/2$ and so any solution of the Fibonacci recurrence has the form $f(n) = c_1((1+\sqrt{5})/2)^n + c_2((1-\sqrt{5})/2)^n$. Use the Fibonacci initial conditions for $n = 0$ and $n = 1$

$$\begin{aligned} c_1 + \qquad\qquad c_2 &= 0 \\ (1+\sqrt{5}/2)c_1 + (1-\sqrt{5}/2)c_2 &= 1 \end{aligned}$$

and solve to get $c_1 = 1/\sqrt{5}$ and $c_2 = -1/\sqrt{5}$, as we found above.

We close by considering the nonhomogeneous case, where the relation has the form $f(n+1) = a_n f(n) + a_{n-1} f(n-1) + \cdots + a_{n-k} f(n-k) + b$ for some nonzero b. We only need a small adjustment to make the transition from the homogeneous case.

This classic example illustrates: in 1883, Edouard Lucas posed the Tower of Hanoi problem.

> In the great temple at Benares, beneath the dome which marks the center of the world, rests a brass plate in which are fixed three diamond needles, each a cubit high and as thick as the body of a bee. On one of these needles, at the creation, God placed sixty four disks of pure gold, the largest disk resting on the brass plate, and the others getting smaller and smaller up to the top one. This is the Tower of Brahma. Day and night unceasingly the priests transfer the disks from one diamond needle to another according to the fixed and immutable laws of Bram-ah, which require that the priest on duty must not move more than one disk at a time and that he must

> place this disk on a needle so that there is no smaller disk below it. When the sixty-four disks shall have been thus transferred from the needle on which at the creation God placed them to one of the other needles, tower, temple, and Brahmins alike will crumble into dusk, and with a thunderclap the world will vanish. (Translation of [De Parville] from [Ball & Coxeter].)

We put aside the question of why the priests don't sit down for a while and have the world last a little longer, and instead ask how many disk moves it will take. Before tackling the sixty four disk problem we will consider the problem for three disks.

To begin, all three disks are on the same needle.

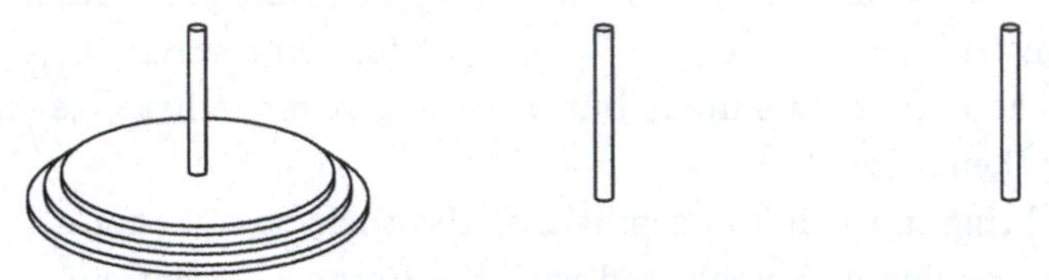

After the three moves of taking the small disk to the far needle, the mid-sized disk to the middle needle, and then the small disk to the middle needle, we have this.

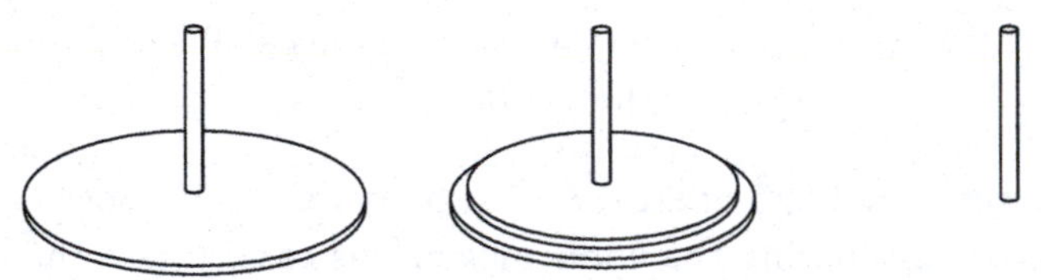

Now we can move the big disk to the far needle. Then to finish we repeat the three-move process on the two smaller disks, this time so that they end up on the third needle, on top of the big disk.

That sequence of moves is the best that we can do. To move the bottom disk at a minimum we must first move the smaller disks to the middle needle, then move the big one, and then move all the smaller ones from the middle needle to the ending needle. Since this minimum suffices, we get this recurrence.

$$T(n) = T(n-1) + 1 + T(n-1) = 2T(n-1) + 1 \qquad \text{where } T(1) = 1$$

Here are the first few values of T.

disks n	1	2	3	4	5	6	7	8	9	10
moves $T(n)$	1	3	7	15	31	63	127	255	511	1023

Of course, these numbers are one less than a power of two. To derive this write the original relation as $-1 = -T(n)+2T(n-1)$. Consider $0 = -T(n)+2T(n-1)$, a linear homogeneous recurrence of order 1. Its associated polynomial is $-\lambda+2$, with the single root $r_1 = 2$. Thus functions satisfying the homogeneous relation take the form $c_1 2^n$.

That's the homogeneous solution. Now we need a particular solution. Because the nonhomogeneous relation $-1 = -T(n) + 2T(n-1)$ is so simple, we can by eye spot a particular solution $T(n) = -1$. Any solution of the recurrence $T(n) = 2T(n-1)+1$ (without initial conditions) is the sum of the homogeneous solution and the particular solution: $c_1 2^n - 1$. Now the initial condition $T(1) = 1$ gives that $c_1 = 1$ and we've gotten the formula that generates the table: the n-disk Tower of Hanoi problem requires $T(n) = 2^n - 1$ moves.

Finding a particular solution in more complicated cases is, perhaps not surprisingly, more complicated. A delightful and rewarding, but challenging, source is [Graham, Knuth, Patashnik]. For more on the Tower of Hanoi see [Ball & Coxeter], [Gardner 1957], and [Hofstadter]. Some computer code follows the exercises.

Exercises

1 How many months until the number of Fibonacci rabbit pairs passes a thousand? Ten thousand? A million?

2 Solve each homogeneous linear recurrence relations.
(a) $f(n) = 5f(n-1) - 6f(n-2)$
(b) $f(n) = 4f(n-2)$
(c) $f(n) = 5f(n-1) - 2f(n-2) - 8f(n-3)$

3 Give a formula for the relations of the prior exercise, with these initial conditions.
(a) $f(0) = 1$, $f(1) = 1$
(b) $f(0) = 0$, $f(1) = 1$
(c) $f(0) = 1$, $f(1) = 1$, $f(2) = 3$.

4 Check that the isomorphism given between S and $\mathbb{R}^k$ is a linear map.

5 Show that the characteristic equation of the matrix is as stated, that is, is the polynomial associated with the relation. (*Hint:* expanding down the final column and using induction will work.)

6 Given a homogeneous linear recurrence relation $f(n) = a_n f(n-1) + \cdots + a_{n-k} f(n-k)$, let $r_1, \ldots, r_k$ be the roots of the associated polynomial. Prove that each function $f_{r_i}(n) = r_k^n$ satisfies the recurrence (without initial conditions).

7 (This refers to the value $T(64) = 18,446,744,073,709,551,615$ given in the computer code below.) Transferring one disk per second, how many years would it take the priests at the Tower of Hanoi to finish the job?

Computer Code

This code generates the first few values of a function defined by a recurrence and initial conditions. It is in the Scheme dialect of LISP, specifically, [Chicken Scheme].

After loading an extension that keeps the computer from switching to floating point numbers when the integers get large, the Tower of Hanoi function is straightforward.

```
(require-extension numbers)

(define (tower-of-hanoi-moves n)
   (if (= n 1)
       1
       (+ (* (tower-of-hanoi-moves (- n 1))
             2)
          1) ) )

; Two helper funcitons
(define (first-few-outputs proc n)
   (first-few-outputs-aux proc n '()) )

(define (first-few-outputs-aux proc n lst)
   (if (< n 1)
   lst
   (first-few-outputs-aux proc (- n 1) (cons (proc n) lst)) ) )
```

(For readers unused to recursive code: to compute $T(64)$, the computer wants to compute $2 * T(63) - 1$, which requires computing $T(63)$. The computer puts the 'times 2' and the 'plus 1' aside for a moment. It computes $T(63)$ by using this same piece of code (that's what 'recursive' means), and to do that it wants to compute $2 * T(62) - 1$. This keeps up until, after 63 steps, the computer tries to compute $T(1)$. It then returns $T(1) = 1$, which allows the computation of $T(2)$ to proceed, etc., until the original computation of $T(64)$ finishes.)

The helper functions give a table of the first few values. Here is the session at the prompt.

```
#;1> (load "hanoi.scm")
; loading hanoi.scm ...
; loading /var/lib//chicken/6/numbers.import.so ...
; loading /var/lib//chicken/6/chicken.import.so ...
; loading /var/lib//chicken/6/foreign.import.so ...
; loading /var/lib//chicken/6/numbers.so ...
#;2> (tower-of-hanoi-moves 64)
18446744073709551615
#;3> (first-few-outputs tower-of-hanoi-moves 64)
(1 3 7 15 31 63 127 255 511 1023 2047 4095 8191 16383 32767 65535 131071 262143 524287 1048575
2097151 4194303 8388607 16777215 33554431 67108863 134217727 268435455 536870911 1073741823
2147483647 4294967295 8589934591 17179869183 34359738367 68719476735 137438953471 274877906943
549755813887 1099511627775 2199023255551 4398046511103 8796093022207 17592186044415
35184372088831 70368744177663 140737488355327 281474976710655 562949953421311 1125899906842623
2251799813685247 4503599627370495 9007199254740991 18014398509481983 36028797018963967
72057594037927935 144115188075855871 288230376151711743 576460752303423487 1152921504606846975
2305843009213693951 4611686018427387903 9223372036854775807 18446744073709551615)
```

This is a list of $T(1)$ through $T(64)$ (the session was edited to put in line breaks for readability).

Appendix

Mathematics is made of arguments (reasoned discourse that is, not crockery-throwing). This section sketches the background material and argument techniques that we use in the book.

This section informally outlines the topics, skipping proofs. For more, [Velleman2] is excellent. Two other sources, available online, are [Hefferon] and [Beck].

Statements

Formal mathematical statements come labelled as a *Theorem* for major points, a *Corollary* for results that follow immediately from a prior one, or a *Lemma* for results chiefly used to prove others.

Statements can be complex and have many parts. The truth or falsity of the entire statement depends both on the truth value of the parts and on how the statement is put together.

Not Where P is a proposition, 'it is not the case that P' is true provided that P is false. For instance, 'n is not prime' is true only when n is the product of smaller integers.

To prove that a 'not P' statement holds, show that P is false.

And For a statement of the form 'P and Q' to be true both halves must hold: '7 is prime and so is 3' is true, while '7 is prime and 3 is not' is false.

To prove a 'P and Q', prove each half.

Or A 'P or Q' statement is true when either half holds: '7 is prime or 4 is prime' is true, while '8 is prime or 4 is prime' is false. In the case that both clauses of the statement are true, as in '7 is prime or 3 is prime', we take the statement as a whole to be true. (In everyday speech people occasionally use 'or' in an exclusive way — "Live free or die" does not intend both halves to hold — but we will not use 'or' in that way.)

To prove 'P or Q', show that in all cases at least one half holds (perhaps sometimes one half and sometimes the other, but always at least one).

If-then An 'if P then Q' statement may also appear as 'P implies Q' or 'P $\implies$ Q' or 'P is sufficient to give Q' or 'Q if P'. It is true unless P is true while Q is false. Thus 'if 7 is prime then 4 is not' is true while 'if 7 is prime then 4 is also prime' is false. (Contrary to its use in casual speech, in mathematics 'if P then Q' does not connote that P precedes Q or causes Q.)

Note this consequence of the prior paragraph: if P is false then 'if P then Q' is true irrespective of the value of Q: 'if 4 is prime then 7 is prime' and 'if 4 is prime then 7 is not' are both true statements. (They are *vacuously true*.) Also observe that 'if P then Q' is true when Q is true: 'if 4 is prime then 7 is prime' and 'if 4 is not prime then 7 is prime' are both true.

There are two main ways to establish an implication. The first way is direct: assume that P is true and use that assumption to prove Q. For instance, to show 'if a number is divisible by 5 then twice that number is divisible by 10' we can assume that the number is $5n$ and deduce that $2(5n) = 10n$. The indirect way is to prove the *contrapositive* statement: 'if Q is false then P is false' (rephrased, 'Q can only be false when P is also false'). Thus to show 'if a natural number is prime then it is not a perfect square' we can argue that if it were a square $p = n^2$ then it could be factored $p = n \cdot n$ where $n < p$ and so wouldn't be prime ($p = 0$ or $p = 1$ don't satisfy $n < p$ but they are nonprime).

Equivalent statements Sometimes, not only does P imply Q but also Q implies P. Some ways to say this are: 'P if and only if Q', 'P iff Q', 'P and Q are logically equivalent', 'P is necessary and sufficient to give Q', 'P $\iff$ Q'. An example is 'an integer is divisible by ten if and only if that number ends in 0'.

Although in simple arguments a chain like "P if and only if R, which holds if and only if S ..." may be practical, to prove that statements are equivalent we more often prove the two halves 'if P then Q' and 'if Q then P' separately.

Quantifiers

Compare these statements about natural numbers: 'there is a natural number x such that x is divisible by x^2' is true, while 'for all natural numbers x, that x is divisible by x^2' is false. The prefixes 'there is' and 'for all' are *quantifiers*.

For all The 'for all' prefix is the *universal quantifier*, symbolized $\forall$.

The most straightforward way to prove that a statement holds in all cases is to prove that it holds in each case. Thus to show that 'every number divisible by p has its square divisible by p^2', take a single number of the form pn and square it $(pn)^2 = p^2n^2$. This is a *typical element* proof. (In this kind of argument be careful not to assume properties for that element other than the ones in the hypothesis. This argument is wrong: "If n is divisible by a prime, say 2, so that $n = 2k$ for some natural number k, then $n^2 = (2k)^2 = 4k^2$ and the square of n is divisible by the square of the prime." That is a proof for the special case $p = 2$ but it isn't a proof for all p. Contrast it with a correct one: "If n is divisible by a prime so that $n = pk$ for some natural number k then $n^2 = (pk)^2 = p^2k^2$ and so the square of n is divisible by the square of the prime.")

There exists The 'there exists' prefix is the *existential quantifier*, symbolized $\exists$.

We can prove an existence proposition by producing something satisfying the property: for instance, to settle the question of primality of $2^{2^5} + 1$, Euler exhibited the divisor 641 [Sandifer]. But there are proofs showing that something exists without saying how to find it; Euclid's argument given in the next subsection shows there are infinitely many primes without giving a formula naming them.

Finally, after answering "Are there any?" affirmatively we often ask "How many?" That is, the question of uniqueness often arises in conjunction with the question of existence. Sometimes the two arguments are simpler if separated so note that just as proving something exists does not show it is unique, neither does proving something is unique show that it exists. (For instance, we can easily show that the natural number halfway between three and four is unique, even thouge no such number exists.)

Techniques of Proof

Induction Many proofs are iterative, "Here's why the statement is true for the number 0, it then follows for 1 and from there to 2 ...". These are proofs by *mathematical induction*. This technique is often not obvious to a person who has not seen it before, even to a person with a mathematical turn of mind. So we will see two examples.

We will first prove that $1 + 2 + 3 + \cdots + n = n(n+1)/2$. That formula has a natural number variable n that is free, meaning that setting n to be 1, or 2, etc., gives a family of cases of the statement: first that $1 = 1(2)/2$, second that $1 + 2 = 2(3)/2$, etc. Our induction proofs involve statements with one free natural number variable.

Each proof has two steps. In the *base step* we show that the statement holds for some intial number $i \in \mathbb{N}$. Often this step is a routine, and short, verification. The second step, the *inductive step*, is more subtle; we will show that this implication holds:

$$\begin{array}{l}\text{If the statement holds from } n = i \text{ up to and including } n = k \\ \text{then the statement holds also in the } n = k+1 \text{ case}\end{array} \qquad (*)$$

(the first line is the *inductive hypothesis*). The Principle of Mathematical Induction is that completing both steps proves that the statement is true for all natural numbers greater than or equal to i.

For the sum of the initial n numbers statement the intuition behind the principle is that first, the base step directly verifies the statement for the case of the initial number $n = 1$. Then, because the inductive step verifies the implication $(*)$ for all k, that implication applied to $k = 1$ gives that the statement is true for the case of the number $n = 2$. Now, with the statement established for both 1 and 2, apply $(*)$ again to conclude that the statement is true for the number $n = 3$. In this way, we bootstrap to all numbers $n \geqslant 1$.

Here is a proof of $1+2+3+\cdots+n = n(n+1)/2$, with separate paragraphs for the base step and the inductive step.

> For the base step we show that the formula holds when $n = 1$. That's easy; the sum of the first 1 natural number equals $1(1+1)/2$.
>
> For the inductive step, assume the inductive hypothesis that the formula holds for the numbers $n = 1$, $n = 2$, ..., $n = k$ with $k \geqslant 1$. That is, assume $1 = 1(1)/2$, and $1+2 = 2(3)/2$, and $1+2+3 = 3(4)/2$, through $1+2+\cdots+k = k(k+1)/2$. With that, the formula holds also in the $n = k+1$ case:
>
> $$1+2+\cdots+k+(k+1) = \frac{k(k+1)}{2} + (k+1) = \frac{(k+1)(k+2)}{2}$$
>
> (the first equality follows from the inductive hypothesis).

Here is another example, proving that every integer greater than or equal to 2 is a product of primes.

> The base step is easy: 2 is the product of a single prime.
>
> For the inductive step assume that each of $2, 3, \ldots, k$ is a product of primes, aiming to show $k+1$ is also a product of primes. There are two possibilities. First, if $k+1$ is not divisible by a number smaller than itself then it is a prime and so is the product of primes. The second possibility is that $k+1$ is divisible by a number smaller than itself, and then by the inductive hypothesis its factors can be written as a product of primes. In either case $k+1$ can be rewritten as a product of primes.

Contradiction Another technique of proof is to show that something is true by showing that it cannot be false. A proof by contradiction assumes that the proposition is false and derives some contradiction to known facts.

The classic example of proof by contradiction is Euclid's argument that there are infinitely many primes.

Suppose that there are only finitely many primes $p_1, \ldots, p_k$. Consider the number $p_1 \cdot p_2 \ldots p_k + 1$. None of the primes on the supposedly exhaustive list divides this number evenly since each leaves a remainder of 1. But every number is a product of primes so this can't be. Therefore there cannot be only finitely many primes.

Another example is this proof that $\sqrt{2}$ is not a rational number.

Suppose that $\sqrt{2} = m/n$, so that $2n^2 = m^2$. Factor out any 2's, giving $n = 2^{k_n} \cdot \hat{n}$ and $m = 2^{k_m} \cdot \hat{m}$. Rewrite.

$$2 \cdot (2^{k_n} \cdot \hat{n})^2 = (2^{k_m} \cdot \hat{m})^2$$

The Prime Factorization Theorem says that there must be the same number of factors of 2 on both sides, but there are an odd number of them $1 + 2k_n$ on the left and an even number $2k_m$ on the right. That's a contradiction, so a rational number with a square of 2 is impossible.

Sets, Functions, and Relations

Sets Mathematicians often work with collections. The most commonly-used kind of collection is a *set*. Sets are characterized by the Principle of Extensionality: two sets with the same elements are equal. Because of this, the order of the elements does not matter $\{2, \pi\} = \{\pi, 2\}$, and repeats collapse $\{7, 7\} = \{7\}$.

We can describe a set using a listing between curly braces $\{1, 4, 9, 16\}$ (as in the prior paragraph), or by using set-builder notation $\{x \mid x^5 - 3x^3 + 2 = 0\}$ (read "the set of all x such that ..."). We name sets with capital roman letters; for instance the set of primes is $P = \{2, 3, 5, 7, 11, \ldots\}$ (except that a few sets are so important that their names are reserved, such as the real numbers $\mathbb{R}$ and the complex numbers $\mathbb{C}$). To denote that something is an *element*, or *member*,) of a set we use '$\in$', so that $7 \in \{3, 5, 7\}$ while $8 \notin \{3, 5, 7\}$.

We say that A is a *subset* of B, written $A \subseteq B$, when $x \in A$ implies that $x \in B$. In this book we use '$\subset$' for the *proper subset* relationship that A is a subset of B but $A \neq B$ (some authors use this symbol for any kind of subset, proper or not). An example is $\{2, \pi\} \subset \{2, \pi, 7\}$. These symbols may be flipped, for instance $\{2, \pi, 5\} \supset \{2, 5\}$.

Because of Extensionality, to prove that two sets are equal $A = B$ show that they have the same members. Often we do this by showing mutual inclusion, that both $A \subseteq B$ and $A \supseteq B$. Such a proof will have a part showing that if $x \in A$ then $x \in B$, and a second part showing that if $x \in B$ then $x \in A$.

When a set has no members then it is the *empty set* $\{\}$, symbolized $\varnothing$. Any set has the empty set for a subset by the 'vacuously true' property of the definition of implication.

Diagrams We picture basic set operations with a *Venn diagram.* This shows $x \in P$.

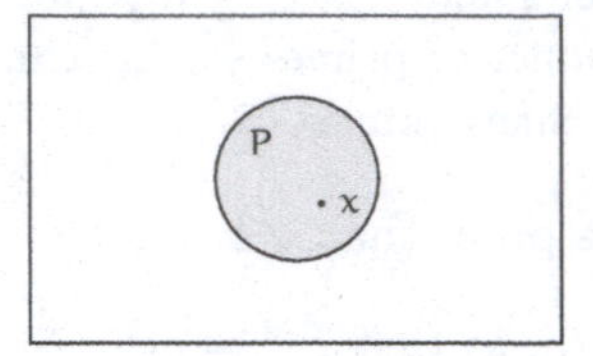

The outer rectangle contains the universe Ω of all objects under discussion. For instance, in a statement about real numbers, the rectangle encloses all members of $\mathbb{R}$. The set is pictured as a circle, enclosing its members.

Here is the diagram for $P \subseteq Q$. It shows that if $x \in P$ then $x \in Q$.

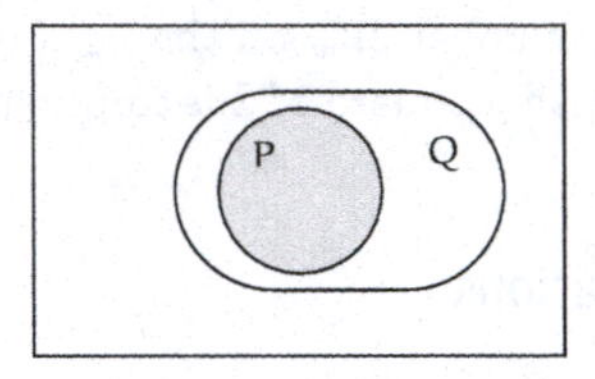

Set Operations The *union* of two sets is $P \cup Q = \{x \mid (x \in P) \text{ or } (x \in Q)\}$. The diagram shows that an element is in the union if it is in either of the sets.

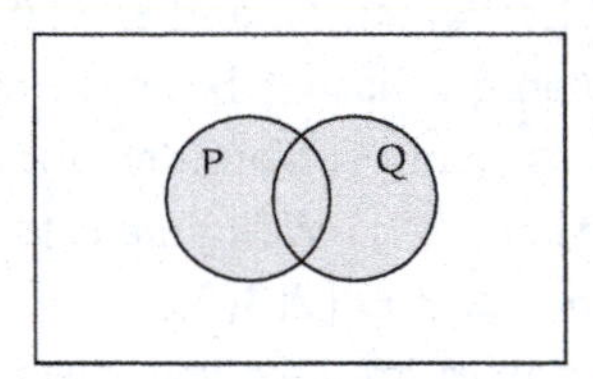

The *intersection* is $P \cap Q = \{x \mid (x \in P) \text{ and } (x \in Q)\}$.

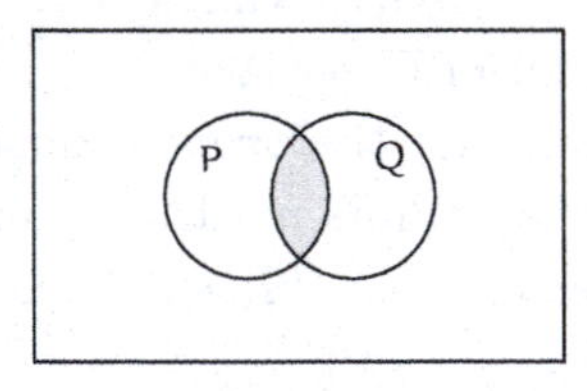

The *complement* of a set P is $P^{\text{comp}} = \{x \in \Omega \mid x \notin P\}$

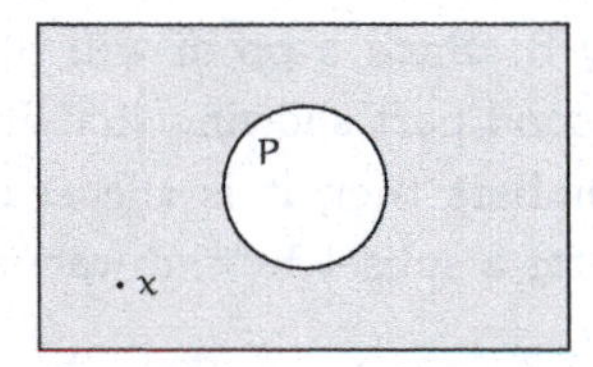

Multisets A *multiset* is a collection that is like a set in that order does not matter, but in which, unlike a set, repeats do not collapse. Thus the multiset $\{2,1,2\}$ is the same as the multiset $\{1,2,2\}$ but differs from the multiset $\{1,2\}$. Note that we use the same $\{\ldots\}$ curly brackets notation as for sets. Also as with sets, we say A is a *multiset subset* if A is a subset of B and A is a multiset.

Sequences In addition to sets and multisets, we also use collections where order matters and where repeats do not collapse. These are *sequences*, denoted with angle brackets: $\langle 2,3,7\rangle \neq \langle 2,7,3\rangle$. A sequence of length 2 is an *ordered pair*, and is often written with parentheses: $(\pi, 3)$. We also sometimes say 'ordered triple', 'ordered 4-tuple', etc. The set of ordered n-tuples of elements of a set A is denoted A^n. Thus $\mathbb{R}^2$ is the set of pairs of reals.

Functions A *function* or *map* $f\colon D \to C$ is is an association between input *arguments* $x \in D$ and output *values* $f(x) \in C$ subject to the the requirement that the function must be *well-defined*, that x suffices to determine $f(x)$. Restated, the condition is that if $x_1 = x_2$ then $f(x_1) = f(x_2)$.

The set of all arguments D is f's *domain* and the set of output values is its *range* $\mathscr{R}(f)$. Often we don't work with the range and instead work with a convenient superset, the *codomain* C. For instance, we might describe the squaring function with $s\colon \mathbb{R} \to \mathbb{R}$ instead of $s\colon \mathbb{R} \to \mathbb{R}^+ \cup \{0\}$.

We picture functions with a *bean diagram.*

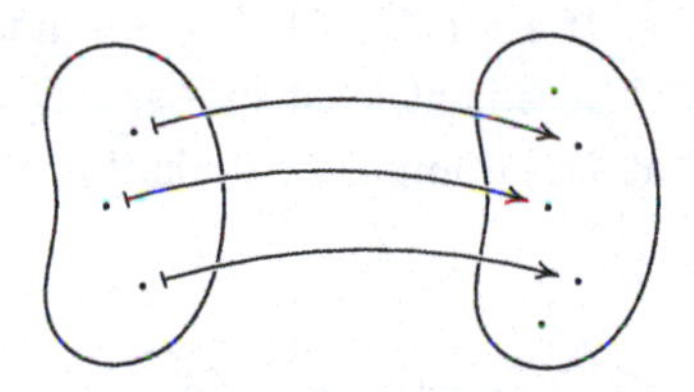

The blob on the left is the domain while on the right is the codomain. The function associates the three points of the domain with three in the codomain. Note that by the definition of a function every point in the domain is associated with a unique point in the codomain, but the converse needn't be true.

The association is arbitrary; no formula or algorithm is required, although in this book there typically is one. We often use y to denote $f(x)$. We also use the notation $x \stackrel{f}{\longmapsto} 16x^2 - 100$, read '$x$ maps under f to $16x^2 - 100$' or '$16x^2 - 100$ is the *image* of x'.

A map such as $x \mapsto \sin(1/x)$ is a combinations of simpler maps, here $g(y) = \sin(y)$ applied to the image of $f(x) = 1/x$. The *composition* of $g\colon Y \to Z$ with $f\colon X \to Y$, is the map sending $x \in X$ to $g(\,f(x)\,) \in Z$. It is denoted $g \circ f\colon X \to Z$. This definition only makes sense if the range of f is a subset of the domain of g.

An *identity map* $\mathrm{id}\colon Y \to Y$ defined by $\mathrm{id}(y) = y$ has the property that for any $f\colon X \to Y$, the composition $\mathrm{id} \circ f$ is equal to f. So an identity map plays the same role with respect to function composition that the number 0 plays in real number addition or that 1 plays in multiplication.

In line with that analogy, we define a *left inverse* of a map $f\colon X \to Y$ to be a function $g\colon \text{range}(f) \to X$ such that $g \circ f$ is the identity map on X. A *right inverse* of f is a $h\colon Y \to X$ such that $f \circ h$ is the identity.

For some f's there is a map that is both a left and right inverse of f. If such a map exists then it is unique because if both g_1 and g_2 have this property then $g_1(x) = g_1 \circ (f \circ g_2)(x) = (g_1 \circ f) \circ g_2(x) = g_2(x)$ (the middle equality comes from the associativity of function composition) so we call it a *two-sided inverse* or just *"the" inverse*, and denote it f^{-1}. For instance, the inverse of the function $f\colon \mathbb{R} \to \mathbb{R}$ given by $f(x) = 2x - 3$ is the function $f^{-1}\colon \mathbb{R} \to \mathbb{R}$ given by $f^{-1}(x) = (x+3)/2$.

The superscript notation for function inverse 'f^{-1}' fits into a larger scheme. Functions with the same codomain as domain $f\colon X \to X$ can be iterated, so that we can consider the composition of f with itself: $f \circ f$, and $f \circ f \circ f$, etc. We write $f \circ f$ as f^2 and $f \circ f \circ f$ as f^3, etc. Note that the familiar exponent rules for real numbers hold: $f^i \circ f^j = f^{i+j}$ and $(f^i)^j = f^{i \cdot j}$. Then where f is invertible, writing f^{-1} for the inverse and f^{-2} for the inverse of f^2, etc., gives that these familiar exponent rules continue to hold, since we define f^0 to be the identity map.

The definition of function requires that for every input there is one and only one associated output value. If a function $f\colon D \to C$ has the additional property that for every output value there is at least one associated input value—that is, the additional property that f's codomain equals its range $C = \mathscr{R}(f)$—then the function is *onto*.

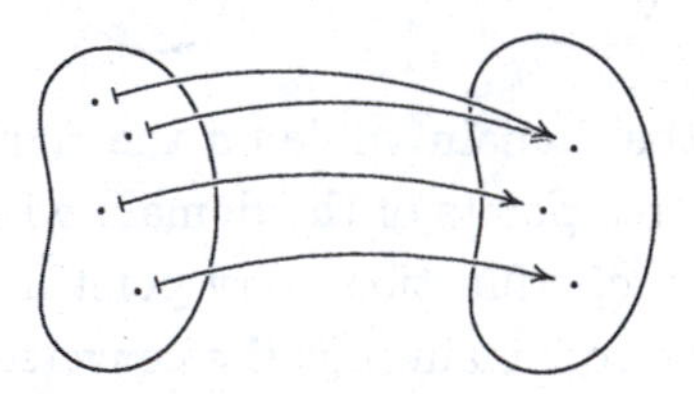

A function has a right inverse if and only if it is onto. (The f pictured above has a right inverse $g\colon C \to D$ given by following the arrows backwards, from right to left. For the codomain point on the top, choose either one of the arrows to follow. With that, applying g first followed by f takes elements $y \in C$ to themselves, and so is the identity function.)

If a function $f\colon D \to C$ has the property that for every output value there is at most one associated input value—that is, if no two arguments share an image so that $f(x_1) = f(x_2)$ implies that $x_1 = x_2$—then the function is *one-to-one*. The bean diagram from earlier illustrates.

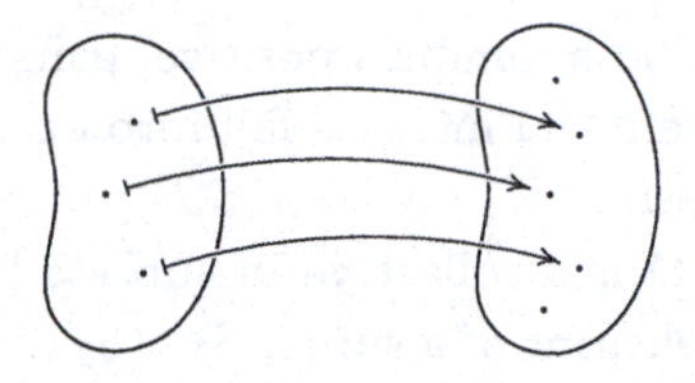

A function has a left inverse if and only if it is one-to-one. (In the picture define $g\colon C \to D$ to follow the arrows backwards for those $y \in C$ that are at the end of an arrow, and to send the point to an arbitrary element in D otherwise. Then applying f followed by g to elements of D will act as the identity.)

By the prior paragraphs, a map has a two-sided inverse if and only if that map is both onto and one-to-one. Such a function is a *correspondence*. It associates one and only one element of the domain with each element of the codomain. Because a composition of one-to-one maps is one-to-one, and a composition of onto maps is onto, a composition of correspondences is a correspondence.

We sometimes want to shrink the domain of a function. For instance, we may take the function $f\colon \mathbb{R} \to \mathbb{R}$ given by $f(x) = x^2$ and, in order to have an inverse, limit input arguments to nonnegative reals $\hat{f}\colon \mathbb{R}^+ \cup \{0\} \to \mathbb{R}$. Then $\hat{f}$ is the *restriction* of f to the smaller domain.

Relations Some familiar mathematical things, such as '$<$' or '$=$', are most naturally understood as relations between things. A *binary relation* on a set A is a set of ordered pairs of elements of A. For example, some elements of the set that is the relation '$<$' on the integers are $(3, 5)$, $(3, 7)$, and $(1, 100)$. Another binary relation on the integers is equality; this relation is the set $\{\ldots, (-1, 1), (0, 0), (1, 1), \ldots\}$. Still another example is 'closer than 10', the set $\{(x, y) \mid |x - y| < 10\}$. Some members of this relation are $(1, 10)$, $(10, 1)$, and $(42, 44)$. Neither $(11, 1)$ nor $(1, 11)$ is a member.

Those examples illustrate the generality of the definition. All kinds of relationships (e.g., 'both numbers even' or 'first number is the second with the digits reversed') are covered.

Equivalence Relations We shall need to express that two objects are alike in some way. They aren't identical, but they are related (e.g., two integers that give the same remainder when divided by 2).

A binary relation $\{(a, b), \ldots\}$ is an *equivalence relation* when it satisfies (1) *reflexivity*: any object is related to itself, (2) *symmetry*: if a is related to b then b is related to a, and (3) *transitivity*: if a is related to b and b is related to c then a is related to c. Some examples (on the integers): '$=$' is an equivalence relation, '$<$' does not satisfy symmetry, 'same sign' is a equivalence, while 'nearer than 10' fails transitivity.

Partitions In the 'same sign' relation $\{(1, 3), (-5, -7), (0, 0), \ldots\}$ there are three kinds of pairs, pairs with both numbers positive, pairs with both negative, and the one pair with both zero. So integers fall into exactly one of three classes, positive, or negative, or zero.

A *partition* of a set Ω is a collection of subsets $\{S_0, S_1, S_2, \ldots\}$ such that every element of S is an element of a subset $S_1 \cup S_2 \cup \cdots = \Omega$ and overlapping parts are equal: if $S_i \cap S_j \neq \varnothing$ then $S_i = S_j$. Picture that Ω is decomposed into non-overlapping parts.

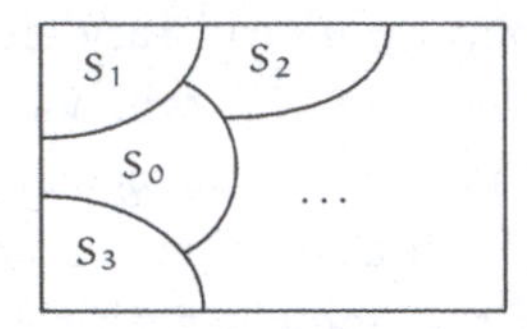

Thus the prior paragraph says that 'same sign' partitions the integers into the set of positives, the set of negatives, and the set containing only zero. Similarly, the equivalence relation '=' partitions the integers into one-element sets.

Another example is the set of strings consisting of a number, followed by a slash, followed by a nonzero number $\Omega = \{n/d \mid n, d \in \mathbb{Z} \text{ and } d \neq 0\}$. Define $S_{n,d}$ by: $\hat{n}/\hat{d} \in S_{n,d}$ if $\hat{n}d = n\hat{d}$. Checking that this is a partition of Ω is routine (observe for instance that $S_{4,3} = S_{8,6}$). This shows some parts, listing in each a couple of its infinitely many members.

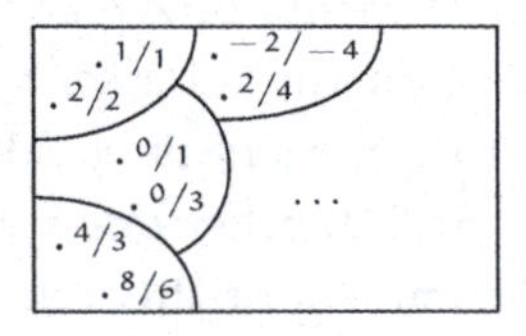

Every equivalence relation induces a partition, and every partition is induced by an equivalence. (This is routine to check.) Below are two examples.

Consider the equivalence relationship between two integers of 'gives the same remainder when divided by 2', the set $P = \{(-1,3), (2,4), (0,0), \ldots\}$. In the set P are two kinds of pairs, the pairs with both members even and the pairs with both members odd. This equivalence induces a partition where the parts are found by: for each x we define the set of numbers related to it $S_x = \{y \mid (x,y) \in P\}$. The parts are $\{\ldots, -3, -1, 1, 3, \ldots\}$ and $\{\ldots, -2, 0, 2, 4, \ldots\}$. Each part can be named in many ways; for instance, $\{\ldots, -3, -1, 1, 3, \ldots\}$ is S_1 and also is S_{-3}.

Now consider the partition of the natural numbers where two numbers are in the same part if they leave the same remainder when divided by 10, that is, if they have the same least significant digit. This partition is induced by the equivalence relation R defined by: two numbers n, m are related if they are together in the same part. For example, 3 is related to 33, but 3 is not related to 102. Verifying the three conditions in the definition of equivalence are straightforward.

We call each part of a partition an *equivalence class*. We sometimes pick a single element of each equivalence class to be the *class representative*.

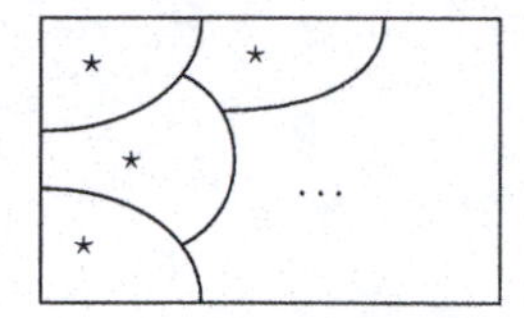

Usually when we pick representatives we have some natural scheme in mind. In that case we call them the *canonical* representatives. An example is that two fractions 3/5 and 9/15 are equivalent. In everyday work we often prefer to use the 'simplest form' or 'reduced form' fraction 3/5 as the class representative.

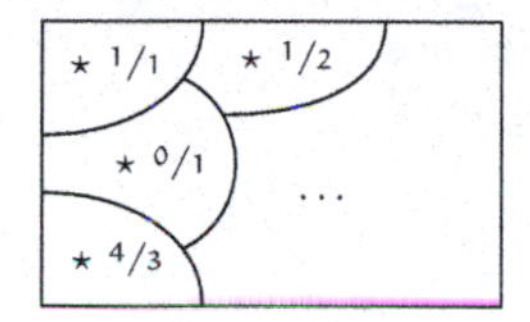

Bibliography

[Ackerson] R. H. Ackerson, *A Note on Vector Spaces*, American Mathematical Monthly, vol. 62 no. 10 (Dec. 1955), p. 721.

[Am. Math. Mon., Jun. 1931] C. A. Rupp (proposer), H. T. R. Aude (solver), problem 3468, American Mathematical Monthly, vol. 37 no. 6 (June-July 1931), p. 355.

[Am. Math. Mon., Feb. 1933] V. F. Ivanoff (proposer), T. C. Esty (solver), problem 3529, American Mathematical Monthly, vol. 39 no. 2 (Feb. 1933), p. 118.

[Am. Math. Mon., Jan. 1935] W. R. Ransom (proposer), Hansraj Gupta (solver), Elementary problem 105, American Mathematical Monthly, vol. 42 no. 1 (Jan. 1935), p. 47.

[Am. Math. Mon., Jan. 1949] C. W. Trigg (proposer), R. J. Walker (solver), Elementary problem 813, American Mathematical Monthly, vol. 56 no. 1 (Jan. 1949), p. 33.

[Am. Math. Mon., Jun. 1949] Don Walter (proposer), Alex Tytun (solver), Elementary problem 834, American Mathematical Monthly, vol. 56 no. 6 (June-July 1949), p. 409.

[Am. Math. Mon., Nov. 1951] Albert Wilansky, *The Row-Sums of the Inverse Matrix*, American Mathematical Monthly, vol. 58 no. 9 (Nov. 1951), p. 614.

[Am. Math. Mon., Feb. 1953] Norman Anning (proposer), C. W. Trigg (solver), Elementary problem 1016, American Mathematical Monthly, vol. 60 no. 2 (Feb. 1953), p. 115.

[Am. Math. Mon., Apr. 1955] Vern Haggett (proposer), F. W. Saunders (solver), Elementary problem 1135, American Mathematical Monthly, vol. 62 no. 4 (Apr. 1955), p. 257.

[Am. Math. Mon., Jan. 1963] Underwood Dudley, Arnold Lebow (proposers), David Rothman (solver), Elementary problem 1151, American Mathematical Monthly, vol. 70 no. 1 (Jan. 1963), p. 93.

[Am. Math. Mon., Dec. 1966] Hans Liebeck, *A Proof of the Equality of Column Rank and Row Rank of a Matrix* American Mathematical Monthly, vol. 73 no. 10 (Dec. 1966), p. 1114.

[Anton] Howard Anton, *Elementary Linear Algebra*, John Wiley & Sons, 1987.

[Arrow] Kenneth J. Arrow, *Social Choice and Individual Values*, Wiley, 1963.

[Austin] David Austin, *How Google Finds Your Needle in the Web's Haystack*, `http://www.ams.org/samplings/feature-column/fcarc-pagerank`, retrieved Feb. 2012.

[Ball & Coxeter] W.W. Rouse Ball, *Mathematical Recreations and Essays*, revised by H.S.M. Coxeter, MacMillan, 1962.

[Beck] Matthias Beck, Ross Geoghegan, *The Art of Proof*, `http://math.sfsu.edu/beck/papers/aop.noprint.pdf`, 2011-Aug-08.

[Beardon] A.F. Beardon, *The Dimension of the Space of Magic Squares*, The Mathematical Gazette, vol. 87, no. 508 (Mar. 2003), p. 112-114.

[Birkhoff & MacLane] Garrett Birkhoff, Saunders MacLane, *Survey of Modern Algebra*, third edition, Macmillan, 1965.

[Blass 1984] A. Blass, *Existence of Bases Implies the Axiom of Choice*, pp. 31 − 33, Axiomatic Set Theory, J. E. Baumgartner, ed., American Mathematical Society, Providence RI, 1984.

[Bridgman] P.W. Bridgman, *Dimensional Analysis*, Yale University Press, 1931.

[Brin & Page] Sergey Brin and Lawrence Page, *The Anatomy of a Large-Scale Hypertextual Web Search Engine*, `http://infolab.stanford.edu/pub/papers/google.pdf`, retrieved Feb. 2012.

[Bryan & Leise] Kurt Bryan, Tanya Leise, *The $25,000,000,000 Eigenvector: the Linear Algebra Behind Google*, SIAM Review, Vol. 48, no. 3 (2006), p. 569-81.

[Casey] John Casey, *The Elements of Euclid, Books I to VI and XI*, ninth edition, Hodges, Figgis, and Co., Dublin, 1890.

[ChessMaster] User ChessMaster of StackOverflow, answer to *Python determinant calculation(without the use of external libraries)*, `http://stackoverflow.com/a/10037087/238366`, answer posted 2012-Apr-05, accessed 2012-Jun-18.

[Chicken Scheme] Free software implementation, Felix L. Winkelmann and The Chicken Team, `http://wiki.call-cc.org/`, accessed 2013-Nov-20.

[Clark & Coupe] David H. Clark, John D. Coupe, *The Bangor Area Economy Its Present and Future*, report to the city of Bangor ME, Mar. 1967.

[Cleary] R. Cleary, private communication, Nov. 2011.

[Clarke] Arthur C. Clarke, *Technical Error*, Fantasy, December 1946, reprinted in Great SF Stories 8 (1946), DAW Books, 1982.

[Con. Prob. 1955] *The Contest Problem Book*, 1955 number 38.

[Cost Of Tolls] *Cost of Tolls*, http://costoftolls.com/Tolls_in_New_York.html, 2012-Jan-07.

[Coxeter] H.S.M. Coxeter, *Projective Geometry*, second edition, Springer-Verlag, 1974.

[Courant & Robbins] Richard Courant, Herbert Robbins, *What is Mathematics?*, Oxford University Press, 1978.

[Cullen] Charles G. Cullen, *Matrices and Linear Transformations*, second edition, Dover, 1990.

[Dalal, et. al.] Siddhartha R. Dalal, Edward B. Fowlkes, & Bruce Hoadley, *Lesson Learned from Challenger: A Statistical Perspective*, Stats: the Magazine for Students of Statistics, Fall 1989, p. 3.

[Davies] Thomas D. Davies, *New Evidence Places Peary at the Pole*, National Geographic Magazine, vol. 177 no. 1 (Jan. 1990), p. 44.

[de Mestre] Neville de Mestre, *The Mathematics of Projectiles in Sport*, Cambridge University Press, 1990.

[De Parville] De Parville, *La Nature*, Paris, 1884, part I, p. 285 – 286 (citation from [Ball & Coxeter]).

[Ebbing] Darrell D. Ebbing, *General Chemistry*, fourth edition, Houghton Mifflin, 1993.

[Ebbinghaus] H. D. Ebbinghaus, *Numbers*, Springer-Verlag, 1990.

[Einstein] A. Einstein, Annals of Physics, v. 35, 1911, p. 686.

[Eggar] M.H. Eggar, *Pinhole Cameras, Perspecitve, and Projective Geometry*, American Mathematical Monthly, August-September 1998, p. 618 – 630.

[Falbo] Clement Falbo, *The Golden Ratio — a Contrary Viewpoint*, College Mathematics Journal, vol. 36, no. 2, March 2005, p. 123 – 134.

[Feller] William Feller, *An Introduction to Probability Theory and Its Applications* (vol. 1, 3rd ed.), John Wiley, 1968.

[Fuller & Logan] L.E. Fuller & J.D. Logan, *On the Evaluation of Determinants by Chiò's Method*, p 49-52, in Linear Algebra Gems, Carlson, et al, Mathematical Association of America, 2002.

[Gardner] Martin Gardner, *The New Ambidextrous Universe*, third revised edition, W. H. Freeman and Company, 1990.

[Gardner 1957] Martin Gardner, *Mathematical Games: About the remarkable similarity between the Icosian Game and the Tower of Hanoi*, Scientific American, May 1957, p. 150 – 154.

[Gardner, 1970] Martin Gardner, *Mathematical Games, Some mathematical curiosities embedded in the solar system*, Scientific American, April 1970, p. 108 – 112.

[Gardner, 1980] Martin Gardner, *Mathematical Games, From counting votes to making votes count: the mathematics of elections*, Scientific American, October 1980.

[Gardner, 1974] Martin Gardner, *Mathematical Games, On the paradoxical situations that arise from nontransitive relations*, Scientific American, October 1974.

[Giordano, Wells, Wilde] Frank R. Giordano, Michael E. Wells, Carroll O. Wilde, *Dimensional Analysis*, UMAP Unit 526, in *UMAP Modules, 1987*, COMAP, 1987.

[Giordano, Jaye, Weir] Frank R. Giordano, Michael J. Jaye, Maurice D. Weir, *The Use of Dimensional Analysis in Mathematical Modeling*, UMAP Unit 632, in *UMAP Modules, 1986*, COMAP, 1986.

[Google Maps] *Directions — Google Maps*, http://maps.google.com/help/maps/directions/, 2012-Jan-07.

[Goult, *et al.*] R.J. Goult, R.F. Hoskins, J.A. Milner, M.J. Pratt, *Computational Methods in Linear Algebra*, Wiley, 1975.

[Graham, Knuth, Patashnik] Ronald L. Graham, Donald E. Knuth, Oren Patashnik, *Concrete Mathematics*, Addison-Wesley, 1988.

[Halmos] Paul R. Halmos, *Finite Dimensional Vector Spaces*, second edition, Van Nostrand, 1958.

[Hamming] Richard W. Hamming, *Introduction to Applied Numerical Analysis*, Hemisphere Publishing, 1971.

[Hanes] Kit Hanes, *Analytic Projective Geometry and its Applications*, UMAP Unit 710, UMAP Modules, 1990, p. 111.

[Heath] T. Heath, *Euclid's Elements*, volume 1, Dover, 1956.

[Hefferon] J Hefferon, *Introduction to Proofs, an Inquiry-Based approach*, http://joshua.smcvt.edu/proofs/, 2013.

[Hoffman & Kunze] Kenneth Hoffman, Ray Kunze, *Linear Algebra*, second edition, Prentice-Hall, 1971.

[Hofstadter] Douglas R. Hofstadter, *Metamagical Themas: Questing for the Essence of Mind and Pattern*, Basic Books, 1985.

[Iosifescu] Marius Iofescu, *Finite Markov Processes and Their Applications*, John Wiley, 1980.

[Kahan] William Kahan, *ChiâĂŹs Trick for Linear Equations with Integer CoefïňĄcients*, http://www.cs.berkeley.edu/~wkahan/MathH110/chio.pdf, 1998, retrieved 2012-Jun-18.

[Kelton] Christina M.L. Kelton, *Trends on the Relocation of U.S. Manufacturing*, UMI Research Press, 1983.

[Kemeny & Snell] John G. Kemeny, J. Laurie Snell, *Finite Markov Chains*, D. Van Nostrand, 1960.

[Kemp] Franklin Kemp *Linear Equations*, American Mathematical Monthly, volume 89 number 8 (Oct. 1982), p. 608.

[Leontief 1951] Wassily W. Leontief, *Input-Output Economics*, Scientific American, volume 185 number 4 (Oct. 1951), p. 15.

[Leontief 1965] Wassily W. Leontief, *The Structure of the U.S. Economy*, Scientific American, volume 212 number 4 (Apr. 1965), p. 25.

[Macdonald & Ridge] Kenneth Macdonald, John Ridge, *Social Mobility*, in *British Social Trends Since 1900*, A.H. Halsey, Macmillian, 1988.

[Math. Mag., Sept. 1952] Dewey Duncan (proposer), W. H. Quelch (solver), Mathematics Magazine, volume 26 number 1 (Sept-Oct. 1952), p. 48.

[Math. Mag., Jan. 1957] M. S. Klamkin (proposer), Trickie T-27, Mathematics Magazine, volume 30 number 3 (Jan-Feb. 1957), p. 173.

[Math. Mag., Jan. 1963, Q237] D. L. Silverman (proposer), C. W. Trigg (solver), Quickie 237, Mathematics Magazine, volume 36 number 1 (Jan. 1963).

[Math. Mag., Jan. 1963, Q307] C. W. Trigg (proposer). Quickie 307, Mathematics Magazine, volume 36 number 1 (Jan. 1963), p. 77.

[Math. Mag., Nov. 1967] Clarence C. Morrison (proposer), Quickie, Mathematics Magazine, volume 40 number 4 (Nov. 1967), p. 232.

[Math. Mag., Jan. 1973] Marvin Bittinger (proposer), Quickie 578, Mathematics Magazine, volume 46 number 5 (Jan. 1973), p. 286, 296.

[Munkres] James R. Munkres, *Elementary Linear Algebra*, Addison-Wesley, 1964.

[Neimi & Riker] Richard G. Neimi, William H. Riker, *The Choice of Voting Systems*, Scientific American, June 1976, p. 21 – 27.

[Oakley & Baker] Cletus O. Oakley, Justine C. Baker, *Least Squares and the* 3 : 40 *Mile*, Mathematics Teacher, Apr. 1977.

[Ohanian] Hans O'Hanian, *Physics*, volume one, W. W. Norton, 1985.

[Onan] Michael Onan, *Linear Algebra*, Harcourt, 1990.

[Online Encyclopedia of Integer Sequences] *Number of different magic squares of order n that can be formed from the numbers* $1, \ldots, n^2$, http://oeis.org/A006052, 2012-Feb-17.

[Petersen] G. M. Petersen, *Area of a Triangle*, American Mathematical Monthly, volume 62 number 4 (Apr. 1955), p. 249.

[Polya] G. Polya, *Mathematics and Plausible Reasoning*, Princeton University Press, 1954.

[Poundstone] W. Poundstone, *Gaming the Vote*, Hill and Wang, 2008. ISBN-13: 978-0-8090-4893-9

[Putnam, 1990, A-5] William Lowell Putnam Mathematical Competition, Problem A-5, 1990.

[Rice] John R. Rice, *Numerical Mathods, Software, and Analysis*, second edition, Academic Press, 1993.

[Rucker] Rudy Rucker, *Infinity and the Mind*, Birkhauser, 1982.

[Ryan] Patrick J. Ryan, *Euclidean and Non-Euclidean Geometry: an Analytic Approach*, Cambridge University Press, 1986.

[Sandifer] Ed Sandifer, *How Euler Did It*, `http://www.maa.org/news/howeulerdidit.html`, 2012-Dec-27.

[Schmidt] Jack Schmidt, `http://math.stackexchange.com/a/98558/12012`, 2012-Jan-12.

[Shepelev] Anton Shepelev, private communication, Feb 19, 2011.

[Seidenberg] A. Seidenberg, *Lectures in Projective Geometry*, Van Nostrand, 1962.

[Strang 93] Gilbert Strang *The Fundamental Theorem of Linear Algebra*, American Mathematical Monthly, Nov. 1993, p. 848 – 855.

[Strang 80] Gilbert Strang, *Linear Algebra and its Applications*, second edition, Harcourt Brace Jovanovich, 1980.

[Taylor] Alan D. Taylor, *Mathematics and Politics: Strategy, Voting, Power, and Proof*, Springer-Verlag, 1995.

[Tilley] Burt Tilley, private communication, 1996.

[Trono] Tony Trono, compiler, *University of Vermont Mathematics Department High School Prize Examinations 1958-1991*, mimeographed printing, 1991.

[USSR Olympiad no. 174] *The USSR Mathematics Olympiad*, number 174.

[Velleman] Dan Velleman, private communication, on multiset in the definition of linearly independent set.

[Velleman2] Daniel J Velleman, *How to Prove It: A Structured Approach*, Cambridge University Press, 2006.

[Weston] J. D. Weston, *Volume in Vector Spaces*, American Mathematical Monthly, volume 66 number 7 (Aug./Sept. 1959), p. 575 – 577.

[Ward] James E. Ward III, *Vector Spaces of Magic Squares*, Mathematics Magazine, vol 53 no 2 (Mar 1980), p 108 – 111.

[Weyl] Hermann Weyl, *Symmetry*, Princeton University Press, 1952.

[Wickens] Thomas D. Wickens, *Models for Behavior*, W.H. Freeman, 1982.

[Wilkinson 1965] *The Algebraic Eigenvalue Problem*, J. H. Wilkinson, Oxford University Press, 1965

[Wikipedia, Lo Shu Square] *Lo Shu Square*, `http://en.wikipedia.org/wiki/Lo_Shu_Square`, 2012-Feb-17.

[Wikipedia, Magic Square] *Magic square*, `http://en.wikipedia.org/wiki/Magic_square`, 2012-Feb-17.

[Wikipedia, Mens Mile] *Mile run world record progression*,
`http://en.wikipedia.org/wiki/Mile_run_world_record_progression`,
2011-Apr-09.

[Wikipedia, Square-cube Law] *The Square-cube law*,
`http://en.wikipedia.org/wiki/Square-cube_law`, 2011-Jan-17.

[Wikipedia, Google Page Rank] *Page Rank*,
`http://en.wikipedia.org/wiki/PageRank`, 2012-Feb-27.

[Wills] Rebecca S. Wills, *Google's Page Rank*, Mathematical Intelligencer, vol. 28, no. 4, Fall 2006.

[Wohascum no. 2] *The Wohascum County Problem Book* problem number 2.

[Wohascum no. 47] *The Wohascum County Problem Book* problem number 47.

[Yaglom] I. M. Yaglom, *Felix Klein and Sophus Lie: Evolution of the Idea of Symmetry in the Nineteenth Century*, translated by Sergei Sossinsky, Birkhäuser, 1988.

[Yuster] Thomas Yuster, *The Reduced Row Echelon Form of a Matrix is Unique: a Simple Proof*, Mathematics Magazine, vol. 57, no. 2 (Mar. 1984), pp. 93-94.

[Zwicker] William S. Zwicker, *The Voters' Paradox, Spin, and the Borda Count*, Mathematical Social Sciences, vol. 22 (1991), p. 187 – 227.

Index